Advances and Applications of Block Copolymers

Advances and Applications of Block Copolymers

Editors

Nikolaos Politakos
Apostolos Avgeropoulos

MDPI • Basel • Beijing • Wuhan • Barcelona • Belgrade • Manchester • Tokyo • Cluj • Tianjin

Editors
Nikolaos Politakos
POLYMAT & Applied
Chemistry Department
University of the Basque Country
Donostia-San Sebastián
Spain

Apostolos Avgeropoulos
Department of Materials
Science Engineering
University of Ioannina
Ioannina
Greece

Editorial Office
MDPI
St. Alban-Anlage 66
4052 Basel, Switzerland

This is a reprint of articles from the Special Issue published online in the open access journal *Polymers* (ISSN 2073-4360) (available at: www.mdpi.com/journal/polymers/special_issues/Advances_and_ Applications_of_Block_Copolymers).

For citation purposes, cite each article independently as indicated on the article page online and as indicated below:

LastName, A.A.; LastName, B.B.; LastName, C.C. Article Title. *Journal Name* **Year**, *Volume Number*, Page Range.

ISBN 978-3-0365-8489-8 (Hbk)
ISBN 978-3-0365-8488-1 (PDF)

© 2023 by the authors. Articles in this book are Open Access and distributed under the Creative Commons Attribution (CC BY) license, which allows users to download, copy and build upon published articles, as long as the author and publisher are properly credited, which ensures maximum dissemination and a wider impact of our publications.

The book as a whole is distributed by MDPI under the terms and conditions of the Creative Commons license CC BY-NC-ND.

Contents

About the Editors vii

Nikolaos Politakos and Apostolos Avgeropoulos
Advances and Applications of Block Copolymers
Reprinted from: *Polymers* **2023**, *15*, 2930, doi:10.3390/polym15132930 1

Irati Barandiaran, Joseba Gomez-Hermoso-de-Mendoza, Junkal Gutierrez, Agnieszka Tercjak and Galder Kortaberria
Nanostructuring Biobased Epoxy Resin with PEO-PPO-PEO Block Copolymer
Reprinted from: *Polymers* **2023**, *15*, 1216, doi:10.3390/polym15051216 5

Konstantinos Artopoiadis, Christina Miskaki, Gkreti-Maria Manesi, Ioannis Moutsios, Dimitrios Moschovas and Alexey A. Piryazev et al.
Thermal and Bulk Properties of Triblock Terpolymers and Modified Derivatives towards Novel Polymer Brushes
Reprinted from: *Polymers* **2023**, *15*, 848, doi:10.3390/polym15040848 19

Nikolaos Politakos
Block Copolymers in 3D/4D Printing: Advances and Applications as Biomaterials
Reprinted from: *Polymers* **2023**, *15*, 322, doi:10.3390/polym15020322 33

Xinyue Zhang, Mingge Zhao and Junhan Cho
Effect of Disparity in Self Dispersion Interactions on Phase Behaviors of Molten A-b-B Diblock Copolymers
Reprinted from: *Polymers* **2022**, *15*, 30, doi:10.3390/polym15010030 65

Sang-In Lee, Min-Guk Seo, June Huh and Hyun-jong Paik
Small-Angle X-ray Scattering Analysis on the Estimation of Interaction Parameter of Poly(*n*-butyl acrylate)-*b*-poly(methyl methacrylate)
Reprinted from: *Polymers* **2022**, *14*, 5567, doi:10.3390/polym14245567 91

Alexandros Ch. Lazanas, Athanasios Katsouras, Michael Spanos, Gkreti-Maria Manesi, Ioannis Moutsios and Dmitry V. Vashurkin et al.
Synthesis and Characterization of Hybrid Materials Derived from Conjugated Copolymers and Reduced Graphene Oxide
Reprinted from: *Polymers* **2022**, *14*, 5292, doi:10.3390/polym14235292 99

Maria Karayianni, Dimitra Koufi and Stergios Pispas
Development of Double Hydrophilic Block Copolymer/Porphyrin Polyion Complex Micelles towards Photofunctional Nanoparticles
Reprinted from: *Polymers* **2022**, *14*, 5186, doi:10.3390/polym14235186 111

Ainhoa Álvarez-Gómez, Jiayin Yuan, Juan P. Fernández-Blázquez, Verónica San-Miguel and María B. Serrano
Polyacrylonitrile-*b*-Polystyrene Block Copolymer-Derived Hierarchical Porous Carbon Materials for Supercapacitor
Reprinted from: *Polymers* **2022**, *14*, 5109, doi:10.3390/polym14235109 131

Yulia L. Kuznetsova, Karina S. Sustaeva, Alexander V. Mitin, Evgeniy A. Zakharychev, Marfa N. Egorikhina and Victoria O. Chasova et al.
Graft Polymerization of Acrylamide in an Aqueous Dispersion of Collagen in the Presence of Tributylborane
Reprinted from: *Polymers* **2022**, *14*, 4900, doi:10.3390/polym14224900 147

Shuhui Ma, Yushuang Hou, Jinlin Hao, Cuncai Lin, Jiawei Zhao and Xin Sui
Well-Defined Nanostructures by Block Copolymers and Mass Transport Applications in Energy Conversion
Reprinted from: *Polymers* **2022**, *14*, 4568, doi:10.3390/polym14214568 161

Miriam Scoti, Fabio De Stefano, Angelo Giordano, Giovanni Talarico and Claudio De Rosa
Crystallization Behavior of Isotactic Propene-Octene Random Copolymers
Reprinted from: *Polymers* **2022**, *14*, 4032, doi:10.3390/polym14194032 193

Miriam Scoti, Fabio De Stefano, Filomena Piscitelli, Giovanni Talarico, Angelo Giordano and Claudio De Rosa
Melt-Crystallizations of α and γ Forms of Isotactic Polypropylene in Propene-Butene Copolymers
Reprinted from: *Polymers* **2022**, *14*, 3873, doi:10.3390/polym14183873 213

Yuan Zhang, Peng Wang, Nan Li, Chunyan Guo and Sumin Li
The Effect of Topology on Block Copolymer Nanoparticles: Linear versus Star Block Copolymers in Toluene
Reprinted from: *Polymers* **2022**, *14*, 3691, doi:10.3390/polym14173691 229

Jia Qu, Qiang Yang, Wei Gong, Meilan Li and Baoyue Cao
Simultaneous Removal of Cr(VI) and Phenol from Water Using Silica-di-Block Polymer Hybrids: Adsorption Kinetics and Thermodynamics
Reprinted from: *Polymers* **2022**, *14*, 2894, doi:10.3390/polym14142894 243

Ameen Arkanji, Viko Ladelta, Konstantinos Ntetsikas and Nikos Hadjichristidis
Synthesis and Thermal Analysis of Non-Covalent PS-*b*-SC-*b*-P2VP Triblock Terpolymers via Polylactide Stereocomplexation
Reprinted from: *Polymers* **2022**, *14*, 2431, doi:10.3390/polym14122431 259

Suhail K. Siddique, Hassan Sadek, Tsung-Lun Lee, Cheng-Yuan Tsai, Shou-Yi Chang and Hsin-Hsien Tsai et al.
Block Copolymer Modified Nanonetwork Epoxy Resin for Superior Energy Dissipation
Reprinted from: *Polymers* **2022**, *14*, 1891, doi:10.3390/polym14091891 273

Katerina Mavronasou, Alexandra Zamboulis, Panagiotis Klonos, Apostolos Kyritsis, Dimitrios N. Bikiaris and Raffaello Papadakis et al.
Poly(vinyl pyridine) and Its Quaternized Derivatives: Understanding Their Solvation and Solid State Properties
Reprinted from: *Polymers* **2022**, *14*, 804, doi:10.3390/polym14040804 285

About the Editors

Nikolaos Politakos

Dr. Nikolaos Politakos is currently a Senior Researcher in POLYMAT, Basque Center for Macromolecular Design & Engineering in Donostia, San Sebastian, Spain. He received his B.Sc. (2005) in Materials Engineering and M.Sc. (2007) in Chemistry and Technology of Materials at the Polytechnic School of the University of Ioannina (UoI), Ioannina, Greece. He obtained his Ph.D. (in 2010) in Materials Engineering from the Polytechnic School of the University of Ioannina (UoI), Ioannina, Greece. He started his postdoctoral research in 2010 at the University of the Basque Country (UPV/EHU) until 2014, working with chemical modifications of polymers and composites with fullerenes. From 2014 to 2017, he worked in CIC biomaGUNE in San Sebastian, synthesizing and characterizing responsive copolymer brushes from surfaces and particles for biomedical applications. From 2017 to 2020, he worked in POLYMAT, conducting research centred around the preparation of 3D sponge-like graphene oxide materials for CO_2 capture and photocatalysis. Finally, in 2020, he joined the Responsive Polymer Therapeutics Group and focussed his research on preparing hydrogels for wound healing. He has published 43 peer-reviewed publications, earning him an h-index of 15 and an i10-index of 19, as well as 602 total citations (according to Google Scholar) at present. He has participated in over 30 national and international conferences and over 20 national and international projects. His research interests include the synthesis of copolymers, biomaterials, chemical modifications, polypeptides, wound healing, smart materials/responsiveness, hybrid materials, hydrophobic materials, and CO_2 adsorption and photocatalysis.

Apostolos Avgeropoulos

Prof. Apostolos Avgeropoulos is a Full Professor at the Department of Materials Science Engineering (DMSE), University of Ioannina (UoI), Greece. He obtained his B.Sc. in Chemistry from the Chemistry Department, University of Athens, in 1992 and his Ph.D. in 1997 in Polymer Science from the same Department and University, under the supervision of Prof. N. Hadjichristidis in collaboration with Prof. E. L. Thomas at DMSE, MIT, USA. He was a post-doctoral researcher at DMSE-MIT for a total of 16 months from 1997 to 2001 and a post-doctoral researcher at the Industrial Chemistry Laboratory, Chemistry Department, University of Athens and the National Center for Scientific Research "DEMOKRITOS" from 1999 to 2002. He was elected as an Adjunct Assistant Professor at DMSE-UoI in 2002. In 2003, he started as a non-tenured Assistant Professor, got tenure in 2007, and was promoted to Associate Professor in 2009. In 2013, he became a full professor in the same department, where he continues his research in polymers science and engineering. His current research interests involve high-molecular-weight copolymers, amphiphilic functionalized block copolymers, nanocrystal/block copolymer composites (structure and structure properties), as well as nano-lithographic and nano-patterning applications of block copolymers with various architectures, among others. He has supervised 145 undergraduate diploma theses as a faculty member, including 46 MSc and 15 PhD students. He is the Director of the Polymer Science and Engineering Laboratory of DMSE-UoI and the Scientific Responsible for the Electron Microscopy Facility of UoI. His overall funding from various agencies is approximately EUR 6,6 M. He has published 3 book chapters, 1 US patent (2 pending), 220 peer-reviewed publications (h-index = 43, 6800 total citations - Google Scholar) at present. He has given 35 invited lectures in domestic and foreign Institutions and has participated in 80 national and international conferences with more than 30 invited talks.

Editorial

Advances and Applications of Block Copolymers

Nikolaos Politakos [1,*] and Apostolos Avgeropoulos [2,3]

1. POLYMAT, Applied Chemistry Department, Faculty of Chemistry, University of the Basque Country, UPV/EHU, Paseo Manuel de Lardizabal 3, 20018 Donostia-San Sebastián, Spain
2. Department of Materials Science Engineering, University Campus-Dourouti, University of Ioannina, 45110 Ioannina, Greece; aavger@uoi.gr
3. Faculty of Chemistry, Lomonosov Moscow State University (MSU), GSP-1, 1-3 Leninskiye Gory, 119991 Moscow, Russia

* Correspondence: nikolaos.politakos@ehu.eus

Polymers are materials that have constantly evolved from the beginning of their discovery until the present day. During the last few decades, polymers have been used in many areas of science and cutting-edge research and have been applied to everyday items. Many polymerization techniques, either synthetic or natural, exist to create molecularly homogeneous materials with specific properties prepared from different types of organic or even inorganic monomers.

The most important class of polymers is block copolymers (BCPs) since they can be prepared through different methods and exhibit various properties combined with their chemically different segments. The importance of block copolymers can be supported by the multitude of related manuscripts published on a daily basis, their use in practical applications in almost all areas, and novel discoveries based on BCPs in many aspects of life. Block copolymers can be used as composites and hybrid materials, providing new properties and, in some cases, responsiveness in terms of temperature, humidity, pH, redox, magnetic, electric, swelling, light, and others. Block copolymers are also very important for self-assembly studies since they can be organized into nanostructures and may be used in applications such as lithography or membranes. In general, BCPs have been used in medicine, chemistry, nanotechnology, physics, water treatment, environmental applications, wound healing, solar cells, photocatalysis, and many other fields.

Nevertheless, block copolymers face new challenges and problems every day. This Special Issue brings together different experimental works and reviews and attempts to cover the majority of the recent advances and applications of block copolymers in the last few years.

This Special Issue has gathered scientific works from research groups examining various copolymers, indicating advances in architecture, functionalities, and applications. The total number of manuscripts (17) published in this Special Issue indicates the importance of BCPs and the fact that many research groups and relevant members of the scientific community are thoroughly interested in the advancement of block copolymers and their applications.

In one study, copolymers with PVP (poly(vinylpyridine)) and its partially quaternized derivatives were studied based on the structure/properties relationship, focusing on solvation and optical properties [1].

Research has also been carried out on fabricating well-ordered epoxy resin nanonetworks modified with poly(butyl acrylate)-b-poly(methyl methacrylate) (PBA-b-PMMA) to enhance energy dissipation by exploiting the self-assembly of polystyrene-b-poly(dimethylsiloxane) (PS-b-PDMS) with gyroid and diamond structures as templates. Selectively etching a PDMS block can provide nanochannels for the polymerization of epoxy resin. A PBA-b-PMMA BCP can be self-assembled into a spherical nanosized micelle in the epoxy matrix, which acts as a toughening agent for forming soft domains [2].

Citation: Politakos, N.; Avgeropoulos, A. Advances and Applications of Block Copolymers. *Polymers* **2023**, *15*, 2930. https://doi.org/10.3390/polym15132930

Received: 9 June 2023
Accepted: 19 June 2023
Published: 2 July 2023

Copyright: © 2023 by the authors. Licensee MDPI, Basel, Switzerland. This article is an open access article distributed under the terms and conditions of the Creative Commons Attribution (CC BY) license (https://creativecommons.org/licenses/by/4.0/).

A process of combining anionic polymerization with ROP copolymers based on PLA was carried out by a group of researchers, providing new insights into polymer design strategies for high-performance PLA-based materials using stereo complexation [3].

In one study, copolymers of PBA-b-PDMAEMA (butyl methacrylate as a hydrophobic monomer and 2-(Dimethylamino) ethyl methacrylate) were prepared by surface-initiated ATRP from SiO_2 and used as a hybrid adsorbent for heavy metals (Cr) and phenol [4].

In another study, linear and star block copolymer nanoparticles of (polystyrene-b-poly(4-vinyl pyridine))n (PS-b-P4VP)n were prepared via RAFT, and their nano-assemblies in toluene were studied. Polymerization kinetics were investigated, and the effect of the arm number of the PS/P4VP chain on the block copolymer nano-assemblies prepared via PISA was explored by changing the length of the PS or the P4VP [5].

Two manuscripts focusing on crystallization are also included in this Special Issue. The first study [6] investigates the crystallization of the α and γ forms of iPP in propene–butene copolymers by analyzing the thermodynamics and kinetics effects, whereas the second study [7] is dedicated to the crystallization behavior of isotactic propene–octene copolymers (iPPC8) synthesized with a metallocene catalyst. The effect of the octene units excluded from the crystals on the crystallization of the α and γ forms of iPP was analyzed and compared to those of the different aforementioned comonomers that are partially included in the crystals of the α and γ forms with different degrees of incorporation.

The synthesis of hybrid materials based on AA grafted in aqueous collagen dispersions was also investigated in one study [8], whereby the determination of the molecular weight characteristics, structure, mechanical properties, and cytotoxicity of the obtained copolymers was carried out, and the authors evaluated the prospects for using the obtained copolymers in scaffold technologies.

Block copolymers can also be a powerful new precursor for fabricating porous carbon structures (PCFs) for electrochemical performance. Different copolymers can provide different nanostructures and thus improve electrochemical performance. Based on the above, a study focusing on the copolymer PAN–b–PS and carbon powder was presented as a comparative study of two different electrode materials for supercapacitors [9].

Additionally, the formation of photo-functional nanoparticles (owing to the intrinsic properties of the porphyrin) through the electrostatic complexation between the poly(2-(dimethylamino) ethyl methacrylate)-b-poly[(oligo ethylene glycol) methyl ether methacrylate] (PDMAEMA-b-POEGMA) double hydrophilic block copolymer or its quaternized strong polyelectrolyte counterpart (QPDMAEMA-b-POEGMA) was investigated [10]. This study examined their solution properties by changing the porphyrin content, photophysical characteristics, and photosensitivity under different temperatures and pHs.

Moreover, a study on mixtures consisting of conjugated polymers and reduced graphene oxide was also published with the major aim of tuning the energy gaps and revolutionizing various electronic applications, such as organic solar cells and OLED displays [11]. Another study is also presented to shed light on the temperature dependence of the Flory–Huggins interaction parameter $\chi(T)$ between the two acrylic monomer species of BA and MMA by small-angle X-ray scattering (SAXS). Well-defined copolymers were prepared via ATRP, and the study's authors measured the order–disorder transition (ODT) temperature within the investigated temperature range [12]. In terms of theory and simulations, the phase behaviors of molten A–b–B copolymers in the weak segregation regime were also investigated using equivalent free energies. In this study, the theoretical calculations were compared with the experimental results [13].

Furthermore, the self-assembly behavior of different terpolymers with the PB1,2 segment modified was studied within the scope of chemical modifications. The as-synthesized polymer brushes from the polydiene segment can have possible applications in nanotechnology [14].

In addition, in another study, the nanostructure of a bio-based epoxy thermosetting formulation was investigated by using epoxy resin modified using different amounts of the PEO–PPO–PEO triblock copolymer. The effect of the copolymer amount on the morphology of the nanostructured thermosetting system was analyzed through atomic force microscopy (AFM) [15].

Finally, two reviews are presented in this Special Issue regarding block copolymers. The first review investigates the regulation of block copolymer self-assembly structures by exploring the factors that affect the self-assembly nanostructure. This review highlights block-copolymer-based mass transport membranes, which play the role of "energy enhancers" in concentration cells, fuel cells, and rechargeable batteries [16].

The second review explores block copolymers' significant contribution to 3D printing in recent years by focusing on bio-applications. It is shown in the study that block copolymers constantly evolve through new synthetic approaches and through the introduction of new blocks with stimulating properties. This review aims to manifest the use of block copolymers as a leading evolving player in 3D/4D printing and show the significant contribution of block copolymers in bio-applications with specific advances [17].

Therefore, it may be concluded that block copolymers play a significant role in the advancement of research in polymer science and engineering. This Special Issue focuses on various matters from different areas of block copolymer research, showing advances and applications in self-assembly, crystallization, synthesis, chemical modification, 3D printing, membranes, energy materials, nanotechnology, hybrid materials, micelles, order–disorder transition, and biomaterials. As guest editors of this Special Issue, we are confident that a significant contribution to the field of block copolymers will be made through this collection of studies.

Conflicts of Interest: The authors declare no conflict of interest.

References

1. Mavronasou, K.; Zamboulis, A.; Klonos, P.; Kyritsis, A.; Bikiaris, D.N.; Papadakis, R.; Deligkiozi, I. Poly(vinyl pyridine) and Its Quaternized Derivatives: Understanding Their Solvation and Solid State Properties. *Polymers* **2022**, *14*, 804. [CrossRef] [PubMed]
2. Siddique, S.K.; Sadek, H.; Lee, T.-L.; Tsai, C.-Y.; Chang, S.-Y.; Tsai, H.-H.; Lin, T.-S.; Manesi, G.-M.; Avgeropoulos, A.; Ho, R.-M. Block Copolymer Modified Nanonetwork Epoxy Resin for Superior Energy Dissipation. *Polymers* **2022**, *14*, 1891. [CrossRef] [PubMed]
3. Arkanji, A.; Ladelta, V.; Ntetsikas, K.; Hadjichristidis, N. Synthesis and Thermal Analysis of Non-Covalent PS-*b*-SC-*b*-P2VP Triblock Terpolymers via Polylactide Stereocomplexation. *Polymers* **2022**, *14*, 2431. [CrossRef] [PubMed]
4. Qu, J.; Yang, Q.; Gong, W.; Li, M.; Cao, B. Simultaneous Removal of Cr(VI) and Phenol from Water Using Silica-di-Block Polymer Hybrids: Adsorption Kinetics and Thermodynamics. *Polymers* **2022**, *14*, 2894. [CrossRef] [PubMed]
5. Zhang, Y.; Wang, P.; Li, N.; Guo, C.; Li, S. The Effect of Topology on Block Copolymer Nanoparticles: Linear versus Star Block Copolymers in Toluene. *Polymers* **2022**, *14*, 3691. [CrossRef] [PubMed]
6. Scoti, M.; De Stefano, F.; Piscitelli, F.; Talarico, G.; Giordano, A.; De Rosa, C. Melt-Crystallizations of α and γ Forms of Isotactic Polypropylene in Propene-Butene Copolymers. *Polymers* **2022**, *14*, 3873. [CrossRef] [PubMed]
7. Scoti, M.; De Stefano, F.; Giordano, A.; Talarico, G.; De Rosa, C. Crystallization Behavior of Isotactic Propene-Octene Random Copolymers. *Polymers* **2022**, *14*, 4032. [CrossRef] [PubMed]
8. Kuznetsova, Y.L.; Sustaeva, K.S.; Mitin, A.V.; Zakharychev, E.A.; Egorikhina, M.N.; Chasova, V.O.; Farafontova, E.A.; Kobyakova, I.I.; Semenycheva, L.L. Graft Polymerization of Acrylamide in an Aqueous Dispersion of Collagen in the Presence of Tributylborane. *Polymers* **2022**, *14*, 4900. [CrossRef] [PubMed]
9. Álvarez-Gómez, A.; Yuan, J.; Fernández-Blázquez, J.P.; San-Miguel, V.; Serrano, M.B. Polyacrylonitrile-*b*-Polystyrene Block Copolymer-Derived Hierarchical Porous Carbon Materials for Supercapacitor. *Polymers* **2022**, *14*, 5109. [CrossRef] [PubMed]
10. Karayianni, M.; Koufi, D.; Pispas, S. Development of Double Hydrophilic Block Copolymer/Porphyrin Polyion Complex Micelles towards Photofunctional Nanoparticles. *Polymers* **2022**, *14*, 5186. [CrossRef] [PubMed]
11. Lazanas, A.C.; Katsouras, A.; Spanos, M.; Manesi, G.-M.; Moutsios, I.; Vashurkin, D.V.; Moschovas, D.; Gioti, C.; Karakassides, M.A.; Gregoriou, V.G.; et al. Synthesis and Characterization of Hybrid Materials Derived from Conjugated Copolymers and Reduced Graphene Oxide. *Polymers* **2022**, *14*, 5292. [CrossRef] [PubMed]
12. Lee, S.-I.; Seo, M.-G.; Huh, J.; Paik, H.-J. Small-Angle X-ray Scattering Analysis on the Estimation of Interaction Parameter of Poly(*n*-butyl acrylate)-*b*-poly(methyl methacrylate). *Polymers* **2022**, *14*, 5567. [CrossRef] [PubMed]
13. Zhang, X.; Zhao, M.; Cho, J. Effect of Disparity in Self Dispersion Interactions on Phase Behaviors of Molten A-b-B Diblock Copolymers. *Polymers* **2023**, *15*, 30. [CrossRef] [PubMed]
14. Artopoiadis, K.; Miskaki, C.; Manesi, G.-M.; Moutsios, I.; Moschovas, D.; Piryazev, A.A.; Karabela, M.; Zafeiropoulos, N.E.; Ivanov, D.A.; Avgeropoulos, A. Thermal and Bulk Properties of Triblock Terpolymers and Modified Derivatives towards Novel Polymer Brushes. *Polymers* **2023**, *15*, 848. [CrossRef] [PubMed]

15. Barandiaran, I.; Gomez-Hermoso-de-Mendoza, J.; Gutierrez, J.; Tercjak, A.; Kortaberria, G. Nanostructuring Biobased Epoxy Resin with PEO-PPO-PEO Block Copolymer. *Polymers* **2023**, *15*, 1216. [CrossRef] [PubMed]
16. Ma, S.; Hou, Y.; Hao, J.; Lin, C.; Zhao, J.; Sui, X. Well-Defined Nanostructures by Block Copolymers and Mass Transport Applications in Energy Conversion. *Polymers* **2022**, *14*, 4568. [CrossRef] [PubMed]
17. Politakos, N. Block Copolymers in 3D/4D Printing: Advances and Applications as Biomaterials. *Polymers* **2023**, *15*, 322. [CrossRef] [PubMed]

Disclaimer/Publisher's Note: The statements, opinions and data contained in all publications are solely those of the individual author(s) and contributor(s) and not of MDPI and/or the editor(s). MDPI and/or the editor(s) disclaim responsibility for any injury to people or property resulting from any ideas, methods, instructions or products referred to in the content.

Article

Nanostructuring Biobased Epoxy Resin with PEO-PPO-PEO Block Copolymer

Irati Barandiaran [1,2], Joseba Gomez-Hermoso-de-Mendoza [1], Junkal Gutierrez [1], Agnieszka Tercjak [1] and Galder Kortaberria [1,*]

[1] Group 'Materials + Technologies' (GMT), Chemical and Environmental Engineering Department, Faculty of Engineering of Gipuzkoa, University of the Basque Country (UPV/EHU), Plaza Europa 1, 20018 Donostia-San Sebastian, Spain
[2] Chemical and Environmental Engineering Department, Faculty of Pharmacy, University of the Basque Country (UPV/EHU), Paseo de la Universidad, 7, 01006 Vitoria-Gasteiz, Spain
* Correspondence: galder.cortaberria@ehu.eus

Abstract: A biobased diglycidyl ether of vanillin (DGEVA) epoxy resin was nanostructured by poly(ethylene oxide-b-propylene oxide-b-ethylene oxide) (PEO-PPO-PEO) triblock copolymer. Due to the miscibility/immiscibility properties of the triblock copolymer in DGEVA resin, different morphologies were obtained depending on the triblock copolymer amount. A hexagonally packed cylinder morphology was kept until reaching 30 wt% of PEO-PPO-PEO content, while a more complex three-phase morphology was obtained for 50 wt%, in which large worm-like PPO domains appear surrounded by two different phases, one of them rich in PEO and another phase rich in cured DGEVA. UV-vis measurements show that the transmittance is reduced with the increase in triblock copolymer content, especially at 50 wt%, probably due to the presence of PEO crystals detected by calorimetry.

Keywords: biobased; epoxy; block copolymer; nanostructuring

Citation: Barandiaran, I.; Gomez-Hermoso-de-Mendoza, J.; Gutierrez, J.; Tercjak, A.; Kortaberria, G. Nanostructuring Biobased Epoxy Resin with PEO-PPO-PEO Block Copolymer. *Polymers* **2023**, *15*, 1216. https://doi.org/10.3390/polym15051216

Academic Editor: Asterios (Stergios) Pispas

Received: 1 February 2023
Revised: 23 February 2023
Accepted: 24 February 2023
Published: 28 February 2023

Copyright: © 2023 by the authors. Licensee MDPI, Basel, Switzerland. This article is an open access article distributed under the terms and conditions of the Creative Commons Attribution (CC BY) license (https://creativecommons.org/licenses/by/4.0/).

1. Introduction

During recent years, bio-based polymers have attracted attention due to the overuse of fossil fuels as well as the increase in greenhouse gas emissions, which causes important environmental issues [1,2]. Those polymers can be obtained from renewable materials, such as lignin [3] or vegetable oils [4,5], among others. Between these different types of polymeric materials, epoxy-based thermosets are the most popular, due to their broad spectrum of properties through the selection of epoxy prepolymers and curing agents, and therefore their use in various applications, such as coatings, adhesives, and composites, among others [6–8]. Over 90% of these epoxide materials are based on bis(4-hydroxyphenylene)-2,2-propane, known as bisphenol A, to which the aromatic ring confers good thermal resistance. Commercialized for more than 50 years, these bisphenol A based thermosets (DGEBA) have been employed in many common products, such as containers, and human health applications, such as filling materials or sealants in dentistry. However, bisphenol A can also mimic the body's own hormones, and it could lead to severe negative health effects [9,10], besides cited environmental issues. Recently, poly-functional glycidyl ether derivatives based on both biobased and barely toxic extracts, such as vanillin [11–15] and phloroglucinol [16–18], which are extracted from lignins and tannins [1,19], and used as food flavoring or active ingredient in medicine, have been studied as new feedstock for thermosets. Between different biobased resins investigated by other authors, diglycidyl ether of vanillin (DGEVA) have shown good thermomechanical properties [20,21].

On the other hand, the self-assembly of block copolymers (BCP) into different nanoscale structures makes them interesting polymeric macromolecules from both academic and industrial points of view. This class of macromolecules consists of two or more covalently linked polymers, which are thermodynamically incompatible, giving rise to a variety of the microstructures. As it is well known, BCPs can self-assemble to form nanoscale structures with domain spacing that depends strongly on molecular weight, segment size and interaction between the blocks among others. Consequently, microphase separation of BCPs is determined by the degree of polymerization, N, the volume fraction of each block, f, and the Flory–Huggins interaction parameter, χ, which depends on temperature. A typical size of the microphase separated BCP domains is in the range of 10–200 nm.

BCPs can microphase separated in stable structures, such as lamellar, hexagonal-packed cylinder, body-centered cubic, close-packed spherical, or bicontinuous cubic gyroid structures. The ability to control both the length scale and the spatial organization of BCP morphologies makes these polymeric materials attractive candidates for use as templates for the fabrication of novel multifunctional materials with application in many fields of nanotechnology and advanced materials.

BCPs can also act as a nanostructuring agent for different homopolymers and thermosetting systems. As the main drawback of epoxy thermosetting polymers, for their applications as adhesives, surface coatings or composite matrices, is their low fracture toughness. One of the successful pathways to achieve high improvements on the toughness of these systems is incorporation of BCPs [22–26]. Use of the BCPs not only improves the toughness of thermosetting polymers but also leads to nanostructured thermosets, which can act as templates for dispersion and selective localization of low molecular weight organic molecules, such as azobenzene or liquid crystals, or inorganic nanoobjects, such as nanoparticles, carbon nanotubes, nanofibers and others.

Nanostructured thermosetting materials can be formed followed two different mechanisms. In the first one, the epoxy precursor acts as a selective solvent and, consequently, the microphase separation takes place before the curing reaction, and the epoxy network formation process only fixed the final morphology. In the second pathway, the microphase separation of the immiscible block takes place by reaction-induced phase separation (RIPS). Thus, the mixture of BCP and epoxy precursors is miscible before curing and the phase separation takes place during network formation.

Many authors, among which our research group can be mentioned, obtained nanostructured thermosetting systems by employing amphiphilic BCPs. Different amphiphilic BCPs used as nanostructuring agents by different authors can be found in Table 1.

Table 1. Relation of different amphiphilic block copolymers and thermosetting precursors used by different authors.

BCPs	Abbreviation	Thermosetting Precursor	References
poly(hexylene oxide)-b-poly(ethylene oxide)	PHO-b-PEO	DGEBA + PN	[24]
poly(ethylene oxide)-b-poly(ethyl ethylene)	PEO-b-PEE	DGEBA + PA	[27]
poly(ethylene oxide)-b-poly(ethylene-alt-propylene)	PEO-b-PEP	DGEBA + MDA	[28]
poly(ethylene oxide)-b-poly(propylene oxide)	PEO-b-PPO	DGEBA + MDA	[29]
poly(ethylene oxide)-b-poly(propylene oxide)-b- poly(ethylene oxide)	PEO-b-PPO-b-PEO	DGEBA + MDA DGEBA + DDM	[30,31] [32–35]
polyethylene-b-poly(ethylene oxide)	PE-b-PEO	DGEBA + MDA DGEBA + MCDEA	[36] [37]
poly(ethylene oxide)-b-poly(dimethylsiloxane)	PEO-b-PDMS	DGEBA + MDA	[38]
poly(ethylene oxide)-b-poly(ε-caprolactone)	PEO-b-PCL	DGEBA + MOCA	[39]
poly(ethylene oxide)-b-polystyrene	PEO-b-PS	DGEBA + MDA DGEBA + MXDA DGEBA + MCDEA DGEBA + DDM	[40] [41–44] [45,46] [47]

Table 1. Cont.

BCPs	Abbreviation	Thermosetting Precursor	References
poly(ε-caprolactone)-b-polybutadiene-b-poly(ε-caprolactone)	PCL-b-PB-b-PCL	DGEBA + MOCA	[48]
poly(ε-caprolactone)-b-poly(n-butyl acrylate)	PCL-b-PBA	DGEBA + MOCA	[49]
poly(heptadecafluorodecyl acrylate)-b-poly(caprolactone)	PaF-b-PCL	DGEBA + MCDEA	[50]
polydimethylsiloxane-b-poly(ε-caprolactone)-b-polystyrene	PDMS-b-PCL-b-PS	DGEBA + MOCA	[51]
poly(ε-caprolactone)-b-polystyrene	PCL-b-PS	DGEBA + MOCA	[52]

As it can be seen in Table 1, poly(ethylene glycol)-poly(propylene glycol)-poly(ethyleneglycol) (PEO-PPO-PEO), has been widely employed for nanostructuring epoxy matrices, mainly DGEBA resin [30–35]. The popularity of PEO-PPO-PEO is due to its commercial availability, including different ratios of each block as well as the simplicity of the experimental procedure and the absence of any chemical synthesis or reaction with the epoxy system [32]. For PEO-PPO-PEO/epoxy blends, the formation of the self-assembled nanostructure depends on the curing conditions, curing agent and the inner characteristics of the BCP [53]. Regarding the effect of BCP composition, Guo et al. [30] obtained different nanostructured features based on the DGEBA/MDA system and PPO-PEO-PPO copolymers with different block ratios. For the BCP with 30 wt% of PEO block, it did not find macroscopic phase separation up to a content of 50 wt%, exhibiting nanostructures based on spherical PPO domains with an average size of about 10 nm. For blends with the BCP with 80 wt% PEO, blends were not macroscopically phase-separated over the entire composition range because of the much higher PEO content, showing composition-dependent nanostructures on the order of 10–100 nm. Sun et al. [31] studied the same systems by solid-state nuclear magnetic resonance (NMR), finding that the domain size and long period depended strongly on the PEO fraction. They demonstrated that PEO blocks were only partially miscible with the cured network. Upon curing, the cross-linked rigid epoxy resin formed a separated microphase, while some PEO were locally expelled out of the cured network, forming another microphase with PPO. Similar systems but employing diamino diphenyl methane (DDM) as a hardener have also been deeply analyzed by our group [32–35]. Firstly, the miscibility and morphological features were studied, together with cure kinetics, by changing cure temperatures and copolymer amount [32]. Depending on the curing condition, phase separation took place at micro or nanoscale due to competition among kinetic and thermodynamic factors. Two distinct phases were present for every blend studied except for the system with 10 wt% PEO–PPO–PEO and cured at a low temperature. A thermodynamic model describing a thermoset/block copolymer considered as only one entity system was proposed. In a second stage, the effect of copolymer composition (block ratios) and curing conditions was analyzed [33]. A delay of cure rate was found, which increased as copolymer content and PEO molar ratio in the block copolymer increased. Infrared spectroscopy showed that PEO block was mainly responsible for physical interactions between the hydroxyl groups of growing epoxy thermoset and ether bonds of block copolymer that led to the delay in cure kinetics. Regarding structural characterization [34], taking into account DGEBA/DDM systems modifided with PEO or PPO homopolymers for comparison, it was found that, depending on the molar ratio among blocks, micro or macrophase separated morphologies were obtained. For high molar ratio among blocks, microphase-separated structures were obtained for all block copolymer contents and cure temperatures, with the self-assembly of PPO into nanoscopic entities. For low molar ratio among blocks, however, the physical interactions among the PEO block and the epoxy matrix were not favourable enough, due to the lower content of this block. Indeed, the micelles formed initially coalesced, leading to macroscopic phase separation, where different morphologies were obtained depending on copolymer content and cure temperature. Finally, the mechanical properties–morphology relationships were also analyzed [35]. Macrophase-separated systems modified with low PEO/PPO block ratio showed a similar behaviour to that for rubber-modified systems.

Increasing the content of a modifier decreases both flexural modulus and strength, while fracture toughness increases. Microphase-separated systems, on the other hand, did not present significant changes in both flexural modulus and strength for low contents, but the critical stress intensity factor increased due to partial miscibility of the PEO block with the epoxy matrix.

Parameswaranpillai et al. [54] nanostructured a DGEBA/DDM system with PEO-PPO-PEO, finding that the phase separation occurred via self-assembly of PPO blocks, followed by the reaction-induced phase separation of PEO blocks, and confirming that phase separated PEO blocks formed the crystalline phase in the amorphous crosslinked epoxy matrix.

In the present work, as a preliminary study for analyzing the nanostructuring of bio-based epoxy thermosetting formulation, DGEVA epoxy resin has been modified using different amounts of PEO-PPO-PEO triblock copolymer ranging from 10 to 50 wt%. Thermal properties are analyzed in terms of differential scanning calorimetry (DSC) and thermogravimetric analysis (TGA), while optical properties are characterized by UV-vis spectroscopy and corroborated by photographs. Finally, the effect of copolymer amount on the morphology of the nanostructured thermosetting system is analyzed by atomic force microscopy (AFM). The possibility of nanostructuring and the control of generated nanostructures will be further employed in future works for the preparation of ternary systems based on biobased epoxy thermosetts by placing nanofillers at the nanodomains.

2. Materials and Methods

2.1. Materials and Sample Preparation

The biobased epoxy used in this research work was diglycidyl ether of vanillin (DGEVA) supplied by Specific Polymers, Castries, France. The curing agent was 4,4'-diaminodiphenylmethane (DDM), purchased from Sigma-Aldrich, Darmstadt, Germany. The block copolymer used was poly(ethylene oxide-b-propylene oxide-b-ethylene oxide) (PEO-PPO-PEO) triblock copolymer (Pluronic F-127) supplied by Sigma Aldrich, Darmstadt, Germany. Chemical structures of employed materials are shown in Table 2. An amine:epoxy ratio of 1:1 was used for the DGEVA/DDM system, while PEO-PPO-PEO block copolymer content was varied from 10 to 50 wt% to design different BCP-DGEVA/DDM systems.

Table 2. Chemical structures of employed materials.

Material	Chemical Structure
DGEVA	
DDM	
PEO-PPO-PEO	

Sample preparation was carried out in the following way. First, DGEVA resin and PEO–PPO–PEO triblock copolymer were blended at 80 °C under mechanic stirring. Then, a corresponding amount of DDM was added with continuous stirring, in an oil bath at 80 °C, until a homogeneous mixture was achieved. Finally, the mixture was poured to the mold, and samples were degassed in a vacuum oven and cured at 120 °C for 6 h and post-cured under vacuum at 190 °C for 2 h. In both cases, a mechanical vacuum pump device was employed.

2.2. Techniques

Differential scanning calorimetry (DSC) measurements of the individual components, as well as BCP-DGEVA/DDM systems, were performed using a DSC3+ from Mettler Toledo equipment (Columbus, OH, USA). Thermal behavior of individual components and the DGVA/DDM system was evaluated by dynamic scans performed from −80 °C to 250 °C at 10 °C/min scan rate. The miscibility of PEO–PPO–PEO triblock copolymer with the uncured DGEVA/DDM system was investigated by dynamic scans performed from −80 °C to 50 °C at 10 °C/min scan rate. The curing processes of all BCP-DGEVA/DDM systems were analyzed by isothermal scan performed at 80, 100 and 120 °C (followed by a dynamic scan from 25 °C to 200 °C at 10 °C/min). Finally, thermal behavior of BCP-DGEVA/DDM systems was analyzed by dynamic scans performed from −80 °C to 250 °C at 10 °C/min scan rate. All experiments were performed under nitrogen atmosphere, with a flow of 10 mL/min.

Thermogravimetric tests were performed on a TGA 500 from TA Instruments Inc. (New Castle, DE, USA). Samples were heated from 25 to 800 °C at a heating rate of 10 °C/min under nitrogen atmosphere.

Fourier-transformed infrared spectroscopy (FTIR) spectra were recorded with a Nicolet Nexus spectrometer from Thermo Fisher Scientific SL (Bilbao, Spain) with a Golden Gate ATR sampling accessory. Background was recorded before every sample and the spectra were obtained in the range of 4000–650 cm^{-1}, performing 32 scans with a resolution of 4 cm^{-1}.

The morphologies of the cured BCP-DGEVA/DDM systems were studied by atomic force microscopy (AFM) under ambient conditions, using a scanning probe microscope Multimode 8 from Bruker (Billerica, MI, USA). Tapping mode (TM) was employed in air using an integrated tip/cantilever (125 mm in length with a 300 kHz resonant frequency). Measurements were performed with 512 scan lines and target amplitude around 0.9 V. Different regions of the cured BCP-DGEVA/DDM systems were scanned to ensure that the morphology of the investigated materials was a representative one. Samples were cut using an ultramicrotome Leica Ultracut R with a diamond blade.

UV-vis transmittance spectra of the cured BCP-DGEVA/DDM systems were performed with a Shimadzu UV-3600 (Kioto, Japan) spectrophotometer in the range between 200 and 800 nm.

3. Results and Discussion

3.1. Differential Scaning Calorimetry Analysis

DSC dynamic measurements were carried out in order to investigate the miscibility between PEO-PPO-PEO triblock copolymer and DGEVA/DDM.

Figure 1A shows thermograms of blend components (PEO-PPO-PEO and DGEVA/DDM). Moreover, uncured BCP-DGEVA/DDM blends with different BCP amounts were also included in Figure 1B. PEO-PPO-PEO thermogram shows a T_g at around −68.5 °C [32] and a melting peak centered at 58.0 °C, related to the melting of crystalline PEO block. DGEVA resin presents a T_g at −41.7 °C that increased up to −32.5 °C when curing agent was added. This behavior can be related to the partial miscibility between the DGEVA resin and the amine before curing. In addition, the DGEVA/DDM thermogram shows an exothermic peak centered around 140 °C, indicating the curing reaction of the blend. If the dynamic thermograms for BCP-DGEVA/DDM systems are compared, the T_g of the DGEVA resin phase decreases from −36.5 °C to −44.0 °C with increasing BCP content, owing to the miscibility of PEO-PPO-PEO and DGEVA [32,37]. Moreover, at the thermogram of the 50BCP-DGEVA/DDM system, the melting of the crystalline phase of the PEO block is detected and, in contrast to that of neat BCP, the melting happens in two steps, indicating the presence of different types of crystals. This phenomenon will be further discussed in the morphology section shown below.

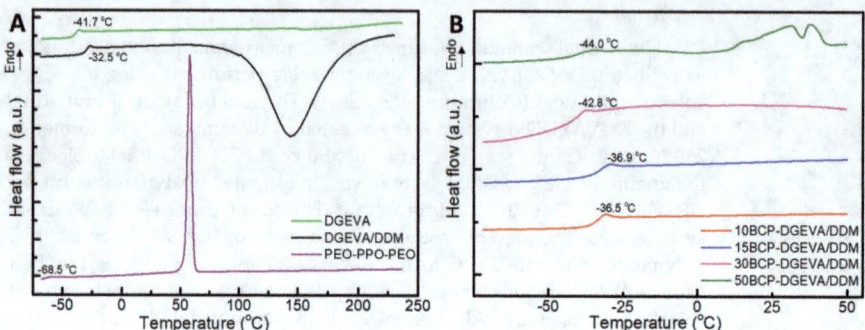

Figure 1. Dynamic DSC thermograms of (**A**) the neat components (DGEVA resin and PEO-PPO-PEO triblock copolymer) and uncured DGEVA/DDM blend, and (**B**) uncured BCP-DGEVA/DDM blends with different BCP contents from 10 to 50 wt%. Note: thermograms have been shifted along the Y-axes for a better visualization.

The curing behavior of all BCP-DGEVA/DDM blends was analyzed by isothermal thermograms at 80, 100 and 120 °C (Figure 2).

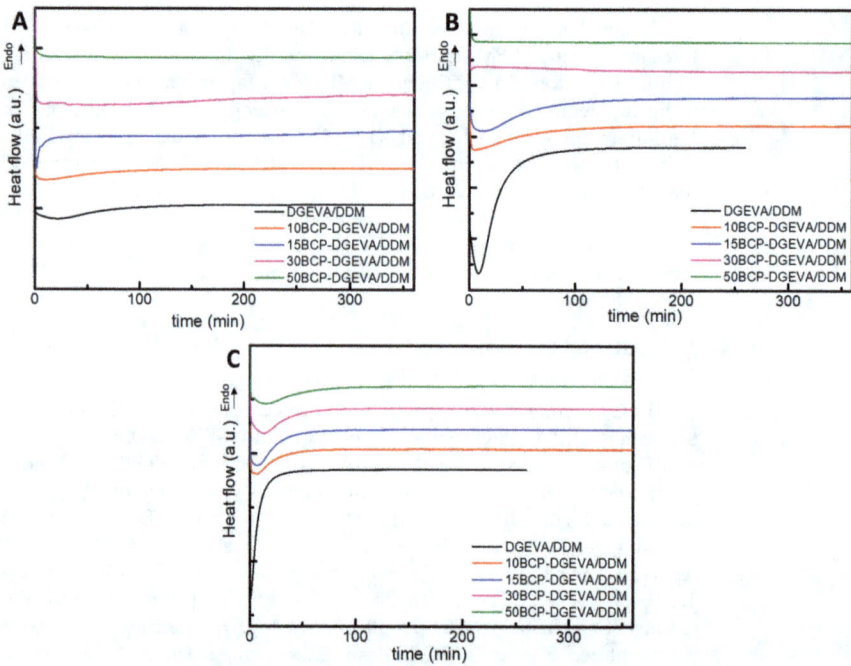

Figure 2. Isothermal DSC thermograms of investigated BCP-DGEVA/DDM systems with BCP content from 0 to 50 wt% at (**A**) 80, (**B**) 100, and (**C**) 120 °C. Note: thermograms have been shifted along the Y-axes for a better visualization.

As can be observed, the reaction rate increased with the increasing of curing temperature, while BCP addition delayed the exothermic peak of the isothermal thermograms at all temperatures, due to the dilution effect of PEO-PPO-PEO [32,46]. For systems with highest amount of BCP, the exothermic peak almost disappeared at 80 and 100 °C, as the full conversion was not reached in the analyzed time scale at 80 and 100 °C. For these

BCP-DGEVA/DDM systems, the curing process would be completed at the post-curing stage. At 120 °C the curing reaction was completed for all investigated BCP-DGEVA/DDM systems, and all composites were cured at 120 °C.

All BCP-DGEVA/DDM systems cured at 120 °C were studied by dynamic DSC analysis (Figure 3). The neat DGEVA/DDM epoxy system showed a T_g at 115.1 °C. With PEO-PPO-PEO triblock copolymer addition, the T_g of the epoxy matrix decreased from 109.5 °C (10BCP-DGEVA/DDM system) to 91.6 °C (50BCP-DGEVA/DDM system), confirming of the epoxy matrix the miscibility between DGEVA epoxy resin and PEO block of PEO-PPO-PEO [26]. However, the sample with 30 wt% of BCP presented a lower T_g value of for epoxy matrix than the sample with 50 wt% of BCP. Moreover, the 50BCP-DGEVA/DDM system presented an additional T_g at −50.2 °C, which could be attributed to the T_g of PEO block of PEO-PPO-PEO. Dynamic thermograms of BCP-DGEVA/DDM systems with BCP content from 10 to 50 wt% also presented endothermic peaks, related to the melting of PEO block of the BCP, which is represented by two peaks that tend to become closer as the BCP content is increased, indicating different types of crystals.

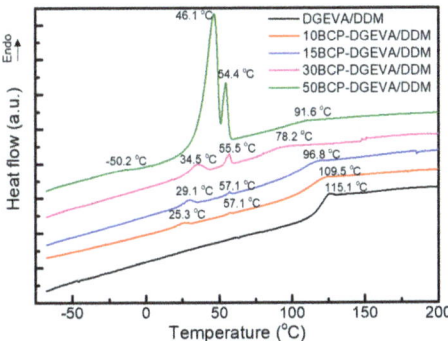

Figure 3. Dynamic DSC thermograms of investigated BCP-DGEVA/DDM systems with BCP content from 0 to 50 wt% cured at 120 °C. Note: thermograms have been shifted along the Y-axes for a better visualization.

As has been pointed out by other authors [26,33], the T_g of a blend depends on the weight fraction of the components. For this reason, it could be expected that the higher the BCP content in the mixture, the lower the T_g of the system will be. However, in this case, although the BCP amount is higher for the 50BCP-DGEVA/DDM system than for the 30BCP-DGEVA/DDM one, the value of the T_g increases. This could indicate that part of the PEO block phase separates from the DGEVA/DDM matrix, as found in previous works of our group [26]. These melting peaks increased significantly in the case of the 50BCP-DGEVA/DDM system. The increase in melting peaks, together with the presence of the T_g of PEO block of BCP and the higher T_g of epoxy matrix when compared with that for the 30BCP-DGEVA/DDM system, could indicate that the phase separated BCP content could be remarkably higher in this case than for the rest of the systems.

3.2. Thermogravimetric Analysis

Thermogravimetric analysis of PEO-PPO-PEO triblock copolymer and developed BCP-DGEVA/DDM systems was also carried out (Figure 4). If the thermal degradation curves of PEO-PPO-PEO and neat DGEVA/DDM are compared, for both samples the main degradation step occurs between 350 and 425 °C. As a result, for BCP-DGEVA/DDM systems the main degradation occurs in the same temperature range, showing that BCP addition does not affect the thermal stability of the system. On the other side, formed char amount (32 wt% for neat epoxy system) proportionally decreases with BCP content from 29 wt% for the 10BCP-DGEVA/DDM system to 14 wt% for the 50BCP-DGEVA/DDM one.

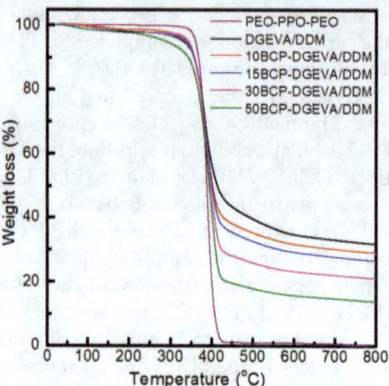

Figure 4. TGA curves of PEO-PPO-PEO block copolymer and BCP-DGEVA/DDM systems with BCP content from 0 to 50 wt% cured at 120 °C.

3.3. Fourier-Transform Infrared Spectroscopy Analysis

Figure 5A shows FTIR spectra of DGEVA resin and the DGEVA/DDM system. If these two spectra are compared, in the case of DGEVA/DDM system, a broad band centered at 3370 cm^{-1} is detected, attributed to the alcohol groups formed after the reaction of the epoxy groups of DGEVA and the amine groups of DDM [33]. In addition, the peak related to the epoxide group (910 cm^{-1}) disappears at the cured spectrum, proving the curing of the epoxy resin [33,55].

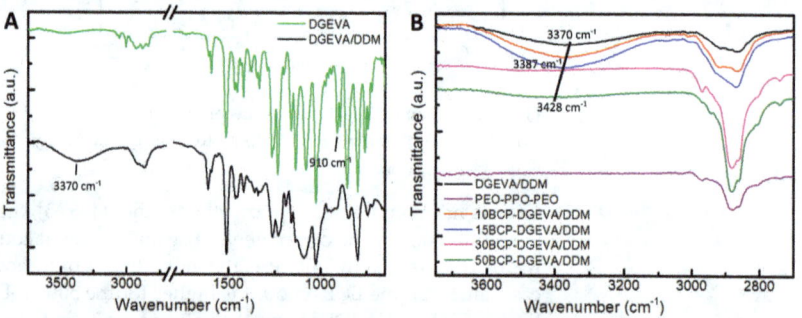

Figure 5. FTIR spectra of (**A**) neat DGEVA resin and DGEVA/DDM system, and (**B**) investigated BCP-DGEVA/DDM systems with BCP content from 0 to 50 wt% cured at 120 °C. Note: spectra have been shifted along the Y-axes for a better visualization.

Regarding the effect of PEO-PPO-PEO triblock copolymer addition, Figure 5B shows that by increasing BCP content, the intensity of the band related to alcohol groups (broad band centered at 3370 cm^{-1}, in DGEVA/DDM) decreases and shifts towards higher wavenumbers (3387 and 3428 cm^{-1} for 15BCP-DGEVA/DDM and 50BCP-DGEVA/DDM, respectively). This could be due to the hydrogen bonding interaction between the OH groups formed in the cured resin and the PEO block of the triblock copolymer [33]. Moreover, in the spectra of systems with higher BCP content (30 and 50 wt%) the bands related to the PEO-PPO-PEO block copolymer present higher intensity.

3.4. Atomic Force Microscopy

The morphologies of the BCP-DGEVA/DDM systems cured at 120 °C were investigated by AFM. As can be observed in Figure 6, all investigated systems show microphase separation at nanoscale. In the case of the samples with BCP content up to 30 wt%, it is remarkable the formation of small nanostructures. For the system with 10 wt% of block copolymer (Figure 6A), a hexagonally packed cylinder morphology is formed, in which the dark spherical domains with diameters ranging from 10 to 15 nm correspond to the PPO block rich phase, while the continuous light phase corresponds to the PEO-epoxy rich one [31]. As other authors have reported for DGEBA epoxy systems [26,37,46], it seems that as a result of the interactions between the PEO block and epoxy resin, the PEO block is miscible with DGEBA epoxy resin, while PPO remains immiscible. In the case of the system based on DGEVA resin, a similar behavior is observed. As shown in Figure 6B, an increase of 5 wt% in PEO-PPO-PEO triblock copolymer content seems not to affect the morphology observed, as the 15BCP-DGEVA/DDM system presents the same morphology than the 10BCP-DGEVA/DDM one. Moreover, for the 30BCP-DGEVA/DDM system, no significant morphological changes are detected, observing a similar hexagonally packed cylinder morphology (marked at the images) than for 10BCP-DGEVA/DDM and 15BCP-DGEVA/DDM systems. On the contrary, when PEO-PPO-PEO concentration rises up to 50 wt%, the morphology changes drastically. In this case, large worm-like domains (PPO block) are observed, surrounded by two different phases, one of them rich in PEO (lower hardness) and the last phase rich in cured DGEVA [56]. This fact could be in agreement with the dynamic DSC results for the 50BCP-DGEVA/DDM system shown in Figure 3, in which an additional T_g attributed to the PEO block of PEO-PPO-PEO was detected.

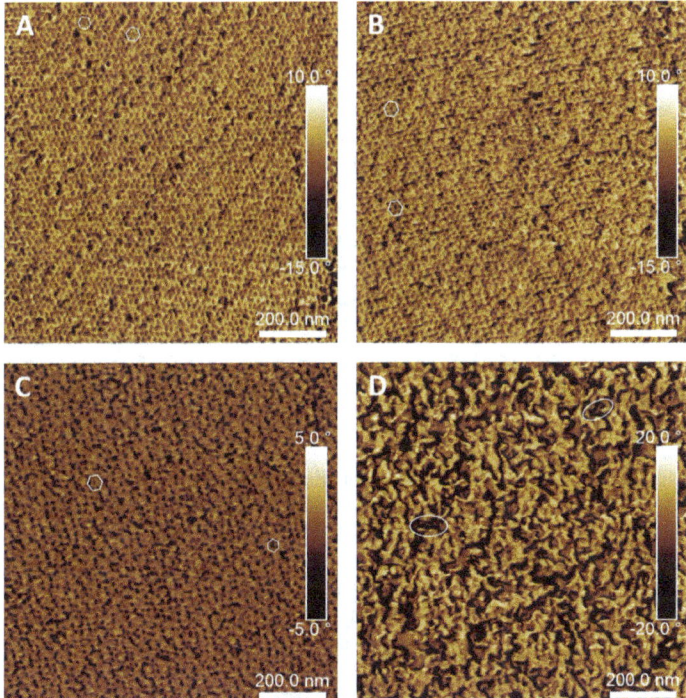

Figure 6. AFM phase images of BCP-DGEVA/DDMP systems cured at 120 °C with (**A**) 10 wt%, (**B**) 15 wt%, (**C**) 30 wt%, and (**D**) 50 wt% of PEO-PPO-PEO.

3.5. UV-Vis Spectroscopy

UV-vis transmittance results of the BCP-DGEVA/DDM systems with different PEO-PPO-PEO contents are shown in Figure 7. The transmittance of the DGEVA/DDM system decreases with the addition of triblock copolymer. The DGEVA/DDM system presents a transmittance value of 77% at 650 nm, while values of 75, 74 and 73% are measured for systems with 10, 15 and 30% of BCP, respectively. When the triblock copolymer content increases to 50 wt%, the transmittance value at 650 nm is reduced to 19%. The presence of PEO crystals observed by DSC for this system (increase in the melting temperature for PEO block) could explain the drastic reduction in the transmittance.

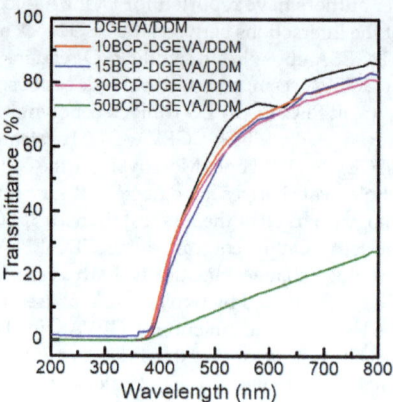

Figure 7. UV-vis transmittance results of BCP-DGEVA/DDM systems with BCP content from 0 to 50 wt% cured at 120 °C.

The digital images of samples shown in Figure 8 corroborate the results obtained by UV-vis transmittance. The systems with PEO-PPO-PEO concentrations up to 30 wt% allow light transmittance, while the 50BCP-DGEVA/DDM system presents much lower transmittance and the image behind cannot be clearly seen.

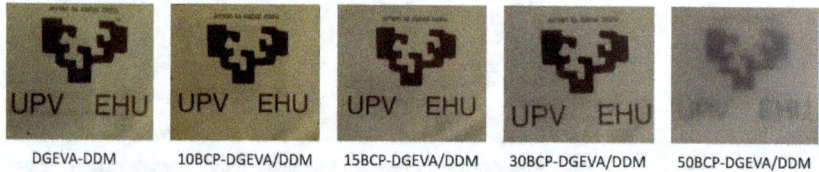

Figure 8. Photographs of all BCP-DGEVA/DDM samples cured at 120 °C.

4. Conclusions

The following conclusions can be extracted from this preliminary work on the nanostructuring of bio-based epoxy matrix by PEO-PPO-PEO block copolymer.

This research work demonstrates that the biobased DGEVA epoxy resin is an adequate resin to be nanostructured with PEO-PPO-PEO triblock copolymer. The curing temperature was set at 120 °C, as at lower temperatures systems with higher BCP content did not reach full conversion, as BCP addition delayed the cure reaction by dilution effect. The investigated systems showed, up to 30 wt% of triblock copolymer, a hexagonally packed cylinder morphology, with spherical domains ranging from 10 to 15 nm. In the 50BCP-DGEVA/DDM system, a change in the morphology was detected, forming a more complex morphology with phase separation of PEO-PPO-PEO triblock copolymer. These results are in agreement with presented DSC thermograms, in which an additional T_g related to PEO crystals was detected, and also with the transmittance data obtained by UV-vis, as the most significant decrease in transmittance was not detected up to a BCP content of 50 wt%, probably due to the presence of PEO crystals.

Author Contributions: Conceptualization, I.B. and G.K.; methodology, J.G.; software, J.G.; validation, I.B., J.G.-H.-d.-M. and A.T.; formal analysis, G.K.; investigation, A.T; resources, G.K.; data curation, I.B.; writing—original draft preparation, G.K. and A.T.; writing—review and editing, A.T.; visualization, J.G.; supervision, G.K.; project administration, A.T.; funding acquisition, G.K. and A.T. All authors have read and agreed to the published version of the manuscript.

Funding: This research was funded by Ministerio de Ciencia, Innovacion y Universidades, grant number PID2021-126417NB-I00 and by the Gobierno Vasco/Eusko Jaurlaritza, grant number IT1690-22.

Institutional Review Board Statement: Not applicable.

Informed Consent Statement: Not applicable.

Data Availability Statement: The data presented are available on request from the corresponding author.

Acknowledgments: J.G.H. thanks Basque Government for PhD Fellowship (PRE_2021_2_0044); Macrobehavior-Mesostructure-Nanotechnology SGIker unit of UPV/EHU is also acknowledged.

Conflicts of Interest: The authors declare no conflict of interest.

References

1. Benyahya, S.; Aouf, C.; Caillol, S.; Boutevin, B.; Pascault, J.P.; Fulcrand, H. Functionalized green tea tannins as phenol-ic prepolymers for bio-based epoxy resins. *Ind. Crops Prod.* **2014**, *53*, 296–307. [CrossRef]
2. Ma, S.; Liu, X.; Jiang, Y.; Zhang, C.; Zhu, J. Bio-based epoxy resin from itaconic acid and its thermosets cured with an-hydride and comonomers. *Green Chem.* **2013**, *15*, 245–254. [CrossRef]
3. Ferdosian, F.; Yuan, Z.; Anderson, M.; Xu, C. Synthesis of lignin-based epoxy resins: Optimization of reaction parame-ters using response surface methodology. *RSC Adv.* **2014**, *4*, 31745–31753. [CrossRef]
4. Tanm, S.G.; Chow, W.S. Thermal properties of anhydride-cured bio-based epoxy blends. *J. Therm. Anal. Calorim.* **2010**, *101*, 1051–1058.
5. Zhu, J.; Chandrashekhara, K.; Flanigan, V.; Kapila, J. Curing and mechanical characterization of a soy-based epoxy resin system. *J. Appl. Polym. Sci.* **2004**, *91*, 3513–3518. [CrossRef]
6. Raquez, J.M.; Deleglise, M.; Lacrampe, M.F.; Krawczak, P. Thermosetting (bio)materials derived from renewable re-sources: A critical review. *Prog. Polym. Sci.* **2010**, *35*, 487–509. [CrossRef]
7. Shi, X.H.; Chen, L.; Liu, B.W.; Long, J.W.; Xu, X.J.; Wang, Y.Z. Carbon fibers decorated by polyelectrolyte complexes toward their epoxy resin composites with high fire safety. *Chin. J. Polym. Sci.* **2018**, *36*, 1375–1384. [CrossRef]
8. Jin, F.L.; Li, X.; Park, S.J. Synthesis and application of epoxy resins: A review. *J. Ind. Eng. Chem.* **2015**, *29*, 1–11. [CrossRef]
9. vom Saal, F.S.; Hughes, C. An extensive new literature concerning low-dose effects of bisphenol A shows the need for a new risk assessment. *Environ. Health Perspect.* **2005**, *113*, 926–933. [CrossRef]
10. vom Saal, F.S.; Myers, J.P. Bisphenol A and risk of metabolic disorders. *JAMA* **2008**, *300*, 1353–1355. [CrossRef]
11. Fache, M.; Darroman, E.; Besse, V.; Auvergne, R.; Caillol, S.; Boutevin, B. Vanillin, a promising biobased build-ing-block for monomer synthesis. *Green Chem.* **2014**, *16*, 1987–1998. [CrossRef]
12. Fache, M.; Monteremal, C.; Boutevin, B.; Caillol, S. Amine hardeners and epoxy crosslinker from aromatic renewable resources. *Eur. Polym. J.* **2015**, *73*, 344–362. [CrossRef]
13. Fache, M.; Boutevin, B.; Caillol, S. Vanillin, a key-intermediate of biobased polymers. *Eur. Polym. J.* **2015**, *68*, 488–502. [CrossRef]
14. Fache, M.; Auvergne, R.; Boutevin, B.; Caillol, S. New vanillin-derived diepoxy monomers for the synthesis of bi-obased thermosets. *Eur. Polym. J.* **2015**, *67*, 527–538. [CrossRef]
15. Fache, M.; Boutevin, B.; Caillol, S. Epoxy thermosets from model mixtures of the lignin-to-vanillin process. *Green Chem.* **2016**, *18*, 712–725. [CrossRef]
16. Menard, R.; Negrell, C.; Fache, M.; Ferry, L.; Sonnier, R.; David, G. From a bio-based phosphorus-containing epoxy monomer to fully bio-based flame-retardant thermosets. *RSC Adv.* **2015**, *5*, 70856–70867. [CrossRef]
17. Menard, R.; Negrell, C.; Ferry, L.; Sonnier, R.; David, G. Synthesis of biobased phosphorus-containing flame retardants for epoxy thermosets comparison of additive and reactive approaches. *Polym. Degrad. Stab.* **2016**, *120*, 300–312. [CrossRef]
18. Fanny, J.; Hélène, N.; Bernard, B.; Sylvain, C. Synthesis of novel vinyl ester from biobased phloroglucinol. *Green Mater.* **2016**, *4*, 63–71.
19. Aouf, C.; Nouailhas, H.; Fache, M.; Caillol, S.; Boutevin, B.; Fulcrand, B.H. Multifunctionalization of gallic acid. Synthesis of a novel bio-based epoxy resin. *Eur. Polym. J.* **2013**, *49*, 1185–1195.
20. Nikafshar, S.; Zabihi, O.; Hamidi, S.; Moradi, Y.; Barzegar, S.; Ahmadi, M.; Naebe, M. A renewable bio-based epoxy resin with improved mechanical performance that can compete with DGEBA. *RSC Adv.* **2017**, *7*, 8694–8701. [CrossRef]
21. Mauck, J.R.; Yadav, S.K.; Sadler, J.M.; La Scala, J.J.; Palmese, G.R.; Schmalbach, K.M.; Stanzione, J.F. Preparation and characterization of highly bio-based epoxy amine thermosets derived from lignocellulosics. *Macromol. Chem. Phys.* **2017**, *218*, 1700013. [CrossRef]

22. Dean, J.M.; Grubbs, R.B.; Saad, W.; Cook, R.F.; Bates, F.S. Mechanical properties of block copolymer vesicle and micelle modified epoxies. *J. Polym. Sci. Part B Polym. Phys.* **2003**, *41*, 2444–2456. [CrossRef]
23. Rebizant, V.; Venet, A.-S.; Tournilhac, F.; Girard-Reydet, E.; Navarro, C.; Pascault, J.-P.; Leibler, L. Chemistry and mechanical properties of epoxy-based thermosets reinforced by reactive and nonreactive SBMX block copolymers. *Macromolecules* **2004**, *34*, 8017–8027. [CrossRef]
24. Thio, Y.S.; Wu, J.; Bates, F.S. Epoxy toughening using low molecular weight poly(hexylene oxide)-poly(ethylene oxide) diblock copolymers. *Macromolecules* **2006**, *39*, 7187–7189. [CrossRef]
25. Liu, J.D.; Thompson, Z.J.; Sue, H.-J.; Bates, F.S.; Hillmyer, M.A.; Dettloff, M.; Jacob, G.; Verghese, N.; Pham, H. Toughening of epoxies with block copolymer micelles of wormlike morphology. *Macromolecules* **2010**, *43*, 7238–7243. [CrossRef]
26. Cano, L.; Builes, D.H.; Tercjak, A. Morphological and mechanical study of nanostructured epoxy systems modified with amphiphilic poly(ethylene oxide-b-propylene oxide-b-ethylene oxide)triblock copolymer. *Polymer* **2014**, *55*, 738–745. [CrossRef]
27. Hillmyer, M.A.; Lipic, P.M.; Hajduk, D.A.; Almdal, K.; Bates, F.S. Self-assembly and polymerization of epoxy resin-amphiphilic block copolymer nanocomposites. *J. Am. Chem. Soc.* **1997**, *119*, 2749–2750. [CrossRef]
28. Lipic, P.M.; Bates, F.S.; Hillmyer, M.A. Nanostructured thermosets from self-assembled amphiphilic block copolymer/epoxy resin mixtures. *J. Am. Chem. Soc.* **1998**, *120*, 8963–8970. [CrossRef]
29. Mijovic, J.; Shen, M.; Sy, J.W.; Mondragon, I. Dynamics and morphology in nanostructured thermoset network/block copolymer blends during network formation. *Macromolecules* **2000**, *33*, 5235–5244. [CrossRef]
30. Guo, Q.; Thomann, R.; Gronski, W.; Thurn-Albrecht, T. Phase behavior, crystallization, and hierarchical nanostructures in self-organized thermoset blends of epoxy resin and amphiphilic poly(ethylene oxide)-block-poly(propylene oxide)-block-poly(ethylene oxide) triblock copolymers. *Macromolecules* **2002**, *35*, 3133–3144. [CrossRef]
31. Sun, P.; Dang, Q.; Li, B.; Chen, T.; Wang, Y.; Lin, H.; Jin, Q.; Ding, D. Mobility, miscibility, and microdomain structure in nanostructured thermoset blends of epoxy resin and amphiphilic poly(ethylene oxide)-block-poly(propylene oxide)-block-poly(ethylene oxide) triblock copolymers characterized by solid-state NMR. *Macromolecules* **2005**, *38*, 5654–5667. [CrossRef]
32. Larrañaga, M.; Gabilondo, N.; Kortaberria, G.; Serrano, E.; Remiro, P.; Riccardi, C.C.; Mondragon, I. Micro- or nanoseparated phases in thermoset blends of an epoxy resin and PEO–PPO–PEO triblock copolymer. *Polymer* **2005**, *46*, 7082–7093. [CrossRef]
33. Larrañaga, M.; Martin, M.D.; Gabilondo, N.; Kortaberria, G.; Eceiza, A.; Riccardi, C.C.; Mondragon, I. Toward microphase separation in epoxy systems containing PEO–PPO–PEO block copolymers by controlling cure conditions and molar ratios between blocks. Part 1. Cure kinetics. *Colloid Polym. Sci.* **2006**, *284*, 1403–1410. [CrossRef]
34. Larrañaga, M.; Arruti, P.; Serrano, E.; de la Caba, K.; Remiro, P.; Riccardi, C.C.; Mondragon, I. Towards microphase separation in epoxy systems containing PEO/PPO/PEO block copolymers by controlling cure conditions and molar ratios between blocks. Part 2. Structural characterization. *Colloid Polym. Sci.* **2006**, *284*, 1419–1430. [CrossRef]
35. Larrañaga, M.; Serrano, E.; Martin, M.D.; Tercjak, A.; Kortaberria, G.; de la Caba, K.; Riccardi, C.C.; Mondragon, I. Mechanical properties-morphology relationships in nano-/microstructured epoxy matrices modified with PEO-PPO-PEO block copolymers. *Polym. Int.* **2007**, *56*, 1392–1403. [CrossRef]
36. Guo, Q.; Thomann, R.; Gronski, W.; Staneva, R.; Ivanova, R.; Stühn, B. Nanostructures, semicrytalline morphology, and nanoscale confinement effect on the crystallization kinetics in self-organized block copolymer/thermoset blends. *Macromolecules* **2003**, *36*, 3635–3645. [CrossRef]
37. Tercjak, A.; Larrañaga, M.; Martin, M.D.; Mondragon, I. Thermally reversible nanostructured thermosetting blends modified with poly(ethylene-b-ethylene oxide) diblock copolymer. *J. Therm. Anal. Calorim.* **2006**, *86*, 663–668. [CrossRef]
38. Guo, Q.; Chen, F.; Wang, K.; Chen, L. Nanostructured thermoset epoxy resin templated by an amphiphilic poly(ethylene oxide)-block-poly(dimethylsiloxane) diblock copolymer. *J. Polym. Sci. Part B Polym. Phys.* **2006**, *44*, 3042–3052. [CrossRef]
39. Meng, F.; Zheng, S.; Liu, T. Epoxy resin containing poly(ethylene oxide)-block-poly(ε-caprolactone) diblock copolymer: Effect of curing agents on nanostructures. *Polymer* **2006**, *47*, 7590–7600. [CrossRef]
40. Meng, F.; Zheng, S.; Li, H.; Liang, Q.; Liu, T. Formation of ordered nanostructures in epoxy thermosets: A mechanism of reaction-induced microphase separation. *Macromolecules* **2006**, *39*, 5072–5080. [CrossRef]
41. Tercjak, A.; Serrano, E.; Garcia, I.; Mondragon, I. Thermoresponsive meso/nanostructured thermosetting materials based on PS-b-PEO block copolymer-dispersed liquid crystal: Curing behavior and morphological variation. *Acta Mater.* **2008**, *56*, 5112–5122. [CrossRef]
42. Tercjak, A.; Mondragon, I. Relationships between the morphology and thermoresponsive behavior in micro/nanostructured thermosetting matrixes containing a 4′-(hexyloxy)-4-biphenylcarbonitrile liquid crystal. *Langmuir* **2008**, *24*, 11216–11224. [CrossRef]
43. Tercjak, A.; Gutierrez, J.; Peponi, L.; Rueda, L.; Mondragon, I. Arrangement of conductive TiO$_2$ nanoparticles in hybrid inorganic/organic thermosetting materials using liquid crystal. *Macromolecules* **2009**, *42*, 3386–3390. [CrossRef]
44. Tercjak, A.; Gutierrez, J.; Mondragon, I. Conductive properties of inorganic/organic nanostructured systems based on block copolymers. *Mater. Sci. Forum* **2012**, *714*, 153–158. [CrossRef]
45. Gutierrez, J.; Tercjak, A.; Mondragon, I. Transparent nanostructured thermoset composites containing well-dispersed TiO$_2$ nanoparticles. *J. Phys. Chem. C* **2010**, *114*, 22424–22430. [CrossRef]
46. Gutierrez, J.; Mondragon, I.; Tercjak, A. Morphological and optical behavior of thermoset matrix composites varying both polystyrene-block-poly(ethylene oxide) and TiO$_2$ nanoparticle content. *Polymer* **2011**, *52*, 5699–5707. [CrossRef]

47. Francis, R.; Baby, D.K. Toughening of epoxy thermoset with polystyrene-block-polyglycolic acid star copolymer: Nanostructure—Mechanical property correlation. *Ind. Eng. Chem. Res.* **2014**, *53*, 17945–17951. [CrossRef]
48. Meng, F.; Zheng, S.; Zhang, W.; Li, H.; Liang, Q. Nanostructured thermosetting blends of epoxy resin and amphiphilic poly(ε-caprolactone)-block-polybutadiene-block-poly(ε-caprolactone) triblock copolymer. *Macromolecules* **2006**, *39*, 711–719. [CrossRef]
49. Xu, Z.; Zheng, S. Reaction-induced microphase separation in epoxy thermosets containing poly(ε-caprolactone)-block-poly(n-butyl acrylate) diblock copolymer. *Macromolecules* **2007**, *40*, 2548–2558. [CrossRef]
50. Ocando, C.; Serrano, E.; Tercjak, A.; Peña, C.; Kortaberria, G.; Calberg, C.; Grignard, B.; Jerome, R.; Carrasco, P.M.; Mecerreyes, D.; et al. Structure and properties of a semifluorinated diblock copolymer modified epoxy blend. *Macromolecules* **2007**, *40*, 4068–4074. [CrossRef]
51. Fan, W.; Wang, L.; Zheng, S. Nanostructures in thermosetting blends of epoxy resin with polydimethylsiloxane-block-poly(ε-caprolactone)-block-polystyrene ABC triblock copolymer. *Macromolecules* **2009**, *42*, 327–336. [CrossRef]
52. Zhu, L.; Zhang, C.; Han, J.; Zheng, S.; Li, X. Formation of nanophases in epoxy thermosets containing an organic–inorganic macrocyclic molecular brush with poly(ε-caprolactone)-block-polystyrene side chains. *Soft Matter* **2012**, *8*, 7062–7072. [CrossRef]
53. Ji, S.; Wan, L.; Liu, C.C.; Nealey, P.F. Directed self-assembly of block copolymers on chemical patterns: A platform for nanofabrication. *Prog. Polym. Sci.* **2016**, *54-55*, 76–172. [CrossRef]
54. Parameswaranpillai, J.; Sidhardhan, S.K.; Harikrishnan, P.; Pionteck, J.; Siengchin, S.; Unni, A.B.; Magueresse, A.; Grohens, Y.; Hameed, N.; Jose, S. Morphology, thermo-echanical properties and surface hydrofobicity of nanostructured epoxy thermosets modified with PEO-PPO-PEO triblock copolymer. *Polym. Test.* **2017**, *59*, 168–176. [CrossRef]
55. Cherdoud-Chihani, A.; Mouzali, M.; Abadie, M.J.M. Study of crosslinking acid copolymer/DGEBA systems by FTIR. *J. Appl. Polym. Sci.* **2003**, *87*, 2033–2051. [CrossRef]
56. Builes, D.H.; Hernandez, H.; Mondragon, I.; Tercjak, A. Relationship between the morphology of nanostructured unsaturated polyesters modified with PEO-b-PPO-b-PEO triblock copolymer and their optical and mechanical properties. *J. Phys. Chem. C* **2013**, *117*, 3563–3571. [CrossRef]

Disclaimer/Publisher's Note: The statements, opinions and data contained in all publications are solely those of the individual author(s) and contributor(s) and not of MDPI and/or the editor(s). MDPI and/or the editor(s) disclaim responsibility for any injury to people or property resulting from any ideas, methods, instructions or products referred to in the content.

Article

Thermal and Bulk Properties of Triblock Terpolymers and Modified Derivatives towards Novel Polymer Brushes

Konstantinos Artopoiadis [1], Christina Miskaki [1], Gkreti-Maria Manesi [1], Ioannis Moutsios [1,2], Dimitrios Moschovas [1], Alexey A. Piryazev [3], Maria Karabela [1], Nikolaos E. Zafeiropoulos [1], Dimitri A. Ivanov [2,3,4] and Apostolos Avgeropoulos [1,4,*]

[1] Department of Materials Science Engineering, University of Ioannina, University Campus-Dourouti, 45110 Ioannina, Greece
[2] Institut de Sciences des Matériaux de Mulhouse—IS2M, CNRS UMR7361, 15 Jean Starcky, 68057 Mulhouse, France
[3] Institute of Problems of Chemical Physics, Russian Academy of Sciences, Chernogolovka, 142432 Moscow, Russia
[4] Faculty of Chemistry, Lomonosov Moscow State University (MSU), GSP-1, 1-3 Leninskiye Gory, 119991 Moscow, Russia
* Correspondence: aavger@uoi.gr; Tel.: +30-2651009001

Abstract: We report the synthesis of three (3) linear triblock terpolymers, two (2) of the ABC type and one (1) of the BAC type, where A, B and C correspond to three chemically incompatible blocks such as polystyrene (PS), poly(butadiene) of exclusively (~100% vinyl-type) -1,2 microstructure ($PB_{1,2}$) and poly(dimethylsiloxane) (PDMS) respectively. Living anionic polymerization enabled the synthesis of narrowly dispersed terpolymers with low average molecular weights and different composition ratios, as verified by multiple molecular characterization techniques. To evaluate their self-assembly behavior, transmission electron microscopy and small-angle X-ray scattering experiments were conducted, indicating the effect of asymmetric compositions and interactions as well as inversed segment sequence on the adopted morphologies. Furthermore, post-polymerization chemical modification reactions such as hydroboration and oxidation were carried out on the extremely low molecular weight $PB_{1,2}$ in all three terpolymer samples. To justify the successful incorporation of –OH groups in the polydiene segments and the preparation of polymeric brushes, various molecular, thermal, and surface analysis measurements were carried out. The synthesis and chemical modification reactions on such triblock terpolymers are performed for the first time to the best of our knowledge and constitute a promising route to design polymers for nanotechnology applications.

Keywords: low molecular weight triblock terpolymers; hydroboration/oxidation reactions; modification of $PB_{1,2}$; high χ terpolymers; self-assembly; asymmetric interactions; double gyroid; polymer brushes; wettability

1. Introduction

The ability of polymers to self-assemble in bulk or thin films using solvents with different selectivity enables the formation of various well-defined morphologies at the nanoscale. This fundamental characteristic is of major importance, leading to their possible utilization in nano-lithographic applications as reported in the literature [1–3].

Triblock terpolymers adopt various ordered three-phase morphologies not evident in linear diblock and triblock copolymers as expected due to the lack of the third segment. It is documented that the morphologies in ABC systems are significantly affected by the block sequence (ABC vs. BCA vs. CAB), the volume fraction ratios, and the three different values of the Flory–Huggins interaction parameters (χ_{AB}, χ_{BC}, χ_{BA}) [4–6]. Unique pattern geometries are adopted after solution and/or spin casting methods, including nanoscale rings that provide electronic [7–10], optical [11–14], and magnetic [15,16] properties that

are highly desirable for several applications. Despite this, due to the dissimilar properties of the adjacent segments, such as solubility and interfacial or surface energy, only limited research has been conducted in this field.

To tune the self-assembly and surface properties, the scientific community has shifted its interest towards polymer brushes [17–27], which are comprised of a polymeric backbone with attached active sites that may provide the desired surface properties [18]. A route to obtaining specific surface properties is to combine different segments bearing functional groups and then perform post-polymerization modification reactions. The modified segment plays the polymer brush role. Note that polydienes constitute a class of materials that can be easily modified, as already reported in the literature [28–30].

ABC and BAC terpolymers consisting of polystyrene or PS (A), polybutadiene of exclusively (~100%) -1,2 microstructure or $PB_{1,2}$ poly(dimethylsiloxane) or PDMS (C) showcase promising potential for pattern fabrication. PS and PDMS are highly immiscible blocks, leading to a high χ parameter and, therefore, extremely low dimensions. Their utilization in film formation for different applications has been thoroughly elaborated [31–36], but the dissimilar surface energies between the vinyl and siloxane segments favor the parallel orientation. To overcome this limitation, several strategies have been employed, such as the use of homopolymer brushes (hydroxyl-terminated PS and/or PDMS), pre-patterned surfaces, solvent annealing, etc. [1,37–41].

New properties are emerging by incorporating an additional low molecular weight segment, namely poly(butadiene). The role of poly(butadiene) of exclusively (~100% vinyl-type) -1,2 microstructure is dual:

(a). Due to the complexity of the system, very low domain periodicities for the PB domains can be formed without miscibility constraints.
(b). Poly(butadiene) can be chemically modified leading to the addition of the -OH functional group in each PB monomeric unit, making the modified PB a sacrificial segment during the film formation, a fact that renders the use of homopolymer brushes unnecessary.

To the best of our knowledge, the synthesis of such terpolymers, their chemical modification, and their self-assembly properties are novel discoveries that have not yet been published in the literature.

Taking into consideration the previous study conducted by Avgeropoulos' group [41] where, among others, the bulk properties of PS-*b*-$PB_{1,4}$-*b*-PDMS or $PB_{1,4}$-*b*-PS-*b*-PDMS were reported, it is straightforward that low molecular weight triblock terpolymers are not frequently encountered in the relative literature and the importance of high χ/low N (N: degree of polymerization) terpolymers in microelectronics could be proven most vital.

Herein, we report the synthesis, characterization, and post-polymerization chemical modification reactions of PS-*b*-PB-*b*-PDMS and PB-*b*-PS-*b*-PDMS terpolymers where the PB is exclusively (~100% vinyl-type) -1,2 microstructure. In total, three (3) samples were synthesized (two of the PS-*b*-$PB_{1,2}$-*b*-PDMS sequence and one of the $PB_{1,2}$-*b*-PS-*b*-PDMS) using anionic polymerization and sequential addition of the three monomers. Note that only through anionic polymerization can such terpolymers with specific molecular characteristics be synthesized. Based on the sub-10 nm requirements, the total average molecular weight values ($\overline{M}_n$) for the terpolymers were chosen to be in the range of 9 kg/mol to 18 kg/mol, while the $\overline{M}_n$ of the $PB_{1,2}$ did not exceed 4 kg/mol in any case.

The molecular characterization of the synthesized samples was accomplished via size exclusion chromatography (SEC), vapor pressure osmometry (VPO), and proton nuclear magnetic resonance spectroscopy (^{1}H-NMR), indicating the synthesis of well-defined, narrowly dispersed terpolymers. In addition, differential scanning calorimetry (DSC) experiments were carried out to determine the characteristic thermal properties of the involved segments. The morphological characterization of all unmodified samples in bulk was performed with transmission electron microscopy (TEM) and small angle X-ray scattering (SAXS), revealing the formation of well-developed structures with dimensions as low as 16 nm in certain cases.

Also, the post-polymerization chemical modification reactions of the $PB_{1,2}$ block are reported, and additional characterization in bulk and solution was employed to evaluate the incorporation of –OH groups in all polydiene vinyl-type monomeric units. The successful chemical modification was verified through thermal analysis (thermogravimetric analysis, or TGA and DSC), infrared spectroscopy, ^{1}H-NMR, and contact angle measurements—before and after post-treatment for comparison reasons.

2. Materials and Methods

2.1. Materials

The purification procedures of the involved reagents, including solvents [benzene (99%), tetrahydrofuran (99%)], monomers [styrene (99%), hexamethylcyclotrisiloxane (D_3) (98%), 1,3-butadiene (99%)], termination agent [methanol (98%)], 1,2-dipiperidinoethane [dipip (98%)], and initiator [*secondary*-BuLi (*sec*-Buli, 1.4 M in cyclohexane)] are documented elsewhere [29,41]. 9-borabicyclo[3.3.1]nonane or 9-BBN (0.5 M in tetrahydrofuran), hydrogen peroxide (H_2O_2) and sodium hydroxide (NaOH) were used as received. All reagents were supplied by Sigma-Aldrich (Sigma-Aldrich Co., St. Louis, MO, USA).

2.1.1. Synthesis of PS-*b*-$PB_{1,2}$-*b*-PDMS

Styrene (3.5 g, 33.6 mmoles) reacted with *sec*-BuLi (1.0 mmole) in the presence of benzene for 18 h at room temperature. A small quantity was retracted from the reactor to specify the molecular characteristics of the first segment. Subsequently, 1,2-dipiperidinoethane (3.0 mmoles) was introduced to increase the polarity of the solution prior to the addition of the 1,3-butadiene (1.5 g, 27.7 mmoles). The use of the polar compound results in only (~100%) 1,2-microstructure. The reaction was left to proceed at 4 °C for 22 h, and a second aliquot was retrieved to determine the molecular characteristics of the diblock precursor. Then, hexamethylcyclotrisiloxane (D_3, 4 g, 54.0 mmoles) was introduced and allowed to react for 18 h at ambient conditions. After the ring opening of the cyclic monomer was achieved, tetrahydrofuran was added to the solution, and the reaction proceeded for 4 h at room temperature and then was placed at −20 °C for seven days under continuous stirring. The final triblock terpolymer was precipitated in a non-solvent (methanol) and vacuum-dried. The described quantities correspond to sample 1 (see Table 1). Alternations on the quantity of monomers and initiator led to the preparation of sample 2. The schematics concerning the chemical reactions are given in Scheme 1a.

Table 1. Molecular characteristics of the synthesized linear triblock terpolymers as received from SEC, VPO, and ^{1}H-NMR characterization techniques.

Sample	A-*b*-B-*b*-C	$\overline{M}_n^{A\,(a)}$ (g/mol)	$\overline{M}_n^{B\,(a)}$ (g/mol)	$\overline{M}_n^{C\,(a)}$ (g/mol)	$\overline{M}_n^{tot\,(a)}$ (g/mol)	$Đ^{SEC\,(b)}$	$f_A^{(c)}$	$f_B^{(c)}$	$f_C^{(c)}$	1,2-Microstructure [c] (%)
1	PS-*b*-$PB_{1,2}$-*b*-PDMS	4.000	4.000	10.000	18.000	1.06	0.22	0.22	0.56	100
2	PS-*b*-$PB_{1,2}$-*b*-PDMS	4.000	1.500	4.700	10.200	1.04	0.39	0.15	0.46	100
3	$PB_{1,2}$-*b*-PS-*b*-PDMS	1.400	4.100	3.400	8.900	1.07	0.16	0.46	0.38	100

[a] VPO in toluene at 45 °C. [b] Dispersity (Đ) calculated from SEC in THF at 35 °C. [c] Mass fractions for the three blocks and 1,2-microstructure percentages for the PB segments calculated from ^{1}H-NMR in CDCl$_3$ at 25 °C.

2.1.2. Synthesis of $PB_{1,2}$-*b*-PS-*b*-PDMS

To synthesize the specific sequence, a slightly different synthetic route was employed from the one previously described. A few oligomeric units of styrene and *sec*-BuLi (0.14 mmoles) were introduced at ambient conditions to the reactor with benzene, enabling the formation of a nucleophile macroinitiator. Then, 1,2-dipiperidinoethane (0.30 mmoles) was added to promote the addition exclusively between the first and second carbon atoms of the 1,3-butadiene (2 g, 37.0 mmoles). The monomer was distilled in the apparatus, and the solution was placed at 4 °C for 22 h until complete consumption of the 1,3-butadiene. A small quantity of polar solvent (THF, ~1 mL) was introduced to change the reaction kinetics. Following that, styrene (3.5 g, 33.6 mmoles) was added and allowed to react at room temperature for 18 h until complete conversion, as already stated. The third monomer (D_3, 4.5 g, 60.0 mmoles) was introduced into the solution in a process similar to the one previously described. In each synthetic step, small aliquots were taken to study the molecular characteristics. The synthetic routes employed for the preparation of the specific triblock terpolymers are presented in Scheme 1b.

2.1.3. Chemical Modification Reactions of $PB_{1,2}$

To obtain –OH functional end groups, the poly(butadiene) of exclusively (~100%) -1,2 microstructure is submitted to post-polymerization chemical modification reactions, namely hydroboration and oxidation. In this process, each vinyl bond (-CH=CH_2 per monomeric unit) is converted to -CH_2-CH_2-OH, leading to the desired -OH group in all monomeric units of the PB blocks. To perform the hydroboration reaction 0.15 g (3.3 mmol, 6.6 mL) of 9-BBN were introduced into 1 g of PS-b-$PB_{1,2}$-b-PDMS (sample 1, corresponding to 2.7 mmoles of the $PB_{1,2}$ segment) which was dissolved in tetrahydrofuran (0.2 w/v%) under nitrogen atmosphere at -15 °C. The solution was left to warm up to ambient conditions and was stirred for 24 h to ensure the completion of the hydroboration process. Notably, the 9-BBN is used in ~20% excess compared to the mol of the $PB_{1,2}$ block. Following this, the solution was placed at -25 °C, and 1 mL of properly degassed methanol was added to deactivate all the excess of the borane reagent. After 30 min, NaOH (3.1 mmoles 6 N, 10% excess compared to the borane moles) was introduced to prevent the development of crosslinked networks due to borane by-products. Following, the oxidizing agent H_2O_2 (6.2 mmoles 30% w/v solution, 50% excess compared to the NaOH moles) was added, and the solution was stirred at -25 °C for 2 h before being placed at 55 °C for 1 h, where the separation of the desired organic and aqueous phases occurred. The organic phase was poured into a 0.25 M NaOH solution, and subsequently, the polymer was dissolved in THF/MeOH and washed with 0.25 M NaOH twice using a Buhner funnel [28,29]. Finally, the polymer was thoroughly washed with copious amounts of distilled water and placed in a vacuum oven to remove any volatile compounds. Coherent procedures were employed in all terpolymers. The chemical modification reactions for sample $PB_{1,2}$-b-PS-b-PDMS or sample 3 are provided in Scheme 1c. Similar reactions are followed for the other sequence of PS-b-$PB_{1,2}$-b-PDMS samples.

2.2. Methods

The instrumentations with which molecular characterization experiments were performed, including SEC, VPO, ^{1}H-NMR, and FT-IR, have already reported in our previous works [30,41].

The thermal characterizations through DSC were carried out in an N_2 atmosphere using aluminum pans (Tzero®, TA Instruments Ltd., Leatherhead, UK) in a TA Instruments Q20 DSC (TA Instruments Ltd., Leatherhead, UK) with a heating/cooling rate of 10 K/min. Thermogravimetric analysis was conducted in a Perkin Elmer Pyris-Diamond instrument (PerkinElmer, Inc., Waltham, MA, USA). The samples (~10 mg) were heated from 40 °C to 600 °C with a heating rate of 5 °C/min under a nitrogen atmosphere.

The wetting properties were studied in a contact angle (OCA 25, DataPhysics Instruments GmbH, Filderstadt, Germany) instrument. The samples were dissolved in different solvents (toluene for the pristine materials and a cosolvent mixture of THF/MeOH in a ratio of approximately 2:1 for the modified ones), preparing 1 wt% solution concentrations, and were then spin coated onto silicon wafers at ambient conditions under specific conditions (3750 rpm for 30 s), leading to films with ~40–50 nm thickness. Silicon wafers were first treated with piranha solution (sulfuric acid/hydrogen peroxide: 3/1), leading to purified substrates. In the measurements, 4 µL droplets of DI water were deposited at a rate equal to 0.5 µL/s. Three different measurements in three regions of the same wafer were conducted, and the average value was calculated and presented in the relative results. The deviation in all cases did not exceed ±2°, indicating the consistency of the results due to the uniform film deposition.

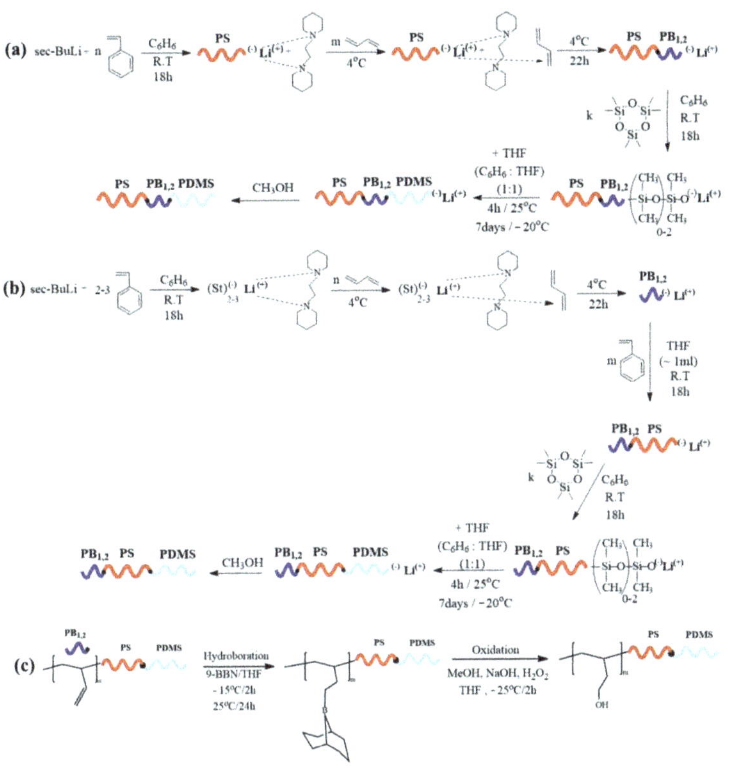

Scheme 1. Synthetic routes corresponding to: (**a**) the synthesis of triblock terpolymers of the PS-*b*-PB$_{1,2}$-*b*-PDMS type through anionic polymerization; (**b**) the synthesis of PB$_{1,2}$-*b*-PS-*b*-PDMS triblock terpolymers through anionic polymerization; and (**c**) post-polymerization chemical modification reactions, including hydroboration and oxidation of the PS-*b*-PB$_{1,2}$-*b*-PDMS terpolymers. Note that similar chemical modification reactions are performed in the case of PB$_{1,2}$-*b*-PS-*b*-PDMS.

The specifics regarding TEM and SAXS measurements are provided elsewhere [42]. The triblock terpolymers were cast from a dilute solution in toluene (5 wt%), and the solvent evaporation was completed in approximately 7 days. Thin sections of the as-cast thin films (ca. 40 nm thick) were obtained in a Leica EM UC7 ultramicrotome [Leica EM UC7 from Leica Microsystems (Wetzlar, Germany)], and subsequently, the sections were picked up on 600-mesh copper grids. The grids were then placed in vapors of a 2% OsO$_4$–water solution (Science Services, Munich, Germany)] for selective staining of the poly(butadiene) segment in order to increase the electron density through crosslinking and to improve the image contrast between PB, PS and PDMS segments.

3. Results and Discussion

Anionic polymerization enabled the synthesis of novel triblock terpolymers of the PS-*b*-PB$_{1,2}$-*b*-PDMS and PB$_{1,2}$-*b*-PS-*b*-PDMS sequences. All samples exhibited narrow dispersity indices, justifying the high standards of the living polymerization. The SEC chromatographs are presented in the Supporting Information (Figure S1a–c) and verify the absence of any by-product during the synthetic procedure of all 3 samples. For clarity reasons, the chromatographs of the PS or PB$_{1,2}$ and intermediate diblock precursors are also given. The molecular characteristics of all blocks for the 3 triblock terpolymer samples were determined through VPO, and the results are summarized in Table 1.

Through the characteristic chemical shifts evident from the ^{1}H-NMR spectra, the mass fractions of the three different segments were calculated and are also given in Table 1. The relative spectra are presented in the Supporting Information (Figure S2a–c). It is important to mention that the existence of exclusively -1,2 microstructure was verified through ^{1}H-NMR, as indicated by the absence of any proton chemical shifts at 5.30 ppm. The complete absence of the two olefinic protons strongly suggests the successful synthesis of pure vinyl elastomers due to the use of 1,2-dipiperidinoethane, which is well elaborated in the literature [29]. From the molecular characterizations, it is therefore concluded that the samples showcase high molecular and compositional homogeneity, a fact that contributes to the formation of well-defined structures during morphological observations.

A DSC analysis was carried out to study the thermal behavior of the terpolymers. The thermographs are shown in the Supporting Information (Figure S3a–c). Two separate glass transition temperatures (T_gs) of approximately −121 °C and 65 °C were recorded, corresponding exclusively to PDMS and PS blocks, respectively. The $\overline{M}_n$ values of the $PB_{1,2}$ in almost all cases were lower than or very close (sample 1) to the entanglement molecular weight (M_e ~3.8 kg/mol) [43], justifying the absence of any glass transition for the PB segments. For the semicrystalline PDMS block, additional transitions were observed at approximately −40 °C and −70 °C, indicating the melting and crystallization of the siloxane crystals, respectively, due to the increased molecular weight of the specific segment (sample 1) compared to the remaining terpolymers. The lower than expected value for the T_g of the PS block is extensively analyzed in the literature [41,42].

3.1. Structure/Properties Relationship

The as-cast OsO$_4$ stained samples for approximately 30 min [44] were morphologically characterized through TEM. The unstained samples due to the similar electron densities between the PS and $PB_{1,2}$ formed two-phase morphologies as expected [41]. Complementary experiments were conducted with SAXS to justify and verify the formation of well-developed structures. It is the very first time in the literature that the self-assembly behavior of such low molecular weight terpolymers with exclusively 100% -1,2-microstucture PB segments is studied.

The interaction parameters between the different components have been theoretically calculated at room temperature and are presented in Table 2. A thorough analysis concerning the estimation of the χ values based on fundamental equations from literature are provided in our previous works [30,41]. The values are critical for the interpretation of the results obtained from the morphological characterization techniques since the dissimilar driving forces promote a different self-assembly behavior for the synthesized terpolymers.

Table 2. Theoretical values of Flory-Huggins interaction parameters for PS/PDMS, PS/$PB_{1,2}$, and $PB_{1,2}$/PDMS.

χ	PS	$PB_{1,2}$	PDMS
PS	-	0.130	0.479
$PB_{1,2}$	0.130	-	0.074

3.2. Sample 1 (Sample Number as Indicated in Table 1)

In this triblock terpolymer similar $\overline{M}_n$ values for PS as well as $PB_{1,2}$ and increased molecular weight of the PDMS block were chosen. This strategy was followed to clarify the effect of the larger elastomeric end block on the overall morphology. The desired composition ratio for the individual segments was approximately 2/2/5, as evident in Table 1. Given that, the predicted morphology should include a PDMS matrix with the olefinic minority blocks forming a complex network. Although enthalpy-driven interactions would prevail over the entropy factors due to the major difference in molecular characteristics. The morphology obtained for this sample was three-phase four-layer lamellae, despite the non-symmetric composition ratio between the three blocks. As evident in the TEM image

(Figure 1a) the white regions correspond to the PS, the black sheets to the $PB_{1,2}$ and the gray areas to the PDMS segments. The dark color of the $PB_{1,2}$ is attributed to the OsO_4 staining. One could easily observe the molecular weight influence on the sheet thickness. Specifically, the dimensional approximation for the first three lamellar sheets (black/white/black) is attributed to the identical molecular weights between the two olefinic segments. The enhanced thickness of grey areas is induced by the almost doubled molecular weight of the PDMS block. The PDMS blocks due to the high incompatibility with the PS segments probably dictate the stretching of the $PB_{1,2}$ blocks. The SAXS results further support the existence of lamellar morphology. The characteristic peak ratio of 1:2:3:4 is evident in Figure 1b, and through the first permitted reflection, the domain spacing was found to be equal to 23 nm (supporting the calculation from the TEM image of 21 nm).

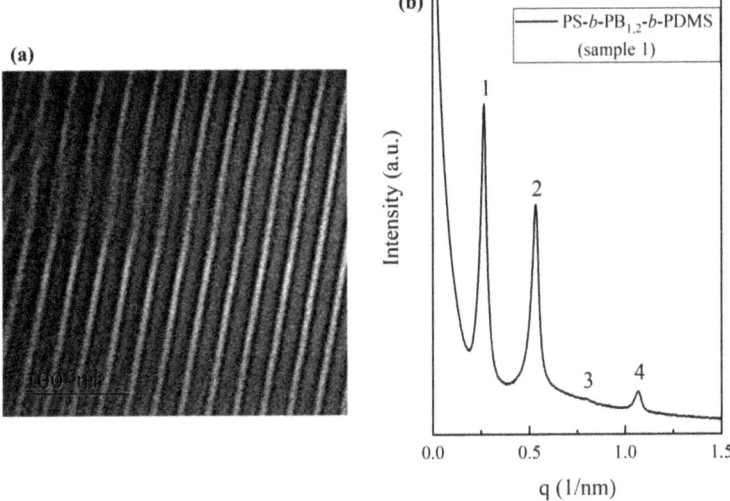

Figure 1. (**a**) Bright field TEM image of the PS-*b*-$PB_{1,2}$-*b*-PDMS or sample 1, after staining with vapors of OsO_4 for approximately 30 min. The white areas correspond to the PS, the grey to the PDMS, and the black areas to the $PB_{1,2}$, and (**b**) the relative 1-D SAXS pattern with the characteristic reflection peaks (1:2:3:4) corresponding to lamellar morphology verifying the TEM image.

3.3. Sample 2 (Sample Number as Indicated in Table 1)

Taking into consideration the mass fraction ratio presented in Table 1, the specific sample shows a composition ratio equal to 4/1/5 between the three blocks. The sample adopted the three-phase four-layer lamellae morphology even though the composition ratio between the components deviates significantly from the 1/1/1 that favors the formation of lamellar sheets (as already reported for Sample 1 as well). This behavior has also been observed in our previous work [41] for the PS-*b*-$PB_{1,4}$-*b*-PDMS sequence with approximate molecular characteristics and composition ratio. Even though the molecular weight of the blocks is quite low, especially for the intermediate elastomeric block, the system formed the structure that requires the lowest free energy. In Figure 2a, the TEM image of sample 2 is illustrated, where the black sheets correspond to the $PB_{1,2}$ due to the enhanced electron density induced after staining, the white to the PS, and the grey to the PDMS. One would expect a different morphology to be derived from the asymmetric compositions, but the asymmetric interactions ($\chi_{PS/PB} \neq \chi_{PB/PDMS} \neq \chi_{PS/PDMS}$) have a dominant role in the formation of a less frustrated structure [45,46]. The SAXS profile corroborated the results obtained from real-space imaging (Figure 2b). Specifically, a peak ratio of 1:2:3, which is in accordance with the lamellar morphology, was observed. Through the first permitted reflection, the domain spacing was calculated to be 16 nm, which is in good proximity with the one calculated through TEM (~14 nm). The fact that the triblock terpolymer was able to self-assemble in a three-phase structure despite the low molecular characteristics of the individual blocks is of paramount importance. It is also obvious that designing high χ terpolymers results in different structures due to the three possible pairs of interactions.

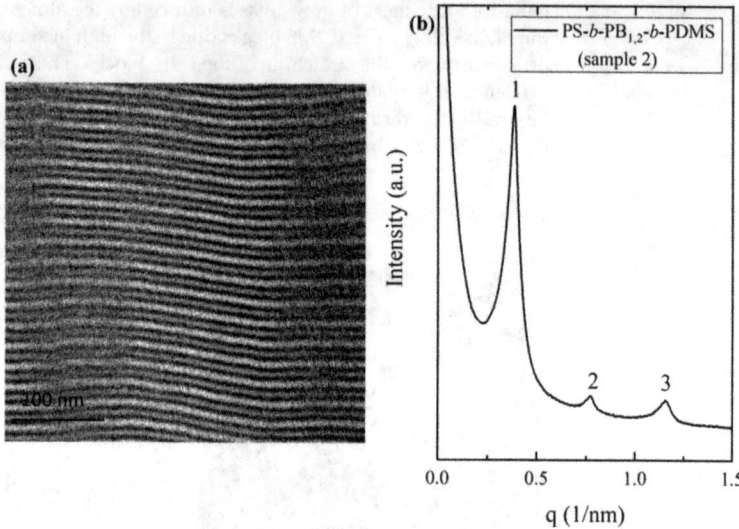

Figure 2. (**a**) Bright field TEM image of the PS-*b*-PB$_{1,2}$-*b*-PDMS or sample 2, after staining with vapors of OsO$_4$ for approximately 30 min. The white areas correspond to the PS, the grey to the PDMS, and the black areas to the PB$_{1,2}$, and (**b**) the relative 1-D SAXS pattern with the characteristic reflection peaks (1:2:3) corresponding to lamellar morphology verifying the TEM image.

3.4. Sample 3 (Sample Number as Indicated in Table 1)

It is well established that the ability to manipulate the block sequences in triblock terpolymers provides different nanostructures [45,46]. This extremely low molecular weight terpolymer (sample 3) showcases a composition ratio of 1/5/4, and in contrast to sample 2, the block sequence of the two segments is inversed. One might expect the formation of either three-phase, four-layer lamellae similar to the previous case or the partial mixing of PB$_{1,2}$/PS due to the low $\overline{M}_n$ values that would eventually lead to a two-phase morphology due to the strong immiscibility between PB$_{1,2}$/PS and PDMS. Strikingly, real-space imaging and scattering results revealed the formation of a network phase where the PS corresponds to the matrix and PB$_{1,2}$ and PDMS constitute the two independent networks (Figure 3a). We speculate that this self-assembly behavior can be explained in terms of asymmetric interactions based on the following relationship: $\chi_{PS/PDMS} \gg \chi_{PS/PB} > \chi_{PB/PDMS}$. Specifically, the most favorable interactions are between the outer segments, meaning PB$_{1,2}$ and PDMS. The favorable enthalpic interactions in combination with the inversed sequence (compared to sample 2) induced packing frustration that allowed the formation of the cubic network phase. The results were supported by the SAXS pattern, as can be clearly identified in Figure 3b. The reflections at the relative *q* values of $\sqrt{6}:\sqrt{8}:\sqrt{14}:\sqrt{16}:\sqrt{18}:\sqrt{24}$ are in good agreement with the Ia$\overline{3}$d space group corresponding to the double-gyroid phase. The appearance of an additional peak at the low *q* region, meaning $\sqrt{2}$, as can be observed in the relative SAXS pattern, is attributed to the distortion or deformation of the lattice [47]. The domain periodicity was calculated using the first permitted peak, which is equal to 18 nm.

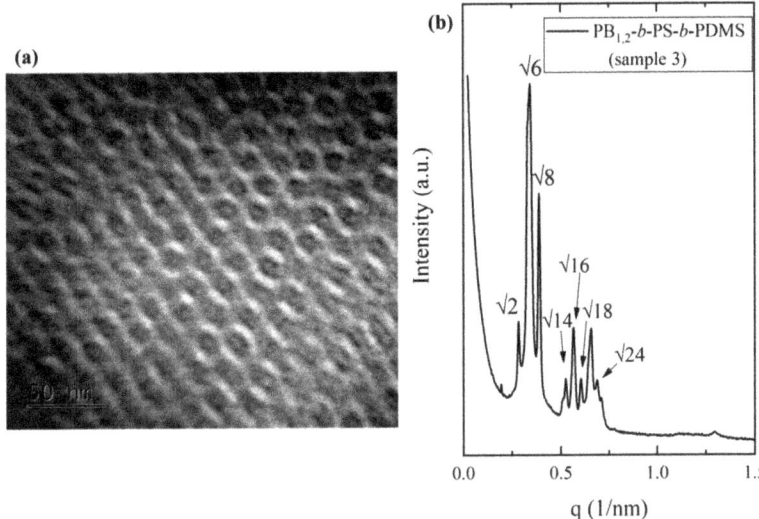

Figure 3. (a) Bright field TEM image of the $PB_{1,2}$-b-PS-b-PDMS or sample 3, after staining with vapors of OsO_4 for approximately 30 min. The white areas correspond to the PS, the grey to the PDMS, and the black areas to the $PB_{1,2}$, and (b) the 1-D SAXS pattern with the characteristic reflections at the relative q values of $\sqrt{6}:\sqrt{8}:\sqrt{14}:\sqrt{16}:\sqrt{18}:\sqrt{24}$ corresponding to double gyroid morphology-verifying the TEM image.

3.5. Chemical Modification Reactions

As evident from Table 1, the total average molecular weight of $PB_{1,2}$ was kept extremely low in all terpolymers to facilitate the physical adsorption of a very thin layer on the surface of the preferred solid substrate after the chemical modification increased its binding capacity. It would therefore be used as a polymeric brush on a corresponding surface. The chemical modification reactions were exclusively performed on the $PB_{1,2}$ segment of each triblock terpolymer. Note that the molecular characteristics of the PS and PDMS blocks remained unchanged while the introduction of the –OH group after the modification reactions (-CH_2-CH_2-OH instead of -CH=CH_2 per monomeric unit) induced an alternation on the mass fraction ratios. This alternation on the molecular characteristics is attributed to the enhanced molecular weight of the modified polybutadiene monomeric unit for the different terpolymers. The results concerning the molecular characteristics of the modified samples are summarized in Table 3.

Table 3. Molecular characteristics of the modified triblock terpolymers.

Modified Sample	A-b-B-b-C	$\overline{M}_n^A$ (g/mol)	$\overline{M}_n^B$ (g/mol)	$\overline{M}_n^C$ (g/mol)	$\overline{M}_n^{tot}$ (g/mol)	f_A	f_B	f_C
1	PS-b-$PB_{1,2}$-b-PDMS (modified)	4.000	5.300	10.000	19.300	0.21	0.27	0.52
2	PS-b-$PB_{1,2}$-b-PDMS (modified)	4.000	1.900	4.700	10.600	0.38	0.18	0.44
3	$PB_{1,2}$-b-PS-b-PDMS (modified)	1.900	4.100	3.400	9.400	0.20	0.44	0.36

Various techniques were employed to justify the successful chemical modification reactions, such as ^{1}H-NMR (Figure S4, Supporting Information), infrared spectroscopy (Figure S5, Supporting Information), TGA (Figure S6, Supporting Information), and contact angle for modified sample 1 (sample number as indicated in Table 3). All modified samples demonstrated a similar behavior after being characterized with the different methods. The comparative results, together with the appropriate interpretation concerning modified sample 1 (sample number as indicated in Table 3) before and after the chemical modifications, are provided for better clarity.

The molecular characterization through ^{1}H-NMR constitutes a reliable tool for the verification of the successful modification through the analysis of the characteristic chemical shifts. It is clear that the original chemical shifts at 5.6 ppm for the PB segment in the pristine terpolymers have been eliminated after the chemical modification, and a new characteristic shift at 2.1 ppm is evident, corresponding to the proton of the –OH group (Figure S4, Supporting Information). IR experiments further supported the ^{1}H-NMR results due to the appearance of characteristic peaks at 3000–3500 cm^{-1} that are attributed to the existence of –OH groups. The IR spectra (Figure S5, Supporting Information) presented for both unmodified and modified samples are transmittance versus wave number (cm^{-1}), and one can clearly observe the differentiation between the two spectra where the different vibrations are illustrated for better understanding.

DSC experiments were also carried out to study the thermal behavior of the terpolymers after the chemical modification reactions. As already discussed, in the pristine materials, in which the molecular characteristics for the PB$_{1,2}$ were significant low ($\overline{M}_n$ < 4.0 kg/mol), the glass transition temperature was absent. As a result, no alternation on the thermographs was observed, even though the physical properties of the terpolymers due to the introduction of the –OH groups are expected to differ. Additionally, TGA experiments revealed a significant weight loss at approximately 100 °C, attributed to the water molecules in the modified sample (Figure S6, Supporting Information).

To find out whether the hydrophobicity of the modified samples is altered compared to the initial terpolymers, we have measured the surface water contact angle of the films, and the results concerning sample 1 before and after the modification are shown in Figure 4, and all data are summarized in Table 4. The results clearly indicate that the contact angle decreases (from approximately 103° to 85°) after the samples are chemically modified in all cases, a fact that is attributed to the more hydrophilic nature of the final modified terpolymer due to the incorporation of –OH groups (Figure 4).

Sample 1

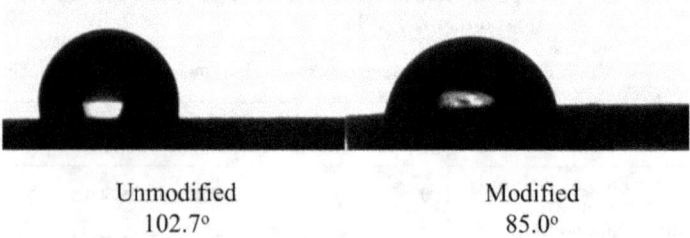

Unmodified
102.7°

Modified
85.0°

Figure 4. Images of water contact angles corresponding to the PS-*b*-PB$_{1,2}$-*b*-PDMS terpolymer (sample 1). The left image corresponds to the unmodified, pristine terpolymer, and the right image to the finally modified case.

Table 4. Water contact angle before and after chemical modification reactions for all different terpolymers.

Sample	A-b-B-b-C	Contact Angle (Degree) Unmodified ($\pm 2°$)	Contact Angle (Degree) Modified ($\pm 2°$)
1	PS-b-PB$_{1,2}$-b-PDMS	102.7	85.0
2	PS-b-PB$_{1,2}$-b-PDMS	103.2	83.1
3	PB$_{1,2}$-b-PS-b-PDMS	101.3	88.2

4. Conclusions

Well-defined triblock terpolymers of the ABC and BAC types consisting of polystyrene (A), poly(butadiene) of exclusively (~100%) -1,2 microstructure (B) and poly(dimethylsiloxane) (C) were synthesized with anionic polymerization. The total average molecular weights were kept low to achieve as small dimensions as possible by adopting different microstructures due to the diverse composition ratios. The self-assembly behavior indicated the impact of asymmetric compositions, different interactions between the adjacent blocks, and segment sequence on the formed structures. Post-polymerization chemical modification reactions such as hydroboration and oxidation were conducted on the PB$_{1,2}$ segments in all different terpolymers. Different molecular, thermal, and surface analysis measurements were performed to verify the successful incorporation of –OH groups in all monomeric units in the polydiene segments and therefore the preparation of polymeric brushes. The results are considered highly promising for nanotechnology applications since the necessity of homopolymer brushes during the film preparation is eliminated due to the physical absorption of the modified polydiene on the substrates.

Supplementary Materials: The following supporting information can be downloaded at: https://www.mdpi.com/article/10.3390/polym15040848/s1. Figure S1: SEC chromatographs corresponding to: (a) PS homopolymer precursor (blue), PS-b-PB$_{1,2}$ intermediate diblock precursor (red) and final triblock terpolymer of the PS-b-PB$_{1,2}$-b-PDMS type or sample 1 (black), (b) PS homopolymer precursor (blue), PS-b-PB$_{1,2}$ intermediate diblock precursor (red) and final triblock terpolymer of the PS-b-PB$_{1,2}$-b-PDMS type or sample 2 (black) and (c) (a) PB$_{1,2}$ homopolymer precursor (blue), PB$_{1,2}$-b-PS intermediate diblock precursor (red) and final triblock terpolymer of the PB$_{1,2}$-b-PS-b-PDMS type or sample 3 (black); Figure S2: ^{1}H-NMR spectra corresponding to: (a) PS-b-PB$_{1,2}$-b-PDMS type or sample 1, (b) PS-b-PB$_{1,2}$-b-PDMS type or sample 2 and (c) PB$_{1,2}$-b-PS-b-PDMS type or sample 3 indicating the chemical shifts attributed to the characteristic protons of the three different monomeric units of PS, PB and PDMS respectively; Figure S3: DSC thermographs corresponding to: (a) PS-b-PB$_{1,2}$-b-PDMS type or sample 1, (b) PS-b-PB$_{1,2}$-b-PDMS type or sample 2, and (c) PB$_{1,2}$-b-PS-b-PDMS type or sample 3; Figure S4: ^{1}H-NMR spectra of the PS-b-PB$_{1,2}$-b-PDMS triblock terpolymer (sample 1) before (up) and after (down) chemical modification reactions; Figure S5: FT-IR spectra of the PS-b-PB$_{1,2}$-b-PDMS triblock terpolymer (sample 1) before (black spectrum) and after (red spectrum) chemical modification reactions; Figure S6: TGA thermograms of the PS-b-PB$_{1,2}$-b-PDMS triblock terpolymer (sample 1) before (black spectrum) and after (red spectrum) chemical modification reactions.

Author Contributions: Conceptualization, A.A.; methodology, A.A.; validation, K.A., C.M., G.-M.M., I.M. and A.A.; formal analysis, K.A., C.M., G.-M.M., I.M. and D.M.; investigation, K.A., C.M., G.-M.M. and I.M. data curation, K.A., C.M., G.-M.M., I.M., D.M., A.A.P., M.K., N.E.Z., D.A.I. and A.A.; writing—original draft preparation, K.A., G.-M.M. and I.M.; writing—review and editing, K.A., G.-M.M., I.M. and A.A.; visualization, A.A.; supervision, A.A.; resources, D.A.I. and K.A., D.M., M.K., N.E.Z. and A.A.; project administration, K.A., D.M., M.K., N.E.Z. and A.A.; funding acquisition, A.A. All authors have read and agreed to the published version of the manuscript.

Funding: The research work was fully supported by the Hellenic Foundation for Research and Innovation (H.F.R.I.) under the "First Call for H.F.R.I. Research Projects to Support Faculty Members and Researchers and the Procurement of High-Cost Research Equipment Grant" (Project Number: 3970, acronym: NANOPOLYBRUSH) for K.A., D.M., M.K., N.E.Z., and A.A. This research was also partially funded by the Ministry of Science and Higher Education of the Russian Federation within State Contract, grant no. 075-15-2022-1105.

Institutional Review Board Statement: Not applicable.

Data Availability Statement: The data presented in this study are available upon request from the corresponding author.

Acknowledgments: K.A., C.M., G.-M.M., I.M., D.M., M.K., N.E.Z. and A.A. would like to acknowledge the Network of Research Supporting Laboratories at the University of Ioannina for letting them use the Electron Microscopy Facility and the Nuclear Magnetic Resonance Spectroscopy Center.

Conflicts of Interest: The authors declare no conflict of interest.

References

1. Angelopoulou, P.P.; Moutsios, I.; Manesi, G.-M.; Ivanov, D.A.; Sakellariou, G.; Avgeropoulos, A. Designing high χ copolymer materials for nanotechnology applications: A systematic bulk vs. thin films approach. *Prog. Polym. Sci.* **2022**, *135*, 101625. [CrossRef]
2. Park, C.; Yoon, J.; Thomas, E.L. Enabling nanotechnology with self-assembled block copolymer patterns. *Polymer* **2003**, *44*, 6725–6760. [CrossRef]
3. Bates, F.S. Polymer-Polymer Phase Behavior. *Science* **1991**, *251*, 898–904. [CrossRef]
4. Bates, F.S.; Fredrickson, G.H. Block Copolymer Thermodynamics: Theory and Experiment. *Ann. Rev. Phys. Chem.* **1990**, *41*, 525–557. [CrossRef]
5. Sinturel, C.; Bates, F.S.; Hillmyer, M.A. High χ-Low N Block Polymers: How Far Can We Go? *ACS Macro Lett.* **2015**, *4*, 1044–1050. [CrossRef]
6. Tureau, M.S.; Epps, T.H., III. Nanoscale Networks in Poly[isoprene-block-styrene-block-(methyl methacrylate)] Triblock Copolymers. *Macromol. Rapid Commun.* **2009**, *30*, 1751–1755. [CrossRef]
7. Lee, J.; Park, J.; Jung, J.J.; Lee, D.; Chang, T. Phase Behavior of Polystyrene-b-polyisoprene-b-poly(methyl methacrylate) Triblock Terpolymer upon Solvent Vapor Annealing. *Macromolecules* **2019**, *52*, 5122–5130. [CrossRef]
8. Lee, S.; Subramanian, A.; Tiwale, N.; Kisslinger, K.; Mumtaz, M.; Shi, L.-Y.; Aissou, K.; Nam, C.-Y.; Ross, C.A. Resolving Triblock Terpolymer Morphologies by Vapor-Phase Infiltration. *Chem. Mater.* **2021**, *32*, 5309–5316. [CrossRef]
9. Zapsas, G.; Patil, Y.; Bilalis, P.; Gnanou, Y.; Hadjichristidis, N. Poly(vinylidene fluoride)/Polymethylene-Based Block Copolymers and Terpolymers. *Macromolecules* **2019**, *52*, 1976–1984. [CrossRef]
10. Hong, S.; Higuchi, T.; Sugimori, H.; Kaneko, T.; Abetz, V.; Takahara, A.; Jinnai, H. Highly oriented and ordered double-helical morphology in ABC triblock terpolymer films up to micrometer thickness by solvent evaporation. *Polym. J.* **2012**, *44*, 567–572. [CrossRef]
11. Wang, H.-F.; Chiu, P.-T.; Yang, C.-Y.; Xie, Y.-C.; Lee, J.-Y.; Tsai, J.C.; Prasad, I.; Jinnai, H.; Thomas, E.L.; Ho, R.-M. Networks with controlled chirality via self-assembly of chiral triblock terpolymers. *Sci. Adv.* **2020**, *6*, eabc3644. [CrossRef]
12. Lee, W.Y.; Chapman, D.V.; Yu, F.; Tait, W.R.T.; Thedford, R.P.; Freychet, G.; Zhernekov, M.; Estroff, L.A.; Wiesner, U.B. Triblock Terpolymer Thin Film Nanocomposites Enabling Two-Color Optical Super-Resolution Microscopy. *Macromolecules* **2022**, *55*, 9452–9464. [CrossRef]
13. Cao, X.; Mao, W.; Mai, Y.; Han, L.; Che, S. Formation of Diverse Ordered Structures in ABC Triblock Terpolymer Templated Macroporous Silicas. *Macromolecules* **2018**, *51*, 4381–4396. [CrossRef]
14. Sun, T.; Tang, P.; Qiu, F.; Yang, Y.; Shi, A.-C. Formation of Single Gyroid Nanostructure by Order-Order Phase Transition Path in ABC Triblock Terpolymers. *Macromol. Theory Simul.* **2017**, *26*, 1700023. [CrossRef]
15. Aissou, K.; Mumtaz, M.; Marcasuzza, P.; Brochon, C.; Cloutet, E.; Fleury, G.; Hadziioannou, G. Highly Ordered Nanoring Arrays Formed by Templated Si-Containing Triblock Terpolymer Thin Films. *Small* **2017**, *13*, 1603184. [CrossRef]
16. Choi, H.K.; Aimon, N.M.; Kim, X.Y.; Gwyther, J.; Manners, I.; Ross, C.A. Hierarchical Templating of a $BiFeO_3$-$CoFe_2O_4$ Multiferroic Nanocomposite by a Triblock Terpolymer Film. *ACS Nano* **2014**, *8*, 9248–9254. [CrossRef]
17. Brittain, W.J.; Minko, S. A structural definition of polymer brushes. *J. Polym. Sci. Part A Polym. Chem.* **2007**, *45*, 3505–3512. [CrossRef]
18. Zhao, B.; Brittain, W.J. Polymer brushes: Surface-immobilized macromolecules. *Prog. Polym. Sci.* **2000**, *25*, 677–710. [CrossRef]
19. Feng, C.; Huang, X. Polymer Brushes: Efficient Synthesis and Applications. *Acc. Chem. Res.* **2018**, *51*, 2314–2323. [CrossRef]
20. Edmonson, S.; Osborne, L.V.; Huck, W. Polymer Brushes via surface-initiated polymerizations. *Chem. Soc. Rev.* **2004**, *33*, 14–22. [CrossRef]
21. Chen, W.-L.; Cordero, R.; Tran, H.; Ober, C.K. 50th Anniversary Perspective: Polymer Brushes: Novel Surfaces for Future Materials. *Macromolecules* **2017**, *50*, 4089–4113. [CrossRef]
22. Olivier, A.; Meyer, F.; Raquez, J.-M.; Damman, P.; Dubois, P. Surface-initiated controlled polymerization as a convenient method for designing functional polymer brushes: From self-assembled monolayers to patterned surfaces. *Prog. Polym. Sci.* **2012**, *37*, 157–181. [CrossRef]
23. Zhang, M.; Muller, A.H.E. Cylindrical Polymer Brushes. *J. Polym. Sci. Part A Polym. Chem.* **2005**, *43*, 3461–3481. [CrossRef]

24. Alaboalirat, M.; Qi, L.; Arrington, K.J.; Qian, S.; Keum, J.K.; Mei, H.; Littrell, K.C.; Sumpter, B.G.; Carillo, J.-M.Y.; Verduzco, R. Amphiphilic Bottlebrush Block Copolymers: Analysis of Aqueous Self-Assembly by Small-Angle Neutron Scattering and Surface Tension Measurements. *Macromolecules* **2018**, *52*, 465–476. [CrossRef]
25. Ludwigs, S.; Böker, A.; Voronov, A.; Rehse, N.; Magerle, R.; Krausch, G. Self-assembly of functional nanostructures from ABC triblock copolymers. *Nat. Mater.* **2003**, *2*, 744–747. [CrossRef]
26. Luzinov, I.; Tsukruk, V.V. Ultrathin Triblock Copolymer Films on Tailored Polymer Brushes. *Macromolecules* **2002**, *35*, 5963–5973. [CrossRef]
27. Huizhen Mah, A.; Mei, H.; Basu, P.; Laws, T.S.; Ruchhoeft, P.; Verduzco, R.; Stein, G.E. Swelling responses of surface-attached bottlebrush polymer networks. *Soft Matter.* **2018**, *14*, 6728–6736.
28. Mao, G.; Wang, J.; Clingman, S.R.; Ober, C.K.; Chen, J.T.; Thomas, E.L. Molecular Design, Synthesis, and Characterization of Liquid Crystal-Coil Diblock Copolymers with Azobenzene Side Groups. *Macromolecules* **1997**, *30*, 2556–2567. [CrossRef]
29. Politakos, N.; Weinman, C.J.; Paik, M.Y.; Sundaram, H.S.; Ober, C.K.; Avgeropoulos, A. Synthesis, molecular, and morphological characterization of initial and modified diblock copolymers with organic acid chloride derivatives. *J. Polym. Sci. Part A Polym. Chem.* **2011**, *49*, 4292–4305. [CrossRef]
30. Politakos, N.; Moutsios, I.; Manesi, G.-M.; Artopoiadis, K.; Tsitoni, K.; Moschovas, D.; Piryazev, A.A.; Kotlyarskiy, D.S.; Kortaberria, G.; Ivanov, D.A.; et al. Molecular and Structure-Properties Comparison of an Anionically Synthesized Diblock Copolymer of the PS-*b*-PI Sequence and Its Hydrogenated or Sulfonated Derivatives. *Polymers* **2021**, *13*, 4167. [CrossRef]
31. Jung, Y.S.; Ross, C.A. Orientation-Controlled Self-Assembled Nanolithography Using a Polystyrene-Polydimethylsiloxane Block Copolymer. *Nano Lett.* **2007**, *7*, 2046–2050. [CrossRef]
32. O' Driscoll, B.M.D.; Kelly, R.A.; Shaw, M.; Mokarian-Tabari, P.; Liontos, G.; Ntetsikas, K.; Avgeropoulos, A.; Petkov, N.; Morris, M.A. Achieving structural control with thin polystyrene-b-polydimethylsiloxane block copolymer films: The complex relationship of interface chemistry, annealing methodology and process conditions. *Eur. Polym. J.* **2013**, *49*, 3445–3554. [CrossRef]
33. Lo, T.-Y.; Krishnan, M.R.; Lu, K.-Y.; Ho, R.-M. Silicon-containing block copolymers for lithographic applications. *Prog. Polym. Sci.* **2018**, *77*, 19–68. [CrossRef]
34. Durand, W.J.; Blachut, G.; Maher, M.J.; Sirard, S.; Tein, S.; Carlson, M.C.; Asano, Y.; Zhou, S.X.; Lane, A.P.; Bates, C.M.; et al. Design of High-χ Block Copolymers for Lithography. *J. Polym. Sci. Part A Polym. Chem.* **2015**, *53*, 344–352. [CrossRef]
35. Jung, Y.S.; Chang, J.B.; Verploegen, E.; Berggren, K.K.; Ross, C.A. A path to ultranarrow patterns using self-assembled lithography. *Nano Lett.* **2010**, *10*, 1000–1005. [CrossRef]
36. Hu, H.; Gopinadhan, M.; Osuji, C.O. Directed self-assembly of block copolymers: A tutorial review of strategies for enabling nanotechnology with soft matter. *Soft Matter* **2014**, *10*, 3867–3889. [CrossRef]
37. Jeong, J.W.; Park, W.I.; Do, L.-M.; Park, J.-H.; Kim, T.-H.; Chae, G.; Jung, Y.S. Nanotransfer Printing with sub-10 nm Resolution Realized using Directed Self-Assembly. *Adv. Mater.* **2012**, *24*, 3526–3531. [CrossRef]
38. Liu, C.-C.; Han, E.; Onses Serdar, M.; Thode, C.J.; Ji, S.; Gopalan, P.; Nealey, P.F. Fabrication of Lithographically Defined Chemically Patterned Polymer Brushes and Mats. *Macromolecules* **2011**, *44*, 1876–1885. [CrossRef]
39. Ipekci, H.H.; Arkaz, H.H.; Serdar Onses, M.; Hancer, M. Superhydrophobic coatings with improved mechanical robustness based on polymer brushes. *Surf. Coat. Technol.* **2016**, *299*, 162–168. [CrossRef]
40. Borah, D.; Rasappa, S.; Senthamaraikannan, R.; Kosmala, B.; Shaw, M.T.; Holmes, J.D.; Morris, M.A. Orientation and Alignment Control of Microphase-Separated PS-*b*-PDMS Substrate Patterns via Polymer Brush Chemistry. *ACS Appl. Mater. Interfaces* **2013**, *5*, 88–97. [CrossRef]
41. Miskaki, C.; Moutsios, I.; Manesi, G.-M.; Artopoiadis, K.; Chang, C.-Y.; Bersenev, E.A.; Moschovas, D.; Ivanov, D.A.; Ho, R.-M.; Avgeropoulos, A. Self-Assembly of Low-Molecular-Weight Assymetric Linear Triblock Terpolymers: How Low Can We Go? *Molecules* **2020**, *25*, 5527. [CrossRef] [PubMed]
42. Liontos, G.; Manesi, G.-M.; Moutsios, I.; Moschovas, D.; Piryazev, A.A.; Bersenev, E.A.; Ivanov, D.A.; Avgeropoulos, A. Synthesis, Molecular Characterization, and Phase Behavior of Miktoarm Star Copolymers of the AB_n and A_nB (n = 2 or 3) Sequences, Where A Is Polystyrene and B Is Poly(dimethylsiloxane). *Macromolecules* **2022**, *55*, 88–99. [CrossRef]
43. Fetters, J.L.; Lohse, J.D.; Milner, T.S.; Graessley, W.W. Packing Length Influence in Linear Polymer Melts on the Entanglement, Critical, and Reptation Molecular Weights. *Macromolecules* **1999**, *32*, 6847–6851. [CrossRef]
44. Avgeropoulos, A.; Paraskeva, S.; Hadjichristidis, N.; Thomas, E.L. Synthesis and Microphase Separation of Linear Triblock Terpolymers of Polystyrene, High 1,4-Polybutadiene, and High 3,4-Polyisoprene. *Macromolecules* **2002**, *35*, 4030–4035. [CrossRef]
45. Zheng, W.; Wang, Z.-G. Morphology of ABC triblock copolymers. *Macromolecules* **1995**, *28*, 7215–7223. [CrossRef]
46. Nakazawa, H.; Ohta, T. Microphase separation of ABC-type triblock copolymers. *Macromolecules* **1993**, *26*, 5503–5511. [CrossRef]
47. Feng, X.; Burke, C.J.; Zhuo, M.; Guo, H.; Yang, K.; Reddy, A.; Prasad, I.; Ho, R.M.; Avgeropoulos, A.; Grason, G.M.; et al. Seeing mesoatomic distortions in soft-matter crystals of a double-gyroid block copolymer. *Nature* **2019**, *575*, 175–179. [CrossRef]

Disclaimer/Publisher's Note: The statements, opinions and data contained in all publications are solely those of the individual author(s) and contributor(s) and not of MDPI and/or the editor(s). MDPI and/or the editor(s) disclaim responsibility for any injury to people or property resulting from any ideas, methods, instructions or products referred to in the content.

Review

Block Copolymers in 3D/4D Printing: Advances and Applications as Biomaterials

Nikolaos Politakos

POLYMAT, Applied Chemistry Department, Faculty of Chemistry, University of the Basque Country, UPV/EHU, Paseo Manuel de Lardizabal 3, 20018 Donostia-San Sebastián, Spain; nikolaos.politakos@ehu.eus

Abstract: 3D printing is a manufacturing technique in constant evolution. Day by day, new materials and methods are discovered, making 3D printing continually develop. 3D printers are also evolving, giving us objects with better resolution, faster, and in mass production. One of the areas in 3D printing that has excellent potential is 4D printing. It is a technique involving materials that can react to an environmental stimulus (pH, heat, magnetism, humidity, electricity, and light), causing an alteration in their physical or chemical state and performing another function. Lately, 3D/4D printing has been increasingly used for fabricating materials aiming at drug delivery, scaffolds, bioinks, tissue engineering (soft and hard), synthetic organs, and even printed cells. The majority of the materials used in 3D printing are polymeric. These materials can be of natural origin or synthetic ones of different architectures and combinations. The use of block copolymers can combine the exemplary properties of both blocks to have better mechanics, processability, biocompatibility, and possible stimulus behavior via tunable structures. This review has gathered fundamental aspects of 3D/4D printing for biomaterials, and it shows the advances and applications of block copolymers in the field of biomaterials over the last years.

Keywords: block copolymers; 3D printing; 4D printing; biomaterials; scaffolds; tissue engineering

Citation: Politakos, N. Block Copolymers in 3D/4D Printing: Advances and Applications as Biomaterials. *Polymers* **2023**, *15*, 322. https://doi.org/10.3390/polym15020322

Academic Editor: George Z. Papageorgiou

Received: 29 November 2022
Revised: 3 January 2023
Accepted: 4 January 2023
Published: 8 January 2023

Copyright: © 2023 by the author. Licensee MDPI, Basel, Switzerland. This article is an open access article distributed under the terms and conditions of the Creative Commons Attribution (CC BY) license (https://creativecommons.org/licenses/by/4.0/).

1. Introduction

Scientific progress is changing rapidly from day to day, leading to new advances in many different aspects of research. Many areas of investigation are transforming, making the advances important every day. More and more new techniques and advances are imposed in scale-up on an industrial scale. An important method in the manufacturing of materials is considered 3D printing [1]. It is an area of investigation constantly evolving and moving towards, in terms of improving the existing pieces of equipment, using new materials, and preparing objects that are more complex. This evolution in the research in materials used in this technique creates the need to develop novel materials that can be printable and have properties for specific needs and areas.

3D was introduced as a manufacturing technique in the 80s and immediately gained attention. Now it is becoming one of the leading methods of fabrication, especially for hierarchically complex architectures, which are not always achievable by conventional manufacturing technologies [1]. This technique has proven to be versatile, and this is mainly because of its operational simplicity. 3D printing can be used in many different areas, such as electronics, education, robotics, sensing, the pharmaceutical industry, design, aerospace, automation, and biomedical engineering [1].

Nowadays, 3D printing production can use different materials, such as metals, ceramics, and polymers without the need for molds or machining typical for conventional formative and subtractive fabrication [2]. The cost of a 3D printer can be less than $500, enabling desktop fabrication of 3D objects even at home [2]. Since the 1980s, there has been a considerable jump in scientific and technological impact in manufacturing [2]. 3D printing is the primary reason for this. Apart from the scientific part, the economic impact is also tremendous since, with 3D printing, there was a change in product development due to lower cost and acceleration in manufacturing. The overall market has significant growth rates and surpassed the value of 5 billion USD in 2015 [2].

Nevertheless, interest in 3D printing has increased over the last 25 years, but bioprinting is still young. It has constantly been developing over the last 10 years, as seen from increased patents and publications. 3D printing has great potential to become the major player in manufacturing complex materials for many applications in the following years. This growth is accomplished by constantly researching new materials with exciting properties and use in wide applications [2–7].

This review aims to explore block copolymers' major contribution to 3D printing in recent years by focusing on bio-applications. It is well known that one of the major materials that are currently in use for 3D printing are polymers. Most of the polymers used in 3D printing in bioprinting are homopolymers. Block copolymers constantly evolve through new synthetic approaches and introducing new blocks with stimulating properties. This review aims to manifest the use of block copolymers as a leading evolving player in 3D/4D printing. This review will try to explore the significant contribution of block copolymers in bio-applications with specific advances.

2. Aspects on 3D and 4D Printing

3D printing is the additive fabrication process in which a material is joined or solidified to form complex geometric structures controlled by a computer. The term was introduced by Emanuel Sachs et al. [8] in 1989 with the invention of binder jetting 3D printing technology. The truth is that the concept has existed since 1986 with the use of stereolithography [9]. 3D printing can be categorized into two general categories in terms of preparation of the structures. The first is considered the "bulk" method, and the second is the "extrusion" method. Bulk methods are based on technologies that can print directly in an extensive reservoir of raw material by selectively curing, sintering, or binding material in a layer-by-layer fashion [9]. Extrusion methods are distinguished by the process of the material that is selectively deposited on a print bed from a separate reservoir. These techniques can be seen in Figure 1.

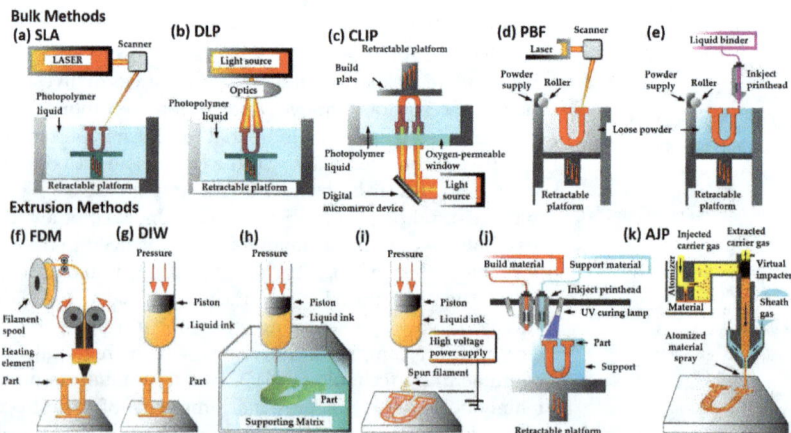

Figure 1. Schematics of bulk and extrusion 3D printing methods. (**a**–**e**) Bulk methods: (**a**) top-down laser stereolithography (SLA), (**b**) top-down digital projection lithography (DLP), (**c**) continuous liquid interface production (CLIP), (**d**) powder bed fusion (PBF), and (**e**) binder jet. (**f**–**k**) Extrusion methods: (**f**) fused deposition modeling (FDM), (**g**) direct ink writing (DIW), (**h**) sacrificial/embedded printing, (**i**) electrospinning, (**j**) direct inkjet printing, and k) aerosol jet printing (AJP). Reprinted with permission from [9].

Bulk methods are considered Figure 1a top-down laser stereolithography (SLA), Figure 1b top-down digital projection lithography (DLP), Figure 1c continuous liquid interface production (CLIP), Figure 1d powder bed fusion (PBF), and Figure 1e binder jet. On the other hand, extrusion methods are Figure 1f fused deposition modeling (FDM), Figure 1g direct ink writing (DIW), Figure 1h sacrificial/embedded printing, Figure 1i electrospinning, Figure 1j direct inkjet printing and Figure 1k aerosol jet printing (AJP). Table 1 gathered information about each technique as long, including basic characteristics of resolution and speed [9]. As can be seen from Table 1, numerous techniques are being used depending on what needs exist for manufacturing different structures. The choice of the method is based mainly on the type of material that is going to be used and the limitations that it has in terms of processability. According to different techniques, the printing speed and the structure's final resolution can be affected.

3D printing is a technique where, in order to show optimized results, efficiency and precision are essential. Specific parameters must be taken into consideration. These parameters can be based on three main ones [10]:

1. Geometry related (nozzle size and filament size)
2. Process related (melting temperature, bed temperature, and printing speed)
3. Structural related (layer thickness, infill geometry-density, raster angle-gap)

Table 1. 3D printing techniques with its basic characteristics [9].

3D Printing Technique	Material	Resolution / Speed	Description
Stereolithography (SLA)	Photocurable resin	50–200 µm [11] / 1000 mL/h [12]	Use of ultraviolet (UV) laser to polymerize a photocurable resin layer by layer
Digital projection lithography (DLP)	Photocurable resin	1 µm [13] / ≈50 mm/h [14–17]	Use of UV light to selectively polymerize a liquid resin with a spatial light modulating element
Continuous liquid interface production (CLIP)	Photocurable resin	10–100 µm [18] / 500 mm/h [19]	Use of a similar projection method to DLP, with the addition of an oxygen permeable window
Two-photon polymerization (2PP)	Photocurable resin	100 nm [11] / 80 nm/s–2 cm/s [13]	Use near-infrared femtosecond laser pulses to polymerize a nanoscale voxel at the focal point of the laser
Powder bed fusion (PBF)	Polymer Metal Ceramic	20–100 µm [14–20] / 1000 mL/h [15–21]	Uses a high-power photon or electron source to fuse the selectively powder layer by layer, while fresh powder is spread onto the previously bonded layer
Binder jet	Polymer Metal Ceramic compatible liquid binder	50–400 µm [22] / 25 mm/h [23]	Jets tiny droplets of binder onto a polymer, metal, or ceramic powder using an inkjet printhead
Fused deposition modeling (FDM)	Thermoplastic filament	100 µm [12] / 100 mL/h [12]	Uses rollers to push thermoplastic filament through a heated metal nozzle
Direct ink writing (DIW)	Viscoelastic ink Shear thinning fluid	1–250 µm [11] / 100 mL/h [12]	Extrudes liquid ink through a nozzle or needle
Sacrificial/embedded printing	Ink compatible with DIW process. Support: shear thinning fluid, or high viscosity reservoir and low viscosity filler combination	1–250 µm [11] / 1300 mL/h [24]	The nozzle of an ink dispensing system is inserted into a matrix of soft material. The supporting structure allows the ink to be 3D printed by tracing a 3D trajectory
Electro-hydrodynamic printing (EHD)	Polymer-based solution	100 nm–20 µm [25] / 20–1500 mL/h [26]	Use a voltage between the nozzle and substrate to eject fluid from the nozzle
Direct inkjet printing	Low-viscosity fluid	240 nm–5 µm [12] / 500 mL/h [27]	Deposition of droplets by means of a valve inside the printhead, formed by electrostatic, thermal, or piezoelectric plates
Aerosol jet printing (AJP)	Metal inks biological inks adhesives polymers	10 µm [28] / 1200 mL/h [29]	Uses aerodynamic focusing to guide a narrow spray of atomized fluid onto a substrate

3D printing manufactures objects with properties and shapes that do not change during their entire product life. 4D printing changes the aspect mentioned above by fabricating "smart" structures that can alter their shape after external stimulus, such as light, shear, water, touch, pH, electromagnetic field, or temperature [30–32]. Easily it can be hypothesized that this type of printing can open new roads to the fabrication of materials with use in applications with high demands. 4D printing can create dynamic structures that can easily be used as biological structures, hydrogels, or stimuli-responsive shapes [2].

One of the first 4D printing was adopted by Skylar Tibbits in fabricating materials that expand at different rates [33]. These materials stretch and fold to form different

shapes when activated by the stimulus [33]. Currently, three different approaches can be recognized:
1. Smart materials that change their shape upon stimuli.
2. 3D printing materials that can act as supporting structures for growth of organic cells.
3. Self-assembly of micro-sized smart particles that, upon stimulus change their pattern.

4D printing has differences in the technologies used for printing, mainly in using smart polymers that are used with traditional 3D printing techniques such as stereolithography, fused deposition modeling, inkjet printing, and others [34]. With slight modifications to the traditional 3D printing technologies, an adaptative 4D technology can be used [35]. More specifically, the FDM technique was modified by adding an air circulation system for 4D printing (cooling down the system below its glass transition temperature). Inkjet printing is also used in manufacturing cells since it is highly biocompatible [35]. DIW can also be used with specific bio-inks to prepare biomedical scaffolds and tissue engineering. SLA is also a technique that is used for the preparation of networks. Modification of SLA with UV-LED parts and projection micro stereolithography (PSL) can be used for structures with shape memory ability that can be used in drug delivery systems [33].

One of the most critical steps toward the evolution of 3D printing is the introduction of functional polymers. These polymers have a specific structure and properties that can be altered and perform different actions (physical, chemical, or biological) based on external stimuli (pH, light, temperature, mechanical loading, or voltage) [36]. The changes happening to these materials are through thermodynamics (change in hydrophobicity), protonation or deprotonation, polarization, rearrangement, and cleavage of bonds [36]. These small changes can lead to macroscopic changes in these materials, is swelling or aggregation, changing of color, or shrinking [37–39]. Specific reference should be made to self-healing polymers that can recover from partial destruction, which can be attributed to the dynamic nature of their bonds and the reversibility of the damage through reformation [40].

3D printing is becoming a versatile technique for manufacturing advanced functional materials with specific properties for many applications. Last few years, this technique has become more approachable to more people and industry. In the following years, 3D printing will have a significant role in energy, food production, medicine, and engineering. In particular, in the medicinal field, it can bring revolution to orthopedics, implants, tissue engineering, scaffolds, drug delivery, and regenerative medicine [1].

3. Bioprinting

Biomaterials are an emerging field. These materials are constantly evolving and developing, and they have a great share in the global market with an increased growth rate [41]. 3D printing is a technique that can be used in the field of biomaterials, providing top-of-the-edge constructions for applications with specific properties. In the last years, 3D printing has increased in popularity as a bioprinting technique because has great potential. Bioprinting can help scientists to fabricate tissues, organs, biological systems, drug delivery systems, and other in vitro systems that mimic their natural counterparts [42]. Bioprinting is defined by Groll et al. [43] as the process of using computer-aided transfer processes to pattern and assembles biologically relevant materials, including molecules, cells, tissues, and biodegradable biomaterials based on prescribed 2D or 3D originations resulting in the formation of the engineered biofunctional construct [43].

In bioprinting, a plethora of materials can be used to achieve the goal, such as cells, drugs, genes, growth factors, and of course, polymers (synthetic or natural). Bioprinting is an interdisciplinary field with a combination of knowledge from chemistry, biology, medicine, computer science, and materials science. In order to achieve the goal, many different parameters must be kept in mind. The bioprinter is the most important part of constructing the biomaterial apart from the material itself. This makes it clear that parameters such as build speed, user-friendliness, full automation capability, ease of sterilization, affordability, versatility, and compactness are important [42]. Conventional manufacturing techniques are not at the top of preparing biomaterials with complete geometry control (pore size, porous network, and porosity). Bioprinting as a fabrication method seems to have the potential to produce complex living and nonliving biological products from cells, molecules, extracellular matrices, and biomaterials [41].

3.1. Techniques in Bioprinting

In the field of bioprinting, four main strategies exist for manufacturing objects via 3D. The fabrication technologies include lithography-based [44–46], drop-on-demand [42,47,48], laser-induced forward transfer [49,50], and extrusion-based [51]. In Table 2, these techniques and their advantages and disadvantages can be found. Of these strategies, the most commonly used is extrusion-based, mainly because of its versatility, affordability, and capability to construct complex structures [52]. Tissues, living cells, organ modules, and organ-on-a-chip are some of the examples that can be 3D printed. The advantages of extrusion n-based are the commercial affordability, capability for printing high cell density bioink, and creating 3D complex structures of materials and multiple cell types [53–55]. For bioprinting, the techniques are less than the standard 3D printing. The most crucial parameter that someone needs to consider is the viability of the cells and, of course, the resolution. Then, some disadvantages are parameters related to the cost and processability, such as poor quality, limitation to specific materials, printing speed, and particular viscosities. Nevertheless, from the abovementioned techniques, it can be seen that now it can print blood vessels, bone, and cartilage, and for other applications, a specific method is used.

The different techniques used in biofabrication have specific applications based on their limitations, advantages, and disadvantages. 3D printing can be used in scaffolds, hydrogels, tissue engineering, and cell growth. Extrusion-based methods are currently used in pharmaceuticals, scaffolds, bone tissue engineering, and cardiovascular medical devices. Finally, other indirect methods (stereolithography or selective laser sintering) can be used in pharmaceutical, biomedical manufacturing, bone tissue engineering, and drug delivery [56].

Table 2. Comparison of four types of bioprinting methods. Combining information from [52].

Printing Method	Advantages	Disadvantages	Applications	Refs.
Drop on demand	Low cost Fast print speed High resolution High cell viability (>85%)	Low cell density (<10^6 cells/mL) Poor quality of vertical structures Bioink with specific range of viscosity	Blood vessel Bone Cartilage Neuron	[57–60]
Lithography based	Low cost High cell viability (>85%) High resolution Good vertical structure fidelity Fast printing speed	Limited to photopolymerization Medium on cell density (<10^8 cells/mL)	Blood vessel Cartilage Bone	[61–70]
Laser assisted	High cell viability (>95%) High resolution Fair vertical structure fidelity	Expensive Medium printing speed Bioink with specific range of viscosity Medium cell density (<10^8 cells/mL)	Blood vessel Bone Skin Adipose	[71–74]
Extrusion based	Moderate resolution Cell-laden bioink Good vertical structure fidelity Supports high viscosity bioink	Fairly expensive slow printing speed low cell viability (40–80%)	Blood vessel Bone Cartilage Neuron Muscle Tumor	[61–69]

3.2. Criteria and Limitations

3D printing has its limitations in terms of materials and technical parameters used during the formulation of the objects for applications in bio. It is important to notice that specific things must be considered since it has to deal with sensitive biological materials. The most important is to control temperature (range between 5–50 °C), pH (range between 6.5–7.8), irradiation, organic solvents (avoided), and pressure (high pressure can destroy cells) to ensure the viability of the structures prepared [75].

In the case of tissue engineering, for example, many criteria should be considered, especially for the case of the hydrogels used (Figure 2). These criteria are [76]:

1. Mechanical properties: tailored to meet specific end-user requirements.
2. Biodegradability/biosorbability: ideally bioresorbability and tunable degradation upon formation of functional tissues.
3. Porosity: porosity or hierarchical transport properties is vital for efficient nutrient and metabolic waste transport and optimal cell migration.
4. Swelling: crucial function in materials diffusion and transport within and through the hydrogel (cell stability and molecule release for drug delivery).
5. Biocompatibility: integrated into the biological system without harming or rejected (minimal or no immune reactions).
6. Cell adhesion: display adhesion property for cell binding.
7. Vascularization: capillary network responsible for nutrients transport to the cells.
8. Bioactivity: trigger/facilitate a biological response within a living system (tissue interactivity and binding ability, excellent osteoconductivity and osteoinductivity, and cell differentiation, attachment, and ingrowth).

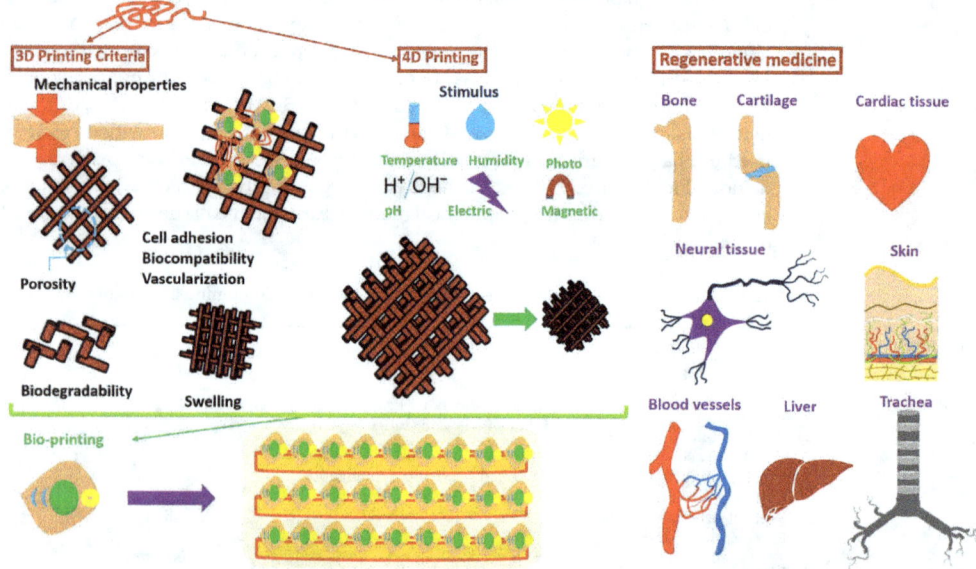

Figure 2. Schematic illustration of the criteria that should be considered for tissue engineering, differences of 3D Vs 4D (bio)-printing and main organs/tissues in regenerative medicine currently under investigation.

3.3. Bioprinting in Regenerative Medicine

Many polymeric materials exist for 3D printing organs or tissues and are under investigation. The research in organs/tissues is concentrated in bone, cartilage, heart valve, cardiac tissue, neural tissue, blood vessels, trachea, liver, and skin in regenerative medicine is vast (Figure 2). Materials that are currently used are based on natural polymers (agarose, alginate, chitosan, collagen, and gelatin) or synthetic ones (homopolymers and copolymers) [41]. For the case of the synthetic polymers, it can be found that the majority of them are based on Polycaprolactone (PCL), Polyurethane (PU), Polyethylene Glycol (PEG), Polylactic-co-glycolic Acid (PLGA), Pluronic Acid (or Poloxamer), and Polydimethylsiloxane (PDMS) [77]. More specifically, in the above mentioned areas, research is concentrated on some polymers (synthetic or natural) with specific properties.

3.3.1. Bone

Bone is one of the most studied areas in 3D printing. Synthetic biodegradable polymers that are currently under investigation are based on poly(caprolactone) (PCL), poly(glycolic acid) (PGA), poly(lactic acid) (PLA), and their copolymers. These materials can undergo hydrolysis, and they produce non-toxic materials. Scaffolds from these materials can be prepared with similar properties to the natural bone [78].

3.3.2. Cartilage

As in the case of bone, this is one of the most studied areas. Bioink can be used in order to mimic shapes from simple to more complex. Works are using thermo-sensitive methacrylate pHPMA-lac-PEG triblock copolymer with chondrocytes showing nice mechanical properties, stability in the long term, and processability via printing [79]. In another work, PCL and alginate hydrogel were used with incorporated embryonic chick cartilage cells proving increased cell viability, proliferation, and differentiation [80]. Other materials that can be used are Poly(ethylene glycol) terephthalate/poly(butylene terephthalate) (PEGT/PBT) block copolymer (grid-like structured scaffold) or an inorganic-organic hybrid of silica poly(tetrahydrofuran)/PCL for fabricating scaffolds for articular cartilage regeneration [78].

3.3.3. Cardiac Tissue

In this case, synthetic polymers that can be used are based on alginate with human cardiac-derived cells, increasing cell viability and retention of the cardiac lineage. Moreover, PEUU was also used in in vivo studies showing increased vessel formation and integration of cells [41].

3.3.4. Heart Valve

Materials used in 3D printing involving PLA, fibrin, collagen, and PGA. In this case, mechanical properties are essential. Methacrylated hyaluronic acid hydrogels have been used, showing cell viability and glycosaminoglycan matrix formation. Bioprinted alginate/gelatin hydrogels were also prepared with high viability and sound biomechanics [41,78].

3.3.5. Neural Tissue

Astrocytes and neurons from embryonic rats were bioprinted in a 3D collagen gel acting as a scaffold. Finally, an artificial neural tissue with murine neural stem cells, collagen hydrogels, and fibrin gels was fabricated, showing high cell viability [41].

3.3.6. Blood Vessels

This area is highly important since there is a high demand for artificial blood vessels. Ideally, an artificial blood vessel should be anti-thrombogenic, durable, biocompatible, and have comparable structural properties to the native ones. 3D bioprinted collagen was used with fugitive ink composed of gelatin with endothelial cells, showing a high potential for practical use. Another natural polymer is based on fibrin by polymerizing fibrinogen and thrombin in creating scaffolds. Finally, a 3D printed aorta based on poly(propylene fumarate) was fabricated using digital light stereolithography, with bioactive properties in vivo and comparable mechanical strength [78].

3.3.7. Trachea

Here, a work based on PCL powder with hydroxyapatite was used for preparing scaffolds. MSC-seeded fibrin coated the scaffold to improve bioactivity; in vivo studies showed reconstruction of the trachea and mechanically stable structures. Other works involve the methacrylated silk fibroin crosslinked via UV. This cell-loaded hydrogel scaffold showed excellent results in vitro [78].

3.3.8. Liver

The liver is one of the most fundamental organs in the human body with essential functions. A work is related with gelatin hydrogel with hepatocyte as an extracellular matrix (ECM), crosslinked with glutaraldehyde. In vitro experiments showed the survival of the cells over two months. A chitosan-gelatin scaffold was prepared to mimic the natural liver in architecture. This hybrid scaffold was highly porous with a well-organized structure [78].

3.3.9. Skin

The skin is an area in which many applications of 3D printing can be found, especially in the case of skin burns. Collagen is one of the obvious candidates for use since it is a big part of the natural skin. With bioprinting using fibroblasts and keratinocytes, fully functional skin was prepared. In vivo experiments revealed the epidermis' formation and blood vessels' presence in the wound area [41].

As 3D printing techniques evolve and more complex structures are prepared, this opens the road for printing different organs and more complex systems of the human body. The printing of bone and skin can move towards printing blood vessels and neurons, as already seen from the numerous kinds of research in biomaterials.

3.4. Bioinks

Bioinks are one of the most important ways of fabricating structures in bioprinting, especially with stereolithography. The bioinks used in printing must have certain aspects, such as biocompatibility, printability, and robustness [81]. The main types of bioinks are microcarriers, cell aggregates, decellularized components, and most importantly, hydrogels [42]. Hydrogels are considered one of the best choices as bioinks can easily be loaded with cells, can be crosslinked, have a high swelling degree, printability, and biocompatibility [81]. Hydrogels can be widely categorized into natural and synthetic ones. Natural ones that are currently used or investigated are based mainly on polysaccharides (agarose, alginate, and chitosan) and components of ECM (fibronectin, laminin, collagen, and gelatin) [81]. When control over mechanical and chemical properties is needed, then synthetic hydrogels can be used. There are notable works for a lot of synthetic hydrogels based mainly on PLA, PCL, PEG and others.

3.5. 4D Printing as Biofabrication Method

4D printing can be used to exploit its capabilities compared to its analogous 3D. The main difference between 3D and 4D is the nature of the material used. For the 4D, the material should have a "smart" behavior. These smart behaviors can be categorized into five main categories.

1. Shape memory (shape is changed in response to an external stimulus).
2. Self-assembly (external stimulus obligates chains into assembly in specific shape).
3. Self-actuating (Automated actuation upon exposure to an external stimulus).
4. Self-sensing (detection of external stimulus and quantification).
5. Self-healing (the damaged structure is auto-repaired)

It can be said that 4D has several advantages compared to 3D. The introduction of the fourth dimension of time, spatial and temporal control over the fabrication process and the final product makes the structure's dynamic [42]. The 4D printing process is influenced by five main factors: (i) Type of stimulus, (ii) type of manufacturing process, (iii) interaction mechanism (stimulus and material), (iv) type of responsive material, and (v) mathematical model of the transformation of the material [42]. The two areas where 4D bioprinting is mostly used in regenerative medicine and tissue engineering for fabricating complex tissue/organ geometries and controlling the tissue microarchitecture with 3D printing. 3D printing is considered as a more traditional method that can also be used for introducing cells to the structure, but 4D is a more dynamic method that can have the "smart" property of the material introduced into the structures (Figure 2).

The "smart" behavior of a material printed in 4D can be generally divided into five significant responsiveness: (i) Temperature, (ii) pH, (iii) humidity, (iv) photo, and (v) magnetism.

Temperature responsiveness: This is one of the most frequently studied with many materials under investigation. Generally, it can be used in tissue engineering. The most common classes are shape memory polymers and responsive polymer solutions. In the first polymeric class, materials such as poly(caprolactone) dimethacrylate, soybean oil epoxidized acrylate, polycaprolactone triol (Ptriol), poly (ether urethane) (PEU), and poly (lactic acid) can be found. For the responsive polymer solutions that can be used in drug delivery and tissue engineering, polymers that can be used are poly(N isopropylacrylamide), poly(N vinyl caprolactam), gelatin and GelMA, collagen and ColMA, methylcellulose, agarose, pluronic and poly(ethylene glycol)-based block polymers [42].

pH responsiveness: This type of responsive behavior is based on polymers that are polyelectrolytes with weak acidic or basic groups. Accepting or releasing protons is the key to their responsiveness to changes in pH. Functional groups like phosphate, tertiary amine, pyridine, carboxyl, and sulfonic are responsible for changes in pH, resulting in structural changes. Drug delivery and sensing are where more applications can have these materials. Natural polymers with this responsiveness are chitosan, gelatin, dextran, alginic acid, and hyaluronic acid. In terms of synthetic ones, polymers such as poly(histidine) (PHIS), poly(acrylic acid) (PAA), poly(acrylamide/maleic acid), Poly(dimethylaminoethyl methacrylate) (PDMAEMA), Poly(2-hydroxyethyl methacrylate) (PHEMA), Sulfonamide/polyethyleneimine (PEI), Poly(N-isopropyl acrylamide-co-butyl methacrylate-co-acrylic acid) and poly(aspartic acid) (PASA) [42,82] exist.

Humidity responsiveness: Changes in humidity can alter the volume of the material. Systems with this responsiveness can transform the absorption or desorption of humidity into a driving force for movement. Polymers that can be used are poly(ethylene glycol) diacrylate (PEGDA), cellulose, and polyurethane copolymers. These materials can have applications such as stents, drug delivery, sensing, soft robotic, and tissue engineering [42,82].

Photo responsiveness: Exposure to light can undergo a physical or chemical transformation for some polymers. Generally, changes in hydrophobicity, charge, polarity, bond strength, or conformation can lead to changes in solubility, degradability, wettability, and optical properties. Using this type of responsiveness has an advantage in terms of zero contact. Polymers with specific side groups, such as diarylethenes, spiropyrans, azobenzenes, and fulgides, are commonly used. In tissue engineering, examples exist with printing hydrogels that swell or shrink upon the external stimulus of light. Some of these systems are based on PNiPAM (functionalized with spirobenzopyran) and a hydrogel of 4,4′-azodibenzoic acid, cyclodextrin, and dodecyl (C12)-modified poly(acrylic acid) [42,82].

Magnetic responsiveness: In this type of responsiveness, polymers are generally functionalized (physically entrapped or covalently bonded) with magnetic nanoparticles that have in their structure magnetic elements such as nickel, iron, cobalt, or oxides. Polymeric scaffolds prepared from these materials can undergo physical changes, structural or mechanical, as a general area of applications is more targeted to drug delivery [42,82].

Electrical responsiveness: This type of responsiveness is based on electric or mechanoelectrical external stimulus. These materials depend on their electrically conductive character to respond to the stimulus. Swelling or shrinkage of the material can undergo under an electric field. Some of these polymers can have applications in drug delivery, sensing, implant devices, artificial muscles, neural tissue engineering, and others. Hydrogels based on these types of polymers are generally based on polyelectrolytes. Some examples are sulfonated-polystyrene (PSS), poly(2-(acrylamide)-2-methylpropanesulfonic acid) (PAMPS), polythiophene (PT), poly(2-hydroxyethyl methacrylate) (PHEMA), poly(3,4-ethylenedioxythiophene): polystyrene sulfonate (PEDOT:PSS) and Poly(L-lactic acid)/Poly(3,4-ethylenedioxythiophene) (PLLA/PEDOT) [82].

4. Advances and Applications of Block Copolymers in 3D/4D Printing in the Area of Biomaterials

Block copolymers are a class of materials that can be used in different aspects of 3D printing. Copolymers are a class of materials with versatile properties and there are many different ways of preparation. Using copolymers instead of homopolymers in 3D printing is based mainly on the versatility and tunability of the properties the researcher wants to achieve. The temperature of the processing during 3D printing is very important. This temperature can be tuned by using a copolymer instead of a homopolymer. Crystallinity is another parameter that needs to be controlled, and copolymers can adjust this. Parameters such as wetting behavior, hydrophilicity, and solubility can also be altered by adding a more hydrophobic block or, reversely, a more hydrophilic one. Finally, mechanical properties can also be different when two or more blocks are combined. The use of copolymers uses the best properties of two synthetic polymers or polypeptides or natural polymers to adjust the material prima to easy processing during 3D printing, improved properties, and better stability and stability biocompatibility. There are numerous examples of copolymers of different architectures, such as AB, ABA, ABC, and star copolymers, or in combination with other inorganic materials or natural polymers. This review will focus on the advances and applications of biomaterials in 3D printing over the last ten years. Copolymers will be divided into (i) AB, (ii) ABA, and (iii) other architectures/combinations with inorganic materials or natural polymers.

4.1. AB Block Copolymers

The research in the area of copolymers type AB in the last years is mainly focused on two essential polymers that also have the lion share in the 3D printing technology, with PEG being the third one. These two polymers are poly(lactic acid) and poly(caprolactone), two polyesters with degradation properties.

Block copolymers of poly(L-lactide-co-ε-caprolactone) (PLACL), Poly(D, L-lactide-co-glycolide) (PLGA), and poly(D, L-lactide-co-glycolide) (PDLGA) were prepared via extrusion-based 3D printing in terms of extensively studying and deeply understanding how printing parameters affect degradation, printability, and properties of lactide-based medical grade polymers. In this work, authors vary printing parameters (pressure, speed, and temperature) to evaluate the relationship between composition, degradation, printability, polymer microstructure, and rheological behavior. It is crucial to notice that the innovation of this work is based on the evaluation of the polymer changes occurring during printing that many do not consider during the preparation of scaffolds. This work found that polymers had good printability at a relatively good speed and high resolution until a certain degree of degradation. Polymers of this chemical composition can thermally decompose from the first minute leading to a decrease in the average block length. PLACL had better printability at higher molecular weights with less degradation than PLGA and PDLGA, which can be explained in terms of viscosity [83].

Scaffolds of PCL with poly(1,3-propylene succinate) (PCL-PPSu), including antimicrobial silver particles printed by a custom-built computer numerical control 3D printer, were prepared. These scaffolds were prepared as promising biomaterials for regenerative skin therapies. Essential points in this study are also the low-processing temperature, the enhanced degradation, and the antimicrobial properties. These scaffolds were quite porous with well-defined interconnected porosity, as seen from SEM images (Figure 3). Structures showed higher enzymatic and hydrolytic degradation rates and better hydrophilicity than PCL. Experiments with *E. coli* and *C. albicans* showed reduced microbial adhesion compared to *P. aeruginosa* and *S. aureus*. These materials can be attractive biomaterials for tissue engineering and wound healing applications [84].

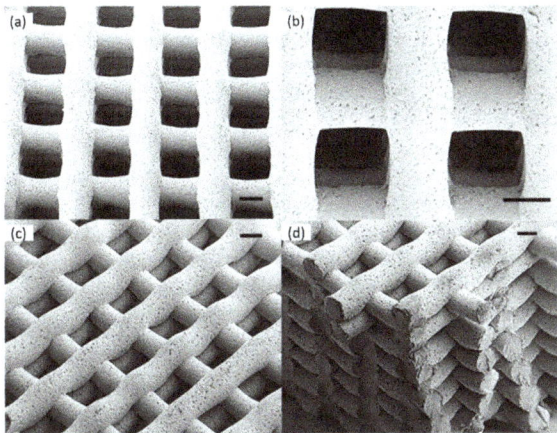

Figure 3. SEM images of 3D printed copolymer impregnated with silver nitrate scaffolds at different magnifications (**a–d**) (scale bars: 200 μm). Reprinted with permission from [84].

Another work used poly(lactide-co-ε-caprolactone) (PLACL) as the ink for printing scaffolds with a desktop 3D printer. These scaffolds have great potential in regenerating soft tissues such as muscle, tendon, nerve, cartilage, and myocardium. An important issue is the biocompatibility and cost-effectiveness of these materials in tissue engineering scaffolds. Here, the authors promoted the incorporation of PCL into PLA as an effective strategy to control stiffness and elasticity. PCL is introduced as a solution to overcome present defects associated with PLA. Combining these two polymers can lead to adjustable mechanical properties and good biocompatibility, compared to their analogous homopolymers. Scaffolds that have different properties and different optical aspect can be seen in Figure 4 where porous cubes with or without large holes (Figure 4a,d), round tubes (Figure 4b), and cambered plates with holes (mimicking the PEEK-based cranium restoration substitute, Figure 4c) are presented [85].

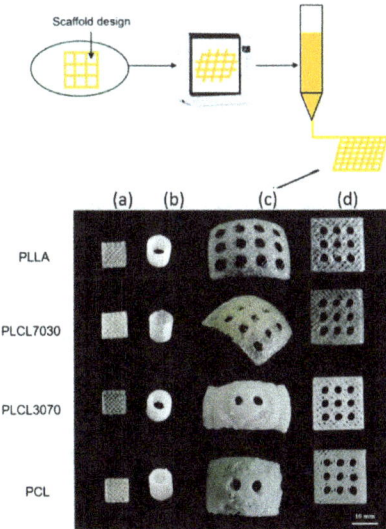

Figure 4. 3D printed micro-structures designed by the computer-aided design model (top) and optical images of different polyester scaffolds. Reprinted with permission from [85].

PCL has also been combined in a block copolymer with polypyrrole (PPy), fabricating a 3D porous nerve guide conduit. This block copolymer was used to fabricate 3D scaffolds with an in-house built electrohydrodynamic-jet 3D printing system. These biodegradable and conductive scaffolds can lead to pretty soft structures with conductive properties and similar mechanics to the native human peripheral nerve (∼6.5 MPa). This work found improved degradation profiles to aid the growth and differentiation of the neuronal cells in vitro. These scaffolds proved to be promising materials for future use and treatment of neurodegenerative disorders [86].

Polylactide, in combination with polycaprolactone, has led to many interesting applications in the area of biomaterials. This combination of poly(D,L-Lactic Acid) (PDLLA) and PCL as photoinks led to digital light 3D printing with the DLP technique of customized and bioresorbable airway stents. This material proved to be a bioresorbable elastomer that can be used in customized medical devices where high precision, elasticity, and degradability are needed. The as-prepared stents have comparable mechanical properties to state-of-the-art silicone stents. An essential point in this work is that these stents would disappear over time, preventing the need for additional procedures. This is something that is significant for children and elderly patients. An in vivo study in healthy rabbits confirmed biocompatibility and showed that the stents stayed in place for at least seven weeks after their incorporation. After this period, they became invisible through radiography, as can be seen in Figure 5 [87].

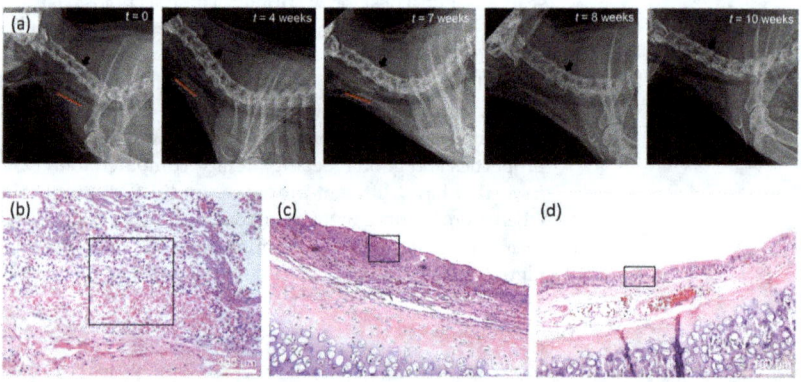

Figure 5. (**a**) Radiographs of a rabbit with the observation period of 10 weeks. Position of the stent is marked with a red line. Black arrows indicate the C4 vertebrae. (**b**–**d**) Inflammatory and tissue morphology changes in the rabbit's trachea 2 (**b**), 6 (**c**), and 10 weeks (**d**) after the stent insertion. Black rectangles represent the parts of the main morphological changes over time including the area of inflammation and necrosis (**b**), squamous metaplastic epithelium (**c**), and pseudostratified columnar (respiratory) epithelium (**d**). Hematoxylin and eosin staining, magnification of 10×10. Reprinted with permission from [87].

Other materials that can be used in 3D printing with the AB architecture are also based on a thermosensitive/moderately hydrophobic poly(2 N-propyl-2-oxazine) (pPrOzi) and thermosensitive/moderately hydrophilic poly(2-ethyl-2-oxazoline) (pEtOx). By using an extrusion-based 3D bioprinter, the preparation of a bio-printable and thermoreversible hydrogel was fabricated for use in tissue engineering applications. These hydrogels have a microporous structure with high mechanical strength (3 kPa). This block copolymer was used in printing various 2D and 3D patterns with high resolution. Furthermore, hADSC stem cells were efficiently encapsulated, and the hydrogel showed cytocompatibility in post-printing cell experiments. This hydrogel can be an alternative in combination with other bioinks to improve printability or be used as a drug delivery platform [88].

Another approach was investigated with a polyester thermoplastic elastomer containing alternating semi-crystalline polybutylene terephthalate as hard segments and amorphous poly(ether terephthalate) as soft segments. This work is more orientated toward improving patient compliance and personalized drug delivery with long-acting drug delivery devices (e.g., implants and inserts). This study highlights processability and drug delivery. These 3D printed materials have outstanding mechanical properties and drug permeability similar to conventionally marketed inserts, such as intravaginal rings, using progesterone (P4) as a model drug. A significant point was the dependence of drug loading (via solvent impregnation) on the external and internal geometry of the 3D-printed structure. The drug was mainly distributed in the outer layers of large structures, hence a high burst effect during release [89].

Finally, work with blocks AB and triblock ABA based on PEG and maleic anhydride led to biocompatible hydrogels. These materials were based on PEG-diol and monomethyl ether PEG that were used to polymerize (with ring-opening polymerization) maleic anhydride and propylene oxide to AB and ABA PEG-poly(propylene maleate) (PPM) and finally to PEG-poly(propylene fumarate) (PPF). In this work, the structures were prepared via continuous digital light processing and were then photochemically printed from an aqueous solution. These 3D printed structures could potentially be applied in soft tissues, such as peripheral nerve regeneration. Experiments with three different cell types (MC3T3, NIH3T3, and Schwann cell lines) showed non-toxic behavior and compatibility [90].

4.2. ABA Block Copolymers

The area of block copolymers type ABA has been studied more in the last years than copolymers type AB. The polymer that is used in the majority of the cases is based on some type of PEG. The FDA considers this polymer safe and inert and has many uses in medicine, biology, and chemistry.

A material that can be used in regenerative medicine, including bioprinting, is based on the synthetic copolymer poly(propylene fumarate)-b-poly(ethylene glycol)-b-poly(propylene fumarate) (PPF-b-PEG-b-PPF). 3D printed materials via digital light processing additive manufacturing can be prepared using this copolymer. These 3D-printed amphiphilic hydrogels show the importance of the nanoscale size and ordering of hydrophobic crosslinked domains in terms of degradation and mechanical properties. It points out a direct correlation between mechanical properties and structure. Compared with other works that studied the number of cross-linking sites concerning mechanical properties/resorbability, this work also considers how these domains are connected. It was shown that size, connectivity, and phase ordering changes resulted from aggregation mechanisms driven by the length of the blocks. The importance of these nanoscale ordering can also be seen in the authors' swelling and in vitro tests [91].

In another work based on polymer poly(N-(2-hydroxypropyl)methacrylamide lactate) A-blocks, (partly derivatized with methacrylate groups), and hydrophilic poly(ethylene glycol) B-blocks (ABA) a hydrogel prepared as a synthetic extracellular matrix for tissue engineering via 3D fiber deposition [92]. According to the authors, this hydrogel can be seen as a potential applicant in bioprinting. The authors showed the preparation of 3D-printed structures with good mechanical properties, stability in photopolymerization, and well-defined vertical porosity. Mechanically, this polymer showed similarities to many natural polymers, including collagen. This similarity makes it an ideal candidate for synthetic extracellular matrix for engineering cartilage. This work has demonstrated the support of encapsulated cells until new tissues are formed, parallel with high chondrocyte viability after one and three days. The highly defined patterning of these printed materials can be observed in Figure 6, where pore shape is maintained, showing good resolution and distinct localization, after loading the fibers (layered and adjacent) with fluorescent microspheres (fluorescent lemon and fluorescent orange) [92].

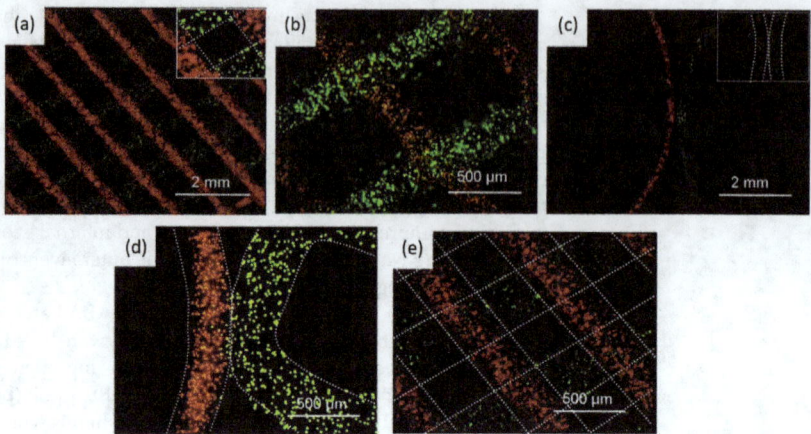

Figure 6. Microscopy pictures of subsequently printed layers of 25 wt% M30P10 hydrogels loaded with Dye-Trak "F" fluorescent microspheres, fluorescent lemon and fluorescent orange, at the concentration of 1×10^6 spheres per mL. (**a**) Two and (**b**) three layer angled constructs (1.5 mm strand spacing) with distinct localization of fluorescent microspheres. (**c**,**d**) Printing of adjacent fibers with circular patterns with maintenance of distinct dye localization. (**e**) Two layers construct with 0.8 mm strand spacing. Reprinted with permission from [92].

An interesting work based on using an ABA copolymer as part of polyurethane is also found in the literature [93]. This work uses the poly(ε-caprolactone)-b-poly(ethylene glycol)-b-poly(ε-caprolactone) triblock copolymer (PCL-b-PEG-b-PCL) to react with hexamethylene diisocyanate to form polyurethane. These hydrogels can be used in 3D printing, especially for tissue engineering scaffolds with high fracture toughness. These structures can have excellent processability and excellent properties in terms of mechanics. The preparation of these materials was done by extrusion using a filament extruder. This work found that a high amount of water (more than 500%) can be uptaken from the materials. In terms of mechanical properties, the high elongation at break, toughness, tensile strength, and tear resistance was proven [93].

PEG is a polymer that is used quite often in ABA architectures. The use of PCL-b-PEG-b-PCL and PLA-b-PEG-b-PLA is another example of this. This work showed that PCL-b-PEG-b-PCL hydrogel with crystallinity could be extruded and printed with adjustments in temperature. 3D structures can effectively be prepared, and good cell compatibility was found with good mechanical properties. According to the authors, these materials can be used as a responsive hydrogel model for 3D bioprinting [94].

3D structures that can have potential applications in tissue engineering are prepared from hydrogels based on an ABA poly(isopropyl glycidyl ether)-b-poly(ethylene oxide)-b-poly(isopropyl glycidyl ether). These structures were prepared via a direct-write printer from a three-stepper motor stage. These structures can have a dual stimuli responsiveness in temperature and shear response [95].

ABA copolymers consisted of a triple stimuli-responsive poly(allyl glycidyl ether)-stat-poly(alkyl glycidyl ether)-b-poly(ethylene glycol)-b-poly(allyl glycidyl ether)-stat-poly(alkyl glycidyl ether) can create new opportunities in the biotechnological field. These materials can have three different responsiveness, based on temperature, pressure (shear-thinning), and UV light, and a direct-write printer prints them. The authors designed these triple responsiveness in order to have easy handling and transfer of the gel (temperature response), printed at ambient conditions (pressure response), and crosslinking for preparing robust structures (UV response). By changing the composition of these materials, the properties of the final object can be tuned, leading to an expansion of a set of materials [96].

An ABA copolymer was prepared from PEG (Block B) and partially methacrylated poly[N-(2-hydroxypropyl) methacrylamide mono/dilactate] (Block A). These hydrogels could be attractive biomaterials since they have biodegradability, tunable thermoresponsiveness/mechanical properties, and cytocompatibility. To improve printability, mechanical properties, and long-term stability, the authors have incorporated methacrylated chondroitin sulfate (CSMA) or methacrylated hyaluronic acid (HAMA). This work showed that the pure ABA copolymer laden with equine chondrocytes showed potential for significant cartilage-like tissue formation in vitro. Finally, incorporating HAMA or CSMA resulted in 3D porous structures with excellent cell viability and improved properties. Overall, are attractive systems for the design of 3D cell-laden constructs for cartilage regeneration [79].

The last work based on PEG as the main component is based on a triblock with PCL as the A block. These hydrogels showed tunable mechanical properties, mainly for soft tissues and high elasticity. Good compatibility with cells to support fibroblast growth in vitro was also studied. The most crucial point of this work is the bioprinting of cells with these hydrogels to form constructs of cell–gel with high viability. The printing procedure has the main component, and a dispenser is mounted onto a robotic stage; then, a motor-controlled stage acts as the printing substrate. A syringe pump drives the extrusion-based dispenser. This work has demonstrated a system with tunable properties, elasticity, and biodegradability. According to the authors, this hydrogel may be fully compatible with other biomaterials, such as proteins, growth factors, peptides, or other bioactive molecules [97].

Apart from PEG as the main component of ABA copolymers, another work involves the acrylic ABA, composed of poly(methyl methacrylate) (PMMA) as block A and poly(n-butyl acrylate) (PnBA) as block B. This copolymer is categorized as a thermoplastic elastomer and is for industrial use. This material is to be used as material in the dental field. The results showed that this copolymer has good physical properties, concluding that it can be used to make provisional restorations for dental 3D printers [98].

An interesting work for materials potentially used as medical implants where biocompatibility and stability are essential is based on a copolymer of poly(styrene-b-isobutylene-b-styrene) (SIBS)/PS homopolymer blend. In this work, filaments of the blends were used for FFF (Fused filament fabrication) 3D printing. 3D constructs can be prepared, as can be seen in Figure 7. The use of PS homopolymer is based on increasing the toughness of the SIBS copolymer due to its limitation in printing because it is soft. This work found that the microarchitecture is tunable, and printability is excellent under specific conditions [99].

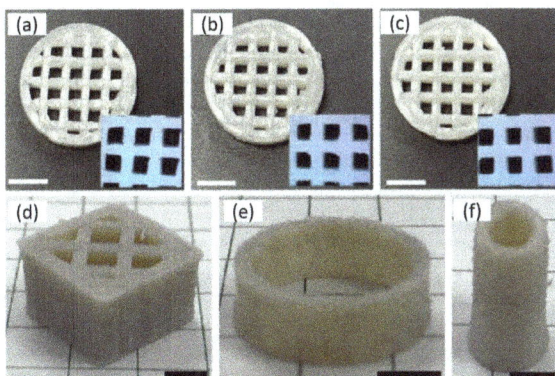

Figure 7. (**a–c**) Photos of 3D-printed cylinders with infill patterns from the SIBS/PS blends: (**a**) SIBS/PS (77/23), (**b**) SIBS/PS (67/33), and (**c**) SIBS/PS (57/43). The insets show higher-magnification images taken with an optical microscope. (**d–f**) 3D-printed objects with different shapes from SIBS/PS (57/43). Scale bars are 5 mm. Reprinted with permission from [99].

A triblock that can be prepared in large quantities is investigated in another work with polymers first processed from a twin-screw extruder for making the filament and then 3D printed (Figure 8). This polymer is based on PCL as the middle block (B) and PLLA or PDLA as the A blocks. The prepared 3D constructs were tested for biocompatibility via MTT assay, using rat bone osteosarcoma cells (UMR-106) to evaluate them as potential biomaterials. It was found that the block length and the composition of the ABA affect the macroscopic properties of the specimens and the mechanical ones. This is important since, varying the final material's properties, customized constructed can be prepared. Finally, the MTT assay found a non-toxic nature of the material [100].

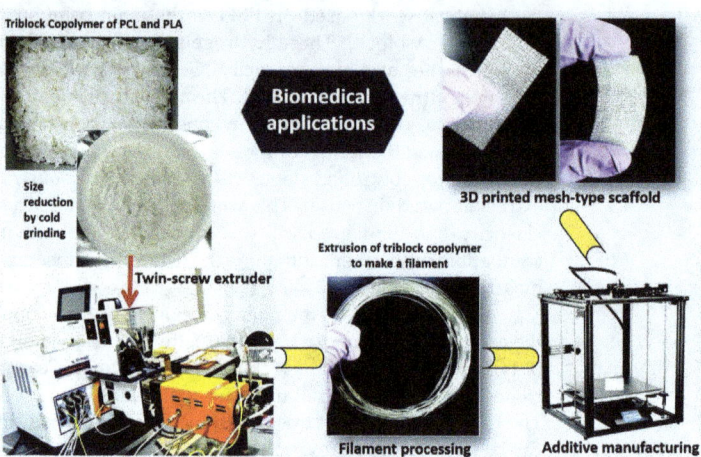

Figure 8. Melt processing of the triblock copolymer (synthesized by the scale-up method) to fabricate the filament which was successfully used for 3D printing a mesh-type scaffold. Reprinted with permission from [100].

An interesting work was proposed using an ABA triblock based on PCL-b-PTMC-b-PCL (trimethylene carbonate as TMC) [101]. The idea of this study was to prepare copolymers with biodegradability and to evaluate physical properties and potential melt-processable thermoplastic elastomeric biomaterials in 3D printing via extrusion of the polymer solution. Mechanical properties showed a tensile strength of 120 MPa, elongation at break 620%, and tensile strength of 16 MPa. With melting points close to 58 °C, these materials can be extruded and are promising as biomaterials for use as implants and scaffolds for tissue engineering. The advantage of these materials is the microporosity that can be tuned for cell culture; however, cell compatibility must be evaluated [101].

A triblock ABA copolymer based on polydimethylsiloxane (PDMS) (B block) and poly(benzyl methacrylate) (PBnMA) (A block) was prepared as a linear-bottlebrush architecture (Figure 9). This copolymer could have much potential in future applications since it has extraordinary properties such as stimuli-reversible, extraordinarily soft, and stretchable elastomer (Figure 10). This material can be used as ink for direct writing in 3D printing structures without post-treatment or external mechanical support. This work compares the material used with existing 3D printable elastomers, showing more than two orders of magnitude softer material and six times more strechable. The authors explain these results from the bottlebrush molecular architecture, which prevents entanglement formation, whereas stretchability ought to have an extensive network strand size. These elastomers can be readily used as a matrix for functional nanoparticle-polymer composites. The self-assembly and thermoreversibility can be an essential advantage for direct-writing printing to avoid solvent evaporation-induced material defects during printing [102].

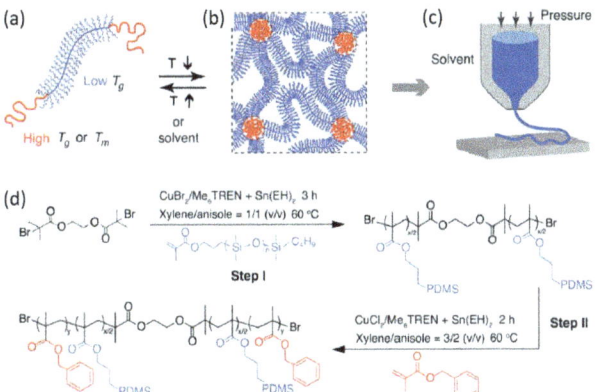

Figure 9. Design concept and synthesis of 3D printable, reversible, ultrasoft, and stretchable elastomers. (**a**) Schematic of a responsive linearbottlebrush-linear triblock copolymer. (**b**) At low temperature, the middle bottlebrush blocks (blue) act as elastic network strands, whereas the high Tg end linear blocks aggregate to form spherical glassy domains. (**c**) Glassy domains dissociate at high temperature or in the presence of solvents, resulting in a solid-to-liquid transition of the network. The stimuli-triggered reversibility allows the elastomers for direct-write 3D printing. (**d**) Synthesis of linear-bottlebrush-linear triblock copolymers using ARGET ATRP. The side chain of the middle bottlebrush block is linear polydimethylsiloxane (PDMS), whereas the end blocks are linear poly(benzyl methacrylate) (PBnMA). A bottlebrush-based triblock polymer is denoted as BnMAy-b-PDMSxw-b-BnMAy, in which y is the number of repeating BnMA units, x is the number of PDMS side chains per bottlebrush, and w represents theMW of PDMS side chains in kg/mol. The weight fraction of the end blocks in the triblock copolymer is kept below 6% to ensure that the bottlebrush-based ABA triblock copolymers self-assemble to a sphere phase. Reprinted with permission from [102].

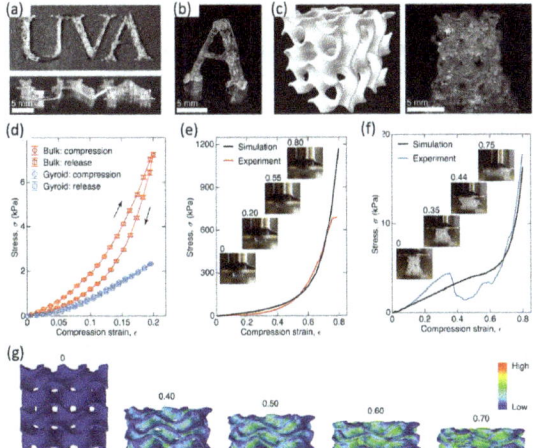

Figure 10. Direct-write printing soft elastomers to create deformable 3D structures. (**a**) 3D printed UVA initials with a stack thickness of 2 mm. Upper: bird's eye view; lower: side view. (**b**) Freestanding, 3D printed letter "A". (**c**) 3D rendering of a cubic gyroid (left) and the corresponding printed product with dimensions $10 \times 10 \times 10$ mm^3 (right). (**d**) For the bulk sample, the compression–release profile exhibits a hysteresis associated with 23% energy dissipation (dashed lines), whereas for the gyroid, there is almost no energy dissipation, as evidenced by the complete overlap between the compression and release profiles (solid lines). The apparent Young's modulus, $E = \sigma/\varepsilon$, of the gyroid is about 8 kPa, nearly half of 20 kPa for the bulk; this is likely because the porous gyroid has a lower

density of about 1/2 of the bulk. The strain rate is 0.005/s. Error bar: standard deviation for $n = 5$. (e) As the compression strain increases from 0.2 to 0.55, the bulk sample exhibits strain-stiffening with the compression stress increasing from 8 to 130 kPa, and further compression with $\varepsilon > 0.55$ results in material fracture (optical images). The stress−strain profile is used to calibrate the FEA simulation (dashed line). (f) Cubic gyroid exhibits a nearly linear elastic deformation up to $\varepsilon = 0.35$, at which the stress is about 5 kPa, 10 times lower than the 50 kPa for the bulk. Slightly above $\varepsilon = 0.35$, the stress exhibits a sharp decrease (solid line). (g) Decrease is associated with structural collapse of the gyroid, as indicated by comparing the snapshots from FEA simulation (dashed line in (f) and upper panel) with the optical images of the gyroid (lower panel) under various extents of compression. The use of UVA initials in 3D printing is under the permission from the UVA Office of Trademark and Licensing. Reprinted with permission from [102].

Finally, a material that can have potential use as bioink is studied in another work, with the use of ABA triblock hydrophilic poly(2-methyl-2-oxazoline) (pMeOx) (block A) and a more hydrophobic poly(2-iso-butyl-2-oxazoline) (piBuOx) (block B). Hydrogel scaffolds of a 20% aqueous solution were printed with a compact bench-top 3D bioprinter from these materials. Rheological experiments show a soft hydrogel above 25 °C with an elastic modulus of 150 Pa, while an increase at almost 600 Pa is observed at 37 °C. The significant point of this work towards the use as bioink was the low cytotoxicity after 24 h at 37 °C observed for both HEK and Calu-3 cells. This work indicates that the fabricated hydrogel may be a potential candidate for printable, functional bioink [103].

4.3. Other Architectures of Block Copolymers and Systems with Other Materials

The preponderance of the works in block copolymers consisted of ABA and AB architectures. Most polymers are based on PLA, PCL, and PEG. In the literature, the last year's other architectures based on the same polymers as AB/ABA architectures or other polymers are currently under investigation. Moreover, using natural polymers, polypeptides, hydroxyapatite, or clay has contributed much to 3D printing in the last few years [104–119].

4.3.1. Other Architectures of Block Copolymers

The use of polypeptides as alternative polymers in 3D printing is also explored [104,108]. This work involves the preparation of highly stable hydrogels with crosslinking under UV. The triblock is an ABC consisting of peptides, glutamic acid, nitrobenzyl-protected cysteine, and isoleucine. This polymer then is deprotected, and a reaction of the cysteine residue with alkyne functionalized four-arm polyethylene glycol (PEG) via nucleophilic thiol–yne chemistry occurred (Figure 11). These hydrogels showed remarkable shear-thinning properties according to the authors. Very important is gellation at lower concentration (3 wt%), desired mechanical properties, low cytotoxicity, and high printability. This material can be an alternative catalyst-free curing method for 3D-printed structures for different biomedical applications [104].

A similar architecture was also investigated to create enzyme-cleavable inorganic-organic hybrid inks with potential applications in scaffolds for bone regeneration. The copolymer used was based on 3-(trimethoxysilyl)propyl methacrylate (TMSPMA) and methyl methacrylate (MMA). The arms were connected at a core consisting of degradable enzymes using a collagenase peptide sequence GLY-PRO-LEU-GLY-PRO-LYS. Three-star copolymers were prepared TMSPMA was randomly distributed, or inner block or outer block (Figure 12). The inorganic-organic hybrid was prepared of a composition of 70 wt% polymer and 30 wt% silica via the sol-gel method and was used for direct 3D extrusion printing without additional carriers or binders. From this work, it was shown that the 3D constructs could be degraded from endogenous tissue-specific enzymes that are involved in the natural remodeling of bone. The specific position of the TMPSMA group controlled the hybrid formation, mechanical properties, degradation, and printability. Better results were found for the inner-star TMSPMA, which led to a more flexible and tougher material than the others [105].

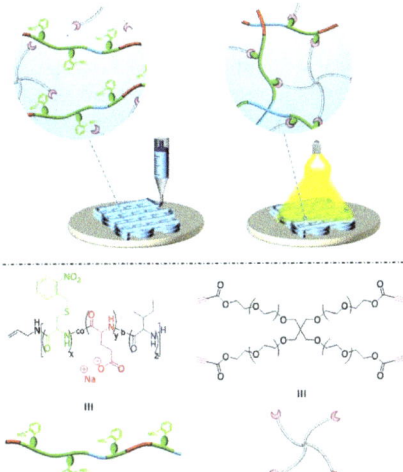

Figure 11. Schematic showing photocleavage of nitrobenzyl protecting groups revealing free thiol functionalities that react with peripheral propiolate functional groups on four-arm PEG. Reprinted with permission from [104].

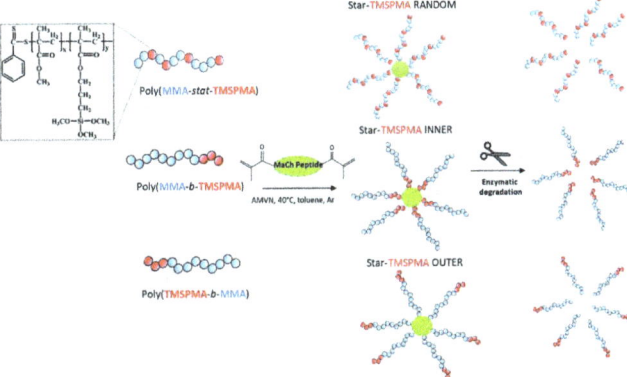

Figure 12. Schematic illustration of the poly(MMA-TMSPMA) star polymers synthesised with arms of three different architectures (random, inner and outer) crosslinked by an MaCh-peptide core and cleavage by collagenase activity. Reprinted with permission from [105].

A different approach in terms of architecture comes from a work where grafted copolymer was prepared [106]. In this work, a PEDOT-g-PCL (poly(3,4-ethylendioxythiophene)-g-poly(ε-caprolactone)) synthesized intending to fabricate electroactive scaffolds for muscle tissue engineering (bioelectronics). Here, it was found that the percentage of PEDOT is a crucial parameter in printability and that only low percentages led to the process with direct ink writing. Biocompatibility of the materials was evaluated with 8220 muscle cells showing encouraging results in compatibility, cell alignment, and myotubes differentiation [106].

Hydrogels that can be used as bioinks with dual sensitivity are very important and can have many potential applications, especially in tissue engineering [107]. In this way, via an microextrusion bioprinter, a triblock copolymer consisted of poly(lactide-co-glycolide)-b-polyethylene glycol-b-poly(lactide-co-glycolide) with acrylate groups in the chain is reported. This hydrogel proved thermo/photo sensitive with excellent shear-thinning properties and fast recovery in the elastic region. The 3D printed scaffolds were very stable after the photopolymerization, which can lead to low-cost and mass production on an industrial scale [107].

Another work of star architecture is based on an amphiphilic copolymer of poly(benzyl-L-glutamate)-b-oligo(L-valine), which forms hydrogels via hydrophobic interactions [108]. This star copolypeptide can be used as ink to rapidly fabricate defined microstructures, opening the road for manufacturing complex scaffolds, which is far more difficult with conventional methods. The fabricated structures were degradable, did not affect the metabolic health of fibroblasts (cell line Balb/3t3), and could have a favorable release of encapsulated molecules [108].

A different approach in the star architecture was introduced by another work, using a dendritic polyester core with a poly(oligo(ethylene glycol) methyl ether acrylate) inner layer and a poly(acrylic acid) as the outer layer [109]. The solution formed a hydrogel by adding metallic ions (zinc, copper(II), aluminum, and ferric ion). 3D structures were printed using a custom-made inkjet printer (Figure 13). It has been reported that the fast gelation of the star upon the addition of the ions can give a potential candidate that can be used in 3D inkjet printing through an in-process cross-linking approach. The authors showed that an 8% polymer solution has enough viscosity to be ejected from the inkjet nozzle. Adding Zn^{2+}, Cu^{2+}, Al^{3+}, and Fe^{3+} formed different hydrogels that also altered the dynamic viscoelasticity. The highest resolution upon printing was found for hydrogel with Fe^{3+} [109].

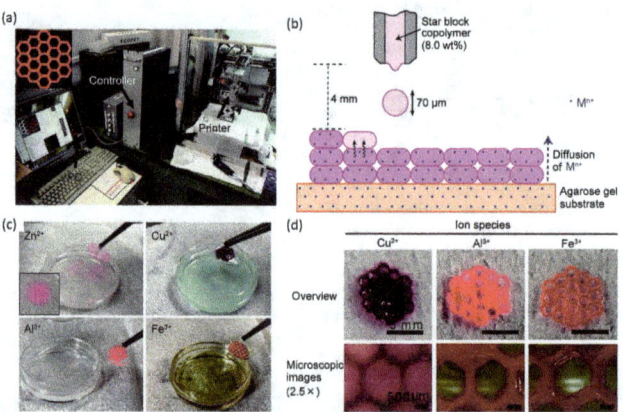

Figure 13. 3D inkjet printing of the star block copolymer hydrogels crosslinked using metallic ions. (**a**) Overview of the inkjet printing system. Inset: horizontal slice of the design of the 3D hydrogel structures that was inputted into the printer. (**b**) Schematic illustration of the printing process. The star block copolymer solution was ejected from the inkjet nozzle towards the agarose gel substrate containing metallic ions. The printed droplets show gelation layer-by-layer through metallic ions supplied from the substrate. (**c**) Transfer of the 3D-printed hydrogels from the substrate. Inset: the gel printed on the substrate containing Zn^{2+} showed extremely low resolution. (**d**) Overview and microscopic images of the 3D-printed hydrogels crosslinked using Cu^{2+}, Al^{3+}, or Fe^{3+}. The hydrogel crosslinked using Zn^{2+} is not shown because it was too brittle to maintain its structure when transferred from the substrate. Reprinted with permission from [109].

An interesting approach is based on the Pluronic F127, poly(ethylene oxide)-b-poly(propylene oxide)-b- poly(ethylene oxide) (PEO–PPO–PEO) with modified end groups of dimethacrylate (FdMA). This component is responsible for reverse thermo-responsiveness; the second is acrylic acid for pH responsiveness. In that way, the FdMA-co-acrylic acid hydrogels will have dual responsiveness. 3D structures via stereolithography are formed, and their environmentally sensitive dimensional behavior was evaluated. As expected, it was found tunable responsiveness in terms of temperature and pH. Combining hydrogels with different compositions, a different response of the structures obtained in temperature and pH, altering their size and geometry accordingly (Figure 14). The novel structures that were 3D printed are expected to contribute to the field of medical devices since they can change their size-volume-geometry "on command", by changes in the temperature and pH [110].

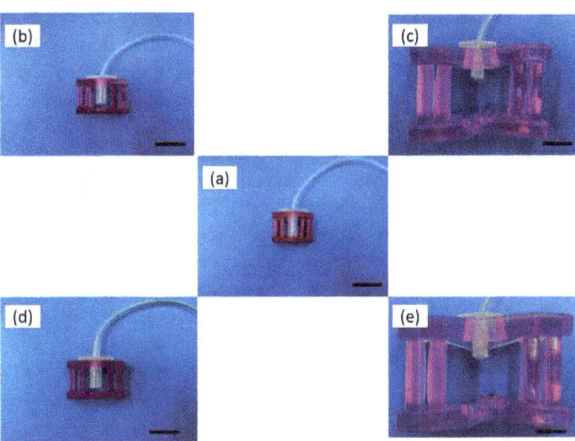

Figure 14. 3D printed valve structure of FA70(80) at different conditions. (**a**) Dry, (**b**) pH 2.0/37 °C (**c**) pH 7.4/37 °C, (**d**) pH 2.0/6 °C and (**e**) pH 7.4/6 °C. (Bar: 2 cm.). Reprinted with permission from [110].

Finally, the pluronic F127 (PEO–PPO–PEO) was used as pentablock, with the end blocks being PLA or PCL. These pentablocks were also modified with 2-isocyanatoethyl methacrylate to improve printability and can be used with acrylic acid and form hydrogels. These structures could have dual responsiveness of pH and temperature due to their blocks and degradability because of PLA and PCL. Adjusting the composition of PLA/PCL, the physicochemical properties of the hydrogels can be tuned, such as water absorption and biodegradation. These structures showed a fast and reversible swelling–deswelling response in pHs 2.0–7.0 or between 10 to 37 °C. The dual behavior for BSA release at different rates has been exploited by changing the release conditions. It has been shown that the PLA/PCL ratio plays an important role. It has been suggested that these materials can be good candidates for colon protein release formulations [111].

4.3.2. Block Copolymers with Other Type of Materials

Research in block copolymers and other materials also has significant and interesting publications [112–119]. Work preparing bioinks for hydrogel implants for controlled drug release is based on different formulations. These systems fabricate cross-linkable chitosan with PLA-PEG and PEGDA. The authors are presenting three different bioink systems: (i) Methacryloyl chitosan/PEGDA, (ii) PLA-PEG (micelle)/PEGDA, and (iii) methacryloyl chitosan/PLA-PEG (micelle)/PEGDA, which are all systems photo-crosslinked during 3D printing. This work studied the biocompatibility–biodegradability of the systems for the controlled release of the hydrophobic drug simvastatin from 3D structures hydrogels for osteogenic stimulation. Tuned mechanical properties and swelling behavior was found for the different systems. The viability of these hydrogels was proved when in contact with NIH3T3 fibroblast cells. The critical point of this work was increased release over 14 weeks with a therapeutic concentration [112].

Another work is based on the preparation of polyurethane, with one of the components being an ABA type or ABCBA pentablock [113]. The first case is PCL-PEG-PCL, and the second one is PLA-PCL-PEG-PCL-PLA. In this way, the final product will also have a biodegradability property that can have vast wide-range utility in various applications as scaffolds. The different compositions and lengths of the blocks led to tunable biodegradability, elastic properties, hydrophilicity, morphology, water uptake, and thermal properties [113].

A work based on combining organic-inorganic components is also found in the literature [114]. This work focuses on a thermoplastic amphiphilic poly(lactic acid-co-ethylene glycol-co-lactic acid) combined with hydroxyapatite. This work aims to create scaffolds with shape-memory effects and macroporosity that can be fabricated by rapid prototyping. These systems were found to have a shape-memory effect at a safe triggering temperature of 50 °C and high elasticity at room temperature. The incorporation of hydroxyapatite until 20% increased the tensile modulus while maintaining shape-memory properties at more than 90%. All the above led to the conclusion that these systems showed that osteoconductivity and osteoinductivity make them ideal candidates for bone regeneration [114].

Another organic-inorganic system is based on the thermoresponsive block copolymer poly(2-methyl-2-oxazoline)-b-poly(2-n-propyl-2-oxazine) (PMeOx-b-PnPrOzi) combined with nanoclay laponite [115]. The work wants to show this formulation's potential application as bioink and evaluate the critical properties relevant to extrusion bioprinting. The authors could show the system's thermoresponsive character and enhanced viscoelastic properties, leading to improved printability (Figure 15). According to the authors, this material can be a versatile support bath for fluid extrusion printing, thermoresponsive self-protection, controlled drug delivery, and a sacrificial template in microfluidic devices [115].

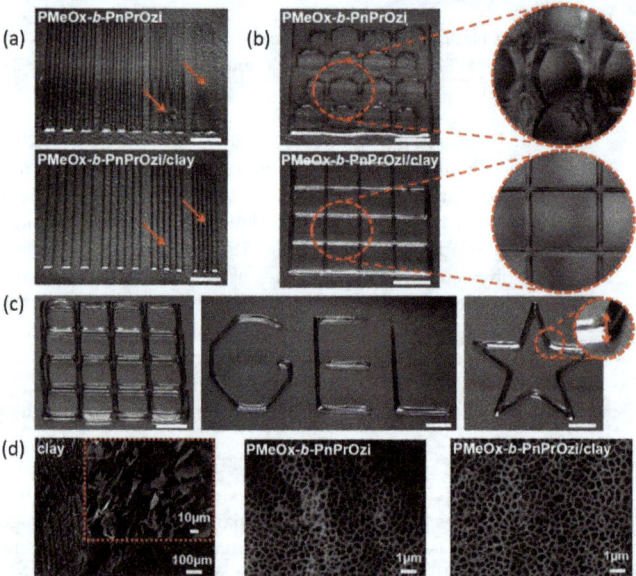

Figure 15. (**a**) 2D printed patterns for assessing the minimal strand-to strand distance and shape fidelity of biomaterial inks (red arrows are guiding for eyes). (**b**) Optical and stereomicroscopic images of the printed constructs composed with three layers of 5 × 5 orthogonal strands with a base area of 20 × 20 mm^2 (amplified in dashed red circles). (**c**) Photographic images of 3D printed six layers of a 5 × 5 woodpile structure, letters "GEL" and a five-pointed star with PMeOx-b-PnPrOzi/clay hydrogel. (**d**) SEM images of clay and cryo-SEM images of PMeOx-b-PnPrOzi and PMeOx-b-PnPrOzi/clay. Scale bars in (**a**–**c**) represent 5 mm. Reprinted with permission from [115].

In order to increase the mechanical properties of polymers to meet the requirements of 3D bioprinting by extrusion, this work proposes the addition of rigid nanoparticles of cellulose nanocrystals (CNC) into a poly(ε-caprolactone/lactide)-b-poly(ethylene glycol)-b-poly(ε-caprolactone/lactide) (PCLA-PEG-PCLA) triblock copolymer. The results showed increased thermal stability, a broader gel state (in terms of temperature) and increased mechanical properties with CNC at higher concentrations. This work proved that incorporating rigid CNC into amorphous gels could improve printability [116].

In another work [117], the authors have a different approach based on nanostructuring in order to increase the biocopmatibility of pluronic gels at printable concentrations. This work fabricates 3D constructs based on the mixing of acrylated and non-acrylated pluronic F127 with hyaluronic acid via UV crosslinking. These gels have a reversible thermo-gelling property and good rheological properties due to F127. A very important point is also the 14 day cell viability that was assessed using a live/dead assay (viable cells stained green and dead cells stained red fluorescent), as shown in Figure 16 [117].

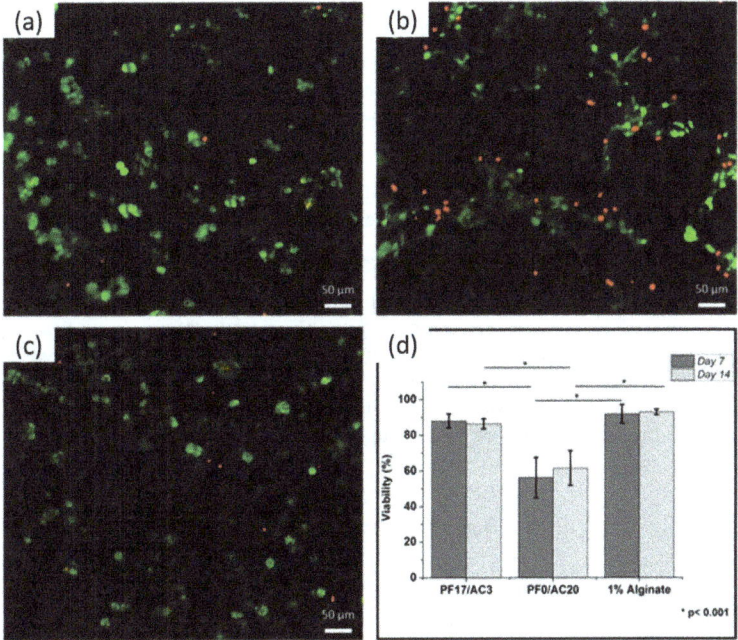

Figure 16. Cell viability of encapsulated bovine chondrocytes at day 14 with live cells in green and dead cells in red. (**a**) Nanostructured PF17/AC3 (**b**) PF0/AC20 and (**c**) 1% alginate. (**d**) Viabilities of encapsulated chondrocytes at day 7 and day 14. Reprinted with permission from [117].

To prepare a hydrogel suitable for cartilage 3D printing applications, a material based on methacrylated chondroitin sulfate and a thermo-sensitive poly(N-(2-hydroxypropyl) methacrylamide-mono/dilactate)/polyethylene glycol triblock copolymer was prepared [118]. The 3D technique for preparing these materials is based on a bioprinter equipped with UV lamps. These hydrogels showed superior rheological properties when compared to their homopolymer analogous. This polymer solution was used as ink to fabricate 3D structures with different porosity. An important point was the good survival and proliferation of chondrogenic cells when incorporated into the hydrogel [118].

Finally, an exciting work involving a two-component hydrogel for preparing scaffolds is presented here [119]. The first component of hydrogel consists of an ABC copolymer in combination with hyaluronic acid (HA) or PEG, both functionalized with N-hydroxysuccinimide. The ABC copolymer was a PEG-b-NiPAAm-b-HPMACys (N-2-Hydroxy-propyl)methacrylamide-Boc-S-acetamidomethyl-L-cysteine) and was used to pre-crosslinked with oxo-ester mediated native chemical ligation and physically crosslinked after deposition on a 37 °C printing plate. The researchers proved that the significant increase in Young modulus (17 to 645 kPa) by covalently grafting to a thermoplastic polymer scaffold (Poly(hydroxymethylglycolide)-co-ε-caprolactone). The authors have successfully demonstrated the increase in mechanical properties of the two-component hydrogels and also high cell viability of chondrocytes in the structure with hyaluronic acid [119].

5. Challenges and Future Works

3D printing, alongside 4D, is currently in constant evolution. New materials, techniques, and 3D structures for various applications are discovered yearly. The future of this technique in the biomaterials area has great potential but also many challenges and obstacles that need to be considered, improved, or changed to the level of the different applications required.

5.1. Challenges

Regarding challenges, the three main directions that need to be considered are: (i) Materials and techniques, (ii) scaffold architecture, and (iii) cell viability/vascularization.

(I) Materials and Techniques

The choice of material is a very crucial issue to deal with for scientists working with 3D printing. The chemical properties, physical properties, potential stimulus, printability, versatility, cost, degradability, and bioavailability will come from the selection of the materials. For hydrogels, for example, the printing resolution needs to be considered. Now it is at 0.3 mm, and for specific scaffolds, a better resolution will be required. Degradation or not is another challenge, especially for scaffolds, where new cells must be adhered to and slowly replaced by the printed scaffold. In the case of by-products after degradation, a careful evaluation must be done to ensure that they are not toxic and do not affect the environment where the 3D-printed object was present. Mechanical properties play a vital role since they are responsible for the printability and stability of the structure. Different parts of the human body need other mechanics for the constructs. In part of the materials, the biocompatibility must always be considered, and the way of sterilization is also a crucial issue for the choice of material.

(II) Scaffold architecture

In the case of scaffold architecture, the physical characteristics should be improved, meaning pore size, distribution and morphology, and topography. It should always be taken in mind that in these scaffolds, cells will adhere and grow and substitute the existing structure. All physical properties should be adequate for the specific cells. During 3D printing, many hydrogels show inhomogeneity due to the uncontrollable crosslinking, leading to poorer mechanical properties, so the way crosslinking is to happen during the preparation of the scaffolds is also quite important.

(III) Cell viability/vascularization

Finally, there are still many unsolved issues in terms of cell viability and vascularization. Cells must be viable and evenly distributed onto the scaffold in tissue engineering. Cells loaded to the printer are also crucial since viability is important during loading, printing, and postprocessing. At present, existing 3D printing technologies can be used to precisely dispense one or two cell-laden hydrogels for the construction of simple or vascularized tissues. They need to recapitulate the intrinsic complexity of vasculature in a natural organ. Building a real blood circulatory network is challenging and one of the future directions. The last point is the 3D printing of personalized constructs or drug delivery for each patient [41,76,77,82].

5.2. Future Works

Overcoming the challenges that now exist in 3D printing technologies, the future can be fruitful for functional materials in biology and medicine. The main lines of investigation that the researchers will focus their research on will be:

(I)   Materials
(II) Combination of natural and synthetic polymers
(III) 3D printers
(IV) Crosslinking
(V)   Testing and simulation of the 3D structures
(VI) New techniques of processing incorporated in 3D

The selection of materials will always be an area of investigation that can lead to the discovery of new combinations or polymers. These formulations will affect the processability, degradation, mechanical properties, and cell viability. Controlling the diversity and the different polymers (natural or synthetic), desirable physiological functions will be captured in specific micro/macro-environments. An investigation will also be made into the 3D printers to improve their capabilities, especially in the bioprinting technology.

In crosslinking, new techniques will be studied to differentiate pre and post-printing chemistries that can improve and better control the crosslinking. In many formulations, the material to be printed consist of different components that can create formulations with agglomerations; therefore, new methods that ensure homogeneous distribution of the different phases must be studied. The testing and simulation of these 3D constructs is also significant. In biomaterials, high standards must be applied in terms of mechanical, physical, chemical, and stimuli. Finally, the investigation will be done with other techniques, such as electrospinning, biofabrication, microfluidics, and 4D printing [41,76,77,82].

6. Conclusions

In the last 10–12 years, 3D/4D printing has changed dramatically. Techniques or polymers used at the beginning are modified or functionalized to have better printability, processability, resolution, and dual properties with or without stimuli. Researchers are constantly investigating new materials and ways of preparing 3D constructs. Every time they fabricate materials that can substitute hard or soft tissues, used as drug delivery vehicles, scaffolds where cells will grow and proliferate and structures printed alongside different types of cells. The structures are every time more complex and try to mimic organs or systems or functions of the human body.

Someone will ask, where is the limit? It can be said that it is almost infinite. Next-generation printed devices are expected to respond after sensing first local changes in live tissues (infection, cancer, neurodegeneration, or inflammation), offering more personalized therapy. In the case of polymers, many architectures and polymers are already in use. Apart from homopolymers (synthetic or natural), block copolymers play a significant role in the evolution and development of 3D/4D printing. Figure 17 shows two graphs of the different architectures and polymers of block copolymers used in the last 10–12 years. Most of the architectures are based on AB and ABA but with natural polymers (e.g., chitosan or hyaluronic acid) or inorganic materials (e.g., hydroxyapatite or clay) and star-like polypeptides to gain their way in more and more applications. In terms of polymers, poly(ethylene glycol), poly(caprolactone), and poly(lactide) are used in half of the works in one way or another. Nevertheless, new polymers are used in at least 18% of the works. One-third of the works also use candidates with potential due to their properties, such as acrylates, acrylamides, peptides, poloxamers, and oxazolines.

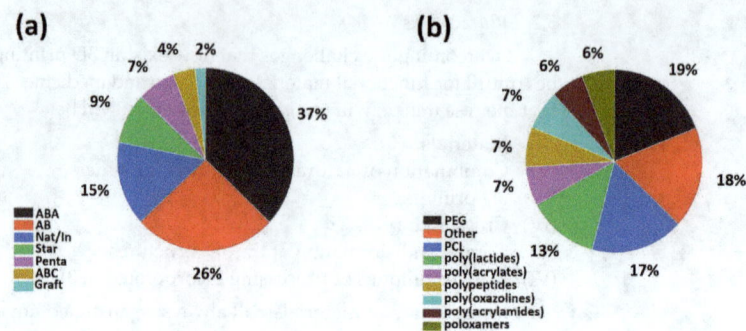

Figure 17. (**a**) Pie chart of the different architectures of block copolymers investigated the last years, and (**b**) pie chart of the different polymers used in block copolymers in the last years.

Finally, it has to be said that the future will bring many discoveries in smart materials for 4D (bio)printing, complete automations in high-resolution printers, computational modeling, and artificial intelligence. The therapy will be transformed into a more personalized one thanks to 3D printing and its constant evolution.

Funding: This research received no external funding.

Institutional Review Board Statement: Not applicable.

Data Availability Statement: Not applicable.

Conflicts of Interest: The author declares no conflict of interest.

Abbreviations

2PP	two-photon polymerization
AB	block copolymers
ABA	triblock copolymers
ABC	triblocl terpolymers
AJP	aerosol jet printing
CLIP	continuous liquid interface production
CNC	cellulose nanocrystals
CSMA	methacrylated chondroitin sulfate
DIW	direct ink writing
DLP	digital projection lithography
EHD	electro-hydrodynamic printing
FDM	fused deposition modeling
FdMA	pluronic F127 dimethacrylate
FFF	fused filament fabrication
HAMA	methacrylated hyaluronic acid
HAMA	hyaluronic acid
HPMACys	N-2-Hydroxy-propyl)methacrylamide-Boc-S-acetamidomethyl-L-cysteine
PAA	poly(acrylic acid)
PAMPS	poly(2-(acrylamide)-2-methylpropanesulfonic acid)
PASA	poly(aspartic acid)
PBF	powder bed fusion
PBnMA	poly(benzyl methacrylate)
PBT	poly(butylene terephthalate)
PCL	polycaprolactone
PCLA	poly(ε-caprolactone/lactide)
PCL-PPSu	poly(1,3-propylene succinate)
PDLGA	poly(D, L-lactide-co-glycolide)

PDMAEMA	poly(dimethylaminoethyl methacrylate)
PDMS	polydimethylsiloxane
PEDOT	poly(3,4-ethylenedioxythiophene)
PEG	polyethylene glycol
PEGDA	poly(ethylene glycol) diacrylate
PEGT	poly(ethylene glycol) terephthalate
PEI	polyethyleneimine
pEtOx	poly(2-ethyl-2-oxazoline)
PEU	poly(ether urethane)
PGA	poly(glycolic acid)
PHEMA	poly(2-hydroxyethyl methacrylate)
PHIS	poly(histidine)
piBuOx	poly(2-iso-butyl-2-oxazoline)
PLA	poly(lactic acid)
PLACL	poly(L-lactide-co-ε-caprolactone)
PLGA	polylactic-co-glycolic Acid
PLLA	poly(L-lactic acid)
pMeOx	poly(2-methyl-2-oxazoline)
PMMA	poly(methyl methacrylate)
PnBA	poly(n-butyl acrylate)
PNiPAM	poly(N-isopropylacrylamide)
PPF	poly(propylene fumarate)
PPG	poly(propylene glycol)
PPM	poly(propylene maleate)
PPO	poly(propylene oxide)
pPrOzi	poly(2 N-propyl-2-oxazine)
PPy	polypyrrole
PSS	sulfonated-polystyrene
PT	polythiophene
PTMC	poly(trimethylene carbonate)
Ptriol	polycaprolactone triol
PU	polyurethane
SIBS	poly(styrene-b-isobutylene-b-styrene)
SLA	laser stereolithography
TMSPMA	3-(trimethoxysilyl)propyl methacrylate

References

1. Nadgorny, M.; Ameli, A. Functional Polymers and Nanocomposites for 3D Printing of Smart Structures and Devices. *ACS Appl. Mater. Interfaces* **2018**, *10*, 17489–17507. [CrossRef] [PubMed]
2. Ligon, S.C.; Liska, R.; Stampfl, J.; Gurr, M.; Mülhaupt, R. Polymers for 3D Printing and Customized Additive Manufacturing. *Chem. Rev.* **2017**, *117*, 10212–10290. [CrossRef] [PubMed]
3. Gebhardt, A. *Rapid Prototyping*; Hanser Verlag: Munich, Germany, 2003.
4. Jacobs, P.F. *Stereolithography and Other RP&M Technologies-from Rapid Prototyping to Rapid Tooling*; SME Publications: Dearborn, MI, USA, 1996.
5. Noorani, R. *Rapid Prototyping. Principles and Applications*; Wiley: Hoboken, NJ, USA, 2006; pp. 34–56.
6. Wohlers, T.; Caffrey, T. *Wohlers Report 2013: Additive Manufacturing and 3D Printing State of the Industry*; Wohlers Associates: Fort Collins, CO, USA, 2013.
7. Pham, D.T.; Gault, R.S. A Comparison of Rapid Prototyping Technologies. *Int. J. Mach. Tool Manuf.* **1998**, *38*, 1257–1287. [CrossRef]
8. Sachs, E.M.; Haggerty, J.S.; Cima, M.J.; Williams, P.A. Three-Dimensional Printing Techniques. U.S. Patent Application No. US5204055A, 20 April 1993.
9. Elder, B.; Neupane, R.; Tokita, E.; Ghosh, U.; Hales, S.; Kong, Y.L. Nanomaterial Patterning in 3D Printing. *Adv. Mater.* **2020**, *32*, 1907142. [CrossRef] [PubMed]
10. Prabhakar, M.M.; Saravanan, A.K.; Lenin, A.H.; Mayandi, K.; Ramalingam, P.S. A short review on 3D printing methods, process parameters and materials. *Mater. Today Proc.* **2021**, *45*, 6108–6114. [CrossRef]
11. Truby, R.L.; Lewis, J.A. Printing soft matter in three dimensions. *Nature* **2016**, *540*, 371–378. [CrossRef]
12. Wallin, T.J.; Pikul, J.; Shepherd, R.F. 3D printing of soft robotic systems. *Nat. Rev. Mater.* **2018**, *3*, 84–100. [CrossRef]
13. Hwang, H.H.; Zhu, W.; Victorine, G.; Lawrence, N.; Chen, S. 3D-Printing of Functional Biomedical Microdevices via Light- and Extrusion-Based Approaches. *Small Methods* **2018**, *2*, 1870021. [CrossRef]

14. B9 Core Series Technical Specifications | B9Creations. Available online: https://www.b9c.com/products/tech-specs (accessed on 3 January 2023).
15. PICO2-Products-Asiga. Available online: https://www.asiga.com/products/printers/pico2/ (accessed on 1 December 2019).
16. Solidator 2-DLP 3D Printer-Fast Stereolithography 3D Printer with a Large Build Volume. Available online: https://www.solidator.com/3D-Printer.html#targetText=The%Solidator%is%an%ultra,13.22%22x8.26%22 (accessed on 3 January 2023).
17. Pro Desktop 3D Printer Technical Specifications | SprintRay Inc. Available online: https://sprintray.com/pro-desktop-3dprinter/pro-desktop-3d-printer-technical-specifications/ (accessed on 3 January 2023).
18. Janusziewicz, R.; Tumbleston, J.R.; Quintanilla, A.L.; Mecham, S.J.; DeSimone, J.M. Layerless fabrication with continuous liquid interface production. *Proc. Natl. Acad. Sci. USA* **2016**, *113*, 11703–11708. [CrossRef]
19. Tumbleston, J.R.; Shirvanyants, D.; Ermoshkin, N.; Janusziewicz, R.; Johnson, A.R.; Kelly, D.; Chen, K.; Pinschmidt, R.; Rolland, J.P.; Ermoshkin, A.; et al. Continuous liquid interface production of 3D objects. *Science* **2015**, *347*, 1349–1352. [CrossRef]
20. Melchels, F.P.W.; Feijen, J.; Grijpma, D.W. A review on stereolithography and its applications in biomedical engineering. *Biomaterials* **2010**, *31*, 6121–6130. [CrossRef]
21. Shusteff, M.; Browar, A.E.M.; Kelly, B.E.; Henriksson, J.; Weisgraber, T.H.; Panas, R.M.; Fang, N.X.; Spadaccini, C.M. One-step volumetric additive manufacturing of complex polymer structures. *Sci. Adv.* **2017**, *3*, eaao5496. [CrossRef]
22. Varotsis, A.B. Introduction to Binder Jetting 3D Printing. Available online: https://www.3dhubs.com/knowledge-base/introduction-binder-jetting-3d-printing#what (accessed on 3 January 2023).
23. Afshar-Mohajer, N.; Wu, C.Y.; Ladun, T.; Rajon, D.A.; Huang, Y. Characterization of particulate matters and total VOC emissions from a binder jetting 3D printer. *Build. Environ.* **2015**, *93*, 293–301. [CrossRef]
24. Hajash, K.; Sparrman, B.; Guberan, C.; Laucks, J.; Tibbits, S. Large-Scale Rapid Liquid Printing. *3d Print. Addit. Manuf.* **2017**, *4*, 123–132. [CrossRef]
25. He, J.; Xu, F.; Dong, R.; Guo, B.; Li, D. Electrohydrodynamic 3D printing of microscale poly (ε-caprolactone) scaffolds with multi-walled carbon nanotubes. *Biofabrication* **2017**, *9*, 15007. [CrossRef]
26. Nagy, Z.K.; Balogh, A.; Démuth, B.; Pataki, H.; Vigh, T.; Szabó, B.; Molnár, K.; Schmidt, B.T.; Horák, P.; Marosi, G.; et al. High speed electrospinning for scaled-up production of amorphous solid dispersion of itraconazole. *Int. J. Pharm.* **2015**, *480*, 137–142. [CrossRef]
27. Singh, M.; Haverinen, H.M.; Dhagat, P.; Jabbour, G.E. Inkjet printing-process and its applications. *Adv. Mater.* **2010**, *22*, 673–685. [CrossRef]
28. Paulsen, J.A.; Renn, M.; Christenson, K.; Plourde, R. *2012 Future of Instrumentation International Workshop (FIIW) Proceedings*; Abbott, D., Ed.; IEEE: Piscataway, NJ, USA, 2012; pp. 47–50.
29. Mahajan, A.; Frisbie, C.D.; Francis, L.F. Optimization of Aerosol Jet Printing for High-Resolution, High-Aspect Ratio Silver Lines. *ACS Appl. Mater. Interfaces* **2013**, *5*, 4856–4864. [CrossRef]
30. Tibbits, S.J.E.; Dikovsky, D.; Hirsch, S. Massachusetts Institute of Technology; Stratasys Ltd. Object of Additive Manufacture with Encoded Predicted Shape Change. U.S. Patent Application No. 2015084422A1, 11 June 2015.
31. Tibbits, S. 4D Printing: Multi-Material Shape Change. *Archit. Design* **2014**, *84*, 116–121. [CrossRef]
32. Momeni, F.; Hassani, S.M.M.; Liu, X.; Ni, J. A Review of 4D Printing. *Mater. Des.* **2017**, *122*, 42–79. [CrossRef]
33. Joshi, S.; Rawat, K.; Karunakaran, C.; Rajamohan, V.; Mathew, A.T.; Koziol, K.; Thakur, V.K.; Balan, A.S.S. 4D printing of materials for the future: Opportunities and challenges. *Appl. Mater. Today* **2020**, *18*, 100490. [CrossRef]
34. Akbari, S.; Zhang, Y.; Wang, D.; Ge, Q. 4D printing and its biomedical applications. In *3D and 4D Printing in Biomedical Applications: Process Engineering and Additive Manufacturing*; Wiley-VCH Verlag GmbH &Co. KGaA: Weinheim, Germany, 2019; Chapter 14; pp. 343–372.
35. Yang, Y.; Chen, Y.; Wei, Y.; Li, Y. 3D printing of shape memory polymer for functional part fabrication. *Int. J. Adv. Manuf. Technol.* **2016**, *84*, 2079–2095. [CrossRef]
36. Liu, F.; Urban, M.W. Recent Advances and Challenges in Designing Stimuli-Responsive Polymers. *Prog. Polym. Sci.* **2010**, *35*, 3–23. [CrossRef]
37. Seuring, J.; Agarwal, S. Polymers with Upper Critical Solution Temperature in Aqueous Solution. *Macromol. Rapid Commun.* **2012**, *33*, 1898–1920. [CrossRef] [PubMed]
38. Dai, S.; Ravi, P.; Tam, K.C. pH-Responsive Polymers: Synthesis, Properties and Applications. *Soft Matter* **2008**, *4*, 435–449. [CrossRef]
39. Davis, D.A.; Hamilton, A.; Yang, J.; Cremar, L.D.; Van Gough, D.; Potisek, S.L.; Ong, M.T.; Braun, P.V.; Martínez, T.J.; White, S.R.; et al. Force-Induced Activation of Covalent Bonds in Mechanoresponsive Polymeric Materials. *Nature* **2009**, *459*, 68–72. [CrossRef]
40. Wei, Z.; Yang, J.H.; Zhou, J.; Xu, F.; Zrínyi, M.; Dussault, P.H.; Osada, Y.; Chen, Y.M. Self-Healing Gels Based on Constitutional Dynamic Chemistry and Their Potential Applications. *Chem. Soc. Rev.* **2014**, *43*, 8114–8131. [CrossRef]
41. Ratheesh, G.; Venugopal, J.R.; Chinappan, A.; Ezhilarasu, H.; Sadiq, A.; Ramakrishna, S. 3D Fabrication of Polymeric Scaffolds for Regenerative Therapy. *ACS Biomater. Sci. Eng.* **2017**, *3*, 1175–1194. [CrossRef]
42. Donderwinkel, I.; van Hest, J.C.M.; Cameron, N.R. Bio-inks for 3D bioprinting: Recent advances and future prospects. *Polym. Chem.* **2017**, *8*, 4451–4471. [CrossRef]

43. Groll, J.; Boland, T.; Blunk, T.; Burdick, J.A.; Cho, D.W.; Dalton, P.D.; Derby, B.; Forgacs, G.; Li, Q.; Mironov, V.A.; et al. Biofabrication: Reappraising the definition of an evolving field. *Biofabrication* **2016**, *8*, 013001. [CrossRef]
44. Zongjie, W.; Raafa, A.; Benjamin, P.; Roya, S.; Sanjoy, G.; Keekyoung, K. Light-activated nitrile imine mediated reaction pathways for the synthesis of bioinks. *Biofabrication* **2015**, *7*, 045009.
45. Barry, J.J.A.; Evseev, A.V.; Markov, M.A.; Upton, C.E.; Scotch-ford, C.A.; Popov, V.K.; Howdle, S.M. In vitro study of hydroxyapatite-based photocurable polymer composites prepared by laser stereolithography and supercritical fluid extraction. *Acta Biomater.* **2008**, *4*, 1603–1610. [CrossRef]
46. Zhu, W.; Ma, X.; Gou, M.; Mei, D.; Zhang, K.; Chen, S. 3D printing of functional biomaterials for tissue engineering. *Curr. Opin. Biotechnol.* **2016**, *40*, 103–112. [CrossRef]
47. Pereira, R.F.; Bártolo, P.J. 3D bioprinting of photocrosslinkable hydrogel constructs. *J. Appl. Polym. Sci.* **2015**, *132*, 42458. [CrossRef]
48. Tuan, R.S.; Boland, G.; Tuli, R. Adult mesenchymal stem cells and cell-based tissue engineering. *Arthritis Res. Ther.* **2003**, *5*, 32. [CrossRef]
49. Hopp, B.; Smausz, T.; Kresz, N.; Barna, N.; Bor, Z.; Kolozsvári, L.; Chrisey, D.B.; Szabó, A.; Nógrádi, A. Survival and Proliferative Ability of Various Living Cell Types after Laser-Induced Forward Transfer. *Tissue Eng.* **2005**, *11*, 1817. [CrossRef]
50. Mandrycky, C.; Wang, Z.; Kim, K.; Kim, D.H. 3D bioprinting for engineering complex tissues. *Biotechnol. Adv.* **2016**, *34*, 422–434. [CrossRef]
51. Murphy, S.V.; Atala, A. 3D bioprinting of tissues and organs. *Nat. Biotechnol.* **2014**, *32*, 773–785. [CrossRef]
52. Cui, X.; Li, J.; Hartanto, Y.; Durham, M.; Tang, J.; Zhang, H.; Hooper, G.; Lim, K.; Woodfield, T. Advances in Extrusion 3D Bioprinting: A Focus on Multicomponent Hydrogel-Based Bioinks. *Adv. Healthcare Mater.* **2020**, *9*, 1901648. [CrossRef]
53. Mironov, V.; Visconti, R.P.; Kasyanov, V.; Forgacs, G.; Drake, C.J.; Markwald, R.R. Organ printing: Tissue spheroids as building blocks. *Biomaterials* **2009**, *30*, 2164–2174. [CrossRef]
54. Jiang, T.; Munguia-Lopez, J.G.; Flores-Torres, S.; Grant, J.; Vi-jayakumar, S.; Leon-Rodriguez, A.D.; Kinsella, J.M. Directing the Self-assembly of Tumour Spheroids by Bioprinting Cellular Heterogeneous Models within Alginate/Gelatin Hydrogels. *Sci. Rep.* **2017**, *7*, 4575. [CrossRef] [PubMed]
55. Jiang, T.; Munguia-Lopez, J.G.; Flores-Torres, S.; Kort-Mascort, J.; Kinsella, J.M. Extrusion bioprinting of soft materials: An emerging technique for biological model fabrication. *Appl. Phys. Rev.* **2019**, *6*, 011310. [CrossRef]
56. Zaszczyńska, A.; Moczulska-Heljak, M.; Gradys, A.; Sajkiewicz, P. Advances in 3D Printing for Tissue Engineering. *Materials* **2021**, *14*, 3149. [CrossRef] [PubMed]
57. Hong, S.; Song, S.J.; Lee, J.Y.; Jang, H.; Choi, J.; Sun, K.; Park, Y.J. Biosci. Cellular behavior in micropatterned hydrogels by bioprinting system depended on the cell types and cellular interaction. *J. Biosci. Bioeng.* **2013**, *116*, 224–230. [CrossRef] [PubMed]
58. Keriquel, V.; Guillemot, F.; Arnault, I.; Guillotin, B.; Miraux, S.; Amedee, J.; Fricain, J.C.; Catros, S. In vivo bioprinting for computer- and robotic-assisted medical intervention: Preliminary study in mice. *Biofabrication* **2010**, *2*, 014101. [CrossRef]
59. Tao, X.; Kyle, W.B.; Mohammad, Z.A.; Dennis, D.; Weixin, Z.; James, J.Y.; Anthony, A. Hybrid printing of mechanically and biologically improved constructs for cartilage tissue engineering applications. *Biofabrication* **2013**, *5*, 015001.
60. Xu, F.; Sridharan, B.; Wang, S.; Gurkan, U.A.; Syverud, B.; Demirci, U. Embryonic stem cell bioprinting for uniform and controlled size embryoid body formation. *Biomicrofluidics* **2011**, *5*, 022207. [CrossRef]
61. Gauvin, R.; Chen, Y.C.; Lee, J.W.; Soman, P.; Zorlutuna, P.; Nichol, J.W.; Bae, H.; Chen, S.; Khademhosseini, A. Microfabrication of complex porous tissue engineering scaffolds using 3D projection stereolithography. *Biomaterials* **2012**, *33*, 3824–3834. [CrossRef]
62. Grogan, S.P.; Chung, P.H.; Soman, P.; Chen, P.; Lotz, M.K.; Chen, S.; D'Lima, D.D. Digital micromirror device projection printing system for meniscus tissue engineering. *Acta Biomater.* **2013**, *9*, 7218–7226. [CrossRef]
63. Dolati, F.; Yu, Y.; Zhang, Y.; De Jesus, A.M.; Sander, E.A.; Ozbolat, I.T. In vitro evaluation of carbon-nanotube-reinforced bioprintable vascular conduits. *Nanotechnology* **2014**, *25*, 145101. [CrossRef]
64. Duan, B.; Hockaday, L.A.; Kang, K.H.; Butcher, J.T. 3D bioprinting of heterogeneous aortic valve conduits with alginate/gelatin hydrogels. *J. Biomed. Mater. Res. Part A* **2013**, *101*, 1255–1264. [CrossRef]
65. Loozen, L.D.; Wegman, F.; Oner, F.C.; Dhert, W.J.A.; Alblas, J. Porous bioprinted constructs in BMP-2 non-viral gene therapy for bone tissue engineering. *J. Mater. Chem. B* **2013**, *1*, 6619–6626. [CrossRef]
66. Malda, J.; Visser, J.; Melchels, F.P.; Jungst, T.; Hennink, W.E.; Dhert, W.J.; Groll, J.; Hutmacher, D.W. 25th Anniversary Article: Engineering Hydrogels for Biofabrication. *Adv. Mater.* **2013**, *25*, 5011–5028. [CrossRef]
67. Zhao, Y.; Yao, R.; Ouyang, L.; Ding, H.; Zhang, T.; Zhang, K.; Cheng, S.; Sun, W. Three-dimensional printing of Hela cells for cervical tumor model in vitro. *Biofabrication* **2014**, *6*, 035001. [CrossRef]
68. Merceron, T.K.; Burt, M.; Seol, Y.J.; Kang, H.W.; Lee, S.J.; Yoo, J.J.; Atala, A. A 3D bioprinted complex structure for engineering the muscle—tendon unit. *Biofabrication* **2015**, *7*, 035003. [CrossRef]
69. Owens, C.M.; Marga, F.; Forgacs, G.; Heesch, C.M. Biofabrication and testing of a fully cellular nerve graft. *Biofabrication* **2013**, *5*, 045007. [CrossRef]
70. Lim, K.; Levato, R.; Costa, F.P.; Castilho, D.M.; Alcala-orozco, R.C.; Dorenmalen, M.A.K.; Meichels, P.W.F.; Gawlitta, D.; Hopper, G.J.; Malda, J.; et al. Bio-resin for high resolution lithography-based biofabrication of complex cell-laden constructs. *Biofabrication* **2018**, *10*, 034101. [CrossRef]
71. Catros, S.; Guillotin, B.; Bačáková, M.; Fricain, J.-C.; Guillemot, F. Effect of laser energy, substrate film thickness and bioink viscosity on viability of endothelial cells printed by Laser-Assisted Bioprinting. *Appl. Surf. Sci.* **2011**, *257*, 5142–5147. [CrossRef]

72. Catros, S.; Fricain, J.C.; Guillotin, B.; Pippenger, B.; Bareille, R.; Remy, M.; Lebraud, E.; Desbat, B.; Amedee, J.; Guillemot, F. Laser-assisted bioprinting for creating on-demand patterns of human osteoprogenitor cells and nano-hydroxyapatite. *Biofabrication* **2011**, *3*, 025001. [CrossRef]
73. Michael, S.; Sorg, H.; Peck, C.-T.; Koch, L.; Deiwick, A.; Chichkov, B.; Vogt, P.M.; Reimers, K. Tissue Engineered Skin Substitutes Created by Laser-Assisted Bioprinting Form Skin-Like Structures in the Dorsal Skin Fold Chamber in Mice. *PLoS ONE* **2013**, *8*, e57741. [CrossRef]
74. Gruene, M.; Unger, C.; Koch, L.; Deiwick, A.; Chichkov, B. Dispensing pico to nanolitre of a natural hydrogel by laser-assisted bioprinting. *Biomed. Eng. Online* **2011**, *10*, 19. [CrossRef] [PubMed]
75. Valot, L.; Martinez, J.; Mehdi, A.; Subra, G. Chemical insights into bioinks for 3D printing. *Chem. Soc. Rev.* **2019**, *48*, 4049–4086. [CrossRef] [PubMed]
76. Advincula, R.C.; Dizon, J.R.C.; Caldona, E.B.; Viers, R.A.; Siacor, F.D.C.; Maalihan, R.D.; Espera Jr, A.H. On the progress of 3D-printed hydrogels for tissue engineering. *MRS Commun.* **2021**, *11*, 539–553. [CrossRef] [PubMed]
77. Liu, F.; Wang, X. Synthetic Polymers for Organ 3D Printing. *Polymers* **2020**, *12*, 1765. [CrossRef] [PubMed]
78. Chung, J.J.; Im, H.; Kim, S.H.; Park, J.W.; Jung, Y. Toward Biomimetic Scaffolds for Tissue Engineering: 3D Printing Techniques in Regenerative Medicine. *Front. Bioeng. Biotechnol.* **2020**, *8*, 586406. [CrossRef] [PubMed]
79. Abbadessa, A.; Mouser, V.H.M.; Blokzijl, M.B.; Gawlitta, D.; Dhert, W.J.A.; Hennink, W.E.; Malda, J.; Vermonden, T. A Synthetic Thermosensitive Hydrogel for Cartilage Bioprinting and Its Biofunctionalization with Polysaccharides. *Biomacromolecules* **2016**, *17*, 2137–2147. [CrossRef]
80. Izadifar, Z.; Chang, T.; Kulyk, W.; Chen, X.; Eames, B.F. Analyzing Biological Performance of 3D-Printed, Cell-Impregnated Hybrid Constructs for Cartilage Tissue Engineering. *Tissue Eng. Part C* **2016**, *22*, 173–188. [CrossRef]
81. Tamay, D.G.; Dursun Usal, T.; Alagoz, A.S.; Yucel, D.; Hasirci, N.; Hasirci, V. 3D and 4D Printing of Polymers for Tissue Engineering Applications. *Front. Bioeng. Biotechnol.* **2019**, *7*, 164. [CrossRef]
82. Morouço, P.; Azimi, B.; Milazzo, M.; Mokhtari, F.; Fernandes, C.; Reis, D.; Danti, S. Four-Dimensional (Bio-)printing: A Review on Stimuli-Responsive Mechanisms and Their Biomedical Suitability. *Appl. Sci.* **2020**, *10*, 9143. [CrossRef]
83. Jain, S.; Fuoco, T.; Yassin, M.A.; Mustafa, K.; Finne-Wistrand, A. Printability and Critical Insight into Polymer Properties during Direct-Extrusion Based 3D Printing of Medical Grade Polylactide and Copolyesters. *Biomacromolecules* **2020**, *21*, 388–396. [CrossRef]
84. Afghah, F.; Ullah, M.; Zanjani, J.S.M.; Sut, P.A.; Sen, O.; Emanet, M.; Okan, B.S.; Culha, M.; Menceloglu, Y.; Yildiz, M.; et al. 3D printing of silver-doped polycaprolactone-poly(propylene succinate) composite scaffolds for skin tissue engineering. *Biomed. Mater.* **2020**, *15*, 035015. [CrossRef]
85. Liu, W.; Feng, Z.; Ou-Yang, W.; Pan, X.; Wang, X.; Huang, P.; Zhang, C.; Kong, D.; Wang, W. 3D printing of implantable elastic PLCL copolymer scaffolds. *Soft Matter* **2020**, *16*, 2141–2148. [CrossRef]
86. Vijayavenkataraman, S.; Kannan, S.; Cao, T.; Fuh, J.Y.H.; Sriram, G.; Lu, W.F. 3D-Printed PCL/PPy Conductive Scaffolds as Three-Dimensional Porous Nerve Guide Conduits (NGCs) for Peripheral Nerve Injury Repair. *Front. Bioeng. Biotechnol.* **2019**, *7*, 266. [CrossRef]
87. Paunović, N.; Bao, Y.; Coulter, F.B.; Masania, K.; Geks, A.K.; Klein, K.; Rafsanjani, A.; Cadalbert, J.; Kronen, P.W.; Kleger, N.; et al. Digital light 3D printing of customized bioresorbable airway stents with elastomeric properties. *Sci. Adv.* **2021**, *7*, eabe9499. [CrossRef]
88. Haider, M.S.; Ahmad, T.; Yang, M.; Hu, C.; Hahn, L.; Stahlhut, P.; Groll, J.; Luxenhofer, R. Tuning the Thermogelation and Rheology of Poly(2-Oxazoline)/Poly(2-Oxazine)s Based Thermosensitive Hydrogels for 3D Bioprinting. *Gels* **2021**, *7*, 78. [CrossRef]
89. Koutsamanis, I.; Paudel, A.; Alva Zúñiga, C.P.; Wiltschko, L.; Spoerk, M. Novel polyester-based thermoplastic elastomers for 3D-printed long-acting drug delivery applications. *J. Control. Release* **2021**, *335*, 290–305. [CrossRef]
90. Dilla, R.A.; Motta, C.M.M.; Snyder, S.R.; Wilson, J.A.; Wesdemiotis, C.; Becker, M.L. Synthesis and 3D Printing of PEG–Poly(propylene fumarate) Diblock and Triblock Copolymer Hydrogels. *ACS Macro Lett.* **2018**, *7*, 1254–1260. [CrossRef]
91. Le Fer, G.; Dilla, R.A.; Wang, Z.; King, J.; Chuang, S.S.C.; Becker, M.L. Clustering and Hierarchical Organization of 3D Printed Poly(propylene fumarate)-block-PEG-block-poly(propylene fumarate) ABA Triblock Copolymer Hydrogels. *Macromolecules* **2021**, *54*, 3458–3468. [CrossRef]
92. Censi, R.; Schuurman, W.; Malda, J.; di Dato, G.; Burgisser, P.E.; Dhert, W.J.A.; van Nostrum, C.F.; di Martino, P.; Vermonden, T.; Hennink, W.E. A Printable Photopolymerizable Thermosensitive p(HPMAm-lactate)-PEG Hydrogel for Tissue Engineering. *Adv. Funct. Mater.* **2011**, *21*, 1833–1842. [CrossRef]
93. Güney, A.; Gardiner, C.; McCormack, A.; Malda, J.; Grijpma, D.W. Thermoplastic PCL-b-PEG-b-PCL and HDI Polyurethanes for Extrusion-Based 3D-Printing of Tough Hydrogels. *Bioengineering* **2018**, *5*, 99. [CrossRef]
94. Cui, Y.; Jin, R.; Zhou, Y.; Yu, M.; Ling, Y.; Wang, L.Q. Crystallization enhanced thermal-sensitive hydrogels of PCL-PEG-PCL triblock copolymer for 3D printing. *Biomed. Mater.* **2021**, *16*, 035006. [CrossRef]
95. Zhang, M.; Vora, a.; Han, W.; Wojtecki, R.J.; Maune, H.; Le, A.B.A.; Thompson, L.E.; McClelland, G.M.; Ribet, F.; Engler, A.C.; et al. Dual-Responsive Hydrogels for Direct-Write 3D Printing. *Macromolecules* **2015**, *48*, 6482–6488. [CrossRef]
96. Karis, D.G.; Ono, R.J.; Zhang, M.; Vora, A.; Storti, D.; Ganter, M.A.; Nelson, A. Cross-linkable multi-stimuli responsive hydrogel inks for direct-write 3D printing. *Polym. Chem.* **2017**, *8*, 4199–4206. [CrossRef]

97. Xu, C.; Lee, W.; Dai, G.; Hong, Y. Highly Elastic Biodegradable Single-Network Hydrogel for Cell Printing. *ACS Appl. Mater. Interfaces* **2018**, *10*, 9969–9979. [CrossRef] [PubMed]
98. Alsandi, Q.; Ikeda, M.; Nikaido, T.; Tsuchida, Y.; Sadr, A.; Yui, N.; Suzuki, T.; Tagami, J. Evaluation of mechanical properties of new elastomer material applicable for dental 3D printer. *J. Mech. Behav. Biomed. Mater.* **2019**, *100*, 103390. [CrossRef]
99. Shen, N.; Liu, S.; Kasbe, P.; Khabaz, F.; Kennedy, J.P.; Xu, W. Macromolecular Engineering and Additive Manufacturing of Poly(styrene-b-isobutylene-b-styrene). *ACS Appl. Polym. Mater.* **2021**, *3*, 4554–4562. [CrossRef]
100. Mulchandani, N.; Masutani, K.; Kumar, S.; Yamane, H.; Sakurai, S.; Kimura, Y.; Katiyar, V. Toughened PLA-b-PCL-b-PLA triblock copolymer based biomaterials: Effect of self-assembled nanostructure and stereocomplexation on the mechanical properties. *Polym. Chem.* **2021**, *12*, 3806–3824. [CrossRef]
101. Güney, A.; Malda, J.; Dhert, W.J.A.; Grijpma, D.W. Triblock Copolymers Based on ε-Caprolactone and Trimethylene Carbonate for the 3D Printing of Tissue Engineering Scaffolds. *Int. J. Artif. Organs* **2017**, *40*, 176–184. [CrossRef]
102. Nian, S.; Zhu, J.; Zhang, H.; Gong, Z.; Freychet, G.; Zhernenkov, M.; Xu, B.; Cai, L.H. Three-Dimensional Printable, Extremely Soft, Stretchable, and Reversible Elastomers from Molecular Architecture-Directed Assembly. *Chem. Mater.* **2021**, *33*, 2436–2445. [CrossRef]
103. Lübtow, M.M.; Mrlik, M.; Hahn, L.; Altmann, A.; Beudert, M.; Lühmann, T.; Luxenhofer, R. Temperature-Dependent Rheological and Viscoelastic Investigation of a Poly(2-methyl-2-oxazoline)-b-poly(2-iso-butyl-2-oxazoline)-b-poly(2-methyl-2-oxazoline)-Based Thermogelling Hydrogel. *J. Funct. Biomater.* **2019**, *10*, 36. [CrossRef]
104. Murphy, R.D.; Kimmins, S.; Hibbitts, A.J.; Heise, A. 3D-extrusion printing of stable constructs composed of photoresponsive polypeptide hydrogels. *Polym. Chem.* **2019**, *10*, 4675–4682. [CrossRef]
105. Li Volsi, A.; Tallia, F.; Iqbal, H.; Georgiou, T.K.; Jones, J.R. Enzyme degradable star polymethacrylate/silica hybrid inks for 3D printing of tissue scaffolds. *Mater. Adv.* **2020**, *1*, 3189–3199. [CrossRef]
106. Dominguez-Alfaro, A.; Criado-Gonzalez, M.; Gabirondo, E.; Lasa-Fernández, H.; Olmedo-Martínez, J.L.; Casado, N.; Alegret, N.; Müller, A.J.; Sardon, H.; Vallejo-Illarramendi, A.; et al. Electroactive 3D printable poly(3, 4-ethylenedioxythiophene)-graft-poly(ε-caprolactone) copolymers as scaffolds for muscle cell alignment. *Polym. Chem.* **2022**, *13*, 109–120. [CrossRef]
107. Zhou, Y.; Cui, Y.; Wang, L.Q. A Dual-sensitive Hydrogel Based on Poly(Lactide-co-Glycolide)-Polyethylene Glycol-Poly(Lactide-co-Glycolide) Block Copolymers for 3D Printing. *Int. J. Bioprint.* **2021**, *7*, 389. [CrossRef]
108. Murphy, R.; Walsh, D.P.; Hamilton, C.A.; Cryan, S.A.; in het Panhuis, M.; Heise, A. Degradable 3D-Printed Hydrogels Based on Star-Shaped Copolypeptides. *Biomacromolecules* **2018**, *19*, 2691–2699. [CrossRef]
109. Nakagawa, Y.; Ohta, S.; Nakamurac, M.; Ito, T. 3D inkjet printing of star block copolymer hydrogels cross-linked using various metallic ions. *RSC Adv.* **2017**, *7*, 55571–55576. [CrossRef]
110. Dutta, S.; Cohn, D. Temperature and pH responsive 3D printed scaffolds. *J. Mater. Chem. B* **2017**, *5*, 9514–9521. [CrossRef]
111. Dutta, S.; Cohn, D. Dually responsive biodegradable drug releasing 3D printed structures. *J. Appl. Polym. Sci.* **2022**, *139*, e53137. [CrossRef]
112. Cisneros, K.; Chowdhury, N.; Coleman, E.; Ferdous, T.; Su, H.; Jennings, J.A.; Bumgardner, J.D.; Fujiwara, T. Long-Term Controlled Release of Simvastatin from Photoprinted Triple-Networked Hydrogels Composed of Modified Chitosan and PLA-PEG Micelles. *Macromol. Biosci.* **2021**, *21*, e2100123. [CrossRef] [PubMed]
113. Il Kim, R.; Lee, G.; Lee, J.-H.; Park, J.J.; Lee, A.S.; Hwang, S.S. Structure–Property Relationships of 3D-Printable Chain-Extended Block Copolymers with Tunable Elasticity and BiodegradabilityZ. *ACS Appl. Polym. Mater.* **2021**, *3*, 4708–4716. [CrossRef]
114. Kutikov, A.B.; Reyer, K.A.; Song, J. Shape-Memory Performance of Thermoplastic Amphiphilic Triblock Copolymer Poly(d, l-lactic acid-co-ethylene glycol-co-d, l-lactic acid) (PELA)/Hydroxyapatite Composites. *Macromol. Chem. Phys.* **2014**, *215*, 2482–2490. [CrossRef]
115. Hu, C.; Hahn, L.; Yang, M.; Altmann, A.; Stahlhut, P.; Groll, J.; Luxenhofer, R. Improving printability of a thermoresponsive hydrogel biomaterial ink by nanoclay addition. *J. Mater. Sci.* **2021**, *56*, 691–705. [CrossRef]
116. Cui, Y.; Jin, R.; Zhang, Y.; Yu, M.; Zhou, Y.; Wang, L.Q. Cellulose Nanocrystal-Enhanced Thermal-Sensitive Hydrogels of Block Copolymers for 3D Bioprinting. *Int. J. Bioprint.* **2021**, *27*, 397. [CrossRef]
117. Müller, M.; Becher, J.; Schnabelrauch, M.; Zenobi-Wong, M. Nanostructured Pluronic hydrogels as bioinks for 3D Bioprinting. *Biofabrication* **2015**, *7*, 035006. [CrossRef]
118. Abbadessa, A.; Blokzijl, M.M.; Mouser, V.H.M.; Marica, P.; Malda, J.; Hennink, W.E.; Vermonden, T. A thermo-responsive and photo-polymerizable chondroitin sulfate-based hydrogel for 3D printing applications. *Carbohydr. Polym.* **2016**, *149*, 163–174. [CrossRef]
119. Boere, K.W.M.; Blokzijl, M.M.; Visser, J.; Elder, J.; Linssen, A.; Malda, J.; Hennink, W.E.; Vermonden, T. Biofabrication of reinforced 3D-scaffolds using two-component hydrogels. *J. Mater. Chem. B* **2015**, *3*, 9067–9078. [CrossRef]

Disclaimer/Publisher's Note: The statements, opinions and data contained in all publications are solely those of the individual author(s) and contributor(s) and not of MDPI and/or the editor(s). MDPI and/or the editor(s) disclaim responsibility for any injury to people or property resulting from any ideas, methods, instructions or products referred to in the content.

Article

Effect of Disparity in Self Dispersion Interactions on Phase Behaviors of Molten A-b-B Diblock Copolymers

Xinyue Zhang, Mingge Zhao and Junhan Cho *

Department of Polymer Science & Engineering, Dankook University, 152 Jukjeon-ro, Suji-gu, Yongin, Gyeonggi-do 16890, Republic of Korea
* Correspondence: jhcho@dankook.ac.kr; Tel.: +82-31-8005-3586

Abstract: Phase behaviors of molten A-b-B diblock copolymers with disparity in self dispersion interactions are revisited here. A free energy functional is obtained for the corresponding Gaussian copolymers under the influence of effective interactions originating in the localized excess equation of state. The Landau free energy expansion is then formulated as a series in powers of A and B density fluctuations up to 4th order. An alternative and equivalent Landau energy is also provided through the transformation of the order parameters to the fluctuations in block density difference and free volume fraction. The effective Flory χ is elicited from its quadratic term as the sum of the conventional enthalpic χ_H and the entropic χ_S that is related to energetic asymmetry mediated by copolymer bulk modulus. It is shown that the cubic term is balanced with Gaussian cubic vertex coefficients in corporation with energetics to yield a critical point at a composition rich in a component with stronger self interactions. The full phase diagrams with classical mesophases are given for the copolymers exhibiting ordering upon cooling and also for others revealing ordering reversely upon heating. These contrasting temperature responses, along with the skewness of phase boundaries, are discussed in relation to χ_H and χ_S. The pressure dependence of their ordering transitions is either barotropic or baroplastic; or anomalously exhibits anomalously both at different stages. These actions are all explained by the opposite responses of χ_H and χ_S to pressure.

Keywords: diblock copolymer; Landau analysis; weak segregation regime; upper order-disorder transition; lower disorder-order transition; barotropicity; baroplasticity

Citation: Zhang, X.; Zhao, M.; Cho, J. Effect of Disparity in Self Dispersion Interactions on Phase Behaviors of Molten A-b-B Diblock Copolymers. *Polymers* **2023**, *15*, 30. https://doi.org/10.3390/polym15010030

Academic Editors: Nikolaos Politakos and Apostolos Avgeropoulos

Received: 31 October 2022
Revised: 7 December 2022
Accepted: 9 December 2022
Published: 21 December 2022

Copyright: © 2022 by the authors. Licensee MDPI, Basel, Switzerland. This article is an open access article distributed under the terms and conditions of the Creative Commons Attribution (CC BY) license (https://creativecommons.org/licenses/by/4.0/).

1. Introduction

Block copolymers have been of great importance for the past several decades because of their self-assembly into arrays of ordered nanoscopic structures such as lamellae, hexagonally packed cylinders, body-centered cubic spheres, double gyroids, other network structures, and Frank-Kasper phases [1–4]. Block copolymers are used in diverse areas and applications such as elastomers, surface modifiers, blend compatibilizers, and templates for directing structured materials towards data storage, nanolithography, and nanopattern transfer [5–9]. Block copolymers in selective solvents can be useful for drug delivery, cancer theranostics, nanoreactors, and stimuli-responsive materials [10,11].

It is well known from phenomenological studies on the corresponding incompressible copolymer systems that their phase behaviors are to be determined by the total number of monomers or chain size N, the component volume fractions ϕ, and the effective Flory interaction parameter χ [12]. However, the copolymer behaviors are considered to be much more complicated than the simple incompressible picture. It is typical that block copolymers exhibit ordering upon cooling, which is referred to as the upper order-disorder transition (UODT) [3,13]. Ordering of block copolymers upon heating has also been found, which is referred to as the lower disorder-order transition (LDOT) [14–19]. Some copolymers have been shown to reveal immiscibility loops [20–23] with both LDOT and UODT. These two types of temperature dependences of the ordering behaviors are driven by different mechanisms. The UODT has an enthalpic origin because it is driven by unfavorable energetics. On the contrary, the LDOT is of an entropic origin that is divided in three-fold ways [24]. Firstly, for copolymers with directional interactions between different monomers, there is entropic penalty in forming such directional pairs. Thus, increase in temperature allows those pairs less to phase separate less [25–27]. Secondly, the disparities in self dispersion interactions or compressibilities between component blocks lead to phase separation to gain more entropy through volume increase [24]. Thirdly, some polymer mixtures without directional interactions or compressibility differences exhibit phase separation because of entropic penalty arisen by asymmetry in monomer structures [28–31].

Block copolymers exhibiting either UODT or LDOT respond to pressure in two different ways. Firstly, their ordered region is enlarged upon pressurization, which is referred to as barotropicity. The unfavorable energetics are augmented by pressurization as a result of the densification of such interactions. Many UODT-type block copolymers such as polystyrene-b-polybutadiene (PS-b-PBD) and PS-b-polyisoprene (PS-b-PI) fall into this category in their responses to pressure [32–36]. Some strongly interacting LDOT-type mixtures exhibits barotropicity due to this densification effect [37]. The transition temperatures change typically by ~20 K over 100 MPa in the absolute sense. Secondly, the ordered region is shrunken upon pressurization, which is observed for some copolymers with substantial disparities in their compressibilities. This phenomenon is referred to as baroplasticity [35,38,39]. Some UODT-type copolymers such as PS-b-poly(n-hexyl methacrylate) (PS-b-PnHMA) [35] and PS-b-poly(ethyl hexly acrylate) (PS-b-PEHA) [40] are baroplastic. LDOT and loop-type block copolymers from PS and ethyl to n-pentyl polymethacrylates also exhibit this property [21,35,39]. The change in transition temperatures varies from ten to several hundred kelvin over 100 MPa in the absolute sense.

Over the years, we have sequentially developed sequentially the random-phase approximation theory [41–43], Landau analysis [44–46], and self-consistent field theory [47–49] for A-b-B block copolymers of all possible types exhibiting UODT, LDOT, barotropicity, and baroplasticity. Narrowing our attention down to Landau approach, the Landau free energy was first obtained as a series in powers of two order parameters, which are A and B density fluctuations, in a direct way [44]. Later in a separate study, an alternative Landau free energy was formulated through the transformation of order parameters [45,46]. The copolymer, with equal self-dispersion interactions for A and B blocks, reveals it is Landau free energy mathematically identical to that of the incompressible counterpart by Leibler. However, an effective Flory χ is shown to carry molecular parameters. Therefore, the symmetric copolymer exhibits a critical point (CP) that is pressure dependent [45,46]. It was argued that the copolymer with disparity in self dispersion interactions yields its Landau energy possessing the nonvanishing and negative cubic term, and the second-order transition is nullified even at the symmetric composition [44,46]. This energetic disparity gives asymmetry in densities or average intermonomer distances for different block domains. The notion that the copolymer phase transition is fully of first order seemed to be in harmony with other known facts. For the liquid-solid transition and isotropic-nematic transition in liquid crystals, their Landau free energy expansions usually possess nonvanishing cubic vertex coefficients [50]. These transitions are only of first order. Here, we revisit the phase behaviors of molten A-b-B diblock copolymers in the weak segregation regime. In the course of formulating the Landau free energy, it is understood that the effective cubic order term is more intricate than previously studied. It is shown that our Landau energy with the deepened conception resurrects the CP, whereas its location is dependent on the disparity in self dispersion interactions. The Landau free energy is derived in two different ways; one is in a direct way with the two order parameters, and the other is through the transformation of the order parameters. Using these two equivalent free energies, the theoretical calculation of the copolymer phase behaviors and transitions are to be compared with experimental results.

2. Theory

2.1. Free Energy Density in the Bulk State

Our system of interest is A-b-B diblock copolymer chains made of A and B monomers in volume V. There are n_c such chains, where each j-block possesses N_j tangent spheres having the identical diameter σ. The close packed volume of j-blocks in the system is given as $V_j = n_c N_j v^*$, where $v^* = \pi\sigma^3/6$ is the monomer volume. Then, the close packed volume fraction of j-block is given as $\phi_j = V_j/\sum V_k = N_j/N_c$, where $N_c = N_A + N_B$ is the copolymer chain size. The overall packing density η is given by $\eta = \sum V_j/V$, and the packing density of j-block is equal to $\eta_j = \phi_j \eta$.

The Helmholtz free energy A of the copolymer melt is given as the sum of ideal A_{id} and non-ideal A_{ni} as $A = A_{id} + A_{ni}$ [41,46,47]. The former A_{id} is given below:

$$\frac{\beta A_{id} v^*}{V} = \frac{\eta}{N_c} \ln \frac{\eta K}{N_c} \qquad (1)$$

where $\beta = 1/k_B T$ as usual, and K is the molecular constant that does not affect any thermodynamic properties. The latter A_{ni} is subdivided into $A_{ni} = A_{HSC} + U_{nb}$, where A_{HSC} implies the excluded volume contribution by hard sphere chains, and U_{nb} represents dispersion (van der Waals) interaction energy between nonbonded monomers. The first contribution A_{HSC} is formulated from Baxter's integral equation theory for adhesive hard spheres under Chiew's connectivity constraint [51–53]. Mathematically stated,

$$\frac{\beta A_{HSC} v^*}{V} = \frac{3}{2}\left[\frac{\eta}{(1-\eta)^2} - \left(1 - \frac{1}{N_c}\right)\frac{\eta}{1-\eta}\right] - \frac{\eta}{N_c}\left[\ln(1-\eta) + \frac{3}{2}\right] \qquad (2)$$

The second contribution U_{nb} is obtained from the Bethe-Peierls-type mean-field energy [54] of locally packed nearest-neighbors around a chosen monomer. There are AA, AB, and BB pairs, whose contact energies are represented by $\bar{\epsilon}_{AA}$, $\bar{\epsilon}_{AB}$, and $\bar{\epsilon}_{BB}$, respectively. Then, U_{nb} can be written as

$$\frac{\beta U_{nb} v^*}{V} = \frac{1}{2} \cdot \beta \cdot \sum_{ij} \phi_i \phi_j \bar{\epsilon}_{ij} \cdot u(\eta) \cdot \eta = \frac{1}{2} \cdot \beta \cdot \sum_{ij} \eta_i \eta_j \bar{\epsilon}_{ij} \cdot \frac{u(\eta)}{\eta} \qquad (3)$$

The density dependence of U_{nb} is determined by $u(\eta) = 4[(\gamma/C)^4 \eta^4 - (\gamma/C)^2 \eta^2]$ with $\gamma = 1/\sqrt{2}$ and $C = \pi/6$. We denote the free energy A per unit volume as $a \equiv A/V$, and its nonideal part as $a_{ni} \equiv A_{ni}/V$.

2.2. Series Expansion of Free Energy Functional

The free energy density functional for an inhomogeneous A-B diblock copolymer melt is written in general as [47]

$$\frac{\beta A_{inh} v^*}{V} = \frac{\eta}{N_c} \ln \frac{\eta K}{N_c} - \frac{\eta}{N_c} \ln\left(\frac{1}{V}\int d\vec{r} \cdot q(\vec{r},1)\right) + \frac{1}{V}\left(\int d\vec{r} \cdot \beta a_{ni}(\vec{r})v^* - \sum_j \int d\vec{r} \cdot i\omega_j(\vec{r}) \cdot \eta_j(\vec{r})\right) \qquad (4)$$

where $a_{ni}(\vec{r})$ is the localized a_{ni} to give the effective short-ranged interactions. The function $\omega_j(\vec{r})$ indicates the external potential conjugate to the local j-density $\eta_j(\vec{r})$. In Equation (4), q is the end-segment distribution function of Gaussian A-b-B chains subject to ω_js which transmits the influence of the local interactions to the chain conformations to describe microphase segregated state.

Fluctuations in various field variables are defined by $\Delta \eta_j(\vec{r}) \equiv \eta_j(\vec{r}) - \eta_j$ and $\Delta \omega_j(\vec{r}) \equiv \omega_j(\vec{r})$, where the spatial average of ω_j is shifted to zero. Then, the logarithm of $Q (\equiv 1/V \cdot \int q d\vec{r})$ in Equation (4) can be expanded as a series in powers of ω_js up to 4th order as

$$\ln Q = \ln\left[\frac{1}{V}\int d\vec{r} \cdot q(\vec{r},1)\right] = \ln \overline{Q} + \sum_{n=2}^{4}\frac{(-1)^n N_c}{n!V}\int \prod_{l=1}^{n}\frac{d\vec{k}_l}{(2\pi)^3} \cdot G_{i_1,\ldots,i_n}^{(n)0}(\vec{k}_1,\ldots,\vec{k}_n)\omega_{i_1}(\vec{k}_1)\ldots\omega_{i_n}(\vec{k}_n) \quad (5)$$

where $\overline{Q}$ is defined by $\overline{Q} = Q(\omega_j \to 0)$, and $G_{ij}^{(2)0}$, $G_{ijk}^{(3)0}$, and $G_{ijkl}^{(4)0}$ are the proper Gaussian correlation functions. It is common to replace $G_{ij}^{(2)0}$ with S_{ij}^0. Equation (5) is written in Fourier form with scattering vectors $\vec{k}$s. In our A-b-B copolymer system, $S_{AA}^0(\vec{k}) = \eta N \cdot d_1(\phi_A, x)$ is used for AA correlations with its gyration radius R_G, where $d_1(\phi_A, x) = 2/x^2 \cdot (e^{-\phi_A x} + \phi_A x - 1)$ is the modified Debye function and $x \equiv k^2 R_G^2$. Likewise, $S_{BB}^0(\vec{k}) = \eta N \cdot d_1(1-\phi_A, x)$ is used for BB correlations. The remaining AB correlations is described by $S_{AB}^0(\vec{k}) = \eta N/2 \cdot [d_1(1,x) - d_1(\phi_A, x) - d_1(1-\phi_A, x)]$.

Now, the free energy is written below as

$$\frac{\beta A_{inh} v^*}{V} \approx \frac{\eta}{N_c}\ln\frac{\eta K}{N_c} + \sum_{n=2}^{4}\frac{(-1)^n N_c}{n!V}\cdot \int \prod_{l=1}^{n}\frac{d\vec{k}_l}{(2\pi)^3}\cdot G_{i_1,\ldots,i_n}^{(n)0}\left(\vec{k}_1,\ldots,\vec{k}_n\right)\omega_i(\vec{k}_1)\ldots\omega_{i_n}(\vec{k}_n)$$
$$+\frac{1}{V}\left(\beta \bar{a}_{ni}v^*V + \frac{1}{2}\sum_{i,j}\int \frac{d\vec{k}}{(2\pi)^3}\cdot \beta D_{ij}a\cdot v^* \Delta \eta_i(\vec{k})\Delta \eta_j(-\vec{k})\right) - \frac{1}{V}\left(\sum_j \int \frac{d\vec{k}}{(2\pi)^3}\cdot \omega_j(\vec{k})\cdot \Delta \eta_j(-\vec{k})\right) \quad (6)$$

where Q is absorbed into K and $\bar{a}_{ni}$ indicates a_{ni} in the homogeneous state. The symbol $D_{ij}a$ denotes the second-order derivatives of a_{ni} as $D_{ij}a \equiv \partial^2 a_{ni}/\partial \eta_i \partial \eta_j$ to give the effective local interactions in two-body level. For compressible systems, the Gaussian correlation functions are diluted by η because of free volume.

The Landau free energy is formulated from Equation (6) by replacing $\Delta \eta_j(\vec{k})$ and $\omega_j(\vec{k})$ with their ensemble averages. For simplicity, we will use the same symbols for their averages. To minimize the Landau free energy, it is required that $\delta(A_{inh}/V)/\delta \omega_j = 0$, which yields the relations between $\omega_j(\vec{k})$s and $\Delta \eta_j(\vec{k})$s. The Landau free energy is then re-written as a series in powers of $\Delta \eta_j(\vec{k})$s as follows:

$$\frac{\beta A_{inh} v^*}{V} = \frac{\eta}{N_c}\ln\frac{\eta K}{N_c} + \beta \bar{a}_{ni} v^* + \frac{1}{2!V}\sum_{ij}\left[\int \frac{d\vec{k}_1}{(2\pi)^3}\frac{d\vec{k}_2}{(2\pi)^3}\Gamma_{ij}^{(2)}\left(\vec{k}_1,\vec{k}_2\right)\Delta\eta_i\left(\vec{k}_1\right)\Delta\eta_j\left(\vec{k}_2\right)\right]$$
$$+\frac{1}{3!V}\sum_{ijk}\left[\int \frac{d\vec{k}_1}{(2\pi)^3}\frac{d\vec{k}_2}{(2\pi)^3}\frac{d\vec{k}_3}{(2\pi)^3}\Gamma_{ijk}^{(3)}\left(\vec{k}_1,\vec{k}_2,\vec{k}_3\right)\Delta\eta_i\left(\vec{k}_1\right)\Delta\eta_j\left(\vec{k}_2\right)\Delta\eta_k\left(\vec{k}_3\right)\right] \quad (7)$$
$$+\frac{1}{4!V}\sum_{ijkl}\left[\int \frac{d\vec{k}_1}{(2\pi)^3}\frac{d\vec{k}_2}{(2\pi)^3}\frac{d\vec{k}_3}{(2\pi)^3}\frac{d\vec{k}_4}{(2\pi)^3}\Gamma_{ijkl}^{(4)}\left(\vec{k}_1,\vec{k}_2,\vec{k}_3,\vec{k}_4\right)\Delta\eta_i\left(\vec{k}_1\right)\Delta\eta_j\left(\vec{k}_2\right)\Delta\eta_k\left(\vec{k}_3\right)\Delta\eta_l\left(\vec{k}_4\right)\right] + O\left(\Delta\eta_j^5\right)$$

The second-order vertex function $\Gamma_{ij}^{(2)}$ is identical to S_{ij}^{-1}, which is given by the sum of Gaussian S_{ij}^{0-1} and effective interaction fields. The higher-order vertex functions are obtained as the combination of Gaussian correlation functions, which can be found elsewhere [12,44,46]. It should be recognized that all the vertex functions require $\sum k_i = 0$.

2.3. Formulation of Landau Free Energy
2.3.1. Method I: Direct Way

Owing to the covalent bonds between A and B blocks, A-b-B diblock copolymer melts exhibit phase separation only on a nanometer scale. These nanoscale mesophases are diverse, but here we consider only the classical ones such as 3-dimensional body-centered cubic spheres (BCC), 2-dimensional hexagonally packed cylinders (HEX), and 1-dimensional lamellae (LAM). The quadratic form of the free energy functional expansion yields the characteristic wavenumber k^* at its minimum, which in turn gives the periodicity of the repeating structures with the domain size D as $D = 2\pi/k^*$. These nanostructures are determined by n characteristic scattering vectors $\vec{K}_j$s, whose magnitudes are $\left|\vec{K}_j\right| = k^*$.

Lamellar mesophase possesses one base vector $\vec{K}_1 = k^* \cdot (1,0,0)$ with $n = 1$. Meanwhile, HEX mesophase possesses three base vectors, $\vec{K}_1 = k^* \cdot (1,0,0)$, $\vec{K}_2 = k^* \cdot \left(-1/2, \sqrt{3}/2, 0\right)$, $\vec{K}_3 = k^* \cdot \left(-1/2, -\sqrt{3}/2, 0\right)$ along with $n = 3$. The last BCC mesophase possesses six base vectors, $\vec{K}_1 = k^*/\sqrt{2} \cdot (1,1,0)$, $\vec{K}_2 = k^*/\sqrt{2} \cdot (-1,1,0)$, $\vec{K}_3 = k^*/\sqrt{2} \cdot (0,1,1)$, $\vec{K}_4 = k^*/\sqrt{2} \cdot (0,1,-1)$, $\vec{K}_5 = k^*/\sqrt{2} \cdot (1,0,1)$, $\vec{K}_6 = k^*/\sqrt{2} \cdot (1,0,-1)$, along with $n = 6$.

Following Leibler's seminal analysis [12,44], the integral in Equation (7) is approximated to the finite sum of integrands at $\vec{K}_j$. Each $\Delta \eta_i(\pm \vec{k}_1)$ is now treated as a plane wave with its amplitude $(1/\sqrt{n})\varsigma_j$ and phase angle $\pm\varphi(i)$ as $\Delta \eta_i(\pm\vec{k}_1) = (1/\sqrt{n})\varsigma_i e^{\pm i\varphi(i)}$. The free energy expansion is greatly simplified to yield the following form as a series in powers of ς_js up to 4th order:

$$\beta \Delta A = \left(\Gamma_{AA}\varsigma_A^2 - 2\Gamma_{AB}\varsigma_A\varsigma_B + \Gamma_{BB}\varsigma_B^2\right) \\ - \left|\alpha_{AAA}\varsigma_A^3 - 3\alpha_{AAB}\varsigma_A^2\varsigma_B + 3\alpha_{ABB}\varsigma_A\varsigma_B^2 - \alpha_{BBB}\varsigma_B^3\right| + \delta_{ijkl}e^{-i\pi \cdot c_B(ijkl)} \cdot \varsigma_i\varsigma_j\varsigma_k\varsigma_l \quad (8)$$

where the necessary treatment of the vertex coefficients of Equation (7) for the three mesophases is given in the Appendix A. In Equation (8), Einstein's summation convention is used when necessary. It is seen that the permutation of indices of α_{AAB} and α_{ABB} yields the identical vertex function values. The cubic coefficients α_{ijk}s for LAM, HEX, and BCC are given respectively as follows:

$$\alpha_{ijk}^{LAM} = 0; \quad \alpha_{ijk}^{HEX} = \frac{12}{3!\left(\sqrt{3}\right)^3}\Gamma_{ijk}(1); \quad \alpha_{ijk}^{BCC} = \frac{48}{3!\left(\sqrt{6}\right)^3}\Gamma_{ijk}(1) \quad (9)$$

In Equation (9), a number h is put into the bracket to indicate the relative angles between the three scattering vectors $\vec{k}_1$, $\vec{k}_2$, and $\vec{k}_3$, where its definition is $h \equiv \left|\vec{k}_1 + \vec{k}_2\right|^2 / (k^*)^2$. Then, the right triangular arrangement of those vectors yields $\left|\vec{k}_1 + \vec{k}_2\right| = \left|\vec{k}_3\right|$ and $h = 1$. The quartic coefficients δ_{ijkl}s are obtained as

$$\delta_{ijkl}^{LAM} = \frac{3!}{4!}\Gamma_{ijkl}(0,0) \quad (10)$$

$$\delta_{ijkl}^{HEX} = \frac{18}{4!\left(\sqrt{3}\right)^4}\left[\Gamma_{ijkl}(0,0) + 4\Gamma_{ijkl}(0,1)\right] \quad (11)$$

$$\delta_{ijkl}^{BCC} = \frac{36}{4!\left(\sqrt{6}\right)^4}\left[\Gamma_{ijkl}(0,0) + 8\Gamma_{ijkl}(0,1) + 2\Gamma_{ijkl}(0,2) + 4\Gamma_{ijkl}(1,2)\right] \quad (12)$$

The set of numbers (h_1, h_2) in Equations (10)–(12) indicates the relative angles between the four scattering vectors $\vec{k}_1$, $\vec{k}_2$, $\vec{k}_3$, and $\vec{k}_4$. We define h_1 and h_2 as $\left|\vec{k}_1 + \vec{k}_2\right|^2 \equiv h_1 \cdot (k^*)^2$ and $\left|\vec{k}_1 + \vec{k}_4\right|^2 \equiv h_2 \cdot (k^*)^2$, respectively. Then, it can be shown that $\left|\vec{k}_1 + \vec{k}_3\right|^2 = (4 - h_1 - h_2) \cdot (k^*)^2$. The Landau free energy is to be minimized with respect to ς_A and ς_B to determine the equilibrium mesophase at a given set of composition, temperature, and pressure.

2.3.2. Method II: Transformation of Order Parameters

Now, let us express our Landau free energy in a more familiar form through the transformation of the order parameters [45,46]. The new order parameters are denoted as $\psi_1(\vec{r})$ and $\psi_2(\vec{r})$, which are defined by the following matrix equation:

$$\begin{bmatrix} \psi_1 \\ \psi_2 \end{bmatrix} = \begin{bmatrix} (1-\phi_A)/\eta & -\phi_A/\eta \\ 1 & 1 \end{bmatrix} \begin{bmatrix} \Delta\eta_A \\ \Delta\eta_P \end{bmatrix} = [M_{ij}] \begin{bmatrix} \Delta\eta_A \\ \Delta\eta_P \end{bmatrix} \quad (13)$$

Using this equation, $\psi_1(\vec{r})$ is given as $\psi_1 = (\Delta\eta_A - \Delta\eta_P)/2\eta$ at $\phi_A = 1/2$. Thus, the profiles of phase segregating A and B blocks are joined to yield a composite profile in phase with A block. The other order parameter $\psi_2(\vec{r})$ is determined to be $\psi_2 = \Delta\eta_A + \Delta\eta_P = -\Delta\eta_f$, which implies the negative fluctuations in free volume fraction. Upon this transformation, the new vertex functions $\bar{\Gamma}$s are obtained from the original vertex functions Γs as

$$\bar{\Gamma}^{(n)}_{i_1...i_n}(\vec{k}_1,...,\vec{k}_n)\psi_{i_1}(\vec{k}_1)\cdots\psi_{i_n}(\vec{k}_n) = \Gamma^{(n)}_{i_1...i_n}(\vec{k}_1,...,\vec{k}_n)\Delta\eta_{i_1}(\vec{k}_1)\cdots\Delta\eta_{i_n}(\vec{k}_n) \quad (14)$$

Then, $\bar{\Gamma}$s are equated to

$$\bar{\Gamma}^{(n)}_{j_1...j_n} = \Gamma^{(n)}_{i_1...i_n} M^{-1}_{i_1 j_1}\cdots M^{-1}_{i_n j_n} \quad (15)$$

where Einstein's summation convention is used for this tensorial equation.

We will consider nanoscale mesophases, whose structures are defined by characteristic scattering vectors $\vec{k}_1 \in \{\pm\vec{K}_n\}$. Regular geometric morphologies are represented by the order parameter ψ_1 that is treated as a plane wave as $\psi_1(\pm\vec{K}_k) = (1/\sqrt{n})\zeta_1 e^{\pm i\varphi_k(1)}$. The remaining ψ_2 is separated into two parts as $-\psi_2 = \Delta\eta_f = -\psi_{2c} - \psi_{2i}$, where the former indicates the excess free volume in phase with the more compressible constituent and the latter represents the excess free volume at the interfaces between domains. While ψ_{2c} is parametrized as $\psi_{2c}(\pm\vec{K}_k) = (1/\sqrt{n})\zeta_{2c} e^{\pm i\varphi_k(2c)}$, ψ_{2i} should have 1/2 period to locate the interfaces as $\psi_{2i}(\pm 2\vec{K}_k) = (1/\sqrt{n})\zeta_{2i} e^{\pm i\varphi_k(2i)}$.

Taking the proper mathematical procedure given in the Appendix B, this alternative Landau free energy is formulated as

$$\beta\Delta A = \bar{\Gamma}_{11}\zeta_1^2 + 2\bar{\Gamma}_{12}\zeta_1\zeta_{2c} + \bar{\Gamma}_{22}\zeta_{2c}^2 + \bar{\Gamma}_{22}(2k*)\zeta_{2i}^2 - \left|a_n\zeta_1^3 + b_1\zeta_1^2\zeta_{2c}\right| + c_4\zeta_1^2\zeta_{2i} + d_n\zeta_1^4 \quad (16)$$

where the coefficient a_n is given respectively for LAM, HEX, and BCC by

$$a_n^{LAM} = 0; \quad a_n^{HEX} = 12/\left(3!3^{3/2}\right)\cdot\bar{\Gamma}_{111}(1); \quad a_n^{BCC} = 48/\left(3!6^{3/2}\right)\cdot\bar{\Gamma}_{111}(1) \quad (17)$$

The coefficient b_1 is given respectively by

$$b_1^{LAM} = 0; \quad b_1^{HEX} = 12/\left(3!3^{3/2}\right)\cdot(3\bar{\Gamma}_{112}(1)); \quad b_1^{BCC} = 48/\left(3!6^{3/2}\right)\cdot(3\bar{\Gamma}_{112}(1)) \quad (18)$$

for LAM, HEX, and BCC. The coefficient c_4 respectively becomes

$$c_4^{LAM} = 2/3!\cdot(3\bar{\Gamma}_{112}(4)); \quad c_4^{HEX} = 6/\left(3!3^{3/2}\right)\cdot(3\bar{\Gamma}_{112}(4)); \quad c_4^{BCC} = 12/\left(3!6^{3/2}\right)\cdot(3\bar{\Gamma}_{112}(4)) \quad (19)$$

The quartic coefficient d_n is given as

$$d_n^{LAM} = \frac{3!}{4!}\bar{\Gamma}_{1111}(0,0) \quad (20)$$

$$d_n^{HEX} = \frac{18}{4!\left(\sqrt{3}\right)^4}\left[\overline{\Gamma}_{1111}(0,0) + 4\overline{\Gamma}_{1111}(0,1)\right] \tag{21}$$

$$d_n^{BCC} = \frac{36}{4!\left(\sqrt{6}\right)^4}\left[\overline{\Gamma}_{1111}(0,0) + 8\overline{\Gamma}_{1111}(0,1) + 2\overline{\Gamma}_{1111}(0,2) + 4\overline{\Gamma}_{1111}(1,2)\right] \tag{22}$$

for LAM, HEX, and BCC, respectively.

Differentiating Equation (16) with respect to ζ_{2c} and ζ_{2i}, and then nullifying those derivatives yield the following conditions:

$$\zeta_{2c} = -\frac{\overline{\Gamma}_{12}}{\overline{\Gamma}_{22}}\zeta_1 \pm \frac{b_1}{2\overline{\Gamma}_{22}}\zeta_1^2; \; \zeta_{2i} = -\frac{c_4}{2\overline{\Gamma}_{22}(2k*)}\zeta_1^2 \tag{23}$$

where + and − signs are assigned to $a_n\zeta_1 + b_1\zeta_{2c} > 0$ and $a_n\zeta_1 + b_1\zeta_{2c} < 0$, respectively. Replacing ζ_{2c} and ζ_{2i} with Equation (23), the free energy becomes in general

$$\beta\Delta A = \left(\overline{\Gamma}_{11} - \frac{\overline{\Gamma}_{12}^2}{\overline{\Gamma}_{22}}\right)\zeta_1^2 - \left|a_n - \frac{b_1\overline{\Gamma}_{12}}{\overline{\Gamma}_{22}}\right|\zeta_1^3 + \left(d_n - \frac{b_1^2}{4\overline{\Gamma}_{22}} - \frac{c_4^2}{4\overline{\Gamma}_{22}(2k*)}\right)\zeta_1^4 \approx \left(\overline{\Gamma}_{11} - \frac{\overline{\Gamma}_{12}^2}{\overline{\Gamma}_{22}}\right)\zeta_1^2 - \left|a_n - \frac{b_1\overline{\Gamma}_{12}}{\overline{\Gamma}_{22}}\right|\zeta_1^3 + d_n\zeta_1^4 \tag{24}$$

Equation (20) is our final suggestion of the alternative Landau free energy to find the equilibrated ordered state as its minimum. It can be seen that the effective cubic and quartic coefficients of the free energy contain not only the Gaussian correlation functions but also interaction-dependent $\overline{\Gamma}_{ij}$. The vertex coefficient $\overline{\Gamma}_{22}$ implies the bulk modulus of the copolymer melt [42]. Thus, the effective quartic coefficient in Equation (24) is further approximated to simply d_n.

2.4. Spinodals and Effective Flory χ

The quadratic form A_2 of the Landau free energy in Equation (9) can be expressed in the matrix form as

$$\beta A_2 = \Gamma_{AA}\zeta_A^2 - 2\Gamma_{AB}\zeta_A\zeta_B + \Gamma_{BB}\zeta_B^2 = \begin{bmatrix}\zeta_A & \zeta_B\end{bmatrix}\begin{bmatrix}\Gamma_{AA} & -\Gamma_{AB} \\ -\Gamma_{AB} & \Gamma_{BB}\end{bmatrix}\begin{bmatrix}\zeta_A \\ \zeta_B\end{bmatrix} \tag{25}$$

The phase stability requires the positive definiteness of A_2. The spinodals are then defined as the border line of stability to require $\det\left[\Gamma_{ij}\right] = 0$ at $k*$ or

$$\Gamma_{AA}/\Gamma_{AB} = \Gamma_{AB}/\Gamma_{BB} \tag{26}$$

The same situation occurs in our alternative Landau free energy in Equation (24), where the spinodals are determined by

$$\overline{\Gamma}_{11} - \overline{\Gamma}_{12}^2/\overline{\Gamma}_{22} = \det\left[\overline{\Gamma}_{ij}\right]/\overline{\Gamma}_{22} = 0 \tag{27}$$

These two different equations for spinodals are simply equivalent because $\det\left[\overline{\Gamma}_{ij}\right] = \eta^2\det\left[\Gamma_{ij}\right]$.

The essence of the phase behavior of diblock copolymer melts is concentrated on effective Flory χ parameter. In our previous works [42,46], χ was properly elicited from the spinodals to consist of two contributions as $\chi = \chi_H + \chi_S$. The former χ_H of our χ indicates the conventional enthalpic contribution gotten from $\overline{\Gamma}_{11}$ in the following way. There are Gaussian and non-Gaussian parts in $\overline{\Gamma}_{11}/\eta$ as

$$\overline{\Gamma}_{11}/\eta = \eta(\Gamma_{AA} - 2\Gamma_{AB} + \Gamma_{BB}) = \eta\left(S_{AA}^{0-1} - 2S_{AB}^{0-1} + S_{BB}^{0-1}\right) + \eta\beta v*(D_{AA}a_{ni} - 2D_{AB}a_{ni} + D_{BB}a_{ni}) \tag{28}$$

where the latter non-Gaussian ones give χ_H as

$$\chi_H = -\frac{1}{2}\beta v * (D_{AA}a_{ni} - 2D_{AC}a_{ni} + D_{CC}a_{ni})\eta = \beta \cdot \frac{1}{2}\Delta\bar{\varepsilon} \cdot |u(\eta)| \tag{29}$$

The symbol $\Delta\bar{\varepsilon}$ ($=\bar{\varepsilon}_{AA} + \bar{\varepsilon}_{BB} - 2\bar{\varepsilon}_{AB}$) implies the exchange energy between $\bar{\varepsilon}_{ij}$'s. Unlike incompressible situations, χ_H possesses density dependence because of $u(\eta)$. Meanwhile, $\bar{\Gamma}_{12}$ ($=\eta/2 \cdot (\Gamma_{AA} - \Gamma_{BB})$) is analyzed to be

$$\bar{\Gamma}_{12} = \frac{\eta}{2}\beta v * (D_{AA}a_{ni} - D_{BB}a_{ni}) = \frac{1}{2}\beta(\bar{\varepsilon}_{AA} - \bar{\varepsilon}_{BB}) \cdot \eta \frac{du}{d\eta} \tag{30}$$

where $\bar{\varepsilon}_{AA} - \bar{\varepsilon}_{BB}$ indicates disparity in self dispersion interactions between constituent blocks. The remaining vertex function $\bar{\Gamma}_{22}$ ($=\sum \Gamma_{ij}/4$) is the average of Γ_{ij}. It was shown in our previous works [42,46] that $\bar{\Gamma}_{22} \approx B_T/\eta^2$, where B_T ($\equiv \eta\,\partial P/\partial\eta)_T$) is the bulk modulus of the copolymer. Therefore, $\bar{\Gamma}_{12}^2/\bar{\Gamma}_{22} \propto [\bar{\varepsilon}_{AA} - \bar{\varepsilon}_{CC}]^2/B_T$ dominantly. As $\bar{\Gamma}_{12}^2/\bar{\Gamma}_{22}$ is always positive, it hampers phase stability. The latter χ_S of our χ represents the entropic contribution to phase stability as

$$\chi_S = \frac{1}{2\eta} \cdot \frac{\bar{\Gamma}_{12}^2}{\bar{\Gamma}_{22}} \tag{31}$$

which is associated with volume fluctuations [42,46]. In general, a component with larger $\bar{\varepsilon}_{jj}$ has a stronger cohesive energy and thus smaller compressibility (larger $\eta_{\phi_j \to 1}$) than the other. Therefore, χ_S vanishes for the copolymers with the same $\bar{\varepsilon}_{jj}$s or compressibility. The determinant $\det[\Gamma_{ij}]$ can then be re-written as

$$\det[\Gamma_{ij}] = \frac{1}{\eta^2}\det[\bar{\Gamma}_{ij}] = \frac{\bar{\Gamma}_{22}}{\eta}\left\{\eta\left(S_{AA}^{0-1} - 2S_{AB}^{0-1} + S_{BB}^{0-1}\right) - 2\chi\right\} \tag{32}$$

This χ is capable of predicting all types of block copolymer phase behaviors.

In response to pressure, χ_H and χ_S behave in the opposite way to each other. Upon pressurization, the increased η augments χ_H, whereas the increased B_T diminishes χ_S. In the case that $|\bar{\varepsilon}_{AA} - \bar{\varepsilon}_{BB}| \to 0$, $\chi_S/\chi \to 0$ and χ_H becomes a dominating contribution to χ. Therefore, pressurization leads the system to a deeper segregation, which is the conventional behavior or barotropicity. In the case that $|\bar{\varepsilon}_{AA} - \bar{\varepsilon}_{BB}|/\bar{\varepsilon}_{AA}$ is more sizable, χ_S/χ gets more substantial. The applied pressure enhances B_T, and then χ_S as well as χ is suppressed by B_T, which is the baroplasticity.

3. Discussions

3.1. Symbolic Arguments on Critical Point

A critical point (CP) or continuous transition point occurs when the spinodal line meets the ODT and OOT lines. The partial minimization of the free energy in Equation (8), with respect to ς_B is obtained by $\partial\Delta A/\partial\varsigma_B = 0$, which yields $\varsigma_B = (\Gamma_{AB}/\Gamma_{BB})\varsigma_A + O(\varsigma_B^2)$. When approaching its CP, $\varsigma_B \to (\Gamma_{AB}/\Gamma_{BB})\varsigma_A$ and higher-order terms can be ignored. Putting this ς_B back into the free energy yields the following symbolic equation:

$$\beta\Delta A = \tau\varsigma_A^2 + \alpha\varsigma_A^3 + \delta\varsigma_A^4 \tag{33}$$

where $\tau \equiv \Gamma_{AA} - \Gamma_{AB}^2/\Gamma_{BB}$ ($\propto \det[\Gamma_{ij}]$) serves as an effective temperature. The condition that $\tau > 0$ indicates the disordered state, above the spinodals for the conventional UODT-type copolymers but below the spinodals for LDOT-type copolymers. The situation that $\tau < 0$ implies the ordered state. The remaining effective coefficients α and δ are given by

$$\alpha \equiv -\left|\alpha_{AAA} - 3\alpha_{AAB}\frac{\Gamma_{AB}}{\Gamma_{BB}} + 3\alpha_{ABB}\left\{\frac{\Gamma_{AB}}{\Gamma_{BB}}\right\}^2 - \alpha_{BBB}\left\{\frac{\Gamma_{AB}}{\Gamma_{BB}}\right\}^3\right| \tag{34}$$

and

$$\delta \equiv \delta_{AAAA} - 4\delta_{AAAB}\frac{\Gamma_{AB}}{\Gamma_{BB}} + 2\{\delta_{AABB} + \delta_{ABAB} + \delta_{ABBA}\}\left\{\frac{\Gamma_{AB}}{\Gamma_{BB}}\right\}^2 - 4\delta_{ABBB}\left\{\frac{\Gamma_{AB}}{\Gamma_{BB}}\right\}^3 + \delta_{BBBB}\left\{\frac{\Gamma_{AB}}{\Gamma_{BB}}\right\}^4 \quad (35)$$

where it is perceived that $\delta_{AABB} = \delta_{BBAA}$, $\delta_{ABAB} = \delta_{BABA}$, and $\delta_{ABBA} = \delta_{BAAB}$. In other cases, such as δ_{ABBB} or δ_{AAAB}, the vertex functions under the permutation of indices are equivalent. It will be seen that δ is dominated by δ_{AAAA} and δ_{BBBB}. A CP is obtainable if the cubic coefficient α vanishes. It is clearly seen in Equation (34) that the energetics come into play in finding the CP through Γ_{ij}s.

In case of using the alternative Landau free energy, the same symbolic expression for the free energy is understood as

$$\beta \Delta A = \overline{\tau}\zeta_A^2 + \overline{\alpha}\zeta_A^3 + \overline{\delta}\zeta_A^4 \quad (36)$$

where the effective coefficients are given as

$$\overline{\tau} = \overline{\Gamma}_{11} - \frac{\overline{\Gamma}_{12}^2}{\overline{\Gamma}_{22}}; \; \overline{\alpha} = -\left|a_n - \frac{b_1 \overline{\Gamma}_{12}}{\overline{\Gamma}_{22}}\right|; \; \overline{\delta} \approx d_n \quad (37)$$

Our alternative Landau free energy in Equation (24) suggests that a CP is obtainable if $a_n - b_1\overline{\Gamma}_{12}/\overline{\Gamma}_{22} = 0$ along with the condition that $\overline{\Gamma}_{11} - \overline{\Gamma}_{12}^2/\overline{\Gamma}_{22} = 0$ or $\det\left[\overline{\Gamma}_{ij}\right] = 0$. It is also observed that the energetics play their role in finding the CP due to $\overline{\Gamma}_{12}$ and $\overline{\Gamma}_{22}$.

The mathematical structure of the effective cubic term in either Equation (34) or Equation (37) demonstrates the existence of CP for an A-b-B copolymer with or without disparity in $\overline{\epsilon}_{ij}$s unlike liquid-solid and nematic-isotropic transitions. The continuous transition for the copolymer with a finite chain size is of course to be destroyed due to concentration fluctuations to that turn to a weak first-order transition [55]. Nonetheless, this mean-field analysis is amenable and neat. It is still of importance because the mean-field behaviors are restored if $N_c \to \infty$ [56]. Furthermore, our Landau free energy works as the starting point for any fluctuation correction analyses.

3.2. Temperature Dependence of Ordering Transitions

3.2.1. UODT System

In this section, we use the Landau free energy in Equation (8) or Equation (24) to discuss various phase behaviors of molten A-b-B copolymers through numerically determining equilibrium mesophases and their stability. Consider first the phase behaviors of PS-b-PBD, which is quite a typical UODT-type block copolymer. In order to probe its phase behavior, our equation-of-state model requires three homopolymer parameters: the self-interaction parameter $\overline{\epsilon}_{jj}$, monomer diameter σ_j, and chain size N_j. The sets of homopolymer parameters for PS and PBD are given in Table 1, where a composite parameter $N_j \pi \sigma_j^3 / 6 M_j$ carrying the ratio of N_j to molecular weight M_j is provided. So, N_j can be determined from the experimental molecular weight of a polymer or N_j is directly given. Cross interactions between different polymers are characterized by $\overline{\epsilon}_{ij}$, which is an adjustable parameter and determined by fitting the phase behaviors of a given block copolymer system or those of the corresponding blends. The ratio $\overline{\epsilon}_{ij}/(\overline{\epsilon}_{ii}\overline{\epsilon}_{jj})^{1/2}$ for PS-b-PBD is determined to be 0.99565 from fitting binodal points of PS/PBD blends [41,57] and also the ordering transitions of PS-b-PBD [41,45,58,59].

Table 1. Molecular Parameters for PS and other polymers that form A-b-B copolymers.

Parameters	PS	PBD	PVME	PI	PEHA
σ_i (Å)	4.039	4.039	3.900 [a]	4.350 [a]	3.840 [a]
$\bar{\varepsilon}_{ii}/k$ (K)	4107.0	4065.9	3644.8	4057.7	3755.7
$N_i \pi \sigma_i^3 / 6 M_i$ (cm³/g) [b]	0.41857	0.49395	0.42906	0.50209	0.48564
$\bar{\varepsilon}_{ij}/\left(\bar{\varepsilon}_{PS}\bar{\varepsilon}_{jj}\right)^{1/2}$	-	0.99565	1.00264	0.99680	0.99880

[a] This discrepancy in monomer diameters is resolved by adopting the conventional Lorentz mixing rule as $\sigma = (\sigma_i + \sigma_j)/2$. [b] This composite parameter gives the ratio of the chain size N_i to molecular weight M_i.

The characteristic squared wavenumber $x^* (= (R_G k^*)^2)$ obtained at the minimum of $\det[\Gamma_{ij}]$ gives the information on the domain size. In Table S1 of Supplemental Materials (SM), x^* for PS-b-PBD is tabulated against ϕ_A. It is seen from this table that x^* is symmetric to $\phi_A = \phi_{PS}$ for the typical UODT systems such as PS-b-PBD, with little to no disparity in self dispersion interactions $\bar{\varepsilon}_{jj}$s.

Prior to the actual phase behaviors, let us briefly take a look at a hypothetical A-b-B diblock copolymer with $N_c = 400$, where each block has the same homopolymer parameters as those of PS and $\bar{\varepsilon}_{ij} = 0.99565(\bar{\varepsilon}_{ii}\bar{\varepsilon}_{jj})^{1/2}$. The exchange energy then becomes $\Delta\bar{\varepsilon}/k = 35.73$ K. In Table S2 of SM, Γ_{ij}s are tabulated for this copolymer at some selected ϕ_As. As is seen in this table, $\Gamma_{AA} = \Gamma_{AB} = \Gamma_{BB}$ at $\phi_A = 0.5$. In this case, $\bar{\Gamma}_{12} = 0$ due to $\bar{\varepsilon}_{AA} = \bar{\varepsilon}_{BB}$. Its phase behavior at ambient pressure is identical to that of the incompressible A-b-B copolymer melt discussed by Leiber. In Table S3 of SM, Γ_{ijk}s are tabulated against the composition ϕ_A. It is noted that Γ_{iii} is negative and large in its magnitude, whereas Γ_{AAB} or Γ_{ABB} is positive and mostly small. It is shown that $\Gamma_{AAA} = \Gamma_{BBB}$ and $\Gamma_{AAB} = \Gamma_{ABB}$ at $\phi_A = 0.5$. Therefore, α in Equation (34) is nullified at this composition to yield the CP, where $N\chi_c = 10.49487$. In case of PS-b-PBD with $N_c = 400$, there is a small difference in $\bar{\varepsilon}_{jj}$s between PS and PBD with $|\bar{\varepsilon}_{PS} - \bar{\varepsilon}_{PBD}|/\bar{\varepsilon}_{PS} = \sim 0.01$. The exchange energy for this copolymer is $\Delta\bar{\varepsilon}/k = 35.66$ K. Based on Γ_{ij}s tabulated for PS-b-PBD in Table S4 of SM, Γ_{ij}s are not identical at $\phi_A = \phi_{PS} = 0.5$. Therefore, α cannot vanish at $\phi_A = 1/2$. The CP of PS-b-PBD is found to be at $\phi_A = 0.50095$ (>1/2) and at $N_c\chi = 10.49494$ because of the small disparity in $\bar{\varepsilon}_{jj}$s.

Using the alternative Landau free energy in Equation (24), the CP of PS-b-PBD system turns out to be $\phi_A = 0.50095$, which is identical to the one using Equation (8) at least up to 9 decimal places. These results prove the equivalence of our two different Landau free energies even though the first method does not provide the profile for the free volume fraction. It needs to be mentioned that the threshold or maximum of the spinodals for PS-b-PBD copolymer is located at $\phi_A = 0.50069$, which is slightly moved to the copolymer with more PBD than at the CP. The shift of the CP is more vivid in the next copolymer exhibiting LDOT.

Starting from the CP of PS-b-PBD copolymer melts, all the transition points at ambient pressure are to be determined by minimizing the Landau free energy given in Equation (8). Using the vertex coefficients as given in Tables S3 to S7 of SM, various transition points are obtained by numerically solving both $\partial\Delta A/\partial\varsigma_A = 0$ and $\partial\Delta A/\partial\varsigma_B = 0$. In Figure 1a, the transition temperatures are plotted against ϕ_A. Because of the small $|\bar{\varepsilon}_{PS} - \bar{\varepsilon}_{PBD}|$, the phase diagram is almost symmetrical. The phase diagram can also be drawn in terms of the well-known relevant parameter for phase segregation, i.e., $N_c\chi$, as shown in Figure 1b. As was mentioned in the previous section, the effective Flory χ is a composite function of various molecular parameters as $\chi = \chi_H + \chi_S$. In Table 2, $N_c\chi$ along with χ_H and χ_S for the symmetric PS-b-PBD copolymer with $N_c = 400$ is tabulated at the selected temperatures and at 0.1 MPa. In this system, χ is almost equal to χ_H. This typical UODT-type copolymer shows the decreasing tendency of χ as $\chi_H \sim 1/T$, as seen in this table.

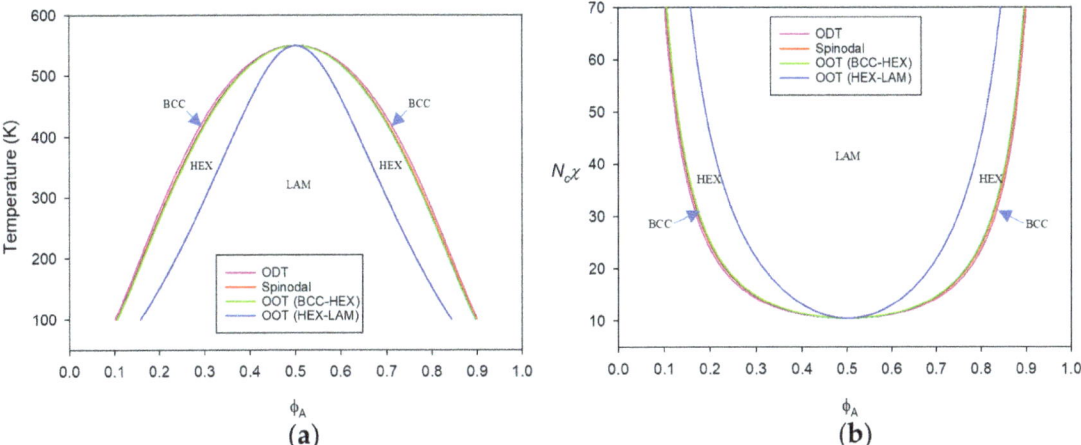

Figure 1. Phase diagram for molten PS-b-PBD with $N_c = 400$ plotted against PS (A) volume fraction ϕ_A in terms of: (**a**) absolute temperature and (**b**) the relevant parameter $N_c\chi$. The disparity in $\bar{\varepsilon}_{jj}$s is $|\bar{\varepsilon}_{PS} - \bar{\varepsilon}_{PBD}|/\bar{\varepsilon}_{PS} = 0.010$ and the exchange energy is $\Delta\bar{\varepsilon}/k = 35.66$ K. As this disparity is quite small, the phase boundaries are almost symmetrical with the CP at $\phi_A = 0.50095$. The arrows indicate BCC mesophase in the narrow region between ODT and spinodals.

Table 2. The relevant parameter $N_c\chi$ and its two contributions, χ_H and χ_S, evaluated at selected temperatures for symmetrical PS-b-PBD with $N_c = 400$ [a].

T (K)	χ_H	χ_S	$N_c\chi$
350	0.04573	4.40866×10^{-5}	18.31053
400	0.03905	4.57456×10^{-5}	15.63757
450	0.03382	4.71030×10^{-5}	13.54714
500	0.02962	4.82407×10^{-5}	11.86625
550	0.02616	4.92196×10^{-5}	10.48438
600	0.02327	5.00868×10^{-5}	9.32755
650	0.02081	5.08801×10^{-5}	8.34431

[a] Pressure is fixed to 0.1 MPa.

In drawing Figure 1, the Landau free energy in Equation (8) is used. If we use the Landau free energy given in Equation (24), where there is only one order parameter amplitude to determine through solving $\partial \Delta A / \partial \zeta_1 = 0$, we get almost the identical phase diagram. The spinodals from the two methods are perfectly identical. The ODT from disorder to BCC is different only by ~0.01 K between the two methods. The calculated differences in OOTs from the two methods are ~0.0007 K and ~0.102 K for BCC-HEX and HEX-LAM OOTs, respectively.

3.2.2. LDOT System

Our second system is a molten diblock copolymer from PS and poly(vinyl methyl ether) (PVME). The corresponding PS/PVME blend is a widely studied blend system that reveals the miscibility between PS and PVME and also the lower critical solution temperature behavior [60]. The origin of their miscibility is considered to be the weak hydrogen bond between the aromatic hydrogen (C-H) and ether oxygen (-O-) [61]. In analyzing copolymer phase behavior, all the molecular parameters for PS-b-PVME are given in Table 1. The cross interaction $\bar{\varepsilon}_{ij}$ for this copolymer is determined to be $\bar{\varepsilon}_{ij} / (\bar{\varepsilon}_{ii}\bar{\varepsilon}_{jj})^{1/2} = 1.00264$ from the binodal points of the corresponding PS/PVME blends, where this $\bar{\varepsilon}_{ij}$ yields $\Delta\bar{\varepsilon}/k = -6.637$ K and the calculated transition temperatures are similar to the experimental values [41,44,60].

The characteristic squared wavenumber $x^* (= (R_G k^*)^2)$ for PS-b-PVME is tabulated against $\phi_A = \phi_{PS}$ in Table S1 of SM. Unlike the typical UODT systems such as PS-b-PBD, it is seen from this table that x^* is slightly asymmetric to ϕ_A. The ratio of $x^*_{\phi_A=0.9}$ to $x^*_{\phi_A=0.1}$ is shown to be $x^*_{\phi_A=0.9}/x^*_{\phi_A=0.1} = 1.0011$, which implies that the domain size of the copolymer richer in PS is shrunken compared with that richer in PVME.

The effect of disparity in self dispersion interactions appears drastically in PS-b-PVME, which exhibits $|\bar{\varepsilon}_{PS} - \bar{\varepsilon}_{PVME}|/\bar{\varepsilon}_{PS} = 0.113$. Since $\bar{\varepsilon}_{PS} > \bar{\varepsilon}_{PVME}$, PS is denser and less compressible than PVME. PS and PVME are compatible with $\Delta\bar{\varepsilon} < 0$ due to the aforementioned weak H-bonds between them. In Table 3, we listed the theoretical χ for PS-b-PVME with $N_c = 20,000$ at $\phi_A = 1/2$ as a function of temperature while fixing pressure to 0.1 MPa. It is seen that the energetic $\chi_H \propto \Delta\bar{\varepsilon}/T$ is negative and decreases with temperature. However, there is comparable entropic $\chi_S \propto [\bar{\varepsilon}_{AA} - \bar{\varepsilon}_{CC}]^2/B_T$, which is always positive and grows with temperature. As a result of these two competing actions, the copolymer is in the disordered state at lower temperatures but reveals nanoscopic phase separation upon heating or LDOT caused by compressibility difference. The phase separation induced in this way requires a large chain size to suppress the combinatorial entropy. In Figure 2, the spinodal points (red line) are plotted against ϕ_A. It is seen that the spinodal line is seriously asymmetric because of the substantial disparity in $\bar{\varepsilon}_{ij}$s. More precisely, the threshold or minimum of the spinodal line is skewed towards more compressible PVME-rich side at ϕ_A = ~0.305. This phenomenon is caused by the fact that the positivity of χ_S always hampers phase stability, which is stronger in the side rich in more compressible PVME. However, this minimum is not the CP. The calculated CP using the free energy in Equation (8) is found to be $\phi_A = 0.50974$, rich in less compressible component PS. This action is caused by the fact that $\bar{\varepsilon}_{PS} > \bar{\varepsilon}_{PVME}$. The stronger binding of PS monomers in turn yields that $\Gamma_{AA} < \Gamma_{AB} < \Gamma_{BB}$, as seen in Table S4 of SM. The system rich in denser component has smaller volume. Therefore, at the CP with a continuous transition, the copolymer system strives to search the composition of comparable volumes of the two components. Henceforth, the critical composition should be $\phi_A > 1/2$ in order to add more volume of less compressible and denser component. This result is in sharp contrast to the phase behavior of the corresponding blend, where the threshold point in the spinodal line is indeed the CP. Using the Landau free energy in Equation (24) yields the CP of PS-b-PVME at $\phi_A = 0.50974$, which is identical to that from Equation (8) up to six decimal places.

Table 3. The relevant parameter $N_c\chi$ and its two contributions, χ_H and χ_S, evaluated at selected temperatures for symmetrical PS-b-PVME with $N_c = 20,000$ [a].

T (K)	χ_H	χ_S	$N_c\chi$
425	−0.00668	0.00656	−2.46405
450	−0.00622	0.00664	8.46668
475	−0.00581	0.00672	18.30529
500	−0.00544	0.00680	27.21505
525	−0.00510	0.00687	35.32983
550	−0.00479	0.00693	42.76065

[a] Pressure is fixed to 0.1 MPa.

Starting from the CP of PS-b-PVME copolymer, all the transition points at ambient pressure are to be determined again using the vertex coefficients as in Tables S2–S7 of SM. Because of the asymmetry in x^*, the δ_{ijkl}s for the copolymer is minutely different from these given in those tables when approaching both extremes at $\phi_A \to 0$ and $\phi_A \to 1$. Figure 2a displays all the phase boundaries as well as spinodals for PS-b-PVME in terms of the absolute temperature. The substantial disparity in $\bar{\varepsilon}_{ij}$s between PS and PVME forces all those lines to skew, as seen in this figure. In Figure 2b, the phase diagram is redrawn in terms of $N_c\chi$, which is quite slanted for molten PS-b-PVME. To get the data in Figure 2, the Landau free energy in Equation (8) is used. Even if the free energy in Equation (24) is used instead, it is observed that we still get almost the identical phase diagram. The spinodals from the two methods are perfectly identical. The ODT from disorder to BCC for the copolymer at ϕ_A = 0.1 using Equation (8) is different by ~0.2 K from that using Equation (24). The calculated difference in OOTs from BCC to HEX using the two methods for the copolymer at the same composition is found to be ~0.2 K. The predicted HEX-LAM OOTs using Equations (8) and (24) are 1040.074 K and 1056.842 K, respectively. In this case, ΔT reaches 16.8 K. However, the agreement between the two methods is satisfactory considering that the difference is less than 2% even in this unreachable temperature region.

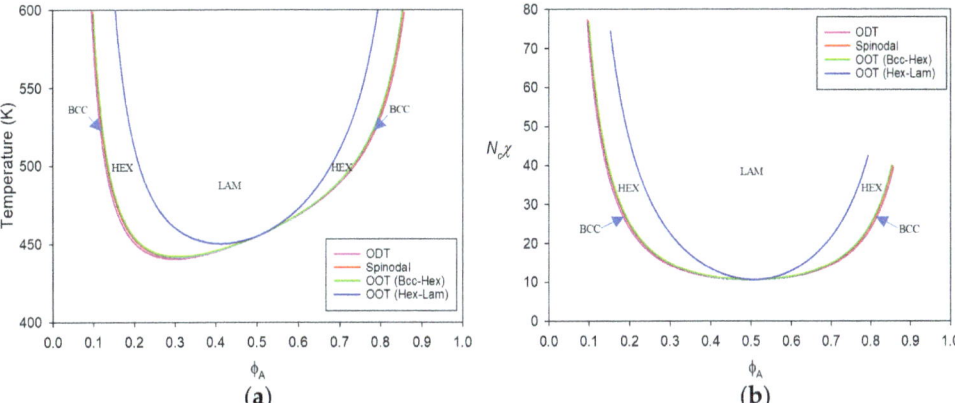

Figure 2. Phase diagram for molten PS-b-PVME with N_c = 20,000 plotted against PS (A) volume fraction ϕ_A in terms of: (**a**) absolute temperature and (**b**) the relevant parameter $N_c\chi$. The disparity in $\bar{\varepsilon}_{jj}$s is sizable as $|\bar{\varepsilon}_{PS} - \bar{\varepsilon}_{PBD}|/\bar{\varepsilon}_{PS}$ = 0.113 and the exchange energy is $\Delta\bar{\varepsilon}/k$ = −6.637 K. In this situation, the phase boundaries are skewed towards more compressible PVME side, but with the CP at ϕ_A = 0.50974. The arrows indicate BCC mesophase in the narrow region between ODT and spinodals.

The compressible nature and disparity in $\bar{\varepsilon}_{jj}$s for PS-b-PVME gives the difference in the order parameter amplitudes. The ratio ς_A/ς_B is shown to be ~1.04 for ODT and BCC-HEX OOT for the copolymer at ϕ_A = 0.1. At other compositions, ς_A/ς_B > 1, which reflects the fact that PS is denser than PVME. As the transition temperature is further increased in case of HEX-LAM OOT for the copolymer at the same composition, ς_A/ς_B is increased to become ~1.11. The density difference between PS and PVME should grow with thermal expansion.

3.3. Pressure Dependence of Ordering Transitions

In this section, we discuss the responses of diblock copolymers to pressure. The first system to consider is PS-b-PI copolymer, whose ordering transition temperatures have been reported by Hajduk et al. [33,62]. The requisite molecular parameters are also given in Table 1. It is seen in this table that $|\bar{\varepsilon}_{PS} - \bar{\varepsilon}_{PI}|/\bar{\varepsilon}_{PS}$ = 0.012, which is similar to that for PS-b-PBD. The cross interaction parameter $\bar{\varepsilon}_{SI}$ = $0.99680(\bar{\varepsilon}_{PS}\bar{\varepsilon}_{PI})^{1/2}$ is determined from fitting the CP (388 K) of PS/PI blend with molecular weights of 2117 and 2594, reported by Rudolf and Cantow [63], and adjusted by comparison with the ODT data for PS-b-PI with M_w = 8000/8500 [33]. Figure 3a depicts the two contributions to χ against pressure for PS-b-PI at $\phi_A = \phi_{PS}$ = 0.442 and at T = 365 K. As is now expected from $|\bar{\varepsilon}_{PS} - \bar{\varepsilon}_{PI}|$ for this copolymer, Flory χ is mostly given by χ_H along with $O(\chi_S) \sim 10^{-5}$. The enthalpic χ_H increases upon pressurization. Although χ_S goes in a reverse way due to the bulk modulus of the copolymer, the effective Flory χ follows χ_H to be strengthened by the applied pressure. In Figure 3b, all the transition points for the copolymer at the same composition are plotted as a function of P. The pressure coefficient, $\Delta T_{trs}/\Delta P$, of the ordering transition is predicted to be ~15 K/100 MPa, which describes well the experimental value of ~17 K/100 MPa for the copolymer with M_w = 16,500 or N_c = 327.4 [33]. This type of pressure response is barotropicity, as already mentioned.

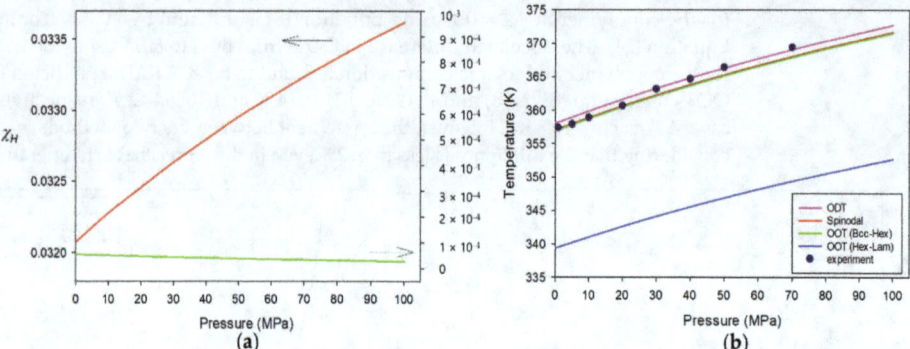

Figure 3. Pressure responses of (**a**) χ_H as well as χ_S at T = 365 K and (**b**) various transitions for molten PS-b-PI with N_c = 327.4 (M_w = 16,500) at $\phi_A = \phi_{PS}$ = 0.442. The symbols in plot (**b**) indicate the experimental ODT data measured by Hajduk et al. The arrows indicate the proper axes for χ_H and χ_S.

Our next system is the copolymer from PS and poly(ethyl hexyl acrylate) (PEHA). The PS-b-PEHA diblock copolymer exhibits a completely reverse response to pressure, as was measured using light scattering (cloud points) and small angle neutron scattering [40]. This copolymer is a member of baroplastic systems, whose nanoscopic phase separation and pressure response were first studied by Mayes and co-workers [38]. Again, all the necessary molecular parameters for PS and PEHA are given in Table 1. It is seen in this table that $|\bar{\epsilon}_{PS} - \bar{\epsilon}_{PEHA}|/\bar{\epsilon}_{PS} = 0.086$, which is quite larger than that for PS-b-PBD. The cross-interaction parameter of $\bar{\epsilon}_{S-EHA} = 0.99880(\bar{\epsilon}_{PS}\bar{\epsilon}_{PEHA})^{1/2}$ is the optimized one to fit the phase behavior of PS-b-PEHA with M_w = 23,000 or N_c = 529.581. In Figure 4a, we display χ and its two contributions, χ_H and χ_S, for the copolymer at $\phi_A = \phi_{PS}$ = 0.42 and at T = 445 K. It is observed in this figure that χ_S is near 20% of χ_H at ambient pressure. As pressure is increased, the enthalpic χ_H is increased due to densification. The entropic χ_S is suppressed by the applied pressure to have $\chi_S/\chi_H \sim 0.12$ at P = 100 MPa. As a result, the effective χ becomes a decreasing function of pressure. Figure 4b depicts all the transition temperatures for the copolymer at this composition. The decrease of its ODT is predicted to be $\Delta T_{trs}/\Delta P = -16$ K/100 MPa, which matches well with the scattering result [40].

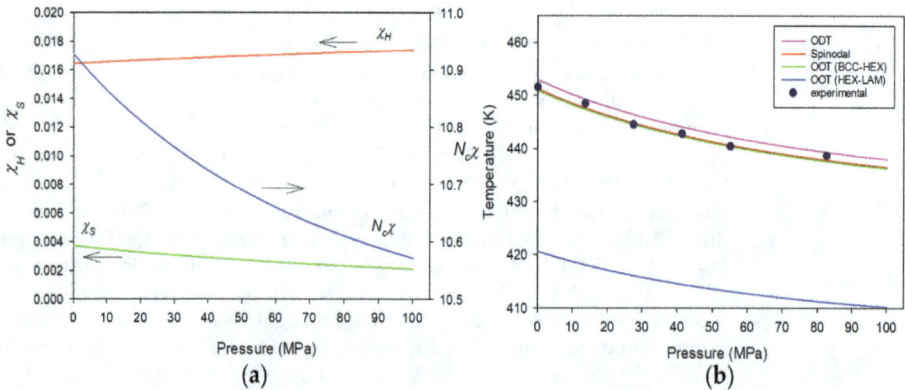

Figure 4. Pressure responses of (**a**) χ_H and χ_S along with $N_c\chi$ at 445 K, and (**b**) various transitions for molten PS-b-PEHA with N_c = 541.8 (M_w = 23,000) at $\phi_A = \phi_{PS}$ = 0.42. The symbols in plot (**b**) indicate the experimental ODT data for the copolymer measured by Lee et al. The arrows indicate the proper axes for χ_H, χ_S, and $N_c\chi$.

Our third system is the diblock copolymer from poly(ethyl ethylene) (PEE) and poly(dimethyl siloxane) (PDMS). The PEE-b-PDMS copolymer is one of UODT-type block copolymers. However, its response to pressure is abnormal in the sense that the copolymer reveals the retreat of its ordering temperatures in the low pressure region and then resurgence of the transition temperatures in the high pressure region [46,64]. This anomalous pressure response of the copolymer can be understood by the subtle balance of χ_H and χ_S. In describing the copolymer, the necessary molecular parameters are given in Table 4. The key elements there are $|\bar{\varepsilon}_{EE} - \bar{\varepsilon}_{DMS}|/\bar{\varepsilon}_{EE} = 0.108$ and $\bar{\varepsilon}_{EE-DMS} = 0.99654(\bar{\varepsilon}_{EE}\bar{\varepsilon}_{DMS})^{1/2}$, determined by fitting the phase behaviors of symmetric PEE-b-PDMS with $M_w = \sim 10{,}700$. Using these parameters, $\Delta\bar{\varepsilon}$ is unfavorable as $\Delta\bar{\varepsilon}/k = 27.158$ K. Figure 5a depicts the effective χ along with its two contributions, χ_H and χ_S, plotted against pressure for PEE-b-PDMS at $\phi_A = \phi_{PEE} = 0.52$ and at 352.5 K. At ambient pressure, it is seen that χ_S is near 20% of χ_H, and at 100 MPa χ_S drops to near 10% of χ_H just as in the case of PS-b-PEHA. However, unlike PS-b-PEHA, the increase of χ_H is more rapid, so that χ_S becomes just 5% of χ_H at 200 MPa. Therefore, χ_H regains the control of the phase behaviors. The transition temperatures of this copolymer are shown in Figure 5b against pressure from 0.1 to 200 MPa. In this figure, the baroplastic, followed by barotropic responses of PEE-b-PDMS, are clearly demonstrated in agreement with the experiment.

Table 4. Molecular parameters for PEE-b-PDMS systems.

Parameters	PEE	PDMS
σ_i (Å)	3.590 [a]	3.952 [a]
$\bar{\varepsilon}_{ii}/k$ (K)	2819.60	2514.50
$N_i \pi \sigma_i^3 / 6M_i$ (cm^3/g)	0.47823	0.41778
$\bar{\varepsilon}_{AB}/(\bar{\varepsilon}_{AA}\bar{\varepsilon}_{BB})^{1/2}$		0.99654

[a] This discrepancy in monomer diameters is resolved by adopting the conventional Lorentz mixing rule as $\sigma = (\sigma_i + \sigma_j)/2$.

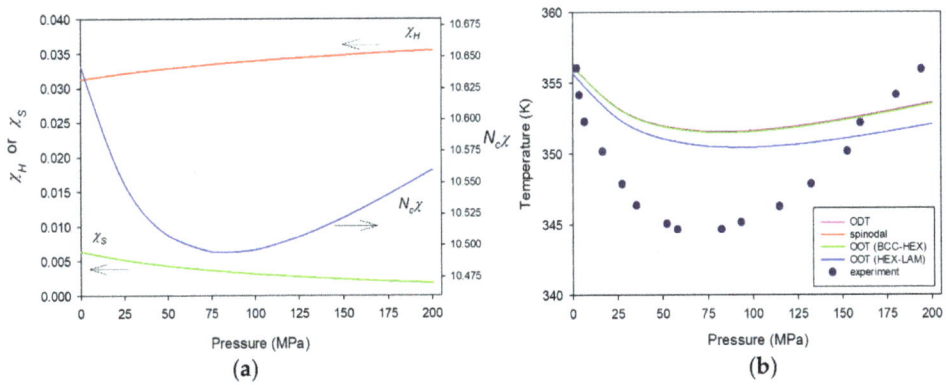

Figure 5. Pressure responses of (**a**) $N_c\chi$ along with its two contributions, χ_H and χ_S, at 352.5 K, and (**b**) various transitions for molten PEE-b-PDMS with $N_c = 282.58$ ($M_w = 10{,}700$) at $\phi_A = \phi_{PEE} = 0.52$. The symbols indicate the experimental ODT data for the copolymer at $\phi_A = 0.50$ measured by Schwahn et al. The arrows indicate the proper axes for χ_H, χ_S, and $N_c\chi$.

We have revisited the Landau free energy for A-b-B diblock copolymer melts with diverse types of phase behaviors from the viewpoint of their response to temperature or pressure. Being analytical with one harmonic for order parameters, the present work deals with the classical nanostructures. For the equilibration of other mesophases, it is necessary to use our self-consistent field theory for the copolymers, which was developed a few years ago [47–49]. All of our works are based on the restricted chain model with the identical monomer diameters σ_js. This restriction can be alleviated to allow for the variation of σ_js. The present mean-field Landau energy with its correct cubic term exhibits continuous transitions or CPs, whose locations are dependent on disparity in $\bar{\varepsilon}_{ij}$s. When the concentration fluctuations are involved, such mean-field CPs are known to be destroyed to yield the weak first-order transition [55]. The necessary fluctuation correction in one-loop order was suggested by Fredrickson and Helfand [56] for the corresponding incompressible A-b-B diblock copolymer systems utilizing Brazovskii's Hamiltonian form. One of the present authors also introduced the similar approach for the copolymer melt in case of small $|\bar{\varepsilon}_{AA} - \bar{\varepsilon}_{BB}|$ [65]. In Appendix C, we provide the fluctuation correction analysis starting from our Landau free energy in Equation (24).

4. Conclusions

Nanoscopic phase behaviors of molten A-b-B diblock copolymers within general disparity in self dispersion interactions are revisited through Landau approach. A continuum space molecular equation of state is first considered to describe such copolymers in the bulk state. A free energy functional is obtained for the inhomogeneous copolymer as that for the corresponding Gaussian chains under the influence of effective two-body interactions from the localized excess equation of state. The free energy functional in the weak segregation regime is then expanded directly as a series in powers of two order parameters, which are fluctuations in A and B block densities ($\langle \Delta \eta_A \rangle$ and $\langle \Delta \eta_B \rangle$), up to 4th order. The order parameters are treated as the sum of plane waves with their amplitude and phase angles. The Fourier-transformed momentum integral for the Landau free energy is approximated to the finite sum of terms at the scattering vectors $\vec{k}$, whose lengths are the characteristic wavenumber k^* signifying the domain sizes of ordered mesophases. A completely alternative Landau free energy is obtained after the transformation of the order parameters to fluctuations in block density difference ($\sim \langle \Delta \eta_A - \Delta \eta_B \rangle / 2\eta$) and negative free volume fraction ($-\langle \Delta \eta_f \rangle$). It is shown that those two different Landau free energies are equivalent to yield almost identical ordering transition temperatures.

The analysis of spinodals from the quadratic term of the Landau energy, which are perfectly identical for the direct and alternative ones, leads to effective Flory χ as the sum of the conventional enthalpic χ_H for exchange energy ($\propto \Delta \bar{\varepsilon} |u(\eta)|$) and entropic χ_S representing disparity in self dispersion interactions mediated by copolymer bulk modulus ($\propto [\bar{\varepsilon}_{AA} - \bar{\varepsilon}_{CC}]^2 / B_T$). The cubic term of the Landau free energy is shown to be balanced with all the Gaussian cubic vertex coefficients Γ_{ijk} in corporation with Γ_{ij} to yield its critical point (CP) depending on asymmetry in self dispersion strengths. The quartic terms of the Landau free energy are given mainly by the Gaussian quartic vertex coefficients Γ_{iiii}s at the proper combinations of the scattering vectors pertinent to the given mesophases.

Taking PS-b-PBD and PS-b-PVME as model systems, the responses of the copolymers to temperature are first discussed. The former copolymer exhibits typical ordering transition upon cooling (UODT), whereas the latter copolymer reveals the reverse ordering transition upon heating (LDOT). The phase boundaries for these copolymers are fully determined by numerically minimizing the Landau energy. The PS-b-PBD copolymer with quite close self-dispersion interactions for both blocks gives a symmetric phase diagram. This phenomenon can be understood because $\chi \approx \chi_H$ and $\chi_S << 1$, in this case. The copolymer phase behaviors at ambient pressure are almost identical to those of the incompressible A-b-B copolymer. In contrast, PS-b-PVME copolymer possesses quite sizable disparity in self dispersion interactions. The substantially large χ_S in this copolymer develops phase segregation tendency upon heating because of the diminished bulk modulus. The phase boundaries are skewed towards the side rich in more compressible PVME. While PS-b-PBD possesses its CP near the symmetric composition, PS-b-PVME pushes its CP towards the copolymer rich in denser component to match domain volumes to fulfill a continuous transition.

The responses of A-b-B copolymers to pressure are investigated by taking PS-b-PI, PS-b-PEHA, and PEE-b-PDMS as model systems. The PS-b-PI copolymer is barotropic, which is typical for many block copolymers. In this case, χ is dominated by χ_H, which is augmented by pressurization due to the increased density. On the contrary, the PS-b-PEHA copolymer is baroplastic with more sizable χ_S. Pressurization suppresses χ_S because of the copolymer bulk modulus, and the decrease in χ_S affects the total χ more than the increase in χ_H. Anomaly is observed for the PEE-b-PDMS copolymer, because it is baroplastic at lower pressure region and then barotropic in higher pressure region. The reason for this complicated pressure response is because there is a subtle competition between χ_H and χ_S, which prevails between the two switches at different stages of pressurization.

Supplementary Materials: The following supporting information can be downloaded at: https://www.mdpi.com/article/10.3390/polym15010030/s1, Table S1: Characteristic squared wavenumber x* at spinodals for PS-b-PBD and PS-b-PVME melts listed at selected compositions ϕ_As; Table S2: Quadratic vertex coefficients Γ_{ij}s for A-b-B with $\bar{\varepsilon}_{AA} = \bar{\varepsilon}_{BB}$ at the indicated compositions ϕ_As; Table S3: Cubic vertex coefficients $\eta^2 N_c \Gamma_{ijk}$s for A-b-B at the indicated compositions ϕ_As; Table S4: Quadratic vertex coefficients Γ_{ij}s for PS-b-PBD and PS-b-PVME at the indicated compositions ϕ_As; Table S5: Quartic vertex coefficients $\eta^3 N_c \delta^{BCC}_{ijkl}/c_{BCC}$s in Equation (12) for BCC-forming PS-b-PBD at the indicated ϕ_As; Table S6: Quartic vertex coefficients $\eta^3 N_c \delta^{HEX}_{ijkl}/c_{HEX}$s in Equation (11) for cylinder-forming PS-b-PBD at the indicated ϕ_As; Table S7: Quartic vertex coefficients $\eta^3 N_c \delta^{LAM}_{ijkl}/c_{LAM}$s in Equation (10) for lamella-forming PS-b-PBD at the indicated ϕ_As.

Author Contributions: Conceptualization, J.C.; methodology, J.C.; software, J.C.; validation, J.C.; formal analysis, J.C.; investigation, X.Z., M.Z. and J.C.; resources, J.C.; data curation, X.Z., M.Z. and J.C.; writing—original draft preparation, J.C.; writing—review and editing, J.C.; visualization, X.Z., M.Z. and J.C.; supervision, J.C.; project administration, J.C.; funding acquisition, J.C. It should be noted that X.Z. and M.Z. contributed equally. All authors have read and agreed to the published version of the manuscript.

Funding: This work was supported by the National Research Foundation of Korea through the Basic Science Research Program (No. 2020R1F1A1056653).

Institutional Review Board Statement: Not applicable.

Informed Consent Statement: Not applicable.

Data Availability Statement: The data presented in this study are available on request from the corresponding author.

Conflicts of Interest: The authors declare no conflict of interest. The funders had no role in the design of the study; in the collection, analyses, or interpretation of data; in the writing of the manuscript; or in the decision to publish the results.

Appendix A. Landau Free Energy in a Direct Way

In this appendix, we provide the detailed procedure in deriving the Landau free energy given in Equation (8). The general integral form of the free energy in Equation (7) is approximated to the finite sum of integrands at n characteristic scattering vectors $\vec{K}_j$s signifying a given structure, where $|\vec{K}_j|$ is equal to its characteristic structural wavenumber k^*. The quadratic form A_2 of the free energy functional in Equation (7) is then given by

$$\beta A_2 = \frac{1}{2}\int d\vec{k}_1 \Gamma_{ij}\left(\vec{k}_1, -\vec{k}_1\right) \Delta\eta_i(\vec{k}_1)\Delta\eta_j(-\vec{k}_1) \approx \frac{1}{2}\sum_{\vec{k}_1 \in \{\pm \vec{K}_n\}} \Gamma_{ij}\left(\vec{k}_1, -\vec{k}_1\right) \Delta\eta_i(\vec{k}_1)\Delta\eta_j(-\vec{k}_1) \quad (A1)$$

There are only 2n such cases. Each $\Delta\eta_i(\pm\vec{k}_1)$ is treated as a plane wave with its amplitude $(1/\sqrt{n})\varsigma_j$ and phase angle $\pm\varphi(i)$ as $\Delta\eta_i(\pm\vec{k}_1) = (1/\sqrt{n})\varsigma_i e^{\pm i\varphi(i)}$. Then, Equation (A1) turns to

$$\beta A_2 = \Gamma_{AA}\left(\vec{k}_1, -\vec{k}_1\right)\varsigma_A^2 + 2\Gamma_{AB}\left(\vec{k}_1, -\vec{k}_1\right)\varsigma_A\varsigma_B e^{i[\varphi(A)-\varphi(B)]} + \Gamma_{BB}\left(\vec{k}_1, -\vec{k}_1\right)\varsigma_B^2 \quad (A2)$$

It is found that all the Γ_{ij}s are positive, as seen in Tables S2 and S3 of SM. The minimization of Equation (A2) gives

$$\beta A_2 = [\Gamma_{AA}(k^*)\varsigma_A^2 - 2\Gamma_{AB}(k^*)\varsigma_A\varsigma_B + \Gamma_{BB}(k^*)\varsigma_B^2] \quad (A3)$$

where the phase angles need to satisfy $\varphi(A) - \varphi(B) = \pi$.

Let us discuss how to obtain higher-order terms in Equation (8) of the main text. This is where the present work takes a step forward with the deepened conception from the previous study [44]. The cubic form A_3 of the free energy functional has the vertex functions of all possible combinations of indices i, j, and k. The same parametrization of $\Delta\eta_i(\vec{k}_1)$ and approximating A_3 to the finite sum of integrands give

$$\beta A_3 = \frac{1}{3!} \sum_{\vec{k}_i \in \{\pm\vec{K}_n\}} \Gamma_{ijk}\left(\vec{k}_1, \vec{k}_2, \vec{k}_3\right) \frac{\varsigma_i\varsigma_j\varsigma_k}{(\sqrt{n})^3} e^{i(\varphi_a(i)+\varphi_b(j)+\varphi_c(k))} \quad (A4)$$

One should bear in mind that the scattering vectors $\vec{k}_1$, $\vec{k}_2$, and $\vec{k}_3$ satisfy $\vec{k}_1 + \vec{k}_2 + \vec{k}_3 = 0$ not to nullify $\Gamma_{ijk}(h)$. These vectors should then form a right triangle to have $h = 1$. The given vertex functions change their signs according to indices. However, the sum of phase angles cannot be decided by their signs, because the sums are not fully independent. All those sums can be determined relative to $\sum \varphi_m(A) \equiv \varphi_a(A) + \varphi_b(A) + \varphi_c(A)$ as follows:

$$\varphi_a(i) + \varphi_b(j) + \varphi_c(k) = \varphi_a(A) + \varphi_b(A) + \varphi_c(A) - \pi \cdot c_B(ijk) = \sum \varphi_m(A) - \pi \cdot c_B(ijk) \quad (A5)$$

where $c_B(ijk)$ is the number of B among indices i, j, and k. Inserting Equation (A5) into A4 yields

$$\beta A_3 = \frac{1}{3!} \sum_{\vec{k}_i \in \{\pm\vec{K}_n\}} e^{i(\varphi_a(A)+\varphi_b(A)+\varphi_c(A))} \cdot \Gamma_{ijk}(1) \frac{\varsigma_i\varsigma_j\varsigma_k}{(\sqrt{n})^3} e^{-i\cdot\pi c_B(ijk)} \quad (A6)$$

where $\Gamma_{ijk}(h)$ is evaluated at $h = 1$. The minimization of Equation (A6) leads in general to

$$\beta A_3 = -\left|\alpha_{AAA}\varsigma_A^3 - 3\alpha_{AAB}\varsigma_A^2\varsigma_B + 3\alpha_{ABB}\varsigma_A\varsigma_B^2 - \alpha_{BBB}\varsigma_B^3\right| \quad (A7)$$

where α_{ijk} is given in Equation (9) of the main text. It should be remembered that there is no way to have $h = 1$ for $\Gamma_{ijk}^{(3)}$ of LAM mesophase. In BCC mesophase, forming a right triangle takes only 4 cases such as $(\vec{K}_1, -\vec{K}_3, -\vec{K}_6)$, $(\vec{K}_1, -\vec{K}_4, -\vec{K}_5)$, $(\vec{K}_2, -\vec{K}_4, \vec{K}_6)$, and $(\vec{K}_2, -\vec{K}_3, \vec{K}_5)$ in order to have $h = 1$.

The quartic form A_4 of the free energy functional is originally made of vertex functions of all possible combinations of indices i, j, k, and l as

$$\beta A_4 = \frac{1}{4!} \sum_{\vec{k}_i \in \{\pm\vec{K}_n\}} \Gamma_{ijkl}\left(\vec{k}_1, \vec{k}_2, \vec{k}_3, \vec{k}_4\right) \frac{\varsigma_i\varsigma_j\varsigma_k\varsigma_l}{(\sqrt{n})^4} e^{i(\varphi_1(i)+\varphi_2(j)+\varphi_3(k)+\varphi_4(l))} \quad (A8)$$

where $\sum \vec{k}_i = 0$ must be satisfied by the integrand. The sums of phase angles are resolved by those of the second and third order terms. However, unlike A_3, the mesophase geometry is directly used for A_4. Still, the values of these sums are determined relative to $\sum \varphi_l(A) \equiv \varphi_a(A) + \varphi_b(A) + \varphi_c(A) + \varphi_d(A)$ for pure A correlations as follows:

$$\varphi_a(i) + \varphi_b(j) + \varphi_c(k) + \varphi_d(l) = \sum \varphi_l(A) - \pi \cdot c_B(ijkl) \quad (A9)$$

where $c_B(ijkl)$ is the number of B among indices $i, j, k,$ and l. Then, A_4 is obtained as

$$\beta A_4 = \frac{1}{4!} \sum_{\vec{k}_i \in \{\pm \vec{K}_n\}} e^{i(\varphi_a(A)+\varphi_b(A)+\varphi_c(A)+\varphi_d(A))} \Gamma_{ijkl}(h_1, h_2) \frac{\varsigma_i \varsigma_j \varsigma_k \varsigma_l}{(\sqrt{n})^4} e^{-i \cdot \pi c_B(ijkl)} \quad (A10)$$

We need to count all the possible cases in accord with LAM, HEX, and BCC mesophases, which are given below in detail.

LAM:

In regards to the quartic form A_4 for lamellae, the only possible option to have the vanishing $\sum \vec{k}_i$ is $\vec{k}_1 - \vec{k}_1 + \vec{k}_1 - \vec{k}_1 = 0$ or $(h_1, h_2) = (0, 0)$. This condition leads to $\sum \varphi_i(A) = \varphi_a(A) - \varphi_a(A) + \varphi_a(A) - \varphi_a(A) = 0$. Then, the quartic form A_4 is expressed as

$$\beta A_4 = \frac{3!}{4!} \Gamma_{ijkl}(0,0) \varsigma_i \varsigma_j \varsigma_k \varsigma_l e^{-i\pi \cdot c_B(ijkl)} \quad (A11)$$

where all such cases are included.

HEX:

For the quartic form A_4 of HEX mesophase, the only possible sets of (h_1, h_2) are $(0,0)$ and $(0,1)$. After counting all such case, we have

$$\beta A_4 = \frac{18}{4! \left(\sqrt{3}\right)^4} \left[\Gamma_{ijkl}(0,0) + 4\Gamma_{ijkl}(0,1) \right] \varsigma_i \varsigma_j \varsigma_k \varsigma_l e^{-i\pi \cdot c_B(ijkl)} \quad (A12)$$

It should be noted that $(h_1, h_2) = (0,0)$ and $(0,1)$ respectively require one $\vec{K}_a$ ($\varphi_a - \varphi_a + \varphi_a - \varphi_a = 0$) and two $\vec{K}_j$s forming either 60° or 120° between them ($\varphi_b - \varphi_b + \varphi_c - \varphi_c = 0$).

BCC:

The quartic form A_4 of the free energy for BCC is given by

$$\beta A_4 = \frac{36}{4!\left(\sqrt{6}\right)^4} \Big\{ \Gamma_{ijkl}(0,0) e^{i(\varphi_a(A)-\varphi_a(A)+\varphi_a(A)-\varphi_a(A))} +$$
$$8\Gamma_{ijkl}(0,1) e^{i(\varphi_b(A)-\varphi_b(A)+\varphi_c(A)-\varphi_c(A))} + 2\Gamma_{ijkl}(0,2) e^{i(\varphi_d(A)-\varphi_d(A)+\varphi_e(A)-\varphi_e(A))} +$$
$$2\Gamma_{ijkl}(1,2) \left(e^{i(\varphi_1(A)+\varphi_2(A)-\varphi_3(A)-\varphi_4(A))} + e^{i(\varphi_1(A)-\varphi_2(A)-\varphi_5(A)-\varphi_6(A))} \right) \Big\} \varsigma_i \varsigma_j \varsigma_k \varsigma_l e^{-i\pi c_B(ijkl)} \quad (A13)$$

There are only one $\vec{K}_a$ required for $(0,0)$ contribution ($\varphi_a - \varphi_a + \varphi_a - \varphi_a = 0$) and two $\vec{K}_j$'s required for $(0,1)$ and $(0,2)$ contributions ($\varphi_b - \varphi_b + \varphi_c - \varphi_c = 0$) to βF_4. In the case of the $(0,1)$ contribution, two $\vec{K}_j$'s form either 60° or 120° between them. In the other case of the $(0,2)$ contribution, two $\vec{K}_j$'s should be selected to form the right angle. However, there are four $\vec{K}_j$'s necessary to describe $(1,2)$ contribution to βF_4, as they are clearly expressed in Equation (A13). Using the result that $\varphi_1(A) - \varphi_3(A) - \varphi_6(A) = \varphi_2(A) - \varphi_4(A) + \varphi_6(A) = 0$ or π, we have

$$\varphi_1(A) + \varphi_2(A) - \varphi_3(A) - \varphi_4(A) = \varphi_1(A) - \varphi_3(A) - \varphi_6(A) = 0 \quad (A14)$$

or

$$\varphi_1(A) + \varphi_2(A) - \varphi_3(A) - \varphi_4(A) = \varphi_1(A) - \varphi_3(A) - \varphi_6(A) + \pi = 2\pi \quad (A15)$$

The same argument applies to the other set of phase angles in Equation (A13), because $\varphi_1(A) - \varphi_4(A) - \varphi_5(A) = \varphi_2(A) - \varphi_4(A) + \varphi_6(A) = 0$ or π. Then, we have

$$\varphi_1(A) - \varphi_2(A) - \varphi_5(A) - \varphi_6(A) = \varphi_1(A) - \varphi_4(A) - \varphi_5(A) = 0 \quad (A16)$$

or

$$\varphi_1(A) - \varphi_2(A) - \varphi_5(A) - \varphi_6(A) = \varphi_1(A) - \varphi_4(A) - \pi - \varphi_5(A) = 0 \quad (A17)$$

Therefore, we have the final expression for the quartic form A_4 as

$$\beta A_4 = \frac{36}{4!\left(\sqrt{6}\right)^4}\left[\Gamma_{ijkl}(0,0) + 8\Gamma_{ijkl}(0,1) + 2\Gamma_{ijkl}(0,2) + 4\Gamma_{ijkl}(1,2)\right]\varsigma_i\varsigma_j\varsigma_k\varsigma_l e^{-i\pi \cdot c_B(ijkl)} \quad \text{(A18)}$$

where $e^{-i\pi \cdot c_B(ijkl)}$ gives 1 or -1 depending on $c_B(ijkl)$ as before.

The Landau free energy is given by $\beta\Delta A = \beta A_2 + \beta A_3 + \beta A_4$ to have its final and general form in Equation (8) for each mesophase with the two amplitudes ς_A and ς_B.

Appendix B. Landau Free Energy through Transformation of Order Parameters

In this appendix, we provide the mathematical procedure to formulate our alternative Landau free energy in a more familiar form through the transformation of order parameters. The transformed order parameters, $\psi_1(\vec{r})$ and $\psi_2(\vec{r})$, are defined in the main text, where the former is a composite density profile in phase with A block and the latter indicates the negative fluctuations in free volume fraction. Upon this change of variables, the new vertex functions $\overline{\Gamma}$s are obtained from the original vertex functions Γs by Equation (15).

We will consider nanoscale mesophases, whose structures are defined by characteristic scattering vectors $\vec{k}_1 \in \left\{\pm\vec{K}_n\right\}$. Those regular geometric morphologies are represented by ψ_1 as the sum of plane waves. The remaining ψ_2 is separated into two parts as $-\psi_2 = -\psi_{2c} - \psi_{2i}$, where $-\psi_{2c}$ indicates the excess free volume at the interfaces between domains and $-\psi_{2i}$ represents the excess free volume in phase with the more compressible constituent. These order parameter parts are parametrized as given in the main text.

The quadratic form A_2 of the free energy expansion can be written as

$$\beta A_2 \approx \frac{1}{2}\sum_{\vec{k}_1 \in \{\pm\vec{K}_n\}}\overline{\Gamma}_{ij}\left(\vec{k}_1, -\vec{k}_1\right)\psi_i(\vec{k}_1)\psi_j(-\vec{k}_1) = \overline{\Gamma}_{11}\zeta_1^2 + 2\overline{\Gamma}_{12}\zeta_1\zeta_{2c}e^{i(\varphi(1)-\varphi(2c))} + \overline{\Gamma}_{22}\zeta_{2c}^2 + \overline{\Gamma}_{22}(2q*)\zeta_{2i}^2 \quad \text{(A19)}$$

where the integral is approximated to the finite sum of the integrand at the characteristic wavevectors. Since $\vec{k}_j - \left(2\vec{k}_j\right) \neq 0$, there is no mixed term such as $\psi_1\psi_{2i}$.

We will discuss how to obtain higher-order terms in Equation (16) of the main text. This is where the present work takes a step forward with the deepened conception from our previous studies [45,46]. The cubic form A_3 is suggested in the following way. The fluctuations in free volume is considered to be small. Therefore, the only possible contributions in our 4-field theory are given by $\psi_1\psi_1\psi_1$, $\psi_1\psi_1\psi_{2c}$, and $\psi_1\psi_1\psi_{2i}$ as

$$\beta A_3 = \tfrac{1}{3!}\sum_{\vec{k}_1 \in \{\pm\vec{K}_n\}}\overline{\Gamma}_{111}(1)\psi_1\left(\vec{k}_1\right)\psi_1\left(\vec{k}_2\right)\psi_1\left(\vec{k}_3\right) + \tfrac{1}{3!}\sum_{\vec{k}_1 \in \{\pm\vec{K}_n\}}\left(\overline{\Gamma}_{112}(1) + \overline{\Gamma}_{121}(1) + \overline{\Gamma}_{211}(1)\right)\psi_1\left(\vec{k}_1\right)\psi_1\left(\vec{k}_2\right)\psi_{2c}\left(\vec{k}_3\right)$$
$$+ \tfrac{1}{3!}\sum_{\vec{k}_1 \in \{\pm\vec{K}_n\}}\left(\overline{\Gamma}_{112}(4) + \overline{\Gamma}_{121}(4) + \overline{\Gamma}_{211}(4)\right)\psi_1\left(\vec{k}_1\right)\psi_1\left(\vec{k}_2\right)\psi_{2i}\left(2,\vec{k}_3\right) \quad \text{(A20)}$$

The number h inside the bracket of $\overline{\Gamma}_{ijk}$ indicates the relative angles between $\vec{k}_1$, $\vec{k}_2$, and $\vec{k}_3$. Therefore, $h = 1$ implies that those three vectors form a right triangle, whereas $h_1 = 4$ means that $\vec{k}_1 = \vec{k}_2$ and $\vec{k}_3 = -2\vec{k}_1$. Equation (A20) is re-written with the parametrized order parameters as

$$\beta A_3 = \frac{1}{3!\left(\sqrt{n}\right)^3}\sum_{\{\vec{K}_a,\vec{K}_b,\vec{K}_c\}}\overline{\Gamma}_{111}(1)\zeta_1^3 e^{i(\varphi_a(1)+\varphi_b(1)+\varphi_c(1))} + \frac{1}{3!\left(\sqrt{n}\right)^3}\sum_{\{\vec{K}_a,\vec{K}_b,\vec{K}_c\}}\left(3\overline{\Gamma}_{112}(1)\right)\zeta_1^2\zeta_{2c}e^{i(\varphi_a(1)+\varphi_b(1)+\varphi_c(2c))}$$
$$+ \frac{1}{3!\left(\sqrt{n}\right)^3}\sum_{\{\vec{K}_a,\vec{K}_a,-2\vec{K}_a\}}\left(3\overline{\Gamma}_{112}(4)\right)\zeta_1^2\zeta_{2i}e^{i(\varphi_a(1)+\varphi_a(1)-\varphi(2i))} \quad \text{(A21)}$$

where $\overline{\Gamma}_{112} = \overline{\Gamma}_{121} = \overline{\Gamma}_{211}$ is used above. In the same way, the quartic form A_4 of the free energy is only given by $\psi_1\psi_1\psi_1\psi_1$ as

$$\beta A_4 = \frac{1}{4!}\sum_{\vec{k}_i \in \{\pm\vec{K}_n\}}\overline{\Gamma}_{1111}\left(\vec{k}_1,\vec{k}_2,\vec{k}_3,\vec{k}_4\right)\psi_1\left(\vec{k}_1\right)\psi_1\left(\vec{k}_2\right)\psi_1\left(\vec{k}_3\right)\psi_1\left(\vec{k}_4\right) = \frac{1}{4!(\sqrt{n})^4}\sum_{\{\vec{K}_a,\vec{K}_b,\vec{K}_c,\vec{K}_d\}}\overline{\Gamma}_{1111}(h_1,h_2)\zeta_1^4 e^{i(\varphi_a(1)+\varphi_b(1)+\varphi_c(1)+\varphi_d(1))} \quad (A22)$$

and other combination of the order parameters are ignored. In Equation (A22), the relative angles between $\vec{k}_j$s are described by h_1 and h_2 defined in the main text.

Let us try to minimize the free energy term by term starting with A_2. The A block is assumed to be less compressible. Then, $\overline{\Gamma}_{12} \sim \Gamma_{11} - \Gamma_{22} < 0$, which requires $\varphi(1) - \varphi(2c) = 0$. The ψ_1 and ψ_{2c} is then totally in phase with each other. The quadratic form A_2 is now written as

$$\beta A_2 = \overline{\Gamma}_{11}\zeta_1^2 + 2\overline{\Gamma}_{12}\zeta_1\zeta_{2c} + \overline{\Gamma}_{22}\zeta_{2c}^2 + \overline{\Gamma}_{22}(2k*)\zeta_{2i}^2 \quad (A23)$$

Later, this relation between phase angles affects the sums of phase angles for the cubic and quartic forms of the free energy. The higher-order free energy terms are structure dependent. Therefore, we probe three classical structures including LAM, HEX, and BCC mesophases.

LAM:

Lamellae possess only one characteristic wave vector $\vec{K}_1$. Since it is only possible to form $\sum \vec{k}_i = 0$ with $\vec{k}_1$, $\vec{k}_1$, and $-2\vec{k}_1$, the cubic form A_3 of the Landau free energy is solely given by $\psi_1\psi_1\psi_{2i}$ as

$$\beta A_3 = \frac{1}{3!\left(\sqrt{1}\right)^3}\sum_{\vec{k}_i \in \{\pm\vec{K}_n\}}(3\overline{\Gamma}_{112}(4))\zeta_1^2\zeta_{2i}e^{i(\varphi_a(1)+\varphi_a(1)-\varphi(2i))} = \frac{2}{3!}(3\overline{\Gamma}_{112}(4))\zeta_1^2\zeta_{2i} \quad (A24)$$

It is shown that $\overline{\Gamma}_{112}$ at $h = 4$ is negative, which requires that $\varphi(1) + \varphi(1) - \varphi(2i) = 0$. The coefficient in Equation (A24) is c_4^{LAM} in Equation (16) of the main text.

The quartic form A_4 of the Landau free energy for LAM is given by

$$\beta A_4 = \frac{1}{4!\left(\sqrt{1}\right)^4}\sum_{\vec{k}_i \in \{\pm\vec{K}_n\}}\overline{\Gamma}_{1111}(0,0)\zeta_1^4 e^{i(\varphi(1)-\varphi(1)+\varphi(1)-\varphi(1))} = \frac{3!}{4!}\overline{\Gamma}_{1111}(0,0)\zeta_1^4 = d_n^{LAM}\zeta_1^4 \quad (A25)$$

where only $(h_1, h_2) = (0,0)$ is possible for our choice of wave vectors.

HEX:

The cubic form A_3 of the Landau free energy for HEX consists of

$$\beta A_3 = \frac{1}{3!(\sqrt{3})^3}\sum_{\vec{k}_i \in \{\pm\vec{K}_n\}}\overline{\Gamma}_{111}(1)\zeta_1^3 e^{i(\varphi_1(1)+\varphi_2(1)+\varphi_3(1))} + \frac{1}{3!(\sqrt{3})^3}\sum_{\vec{k}_i \in \{\pm\vec{K}_n\}}(3\overline{\Gamma}_{112}(1))\zeta_1^2\zeta_{2c}e^{i(\varphi_1(1)+\varphi_2(1)+\varphi_3(2c))}$$
$$+\frac{1}{3!(\sqrt{3})^3}\sum_{\vec{k}_i \in \{\pm\vec{K}_n\}}(3\overline{\Gamma}_{112}(4))\zeta_1^2\zeta_{2i}e^{i(\varphi_a(1)+\varphi_a(1)-\varphi(2i))} \quad (A26)$$

For the first two parts, the situation is a bit complicated. It should be noticed that $\overline{\Gamma}_{112}(1)$'s and $\overline{\Gamma}_{112}(4)$'s are all negative. However, $\overline{\Gamma}_{111}(1)$ changes its sign, which leads to two different cases. In case that $\overline{\Gamma}_{12} < 0$ and $\overline{\Gamma}_{111}(1) < 0$, $\varphi_1(1) + \varphi_2(1) + \varphi_3(1) = 0$ and $\varphi_1(1) + \varphi_2(1) + \varphi_3(2c) = -\varphi_3(1) + \varphi_3(2c) = 0$ in minimizing the free energy. Consider the opposite case that $\overline{\Gamma}_{12} < 0$ and $\overline{\Gamma}_{111}(1) > 0$. If we choose $\varphi_1(1) + \varphi_2(1) + \varphi_3(1) = \pi$ for $\overline{\Gamma}_{111}(1)$ first, then $\varphi_1(1) + \varphi_2(1) + \varphi_3(2c) = -\varphi_3(1) + \pi + \varphi_3(2c) = \pi$ for $\overline{\Gamma}_{112}(1)$. If we choose $\varphi_1(1) + \varphi_2(1) + \varphi_3(2c) = 0$ for $\overline{\Gamma}_{112}(1)$ first, then $\varphi_1(1) + \varphi_2(1) + \varphi_3(1) = -\varphi_3(2c) + \varphi_3(1) = 0$ reversely for $\overline{\Gamma}_{111}(1)$. In all these cases, it is seen that $\varphi_1(1) + \varphi_2(1) + \varphi_3(2c) = \varphi_1(1) + \varphi_2(1) + \varphi_3(1)$. Owing to the negativity of $\overline{\Gamma}_{112}(4)$, it is gotten that $\varphi_a(1) + \varphi_a(1) - \varphi(2i) = 0$. Summarizing all, A_3 should be as follows:

$$\beta A_3 = \frac{1}{3!(\sqrt{3})^3} \sum_{\{\vec{K}_a, \vec{K}_b, \vec{K}_c\}} e^{i(\varphi_a(1)+\varphi_b(1)+\varphi_c(1))} \cdot (\overline{\Gamma}_{111}(1)\zeta_1^3 + 3\overline{\Gamma}_{112}(1)\zeta_1^2\zeta_{2c}) + \frac{6}{3!(\sqrt{3})^3}(3\overline{\Gamma}_{112}(4))\zeta_1^2\zeta_{2i}$$
$$= -\frac{12}{3!(\sqrt{3})^3}\left|\overline{\Gamma}_{111}(1)\zeta_1^3 + 3\overline{\Gamma}_{112}(1)\zeta_1^2\zeta_{2c}\right| + \frac{6}{3!(\sqrt{3})^3}(3\overline{\Gamma}_{112}(4))\zeta_1^2\zeta_{2i} \quad (A27)$$

where the last expression is the outcome of minimizing A_3. The coefficients in Equation (A27) are a_n^{HEX}, b_1^{HEX}, and c_4^{HEX} in Equation (16) of the main text. It should be emphasized that there is a chance to have $\overline{\Gamma}_{111}(1)\zeta_1 + 3\overline{\Gamma}_{112}(1)\zeta_{2c} = 0$, if $\overline{\Gamma}_{111}(1) > 0$.

The quartic form A_4 is considered to contain only the contribution by ψ_1 as

$$\beta A_4 = \frac{1}{4!}\sum_{\vec{q}_i \in \{\pm Q_n\}} \overline{\Gamma}_{1111}(\vec{q}_1, \vec{q}_2, \vec{q}_3, \vec{q}_4)\psi_1(\vec{q}_1)\psi_1(\vec{q}_2)\psi_1(\vec{q}_3)\psi_1(\vec{q}_4)$$
$$= \frac{1}{4!(\sqrt{n})^4}\left\{\sum_{(0,0)}\overline{\Gamma}_{1111}(0,0)\zeta_1^4 e^{i(\varphi_a(1)-\varphi_a(1)+\varphi_a(1)-\varphi_a(1))} + \frac{1}{(\sqrt{n})^4}\sum_{(0,1)}\overline{\Gamma}_{1111}(0,1)\zeta_1^4 e^{i(\varphi_b(1)-\varphi_b(1)+\varphi_c(1)-\varphi_c(1))}\right\} \quad (A28)$$

It can be seen that $\varphi_a(1) - \varphi_a(1) + \varphi_a(1) - \varphi_a(1) = 0$ for $(h_1, h_2) = (0,0)$ contribution and also $\varphi_b(1) - \varphi_b(1) + \varphi_c(1) - \varphi_c(1) = 0$ for $(h_1, h_2) = (0,1)$ contribution. Therefore, A_4 becomes

$$\beta A_4 = \frac{18}{4!(\sqrt{3})^4}\left[\overline{\Gamma}_{1111}(0,0) + 4\overline{\Gamma}_{1111}(0,1)\right]\zeta_1^4 = d_n^{HEX}\zeta_1^4 \quad (A29)$$

by counting all such cases for HEX mesophase.

BCC:

BCC mesophase requires the correct combinations from the six base vectors. The cubic form A_3 is given as

$$\beta A_3 = \frac{1}{3!(\sqrt{6})^3}\sum_{\vec{k}_i \in \{\pm\vec{K}_n\}}\overline{\Gamma}_{111}(1)\zeta_1^3\left\{e^{i(\varphi_1(1)-\varphi_3(1)-\varphi_6(1))} + e^{i(\varphi_1(1)-\varphi_4(1)-\varphi_5(1))} + e^{i(\varphi_2(1)-\varphi_4(1)+\varphi_6(1))} + e^{i(\varphi_2(1)-\varphi_3(1)+\varphi_5(1))}\right\}$$
$$+ \frac{1}{3!(\sqrt{6})^3}\sum_{\vec{k}_i \in \{\pm\vec{K}_n\}}3\overline{\Gamma}_{112}(1)\zeta_1^2\zeta_{2c}\left\{e^{i(\varphi_1(1)-\varphi_3(1)-\varphi_6(2c))} + e^{i(\varphi_1(1)-\varphi_4(1)-\varphi_5(2c))} + e^{i(\varphi_2(1)-\varphi_4(1)+\varphi_6(2c))} + e^{i(\varphi_2(1)-\varphi_3(1)+\varphi_5(2c))}\right\} \quad (A30)$$
$$+ \frac{1}{3!(\sqrt{6})^3}\sum_{\vec{k}_i \in \{\pm\vec{K}_n\}}3\overline{\Gamma}_{112}(4)\zeta_1^3\zeta_{2i}e^{i(\varphi_a(1)+\varphi_a(1)-\varphi(2i))}$$

Regarding the first two parts, we need to do the same argument as that for HEX mesophase. In any cases, it is obtained that $\varphi_a(1) + \varphi_b(1) + \varphi_c(2c) = \varphi_a(1) + \varphi_b(1) + \varphi_c(1)$. The cubic form A_3 can then be written as

$$\beta A_3 = \frac{1}{3!(\sqrt{6})^3}\sum_{\{\vec{K}_a, \vec{K}_b, \vec{K}_c\}}\left\{e^{i(\varphi_1(1)-\varphi_3(1)-\varphi_6(1))} + e^{i(\varphi_1(1)-\varphi_4(1)-\varphi_5(1))} + e^{i(\varphi_2(1)-\varphi_4(1)+\varphi_6(1))} + e^{i(\varphi_2(1)-\varphi_3(1)+\varphi_5(1))}\right\}\left[\overline{\Gamma}_{111}(1)\zeta_1^3 + 3\overline{\Gamma}_{112}(1)\zeta_1^2\zeta_{2c}\right] + \frac{1}{3!(\sqrt{6})^3}\sum_{\{\vec{K}_a, \vec{K}_b, \vec{K}_c\}}3\overline{\Gamma}_{112}(4)\zeta_1^2\zeta_{2i}e^{i(\varphi_a(1)+\varphi_a(1)-\varphi(2i))} \quad (A31)$$

The sums of phase angles in the first bracket should go to either 0 or π simultaneously in order to minimize Equation (A31). The phase angles associated with $\overline{\Gamma}_{112}(4)$ sum up to 0 due to $\overline{\Gamma}_{112}(4) < 0$. Summarizing all, we have the same type of formula for BCC as that for HEX as

$$\beta A_3 = -\frac{48}{3!(\sqrt{6})^3}\left|\overline{\Gamma}_{111}(1)\zeta_1^3 + 3\overline{\Gamma}_{112}(1)\zeta_1^2\zeta_{2c}\right| + \frac{12}{3!(\sqrt{6})^3}\cdot 3\overline{\Gamma}_{112}(4)\zeta_1^2\zeta_{2i} \quad (A32)$$

where the coefficients in Equation (A32) are a_n^{BCC}, b_1^{BCC}, and c_4^{BCC} in Equation (16) of the main text.

The quartic form A_4 for BCC mesophase is given by

$$\beta A_4 = \frac{1}{4!(\sqrt{6})^4}\left[\sum_{(0,0)}\overline{\Gamma}_{1111}(0,0)\zeta_1^4 e^{i(\varphi_a(1)-\varphi_a(1)+\varphi_a(1)-\varphi_a(1))} + \sum_{(0,1)}\overline{\Gamma}_{1111}(0,1)\zeta_1^4 e^{i(\varphi_b(1)-\varphi_b(1)+\varphi_c(1)-\varphi_c(1))}\right.$$
$$\left. + \sum_{(0,2)}\overline{\Gamma}_{1111}(0,2)\zeta_1^4 e^{i(\varphi_d(1)-\varphi_d(1)+\varphi_c(1)-\varphi_c(1))} + \sum_{(1,2)}\overline{\Gamma}_{1111}(1,2)\zeta_1^4\left\{e^{i(\varphi_1(1)+\varphi_2(1)-\varphi_3(1)-\varphi_4(1))} + e^{i(\varphi_1(1)-\varphi_2(1)-\varphi_5(1)-\varphi_6(1))}\right\}\right] \quad (A33)$$

The sums of phase angles become zero for $(h_1, h_2) = (0,0), (0,1),$ and $(0,2)$. For $(h_1, h_2) = (1,2)$ contributions, we provide a table to show all the results for those sums.

Table A1. Sum of phase angles associated with $\overline{\Gamma}_{1111}(1,2)$.

Case	$\varphi_1(1) + \varphi_2(1) - \varphi_3(1) - \varphi_4(1)$ [a]	$\varphi_1(1) - \varphi_2(1) - \varphi_5(1) - \varphi_6(1)$ [b]
$\varphi_a(1) + \varphi_b(1) + \varphi_c(1) = 0$	$\varphi_1(1) + \varphi_2(1) - \varphi_3(1) - \varphi_4(1)$ $= \varphi_1(1) - \varphi_5(1) - \varphi_4(1) = 0$	$\varphi_1(1) - \varphi_2(1) - \varphi_5(1) - \varphi_6(1)$ $= \varphi_1(1) - \varphi_5(1) - \varphi_4(1) = 0$
$\varphi_a(1) + \varphi_b(1) + \varphi_c(1) = \pi$	$\varphi_1(1) + \varphi_2(1) - \varphi_3(1) - \varphi_4(1)$ $= \varphi_1(1) + \pi - \varphi_5(1) - \varphi_4(1) = 2\pi$	$\varphi_1(1) - \varphi_2(1) - \varphi_5(1) - \varphi_6(1)$ $= \varphi_1(1) - \varphi_5(1) - \varphi_4(1) - \pi = 0$

[a] This sum is given from $\vec{K}_1 + \vec{K}_2 - \vec{K}_3 - \vec{K}_4 = 0$ out of the six base vectors of BCC. [b] This sum is given from $\vec{K}_1 - \vec{K}_2 - \vec{K}_5 - \vec{K}_6 = 0$.

Using this table, A_4 is given as

$$\beta A_4 = \frac{36}{4!\left(\sqrt{6}\right)^4}\{\overline{\Gamma}_{1111}(0,0) + 8\overline{\Gamma}_{1111}(0,1) + 2\overline{\Gamma}_{1111}(0,2) + 4\overline{\Gamma}_{1111}(1,2)\}\zeta_1^4 = d_n^{BCC}\zeta_1^4 \quad (A34)$$

where all such cases are corrected counted.

The Landau free energy is given by $\beta\Delta A = \beta A_2 + \beta A_3 + \beta A_4$ with the three amplitudes $\zeta_1, \zeta_{2c},$ and ζ_{2i}. Differentiating ΔA with respect to ζ_{2c} and ζ_{2i} and then nullifying those derivatives give Equation (23). Replacing such ζ_{2c} and ζ_{2i} into ΔA yields the final mathematical expression of our alternative Landau free energy in Equation (24).

Appendix C. Free Energy Expansion through Fluctuation Correction in One-Loop Order

Diblock copolymers belong to Brazovskii universality class. In the mean-field picture, they possess their CP depending on disparity in self dispersion interactions. However, the effect of concentration fluctuations is known to destroy the mean-field CP to yield weak first order transition, as was revealed by Brazovskii [55]. A simplified fluctuation correction analysis adopted by Fredrickson and Helfand for incompressible diblock copolymers [56] is then applied to the compressible Landau energy, especially our alternative version in Equation (24). The free energy expansion $\beta\Delta A$ is first divided by η as

$$\frac{\beta\Delta A}{\eta} = \left(\frac{\overline{\Gamma}_{11}}{\eta} - \frac{\overline{\Gamma}_{12}^2}{\eta\overline{\Gamma}_{22}}\right)\zeta_1^2 - \left|\frac{a_n}{\eta} - \frac{b_1\overline{\Gamma}_{12}}{\eta\overline{\Gamma}_{22}}\right|\zeta_1^3 + \frac{d_n}{\eta}\zeta_1^4 \quad (A35)$$

which is cast back to the integral expression as

$$\frac{\beta\Delta A}{\eta} \approx \sum_{n=2}^{4}\frac{1}{n!}\int \overline{\Gamma}'_n\left(\vec{k}_1,\ldots,\vec{k}_{n-1}\right)\cdot\psi_1\left(\vec{k}_1\right)\ldots\psi_{n-1}\left(\vec{k}_{n-1}\right)\psi_n\left(-\sum_{l=1}^{n-1}\vec{k}_l\right) \quad (A36)$$

where

$$\overline{\Gamma}'_2(k^*) = \frac{\overline{\Gamma}_{11}}{\eta} - \frac{\overline{\Gamma}_{12}^2}{\eta\overline{\Gamma}_{22}} \quad (A37)$$

and the remaining effective vertex coefficients, $\overline{\Gamma}'_3$ and $\overline{\Gamma}'_4$, are obtained in the corresponding fashion. The order parameter ψ_1 is rewritten with the more general order parameter

$\overline{\psi}_1 + \widetilde{\psi}_1$, where $\overline{\psi}_1$ is the mean and $\widetilde{\psi}_1$ the fluctuation part. An average external potential M, which is conjugate to $\overline{\psi}_1 + \widetilde{\psi}_1$, can be given from the functional differentiation of the free-energy expansion with respect to either $\overline{\psi}_1$ or $\widetilde{\psi}_1$. After Brazovskii's approximate closure relation [55] is employed, the inspection of M yields the following self-consistent equation for a function S^{-1}, which is the correction to Γ'_2 owing to the fluctuation effects as

$$\overline{S}^{-1}(k) = \overline{\Gamma}'_2(k) + \overline{\Gamma}'_4(k^*)\overline{\psi}^2 + \frac{1}{2}\overline{\Gamma}'_4(k^*) \cdot \int d\vec{k}_1 \overline{S}(\vec{k}_1) \quad (A38)$$

Because of the profound minimum of $\overline{\Gamma}'_2$ at k^*, it is expanded around k^* up to the quadratic order as $\overline{\Gamma}'_2(k) = \overline{\Gamma}'_2(k^*) + c(k - k^*)^2$, where the symbol c indicates $c = 1/2 \cdot \partial^2 \overline{\Gamma}'_2 / \partial k^2 \big)_{k^*}$. This expression for $\overline{\Gamma}'_2$ is put into Equation (A38) to yield $\overline{S}^{-1}(k) = \overline{S}^{-1}(k^*) + c(k - k^*)^2$, which helps us to evaluate an ultraviolet divergent integral $\int d\vec{k}_1 \overline{S}(\vec{k}_1)$ as

$$\int d\vec{k}_1 \overline{S}(\vec{k}_1) = (3x^*/N_c\pi)/\sqrt{\overline{S}^{-1}(k^*) \cdot \widetilde{c}} \quad (A39)$$

The new symbol $\widetilde{c}$ is given by $\widetilde{c} = N_c x^*/3 \cdot \partial^2 \overline{\Gamma}'_2 / \partial x^2 \big)_{x^*}$, where $x = k^2 R_G^2$ as before. The corresponding x^* is then evaluated at k^*. The $\overline{S}^{-1}$ at k^* then becomes

$$\overline{S}^{-1}(k^*) = \overline{\Gamma}'_2(k^*) + \overline{\Gamma}'_4(k^*)\zeta^2 + \frac{\overline{\Gamma}'_4(k^*) \cdot 3x^*/2\pi}{N_c \sqrt{\overline{S}^{-1}(k^*) \cdot \widetilde{c}}} \quad (A40)$$

where ζ is the amplitude parameter of $\overline{\psi}_1$. The desired free energy with the inclusion of fluctuation correction is now given from the integration of the approximate expression for M as

$$\frac{\beta \Delta A}{\eta} = \frac{1}{2\overline{\Gamma}'_4}\left(\overline{S}^{-2}(k^*) - \overline{S}_D^{-2}(k^*)\right) + \frac{3x^*/2\pi}{N_c\sqrt{\widetilde{c}}}\left(\sqrt{\overline{S}^{-1}(k^*)} - \sqrt{\overline{S}_D^{-1}(k^*)}\right) - \left|\frac{a_n}{\eta} - \frac{b_1 \overline{\Gamma}_{12}}{\eta \overline{\Gamma}_{22}}\right|\zeta^3 + \left(\frac{d_n}{\eta} - \frac{\overline{\Gamma}'_4}{2}\right)\zeta^4 \quad (A41)$$

where $\overline{S}_D^{-1}$ represents $\overline{S}^{-1}$ in the disordered state, and can thus be obtained if ζ in Equation (A40) is taken to be zero.

References

1. Hadjichristidis, N.; Pispas, S.; Floudas, G.A. *Block Copolymers: Synthetic Strategies, Physical Properties, and Applications*; John Wiley & Sons, Inc.: Hoboken, NJ, USA, 2003.
2. Hamley, I.W. *Developments in Block Copolymer Science and Technology*; John Wiley & Sons Ltd.: Hoboken, NJ, USA, 2004.
3. Bates, F.S.; Fredrickson, G.H. Block Copolymer Thermodynamics: Theory and Experiment. *Annu. Rev. Phys. Chem.* **1990**, *41*, 525–557. [CrossRef] [PubMed]
4. Lodge, T.P. Block Copolymers: Long-Term Growth with Added Value. *Macromolecules* **2020**, *53*, 2–4. [CrossRef]
5. Lodge, T.P. Block Copolymers: Past Successes and Future Challenges. *Macromol. Chem. Phys.* **2003**, *204*, 265–273. [CrossRef]
6. Kipp, D.; Ganesan, V. Influence of Block Copolymer Compatibilizers on the Morphologies of Semiflexible Polymer/Solvent Blends. *J. Phys. Chem. B* **2014**, *118*, 4425–4441. [CrossRef]
7. Jin, C.; Olsen, B.C.; Wu, N.L.Y.; Buriak, J.M. Sequential Nanopatterned Block Copolymer Self-assembly on Surfaces. *Langmuir* **2016**, *32*, 5890–5898. [CrossRef]
8. Kim, B.H.; Kim, J.Y.; Jeong, S.J.; Hwang, J.O.; Lee, D.H.; Shin, D.O.; Choi, S.-Y.; Kim, S.O. Surface Energy Modification by Spin-cast, Large-area Graphene Film for Block Copolymer Lithography. *ACS Nano* **2010**, *4*, 5464–5470. [CrossRef]
9. Wi, D.; Kim, J.; Lee, H.; Kang, N.-G.; Lee, J.; Kim, M.-J.; Lee, J.-S.; Ree, M. Finely Tuned Digital Memory Modes and Performances in Diblock Copolymer Devices by Well-defined Lamellar Structure Formation and Orientation Control. *J. Mater. Chem. C* **2016**, *4*, 2017–2027. [CrossRef]
10. Schacher, F.H.; Rupar, P.A.; Manners, I. Functional Block Copolymers: Nanostructured Materials with Emerging Applications. *Angew. Chem. Int. Ed.* **2012**, *51*, 2–25. [CrossRef]
11. Yasen, W.; Dong, R.; Aini, A.; Zhu, X. Recent advances on supramolecular block copolymers for biomedical applications. *J. Mater. Chem. B* **2020**, *8*, 8219–8231. [CrossRef]

12. Leibler, L. Theory of Microphase Separation in Block Copolymers. *Macromolecules* **1980**, *13*, 1602–1617. [CrossRef]
13. Hashimoto, T. *Thermoplastic Elastomers*; Holden, G., Legge, N.R., Quirk, R.P., Schroeder, H.E., Eds.; Hanser: Cincinnati, OH, USA, 1996.
14. Russell, T.P.; Karis, T.E.; Gallot, Y.; Mayes, A.M. A Lower Critical Ordering Transition in a Diblock Copolymer Melt. *Nature* **1994**, *368*, 729–732. [CrossRef]
15. Ruzette, A.-V.G.; Banerjee, P.; Mayes, A.M.; Pollard, M.; Russell, T.P.; Jerome, R.; Slawecki, T.; Hjelm, R.; Thiyagarajan, P. Phase Behavior of Diblock Copolymers between Styrene and n-Alkyl Methacrylates. *Macromolecules* **1998**, *31*, 8509–8516. [CrossRef]
16. Mansky, P.; Tsui, O.K.C.; Russell, T.P.; Gallot, Y. Phase coherence and microphase separation transition in diblock. *Macromolecules* **1999**, *32*, 4832–4837. [CrossRef]
17. Weidisch, R.; Stamm, M.; Schubert, D.W.; Arnold, M.; Budde, H.; Horing, S. Correlation between Phase Behavior and Tensile Properties of Diblock Copolymers. *Macromolecules* **1999**, *32*, 3405–3411. [CrossRef]
18. Hasegawa, H.; Sakamoto, N.; Taneno, H.; Jinnai, H.; Hashimoto, T.; Schwahn, D.; Frielinghaus, H.; Janben, S.; Imai, M.; Mortensen, K. SANS Studies on Phase Behavior of Block Copolymers. *J. Phys. Chem. Solids* **1999**, *60*, 1307–1312. [CrossRef]
19. Fischer, H.; Weidisch, R.; Stamm, M.; Budde, H.; Horing, S. The Phase Diagram of the System Poly(styrene-block-n-butyl methacrylate). *Colloid Polym. Sci.* **2000**, *278*, 1019–1031. [CrossRef]
20. Ryu, D.Y.; Jeong, U.; Kim, J.K.; Russell, T.P. Closed-Loop Phase Behaviour in Block Copolymers. *Nat. Mater.* **2002**, *1*, 114–117.
21. Ryu, D.Y.; Lee, D.H.; Kim, J.K.; Lavery, K.A.; Russell, T.P.; Han, Y.S.; Seong, B.S.; Lee, C.H.; Thiyagarajan, P. Effect of Hydrostatic Pressure on Closed-Loop Phase Behavior of Block Copolymers. *Phys. Rev. Lett.* **2003**, *90*, 235501. [CrossRef]
22. Ryu, D.Y.; Lee, D.H.; Jang, J.; Kim, J.K.; Lavery, K.A.; Russell, T.P. Complex Phase Behavior of a Weakly Interacting Binary Polymer Blend. *Macromolecules* **2004**, *37*, 5851–5855. [CrossRef]
23. Li, C.; Lee, D.H.; Kim, J.K.; Ryu, D.Y.; Russell, T.P. Closed-Loop Phase Behavior for Weakly Interacting Block Copolymers. *Macromolecules* **2006**, *39*, 5926–5930. [CrossRef]
24. Sanchez, I.C.; Panayiotou, C.G. Equation of State Thermodynamics of Polymer Solutions. In *Thermodynamic Modeling*; Sandler, S., Ed.; Marcel Dekker: New City, NY, USA, 1992.
25. ten Brinke, G.; Karasz, F.E. Lower Critical Solution Temperature Behavior in Polymer Blends: Compressibility and Directional-specific Interactions. *Macromolecules* **1984**, *17*, 815–820. [CrossRef]
26. Sanchez, I.C.; Balazs, A.C. Generalization of the Lattice-Fluid Model for Specific Interactions. *Macromolecules* **1989**, *22*, 2325–2331. [CrossRef]
27. Cho, J.; Kwon, Y.K. Mean—Field and Fluctuation Correction Analyses for a Diblock Copolymer Melt Exhibiting an Immiscibility Loop. *J. Polym. Sci. Part B Polym. Phys.* **2003**, *41*, 1889–1896. [CrossRef]
28. Krishnamoorti, R.; Graessley, W.W.; Fetters, L.J.; Garner, R.T.; Lohse, D.J. Anomalous Mixing Behavior of Polyisobutylene with Other Polyolefins. *Macromolecules* **1995**, *28*, 1252–1259. [CrossRef]
29. Mulhearn, W.D.; Register, R.A. Lower Critical Ordering Transition of an All-Hydrocarbon Polynorbornene Diblock Copolymer. *ACS Macro Lett.* **2017**, *6*, 808–812. [CrossRef]
30. Luettmer-Strathmann, J.; Lipson, J.E.G. Miscibility of Polyolefin Blends. *Macromolecules* **1999**, *32*, 1093–1102. [CrossRef]
31. Dudowicz, J.; Freed, K.F.; Douglas, J.F. Beyond Flory-Huggins Theory: New Classes of Blend Miscibility Associated with Monomer Structural Asymmetry. *Phys. Rev. Lett.* **2002**, *88*, 095503. [CrossRef] [PubMed]
32. Hajduk, D.A.; Urayama, P.; Gruner, S.M.; Erramilli, S.; Register, R.A.; Brister, K.; Fetters, L.J. High Pressure Effects on the Disordered Phase of Block Copolymer Melts. *Macromolecules* **1995**, *28*, 7148–7156. [CrossRef]
33. Hajduk, D.A.; Gruner, S.M.; Erramilli, S.; Register, R.A.; Fetters, L.J. High-Pressure Effects on the Order-Disorder Transition in Block Copolymer Melts. *Macromolecules* **1996**, *29*, 1473–1481. [CrossRef]
34. Ladynski, H.; Odorico, D.; Stamm, M. Effect of Pressure on the Microphase Separation of the Symmetric Diblock Copolymer Poly(styrene-b-butadiene). *J. Non-Cryst. Sol.* **1998**, *235*, 491–495. [CrossRef]
35. Ruzette, A.-V.; Mayes, A.M.; Pollard, M.; Russell, T.P.; Hammouda, B. Pressure Effects on the Phase Behavior of Styrene/n-Alkyl Methacrylate Block Copolymers. *Macromolecules* **2003**, *36*, 3351–3356. [CrossRef]
36. Ahn, H.; Ryu, D.Y.; Kim, Y.M.; Kwon, K.W.; Lee, J.M.; Cho, J. Phase Behavior of Polystyrene-b-poly(methyl methacrylate) Diblock Copolymer. *Macromolecules* **2009**, *42*, 7897–7902. [CrossRef]
37. Hammouda, B.; Ho, D.; Kline, S. SANS from Poly(ethylene oxide)/Water Systems. *Macromolecules* **2002**, *35*, 8578–8585. [CrossRef]
38. Gonzalez-Leon, J.A.; Acar, M.H.; Ryu, S.W.; Ruzette, A.-V.; Mayes, A.M. Low-Temperature Processing of Baroplastics by Pressure-Induced Flow. *Nature* **2003**, *426*, 424–428. [CrossRef]
39. Pollard, M.; Russell, T.P.; Ruzette, A.-V.; Mayes, A.M.; Gallot, Y. The Effect of Hydrostatic Pressure on the Lower Critical Ordering Transition in Diblock Copolymers. *Macromolecules* **1998**, *31*, 6493–6498. [CrossRef]
40. Lee, J.; Wang, T.; Shin, K.; Cho, J. High-pressure neutron scattering and random-phase approximation analysis of a molten Baroplastic diblock copolymer. *Polymer* **2019**, *175*, 265–271. [CrossRef]
41. Cho, J. Analysis of Phase Separation in Compressible Polymer Blends and Block Copolymers. *Macromolecules* **2000**, *33*, 2228–2241. [CrossRef]
42. Cho, J. Effective Flory Interaction Parameter and Disparity in Equation-of-State Properties for Block Copolymers. *Polymer* **2007**, *48*, 429–431. [CrossRef]
43. Cho, J. Blends of Two Diblock Copolymers with Opposite Phase Behaviors. *Macromol. Theory Simul.* **2014**, *23*, 442–451. [CrossRef]

44. Cho, J. Microphase Separation upon Heating in Diblock Copolymer Melts. *Macromolecules* **2001**, *34*, 1001–1012. [CrossRef]
45. Cho, J. Analysis of Compressible Diblock Copolymer Melts That Microphase Separate upon Cooling. *Macromolecules* **2001**, *34*, 6097–6106. [CrossRef]
46. Cho, J. A Landau Free Energy for Diblock Copolymers with Compressibility Difference between Blocks. *J. Chem. Phys.* **2003**, *119*, 5711–5721. [CrossRef]
47. Cho, J. Superposition in Flory-Huggins χ and Interfacial Tension for Compressible Polymer Blends. *ACS Macro Lett.* **2013**, *2*, 544–549. [CrossRef]
48. Cho, J. Pressure Effects on Nanostructure Development of ABC Copolymers. *Polymer* **2016**, *97*, 589–597. [CrossRef]
49. Cho, J. Identification of Some New Triply Periodic Mesophases from Molten Block Copolymers. *Polymers* **2019**, *11*, 1081. [CrossRef]
50. Chaikin, P.M.; Lubensky, T.C. *Principles of Condensed Matter Physics*; Cambridge University Press: Cambridge, UK, 1995.
51. Baxter, R.J. Percus–Yevick Equation for Hard Spheres with Surface Adhesion. *J. Chem. Phys.* **1968**, *49*, 2770. [CrossRef]
52. Barboy, B. Solution of the compressibility equation of the adhesive hard-sphere model for mixtures. *Chem. Phys.* **1975**, *11*, 357. [CrossRef]
53. Chiew, Y.C. Percus-Yevick integral-equation theory for athermal hard-sphere chains. *Molec. Phys.* **1990**, *70*, 129–143. [CrossRef]
54. Huang, K. *Statistical Mechanics*, 2nd ed.; John Wiley & Sons: Hoboken, NJ, USA, 1987.
55. Brazovskii, S.A. Phase transition of an isotropic system to a nonuniform state. *Sov. Phys.-JETP* **1975**, *41*, 85.
56. Fredrickson, G.H.; Helfand, E. Fluctuation effects in the theory of microphase separation in block copolymers. *J. Chem. Phys.* **1987**, *87*, 697–705. [CrossRef]
57. Roe, R.J.; Zin, W.C. Determination of the Polymer-Polymer Interaction Parameter for the Polystyrene-Polybutadiene Pair. *Macromolecules* **1980**, *13*, 1221–1228. [CrossRef]
58. Zin, W.C.; Roe, R.J. Phase equilibria and transition in mixtures of a homopolymer and a block copolymer. 1. Small-angle x-ray scattering study. *Macromolecules* **1984**, *17*, 183–188. [CrossRef]
59. Roe, R.J.; Zin, W.C. Phase equilibria and transition in mixtures of a homopolymer and a block copolymer. 2. Phase diagram. *Macromolecules* **1984**, *17*, 189–194. [CrossRef]
60. Han, C.C.; Bauer, B.J.; Clark, J.C.; Muroga, Y.; Matsushita, Y.; Okada, M.; Tran-cong, Q.; Chang, T.; Sanchez, I.C. Temperature, composition and molecularweight dependence of the binary interaction parameter of polystyrene/poly(vinyl methyl ether) blends. *Polymer* **1988**, *29*, 2002–2014. [CrossRef]
61. Green, M.M.; White, J.L.; Mirau, P.; Scheinfeld, M.H. C-H to O Hydrogen Bonding: The Attractive Interaction in the Blend between Polystyrene and Poly(vinyl methyl ether). *Macromolecules* **2006**, *39*, 5971–5973. [CrossRef]
62. Cho, J. Study on the χ Parameter for Compressible Diblock Copolymer Melts. *Macromolecules* **2002**, *35*, 5697–5706. [CrossRef]
63. Rudolf, B.; Cantow, H.-J. Description of Phase Behavior of Polymer Blends by Different Equation-of-State Theories. 1. Phase Diagrams and Thermodynamic Reasons for Mixing and Demixing. *Macromolecules* **1995**, *28*, 6586–6594. [CrossRef]
64. Schwahn, D.; Frielinghaus, H.; Mortensen, K.; Almdal, K. Abnormal Pressure Dependence of the Phase Boundaries in PEE-PDMS and PEP-PDMS Binary Homopolymer Blends and Diblock Copolymers. *Macromolecules* **2001**, *34*, 1694–1706. [CrossRef]
65. Cho, J. Concentration fluctuation effects on the phase behavior of compressible diblock copolymers. *J. Chem. Phys.* **2004**, *120*, 9831–9840. [CrossRef]

Disclaimer/Publisher's Note: The statements, opinions and data contained in all publications are solely those of the individual author(s) and contributor(s) and not of MDPI and/or the editor(s). MDPI and/or the editor(s) disclaim responsibility for any injury to people or property resulting from any ideas, methods, instructions or products referred to in the content.

Communication

Small-Angle X-ray Scattering Analysis on the Estimation of Interaction Parameter of Poly(*n*-butyl acrylate)-*b*-poly(methyl methacrylate)

Sang-In Lee [1,2], Min-Guk Seo [1], June Huh [3,4,*] and Hyun-jong Paik [1,*]

1 Department of Polymer Science and Engineering, Pusan National University, Busan 46241, Republic of Korea
2 LX MMA R&D Center, 188, Munji-ro, Yuseong-gu, Daejeon 34122, Republic of Korea
3 Department of Chemical and Biological Engineering, Korea University, Seoul 02841, Republic of Korea
4 Department of Life Sciences, Korea University, Seoul 02841, Republic of Korea
* Correspondence: junehuh@korea.ac.kr (J.H.); hpaik@pusan.ac.kr (H.-j.P.)

Abstract: The temperature dependence of the Flory–Huggins interaction parameter χ for poly(*n*-butyl acrylate)-*b*-poly(methyl methacrylate) (PBA-*b*-PMMA) was quantified from small-angle X-ray scattering (SAXS) analysis using random phase approximation (RPA) theory. It was found from the χ estimation ($\chi = 0.0103 + 14.76/T$) that the enthalpic contribution, χ_H, a measure for temperature susceptibility of χ, is 1.7–4.5 folds smaller for PBA-*b*-PMMA than for the conventional styrene-diene-based block copolymers, which have been widely used for thermoplastic elastomers. This finding suggests that these fully acrylic components can be a desirable chemical pair for constituting terpolymers applied for thermally stable and mechanically resilient elastomers.

Keywords: acrylic block copolymer; Flory–Huggins interaction parameter; thermoplastic elastomer; small-angle X-ray scattering

1. Introduction

The development of technologies of ligated anionic polymerization (LAP) and controlled radical polymerization (CRP) have allowed the commercialization of novel block copolymers with well-defined molecular architectures [1–4]. Various block copolymers are being produced using LAP and nitroxide-mediated polymerization (NMP) technology, the representative example being Kurarity by Kuraray and Nanostrength by Arkema. In particular, fully acrylic block copolymers [2,5–10], comprised of a rubbery poly(alkyl acrylate) block and glassy poly(alkyl methacrylate) block, have attracted steady interest in the thermoplastic elastomer (TPE) community because of the much higher weather resistance of poly(meth)acrylates when compared to traditional diene-based TPEs [11–13]. Well-designed triblock copolymers consisting of two terminal blocks with glassy segments and a middle block with rubbery segments can form glassy nanodomains embedded in a rubbery matrix, where chains bridging between two different glassy domains, which can efficiently persist the mechanical deformation, play a critical role for the overall mechanical properties of TPE. In this respect, the interaction between block components, usually quantified by the Flory–Huggins interaction parameter χ, is of critical importance in controlling the conformational and morphological behavior, which determines the mechanical performance of TPE.

Among many possible acrylic components, the block copolymers comprised of an *n*-butyl acrylate (BA) block and a methyl methacrylate (MMA) block are one of the promising species for TPE owing to their excellent properties, such as light sensitivity and oxidation stability. Previously, several studies have been carried out to synthesize PBA-PMMA copolymers with various chain architectures [2,5–10,14–16], some of which have reported their methods for the generation of elastomers and thermomechanical properties. Nonetheless, no works have been conducted for the temperature dependence of the interaction parameter $\chi(T)$ between PBA and PMMA despite its importance for thermomechanical properties, which are closely related to conformational and morphological behavior via $\chi(T)$. For instance, it has been theoretically reported that the fraction of bridging conformations between nanodomains, which contributes favorably to the mechanical performance of TPE, is proportional to $\chi^{-1/9}$ [17]. This prediction suggests that introducing a constituent block pair with small χ with weak temperature dependence is advantageous for developing a TPE with enhanced thermal stability.

In this paper, we report the temperature dependence of the Flory–Huggins interaction parameter $\chi(T)$ between the two acrylic monomer species of BA and MMA by small-angle X-ray scattering (SAXS) measurements of a molten PBA-b-PMMA block copolymer using a random phase approximation (RPA) analysis. To do this, we synthesized the PBA-b-PMMA diblock copolymer with a carefully chosen molecular weight by ATRP, which allowed measuring the order-disorder transition (ODT) temperature within the investigated temperature range.

2. Materials and Methods

2.1. Materials

Butyl acrylate (TCI, 99%), methyl methacrylate (DAEJUNG CHEMICALS & METALS, Siheung, Korea 99.5%), and styrene (Junsei, Tokyo, Japan, 99.5%) were purified by passing through a basic alumina column before use. Copper(I) bromide (CuBr, Sigma-Aldrich, Seoul, Korea, 98.0%) and copper(I) chloride (CuCl, Sigma-Aldrich, 98.0%) were purified by stirring with glacial acetic acid, followed by filtering and washing the resulting solid four times with ethanol. The solid was dried under a vacuum for two days. Anisole (DAEJUNG CHEMICALS & METALS, 98.0%), (1-Bromoethyl)benzene (Sigma-Aldrich, 98%), copper(II) bromide (CuBr2, Sigma-Aldrich, 98.0%), copper(II) chloride (CuCl2, Sigma-Aldrich, 98.0%) and N,N,N',N'',N''-pentamethyldiethylenetriamine (PMDETA, Sigma-Aldrich, 98%), and all other chemicals, internal standards, and solvents were used as received.

2.2. ATRP Synthesis

2.2.1. Synthesis of Poly(n-butyl acrylate) (PBA) Macro Initiator

A dried 100 mL Schlenk flask was charged with CuBr (152.97 mg, 1.07 mmol), CuBr$_2$ (4.86 mg, 0.02 mmol). Then, degassed butyl acrylate (30.0 mL, 208.27 mmol), (1-Bromoethyl)benzene (0.15 mL, 1.09 mmol), PMDETA (0.23 mL, 1.09 mmol) and anisole (20.0 mL) were added to the flask via N$_2$ purged syringes. The mixture was reacted in an oil bath at 90 °C. At 40% monomer conversion (measured by gas chromatography), the polymerization was quenched by THF and exposed to air. Then, the reaction mixture was passed through a neutral alumina column to remove the Cu catalyst. After being concentrated using a rotary evaporator, the solution was dropped into methanol/water (8:2) to obtain the polymer as a precipitate. The molecular weight of the resulting polymer in SEC was determined as M_n = 12,460 g/mol and M_w/M_n = 1.084, where M_n and M_w are the number-averaged and the weight-averaged molecular weight, respectively.

2.2.2. Synthesis of Poly(n-butyl acrylate)-b-poly(methyl methacrylate) (PBA-b-PMMA) Diblock Copolymer

A dried 50 mL Schlenk flask was charged with CuCl (122.0 mg, 0.91 mmol) and CuCl$_2$ (7.50 mg, 0.02 mmol). After being sealed with a glass stopper, the flask was evacuated and back-filled with N$_2$ three times. Then, degassed methyl methacrylate and PBA Macro initiator (1 g, 0.09 mmol) resolved in anisole (10.0 mL) were added to the flask via N$_2$ purged syringes. The mixture was reacted in an oil bath at 85 °C. The polymerizations were quenched by THF and exposed to air. Then, the reaction mixtures were passed through a neutral alumina column to remove the Cu catalyst. After being concentrated using a rotary evaporator, the solutions were dropped into methanol to obtain the polymer as precipitates. After filtration and drying under a vacuum, the polymers were isolated as white powder. The molecular weight of the resulting polymer in SEC was determined as M_n = 23,470 g/mol and M_w/M_n = 1.151.

2.2.3. Conversion Analysis

In all the ATRP stages, monomer conversion was confirmed by HP 5890 gas chromatography (GC) equipped with an HP101 column. The number-averaged molecular weight and the dispersity were determined by SEC calibrated with PMMA standards.

2.3. SAXS Measurements

SAXS measurements were performed at the 4C SAXS II beamline of the Pohang Light Source II (PLS II) with 3 GeV power at the POSTECH. A sample-to-detector distance of 3.00 m was used for SAXS. SAXS profiles were measured in situ at each temperature with the SAXS apparatus described elsewhere. The sample was placed in the sample chamber filled with nitrogen gas, and the temperature was controlled with an accuracy of +/−0.03 °C. Since SAXS measurements were conducted in the temperature range of 170–230 °C, precise temperature control was required for block copolymer samples to avoid the possible thermal degradation of PMMA [18,19]. GPC found no mass loss after SAXS measurements of samples. The profiles were desmeared for slit-height and slit-width effects and corrected for absorption, air scattering, and thermal diffuse scattering as described elsewhere [20]. The SAXS measurements were conducted during a heating process that started at 130 °C. During the heating run, the PBA-*b*-PMMA sample was maintained for 1h at a specific temperature to obtain thermal equilibrium as much as possible.

3. Results

Figure 1a presents the SAXS profiles of PBA-*b*-PMMA obtained at various temperatures covering both disordered and ordered regimes. As seen in Figure 1, the broad SAXS curves in the high-temperature regimes become sharply peaked below T = 180 °C, at which point the PBA-*b*-PMMA undergoes the order-disorder transition. This is manifested more clearly in Figure 1b, where the inverse of the maximum scattering intensity (I_m^{-1}) and the half-width at half-maximum (σ_q^2) are plotted against the reciprocal temperature (T^{-1}), where a jump-like discontinuity of I_m^{-1} and σ_q^2 at 180 °C can be seen.

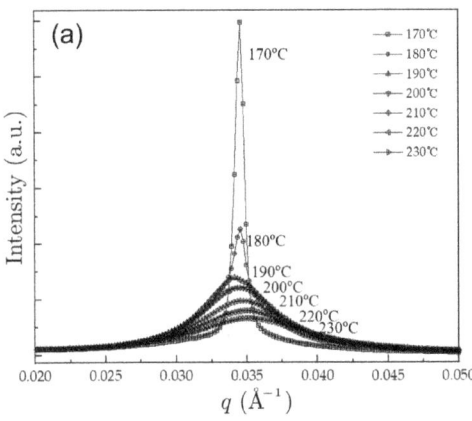
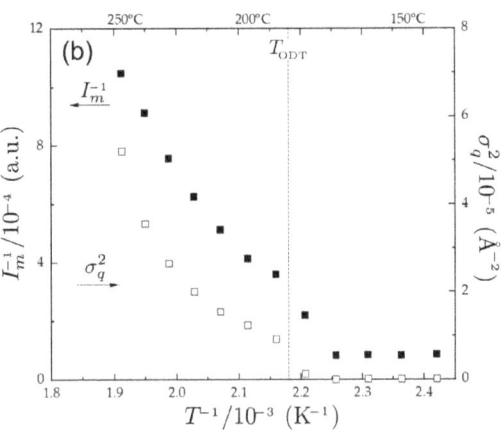

Figure 1. (a) SAXS profiles at various temperatures and (b) the inverse of the maximum scattering intensity (I_m^{-1}) and the half-width at half-maximum (σ_q^2) versus the inverse of the temperature for PBA-*b*-PMMA. The dashed line in (b) indicates the temperature at ODT, T_{ODT}.

Having identified the disordered and the ordered regime by locating the temperature at ODT (T_{ODT}), the temperature-dependence of χ parameters between PBA and PMMA was estimated by RPA analysis at $T > T_{MF}$, where $T > T_{MF}$ is the crossover temperature from a mean-field disordered regime to the non-mean-field regime [21–23]. The RPA states that the scattered intensity $I(q)$ at scattering vector q in the disordered state is linearly proportional to $[\Gamma(q, R_1, R_2) - 2\chi]^{-1}$ where Γ is referred to as the second vertex function related to the single chain density correlation functions in the ideal state, and R_α is the root-mean-square radius of gyration of the component (α = 1: PBA, α = 2: PMMA). The estimation of χ at each temperature was then obtained by fitting $I(q)/I_m$ using the fitting function $F(q; R_1, R_2, \chi) = \mathcal{N}_{min}[\Gamma(q, R_1, R_2) - 2\chi]^{-1}$ with the fitting parameters R_1, R_2 and χ for a given set of molecular parameters (Table 1), where $\mathcal{N}_{min}$ is the minimum value of $\Gamma(q, R_1, R_2) - 2\chi$. The complete formulas for the fitting function with molecular parameters are documented in Appendix A.

Table 1. Molecular parameters of PBA-*b*-PMMA used for fitting with the fitting function $F(q; R_1, R_2, \chi) = \mathcal{N}_{min}[\Gamma(q, R_1, R_2) - 2\chi]^{-1}$ using Equations (A1)–(A3).

Volume Fraction of PBA Block (f_1)	Dispersity of PBA Block (λ_1)	Dispersity of PMMA Block (λ_2)	Total Number of unit Segments ($\bar{N}$)
0.553	1.084	1.579 [1]	208.2 [2]

[1] Parameterized by $\lambda - 1 = (\lambda_1 - 1)w_1^2 + (\lambda_2 - 1)w_2^2$ with $\lambda = 1.151$, $\lambda_1 = 1.084$, and $w_1 = 0.53 = 1 - w_2$, where w_α is the weight fraction of the component α. [2] Parameterized by $\bar{N} = (v_1 N_1 + v_2 N_2)/(v_1 v_2)^{1/2}$ with $N_1 = 97.3$, $N_2 = 110.1$, $v_1 = 118.7$ cm^3/mol and $v_2 = 84.7$ cm^3/mol, where N_α and v_α are the number-averaged degree of polymerization and the molar volume of the component α, respectively.

Figure 2 shows the normalized scattering profiles $I(q)/I_m$ fitted to $F(q; R_1, R_2, \chi)$ at various temperatures in the disordered state, by which the temperature dependence $\chi(T)$ can be obtained assuming the linear relation $\chi = \chi_S + \chi_H/T$ in the mean-field regime where χ_S and χ_H are the temperature-independent constants associated with the entropic and enthalpic contribution to the overall χ [23–26]. It is worth commenting that although the peak fits well to the RPA model in the region $q > 0.03$ Å^{-1}, the agreement in the low q region is poor, which is also observed in the similar previous works on the χ measurements using SAXS [27–29]. While this could be due to the slope in the SAXS data or background mismatch, there are also errors due to the polydispersity of the block copolymer and the conformational asymmetry between blocks [27]. As for the present work, since the corrections for polydispersity and asymmetry were taken into account in our RPA equations, it is likely that this error in the low q region is mainly due to the slope in the SAXS data. Nonetheless, the estimation of χ, which uses the inverse proportionality between χ and the width of scattering profile in the q region at 0.03 Å^{-1} < q < 0.04 Å^{-1}, is unaffected because the peak width is independent of the slope in the SAXS.

Figure 3 shows the temperature dependence of the estimated χ values and the characteristic length $2\pi/q*$, where $q*$ is the dominant wave vector in the scattering profile such that $I_m = I(q*)$. The two plateau regions in the plot of $T^{-1} - 2\pi/q*$ also identify the mean-field ($T > T_{MF}$) and the ordered region ($T < T_{ODT}$), respectively. The resultant temperature dependence $\chi(T)$, which was obtained by the linear regression in the mean-field regime, was found to be

$$\chi = (0.0103 \pm 0.0031) + \frac{(14.76 \pm 1.64)}{T} \quad (1)$$

where the thermodynamic temperature is used for T.

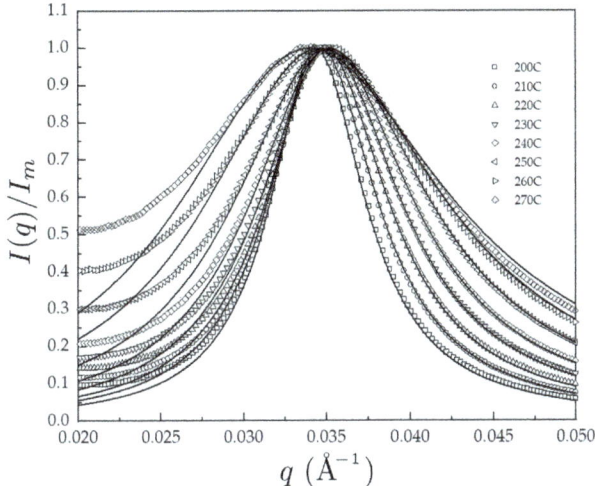

Figure 2. The normalized scattering profiles $I(q)/I_m$ fitted to $F(q; R_1, R_2, \chi)$ at various temperatures in the disordered state for PBA-b-PMMA.

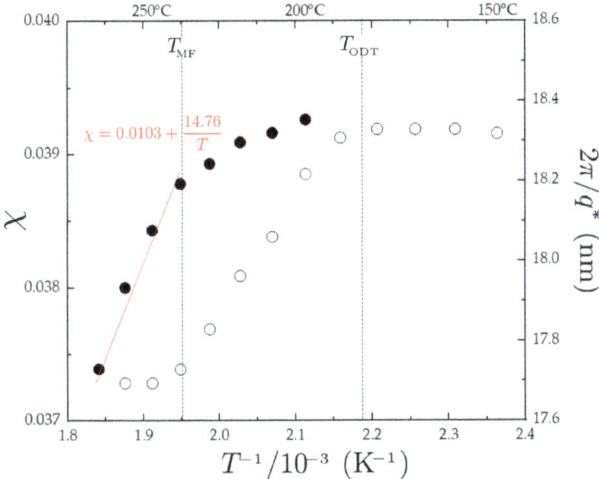

Figure 3. Temperature dependence of χ values and the characteristic length ($2\pi/q*$) for PBA-b-PMMA. The mean-field ($T > T_{\text{MF}}$) and the ordered region ($T < T_{\text{ODT}}$) are identified by two plateau regions in the $T^{-1} - 2\pi/q*$ plot. The red solid line represents the linear regression of (T^{-1}, χ) data in the mean region.

It is of interest to compare our estimation of $\chi(T)$ for PBA-b-PMMA to those for other diene-based block copolymer systems widely used for TPE, such as polystyrene-polybutadiene-polystyrene (SBS) or polystyrene-polyisoprene-polystyrene (SIS) triblock copolymers. Table 2 compares some of the available $\chi(T)$ measured for the polystyrene-b-polybutadiene diblock (PS-b-PB) and the polystyrene-b-polyisoprene diblock (PS-b-PI) to that of our system of PBA-b-PMMA.

Table 2. $\chi(T)$ reported for various diene-based block copolymers compared to that of PBA-b-PMMA.

System	$\chi(T)$	Ref
PS-b-PB [1]	$-0.021 + 25/T$	[30]
PS-b-PB [2]	$-0.027 + 28/T$	[31]
PS-b-PI [3]	$-0.0090 + 25/T$	[30]
PS-b-PI [4]	$-0.0937 + 66/T$	[32]
PBA-b-PMMA	$0.0103 + 14.76/T$	

[1] 1,2 and 1,4 diene microstructures are 95% and 5%, respectively. [2] Diene microstructures are unspecified. [3] 1,4 and 3,4 diene microstructures are 93% and 7%, respectively. [4] 1,2, 1,4 and 3,4 diene microstructures are 38%, 3% and 59%, respectively.

As shown in Table 2, the temperature dependence of χ, which can be characterized by the enthalpic contribution χ_H, is weaker for the PBA-b-PMMA block copolymer than that for diene-based block copolymers. For instance, χ values in the temperature range of 100–200 °C are 0.042–0.050 for PBA-b-PMMA, 0.032–0.046 or 0.032–0.048 for PS-b-PB, and 0.062–0.076 or 0.046–0.083 for PS-b-PI, which suggests less temperature-dependent conformational and morphological behavior for PBA-PMMA-based block copolymers than those for the conventional styrene-diene-based block copolymers.

4. Conclusions

In conclusion, we report the temperature dependence of the Flory–Huggins interaction parameter $\chi(T)$ between BA and MMA components by analyzing SAXS measurements fitted to RPA equations for a molten PBA-b-PMMA diblock at various temperatures. It was found from the χ estimation that the BA-MMA interaction is less susceptible to temperature than the interactions of styrene-diene-based block copolymers such as SIS and SBS triblock copolymers used commercially for TPE applications. Since the fraction of bridging conformation of triblock copolymer, a key for mechanically resilient TPE, is proportional to $\chi^{-1/9}$, this weaker temperature dependence of χ between BA and MMA provides a desirable option when designing terpolymers for developing thermally stable TPE.

Author Contributions: Conceptualization, S.-I.L., J.H. and H.-j.P.; methodology, S.-I.L., M.-G.S., J.H. and H.-j.P.; software, S.-I.L. and J.H.; validation, J.H. and H.-j.P.; formal analysis, S.-I.L. and J.H.; investigation, S.-I.L., M.-G.S., J.H. and H.-j.P.; resources, S.-I.L., M.-G.S., J.H. and H.-j.P.; data curation, S.-I.L. and J.H.; writing—original draft preparation, S.-I.L. and J.H.; writing—review and editing, J.H.; visualization, S.-I.L. and J.H.; supervision, J.H. and H.-j.P.; project administration, J.H. and H.-j.P.; funding acquisition, J.H. and H.-j.P. All authors have read and agreed to the published version of the manuscript.

Funding: This research was funded by the National Research Foundation (NRF) of Korea via grant number 2020R1A4A2002903, 2020R1F1A1065951, 2021R1A2B5B01002081 and by the Creative Materials Discovery Program funded by the Ministry of Science and ICT via grant number NRF-2018M3D1A1058536.

Institutional Review Board Statement: Not applicable.

Data Availability Statement: The data presented in this study are available on request from the corresponding author.

Acknowledgments: S.-I.L. acknowledges the support from LX MMA R&D Center.

Conflicts of Interest: The authors declare no conflict of interest.

Appendix A

Here, we briefly summarize the fitting function $F(q; R_1, R_2, \chi) = \mathcal{N}_{min}[\Gamma(q, R_1, R_2) - 2\chi]^{-1}$, which was used for the fitting SAXS data, $I(q)/I_m$, to extract χ of PBA-b-PMMA. The fitting function is based on the second-order vertex function $\Gamma(q)$ related to the single chain correlation functions between the component α and β in the ideal binary system, $g_{\alpha\beta}$ (α,β = 1 (PBA), 2 (PMMA)) in the RPA equation [33]:

$$\Gamma(q) = \frac{\sum_{\alpha\beta} g_{\alpha\beta}(q)}{|g_{\alpha\beta}|} \tag{A1}$$

The single chain correlation functions between the same species (g_{11}, g_{22}) and between the unlike species (g_{12}) for ideal diblock copolymer are given as

$$g_{11}(q) = \frac{2\tilde{N}f_1^2}{x_1^2}\left[x_1 - 1 + u(\lambda_1, x_1)\right] \tag{A2}$$

$$g_{12}(q) = g_{21}(q)\frac{2\tilde{N}f_1f_2}{x_1x_2}\left[1 - u(\lambda_1, x_1)\right]\left[1 - u(\lambda_2, x_2)\right] \tag{A3}$$

and g_{22} is obtained by switching the indices of species from 1 to 2 in Equation (A2). In Equations (A2) and (A3), $\tilde{N}$ is the number of segments (or the number of reference volumes) in the diblock, f_α is the volume fraction of the component α, $x_\alpha = q^2R_\alpha^2$ is the square of the scaled wave vector where R_α is the root-mean-square radius of gyration of the component α, and $u(\lambda_\alpha, x_\alpha)$ is the function for taking into account the dispersity λ_α of the component α, which is given by [27,34]

$$u(\lambda_\alpha, x_\alpha) = \left[x_\alpha(\lambda_\alpha - 1) + 1\right]^{-\frac{1}{\lambda_\alpha - 1}} \tag{A4}$$

The total number of segments, $\tilde{N}$, can be further expressed by $\tilde{N} = (v_1N_1 + v_2N_2)/(v_1v_2)^{1/2}$, where N_α and v_α are the number-averaged degrees of polymerization and the molar volume of the component α, respectively, and given for the present PBA-b-PMMA system as N_1 = 97.3, N_2 = 110.1, v_1 = 118.7 cm^3/mol and v_2 = 84.7 cm^3/mol. Lastly, the dispersity of the second block component (2:PMMA) in the ATRP, λ_2, was obtained by using the rule for the sum of variances of the distributions: $\lambda - 1 = (\lambda_1 - 1)w_1^2 + (\lambda_2 - 1)w_2^2$ [35]. Here, λ and λ_1 are the overall dispersities of the block copolymer, the dispersity of the first block component (1:PBA), and w_α is the weight fraction of the component α, respectively, which are given for the present PBA-b-PMMA system as λ = 1.151, λ_1 = 1.084, and w_1 = 0.53.

References

1. Braunecker, W.A.; Matyjaszewski, K. Controlled/living Radical Polymerization: Features, Developments, and Perspectives. *Prog. Polym. Sci.* **2007**, *32*, 93–146. [CrossRef]
2. Tong, J.D.; Moineau, G.; Leclere, P.; Brédas, J.L.; Lazzaroni, R.; Jérôme, R. Synthesis, Morphology, and Mechanical Properties of Poly(methyl methacrylate)-b-poly(n-butyl acrylate)-b-poly(methyl methacrylate) Triblocks: Ligated Anionic Polymerization vs. Atom Transfer Radical Polymerization. *Macromolecules* **2000**, *33*, 470–479. [CrossRef]
3. Destarac, M. Controlled Radical Polymerization: Industrial Stakes, Obstacles and Achievements. *Macromol. React. Eng.* **2010**, *4*, 165–179. [CrossRef]
4. Matyjaszewski, K. Atom Transfer Radical Polymerization (ATRP): Current Status and Future Perspectives. *Macromolecules* **2012**, *45*, 4015–4039. [CrossRef]
5. Shipp, D.A.; Wang, J.L.; Matyjaszewski, K. Synthesis of Acrylate and Methacrylate Block Copolymers Using Atom Transfer Radical Polymerization. *Macromolecules* **1998**, *31*, 8005–8008. [CrossRef]
6. Moineau, C.; Minet, M.; Teyssié, P.; Jérôme, R. Synthesis and Characterization of Poly(methyl methacrylate)-block-poly(n-butyl acrylate)-block-poly(methyl methacrylate) Copolymers by Two-Step Controlled Radical Polymerization (ATRP) Catalyzed by NiBr2(PPh3)2, 1. *Macromolecules* **1999**, *32*, 8277–8282. [CrossRef]
7. Garcia, M.F.; de la Fuente, J.L.; Fernandez-Sanz, M.; Madruga, E.L. The Importance of Solvent Polar Character on the Synthesis of PMMA-b-PBA Block Copolymers by Atom Transfer Radical Polymerization. *Polymer* **2001**, *42*, 9405–9412. [CrossRef]
8. Guillaneuf, Y.; Gigmes, D.; Marque, S.R.; Astolfi, P.; Greci, L.; Tordo, P.; Bertin, D. First Effective Nitroxide-Mediated Polymerization of Methyl Methacrylate. *Macromolecules* **2007**, *40*, 3108–3114. [CrossRef]

9. Tran, T.A.; Leonardi, F.; Bourrigaud, S.; Gerard, P.; Derail, C. All Acrylic Block Copolymers Based on Poly(methyl methacrylate) and Poly(butyl acrylate): A Link between the Physico-Chemical Properties and the Mechanical Behaviour on Impact Tests. *Polym. Test.* **2008**, *27*, 945–950. [CrossRef]
10. Roos, S.G.; Müller, A.H.; Matyjaszewski, K. Copolymerization of n-Butyl Acrylate with Methyl Methacrylate and PMMA Macromonomers: Comparison of Reactivity Ratios in Conventional and Atom Transfer Radical Copolymerization. *Macromolecules* **1999**, *32*, 8331–8335. [CrossRef]
11. Holden, G.; Bishop, E.; Legge, N.R. Thermoplastic Elastomers. *J. Polym. Sci. Part C Polym. Symp.* **1969**, *26*, 37–57. [CrossRef]
12. Holden, G.L.; Legge, N.R.; Quirk, R.P.; Schroeder, H.E. *Thermoplastic Elastomers*; Hanser: Munich, Germany, 1996.
13. Fetters, L.J.; Morton, M. Synthesis and Properties of Block Polymers. I. Poly(α-methyl styrene)-b-polyisoprene-b-poly(α-methylstyrene). *Macromolecules* **1969**, *2*, 453–458. [CrossRef]
14. Dufour, B.; Koynov, K.; Pakula, T.; Matyjaszewski, K. PBA–PMMA 3-Arm Star Block Copolymer Thermoplastic Elastomers. *Macromol. Chem. Phys.* **2008**, *209*, 1686–1693. [CrossRef]
15. Nese, A.; Mosnacek, J.; Juhari, A.; Yoon, J.A.; Koynov, K.; Kowalewski, T.; Matyjaszewski, K. Synthesis, Characterization, and Properties of Starlike Poly(n-butyl acrylate)-b-poly(methyl methacrylate) Block Copolymers. *Macromolecules* **2010**, *43*, 1227–1235. [CrossRef]
16. Zhang, J.; Wang, Z.; Wang, X.; Wang, Z. The Synthesis of Bottlebrush Cellulose-Graft-Diblock Copolymer Elastomers via Atom Transfer Radical Polymerization Utilizing a Halide Exchange Technique. *Chem. Commun.* **2019**, *55*, 13904–13907. [CrossRef]
17. Milner, S.T.; Witten, T.A. Bridging Attraction by Telechelic Polymers. *Macromolecules* **1992**, *25*, 5495–5503. [CrossRef]
18. Korobeinichev, O.P.; Paletsky, A.A.; Gonchikzhapov, M.B.; Glaznev, R.K.; Gerasimov, I.E.; Naganovsky, Y.K.; Shundrina, I.K.; Snegirev, A.Y.; Vinu, R. Kinetics of Thermal Decomposition of PMMA at Different Heating Rates and in a Wide Temperature Range. *Thermochim. Acta* **2019**, *671*, 17–25. [CrossRef]
19. Nikolaidis, A.K.; Achilias, D.S. Thermal Degradation Kinetics and Viscoelastic Behavior of Poly(Methyl Methacrylate)/Organomodified Montmorillonite Nanocomposites Prepared via In Situ Bulk Radical Polymerization. *Polymers* **2018**, *10*, 491. [CrossRef]
20. Bodycomb, J.; Yamaguchi, D.; Hashimoto, T. Observation of a Discontinuity in the Value of I_m^{-1} at the Order-Disorder Transition in Diblock Copolymer/Homopolymer and Diblock Copolymer/Diblock Copolymer Blends. *Polym.J.* **1996**, *28*, 821–824. [CrossRef]
21. Bates, F.S.; Rosedale, J.H.; Fredrickson, G.H. Fluctuation Effects in a Symmetric Diblock Copolymer near the Order–Disorder Transition. *J. Chem. Phys.* **1990**, *92*, 6255–6270. [CrossRef]
22. Abu-Sharkh, B.; AlSunaidi, A. Morphology and Conformation Analysis of Self-Assembled Triblock Copolymer Melts. *Macromol. Theory Simul.* **2006**, *15*, 507–515. [CrossRef]
23. Sakamoto, N.; Hashimoto, T. Order-Disorder Transition of Low Molecular Weight Polystyrene-block-polyisoprene: 1. SAXS Analysis of Two Characteristic Temperatures. *Macromolecules* **1995**, *28*, 6825–6834. [CrossRef]
24. Schwahn, D.; Willner, L. Phase Behavior and Flory-Huggins Interaction Parameter of Binary Polybutadiene Copolymer Mixtures with Different Vinyl Content and Molar Volume. *Macromolecules* **2002**, *35*, 239–247. [CrossRef]
25. Russell, T.P.; Hjelm, R.P., Jr.; Seeger, P.A. Temperature Dependence of the Interaction Parameter of Polystyrene and Poly(methyl methacrylate). *Macromolecules* **1990**, *23*, 890–893. [CrossRef]
26. Zhao, Y.; Sivaniah, E.; Hashimoto, T. SAXS Analysis of the Order-Disorder Transition and the Interaction Parameter of Polystyrene-block-poly(methyl methacrylate). *Macromolecules* **2008**, *41*, 9948–9951. [CrossRef]
27. Sakurai, S.; Mori, K.; Okawara, A.; Kimishima, K.; Hashimoto, T. Evaluation of Segmental Interaction by Small-Angle X-ray Scattering Based on the Random-Phase Approximation for Asymmetric, Polydisperse Triblock Copolymers. *Macromolecules* **1992**, *25*, 2679–2691. [CrossRef]
28. Zha, W.; Han, C.D.; Lee, D.H.; Han, S.H.; Kim, J.K.; Kang, J.H.; Park, C. Origin of the Difference in Order-Disorder Transition Temperature between Polystyrene-block-poly(2-vinylpyridine) and Polystyrene-block-poly(4-vinylpyridine) Copolymers. *Macromolecules* **2007**, *40*, 2109–2119. [CrossRef]
29. Tanaka, H.; Hashimoto, T. Thermal Concentration Fluctuations of Block Polymer/Homopolymer Mixtures in the Disordered State. 1. Binary Mixtures of SI/HS. *Macromolecules* **1991**, *24*, 5398–5407. [CrossRef]
30. Anastasiadis, S.H.; Gancarz, I.; Koberstein, J.T. Interfacial Tension of Immiscible Polymer Blends: Temperature and Molecular Weight Dependence. *Macromolecules* **1988**, *21*, 2980–2987. [CrossRef]
31. Hewel, M.; Ruland, W. Microphase Separation Transition in Block Copolymer Melts. *Makromol. Chem. Macromol. Symp.* **1986**, *4*, 197–202. [CrossRef]
32. Mori, K.; Hasegawa, H.; Hashimoto, T. Small-Angle X-Ray Scattering from Bulk Block Polymers in Disordered State. Estimation of χ-Values from Accidental Thermal Fluctuations. *Polym. J.* **1985**, *17*, 799–806. [CrossRef]
33. Leibler, L. Theory of Microphase Separation in Block Copolymers. *Macromolecules* **1980**, *13*, 1602–1617. [CrossRef]
34. Burger, C.; Ruland, W.; Semenov, A.N. Polydispersity Effects on the Microphase-Separation Tran-sition in Block Copolymers. *Macromolecules* **1990**, *23*, 3339–3346. [CrossRef]
35. Lynd, N.A.; Hillmyer, M.A. Influence of Polydispersity on the Self-Assembly of Diblock Copolymers. *Macromolecules* **2005**, *38*, 8803–8810. [CrossRef]

Article

Synthesis and Characterization of Hybrid Materials Derived from Conjugated Copolymers and Reduced Graphene Oxide

Alexandros Ch. Lazanas [1], Athanasios Katsouras [1], Michael Spanos [1,2], Gkreti-Maria Manesi [1], Ioannis Moutsios [1,3], Dmitry V. Vashurkin [4,5], Dimitrios Moschovas [1], Christina Gioti [1], Michail A. Karakassides [1], Vasilis G. Gregoriou [2], Dimitri A. Ivanov [3,4,5], Christos L. Chochos [1,2,*] and Apostolos Avgeropoulos [1,4,*]

1. Department of Materials Science Engineering, University of Ioannina, 45110 Ioannina, Greece
2. National Hellenic Research Foundation (NHRF), 48 Vassileos Constantinou Avenue, 11635 Athens, Greece
3. Institut de Sciences des Matériaux de Mulhouse—IS2M, CNRS UMR7361, 15 Jean Starcky, 68057 Mulhouse, France
4. Faculty of Chemistry, Lomonosov Moscow State University, 9MSU, GSP-1, 1-3 Leninskiye Gory, 119991 Moscow, Russia
5. Institute of Problems of Chemical Physics, Russian Academy of Sciences, Chernogolovka, 142432 Moscow, Russia
* Correspondence: chochos@eie.gr (C.L.C.); aavger@uoi.gr (A.A.); Tel.: +30-26-5100-9001 (A.A.)

Abstract: In this study the preparation of hybrid materials based on reduced graphene oxide (rGO) and conjugated copolymers is reported. By tuning the number and arrangement of thiophenes in the main chain (indacenothiophene or indacenothienothiophene) and the nature of the polymer acceptor (difluoro benzothiadiazole or diketopyrrolopyrrole) semiconducting copolymers were synthesized through Stille aromatic coupling and characterized to determine their molecular characteristics. The graphene oxide was synthesized using the Staudenmaier method and was further modified to reduced graphene oxide prior to structural characterization. Various mixtures with different rGO quantities and conjugated copolymers were prepared to determine the optoelectronic, thermal and morphological properties. An increase in the maximum absorbance ranging from 3 to 6 nm for all hybrid materials irrespective of the rGO concentration, when compared to the pristine conjugated copolymers, was estimated through the UV-Vis spectroscopy indicating a differentiation on the optical properties. Through voltammetric experiments the oxidation and reduction potentials were determined and the calculated HOMO and LUMO levels revealed a decrease on the electrochemical energy gap for low rGO concentrations. The study indicates the potential of the hybrid materials consisting of graphene oxide and high band gap conjugated copolymers for applications related to organic solar cells.

Keywords: hybrid materials; conjugated copolymers; rGO; HOMO; LUMO; indacenothiophene; indacenothienothiophene; cyclic voltammetry

1. Introduction

Over the last two decades graphene [1] has made an impressive impact on current technology due to its remarkable properties. The oxidized form of graphene, namely graphene oxide (GO), has been known for almost two centuries [2], and despite its impressive features, it lacks the electronic properties of graphene due to the mixed sp^3 and sp^2 orbital attributed to the oxygen groups. Many efforts have been conducted to reduce the oxygen groups using reductive substances such as NaBH$_4$ etc. [3,4]. This allotropic form of carbon is generally known as reduced graphene oxide (rGO) and constitutes a fairly good compromise between the outstanding electronic properties of graphene and the excellent solubility of graphene oxide. This fact is allocated to the reduction process which is not exclusively completed (not 100% yield of the process) and the produced rGO still retains some of the aforementioned oxygen groups that enable stable dispersions with hydrophobic interactions, since rGO is not water-soluble similar to its precursor [5].

The above-mentioned characteristics strongly suggest that rGO could readily form films for a great number of applications such as organic photovoltaics [6,7], sensors [8] etc. To that end, many groups have endeavored the combination of conjugated polymers and rGO [9–11] to address the fact that their possible π-π stacking with graphene and rGO enables the fine-tuning of the optoelectronic polymers. Such an approach leads to the opportunity for a flexible and adjustable system which may solve the relatively low photovoltaic efficiencies that these polymers commonly exhibit [12,13]. As a result, a wide range of applications based on composite materials consisting of graphene oxide in polymer matrices have been reported in the literature. These systems have already been used as membranes [14,15], energy devices [16], supercapacitors [17], drug delivery [18] flexible and/or wearable electronics [19] etc.

The non-covalent attachment between the conjugated polymer and the rGO has significant advantages. The rGO is added to the polymer solution and the mixture is agitated through sonication. This approach leads to the stabilization through non-covalent interactions which do not induce intrinsic changes in the chemical structures, electronic or mechanical properties of the rGO. The π-π interactions between the rGO and the conjugated backbone render the hybrid materials capable of being easily processed due to the rGO solubility. The exquisite properties presented on the specific hybrid copolymers have paved the way for their extensive use in photovoltaic devices using facile methods [20–23].

Herein, our focus was the synthesis of high band gap conjugated polymers [24] based on indacenothiophene and indacenodithienothiophene [25] through the Stille aromatic coupling method. These monomers play the role of donors, in *donor-acceptor* units with varying acceptor units. The characterization through Gel Permeation Chromatography (GPC) indicated the molecular features of the copolymers such as number average molecular weight and dispersity (Đ). In addition, we report the synthesis of graphene oxide from graphite powder using a modified Staudenmaier method [2] and its further reduction to obtain the reduced graphene oxide [3]. X-ray Diffraction Analysis (XRD) was carried out to assess the characteristic properties before and after the oxygen incorporation between the graphene oxide sheets. The preparation of mixtures consisting of conjugated polymers and reduced graphene oxide is expected to have a significant role in the fine tuning of their energy gaps and possibly revolutionize various electronic applications such as organic solar cells, OLED displays etc. The materials were mixed in an appropriate dispersing medium (ortho-dichlorobenzene or o-DCB) and sonicated for 6 h prior to their characterization. Specifically, the hybrid dispersions were morphologically characterized using Scanning Electron Microscopy (SEM) and Energy Dispersive Spectroscopy (EDS). Their optoelectronic properties were evaluated with UV-Vis Spectroscopy and Cyclic Voltammetry (CV). Also, the thermal stability of the involved materials (GO, rGO, conjugated copolymers, hybrid materials) was verified through thermogravimetric analysis (TGA).

2. Materials and Methods

2.1. Materials

All monomers (thiophene-based, difluorobenzothiadiazole and diketopyrrolopyrrole) and solvents (toluene, methanol, acetone, hexane, acetonitrile, ortho-dichlorobenzene) were purchased from Sigma-Aldrich (St. Louis, MO, USA) and used without further purification. All reagents for the preparation of the samples [tris (dibenzylideneacetone)dipalladium (0) or Pd_2dba_3, tri (o-tolyl)phosphine or P (o-Tol)$_3$, powdered graphite, sulfuric acid, nitric acid, potassium chlorate powder and sodium borohydride or $NaBH_4$] were purchased from Fluka (Fluka Chemie GmbH, Buchs, Switzerland) and used without additional manipulations.

2.2. Methods

Number average molecular weights ($\overline{M}_n$) and dispersity indices (Đ) were determined with Gel Permeation Chromatography (GPC) at 150 °C on a high temperature PL-GPC 220 system using a PL-GEL 10 μm guard column, three PL-GEL 10 μm Mixed-B columns and o-DCB as the eluent. The instrument (Agilent Technologies, St. Clara, CA, USA) was calibrated with narrow polystyrene standards (Mp ranged from 4830 g/mol to 3,242,000 g/mol).

UV/vis spectra were measured on a Cary 5000 UV-Vis-NIR (Agilent Technologies, St. Clara, CA, USA) in dual beam mode.

Raman spectroscopy studies were conducted through a micro-Raman RM 1000 Renishaw system (Renishaw, Wotton-under-Edge, UK). The power of the laser was 30 mW and a 2 μm focus spot was utilized.

X-ray diffraction (XRD) measurements were carried out in a Bruker D8 Advance (Bruker, Billerica, MA, USA) with Bragg–Brentano geometry with LYNXEYE detector in 2θ range from 2° to 30°. For the X-rays, a Cu-K$_\alpha$ wire was used, resulting in a radiation of 1.5406 Å wavelength.

Thermogravimetric analysis (TGA) was carried out in a Perkin Elmer Pyris-Diamond instrument (PerkinElmer, Inc., Waltham, MA, USA). Approximately 10 mg of each sample (graphene oxide, reduced graphene oxide, pristine conjugated copolymers and final hybrid materials) was heated from 40 °C to 800 °C following a step of 5 °C/min under nitrogen atmosphere.

Scanning Electron Microscope (SEM) images were obtained on gold coated samples (Polaron SC7620 sputter coater, Thermo VG Scientific, Waltham, MA, USA) using a JEOL JSM6510LV scanning electron microscope equipped with an INCA PentaFETx3 (Oxford Instruments, Abingdon, UK) energy dispersive X-ray (EDX) spectroscopy detector. Data acquisition was performed with an accelerating voltage of 20 kV and 60 s accumulation time, respectively. The samples were prepared using an appropriate solvent for solution casting and then dried overnight in a vacuum oven. Prior to the morphological characterization a gold/palladium target was used to sputter the samples utilizing the SC7620 Mini Sputter Coater/Glow Discharge System (Quorum technologies, Sacramento, USA). Also, additional experiments were conducted on a specific copolymer (copolymer-C2 with 5 wt% and 10 wt% rGO) in order to study its morphological behavior in thin film state. The samples were prepared using spin coating onto silicon wafer under ambient conditions (spinning velocity ~3000 rpm for approximately 30 s, polymer solution in o-DCB concentration equal to 3%).

Voltammetric Experiments were performed with a PGSTAT12 electrochemical analyzer (Metrohm Autolab, Utrecht, Netherlands) in a single compartment three-electrode cell. Bare or modified GCEs (IJ Cambria, Swansea, UK) were used as working electrodes and a platinum wire served as the auxiliary electrode. The reference electrode was an Ag/AgCl 3M KCl (IJ Cambria, Swansea, UK) electrode and all potentials hereafter are quoted to the potential of this electrode. Cyclic Voltammograms (CVs) were recorded in 0.1 mol L^{-1} Bu$_4$NBF$_4$ in solvent mixture acetonitrile (ACN):o-DCB at a scan rate of 0.05 V s^{-1}. All potentials are quoted after the Fc/Fc$^+$ redox couple as explained in the experimental section.

2.3. Synthesis of Hybrid Materials

Three (3) conjugated copolymers using thiophene-based and difluoro benzothiadiazole or diketopyrrolopyrrole as monomers were synthesized based on the Stille aromatic coupling in a process that has been already described by our group [26] in order to be used as matrices for the preparation of hybrid materials. In a three-neck round bottom flask, fitted with a condenser, 3 cycles of argon/vacuum were conducted. Afterwards, the donor- and the acceptor-type monomers were added along with Pd$_2$dba$_3$, which is the catalyst (22.90 mg), and P (o-Tol)$_3$ as a ligand (60.87 mg) and 3 cycles of argon/vacuum were carried out again. Finally, toluene (20 mL) was added, and 3 cycles of argon/vacuum were again performed. The reaction mixture is allowed to react for at least 48 h at 120 °C.

Upon completion of the polymerization, the synthesized material was precipitated in 500 mL of methanol and filtered on a paper filter. The filter was placed in a Soxhlet type device and solvent cleaning of different polarity starts. Initially, the polymer was purified with methanol (250 mL) to remove any catalyst residues, then with acetone (250 mL) to remove any unreacted monomers, followed by hexane (250 mL) to remove any oligomers and finally with chloroform (250 mL) to extract the desired copolymer. The chloroform fraction was concentrated on a rotary evaporator to a volume of approximately 50 mL and the copolymer was precipitated in 800 mL of methanol. The precipitated copolymer was collected by vacuum filtration and dried for 24 h on a PTFE filter. Further drying of the polymer in a vacuum oven at 40 °C for 9 h led to a yield of 0.86 g of the final copolymer. The synthetic procedure and chemical structures of the synthesized conjugated copolymers are presented in the Supporting Information for clarification reasons (Scheme S1). The molecular characteristics of the copolymers were specified using GPC and the results are summarized in Table S1 (Supporting Information). The three different molecular weight copolymers are abbreviated as C1, C2 and C3 respectively as evident in Table S1 in the supporting information. In addition, in Figure S1 the GPC curves with respect to the different copolymers are presented.

2.4. Synthesis of Graphene Oxide and Reduction to the Desired rGO

Graphene oxide was synthesized with the modified Staudenmaier method which is widely established [2]. In a typical synthesis, powdered graphite (5 g, purum powder ≤ 0.2 mm; Fluka) was added to a mixture of concentrated sulfuric acid (200 mL, 95–97 wt%) and nitric acid (100 mL, 65 wt%) while cooling in an ice-water bath. Potassium chlorate powder (100 g, purum > 98.0%; Fluka) was added to the mixture gradually while stirring and cooling. The reaction was quenched after 18 h by pouring the mixture into distilled water and the oxidation product was washed until the pH reached a value of 6.0 and finally dried at room temperature. The resulting GO was re-dispersed in 200 mL distilled water under agitation for 24 h. Then, 50 mL of an $NaBH_4$ aqueous solution (8 mg/mL) were added and the dispersion was agitated for another 24 h [3]. Finally, the dispersion was centrifuged for five (5) times and the sediment was left to dry on a glass substrate. Using this procedure rGO was efficiently prepared.

2.5. Preparation of the Hybrid Dispersions

Many solvents have been proposed as appropriate dispersion mediums for rGO [27,28]. High-boiling point solvents showcase an outstanding performance in terms of stability. Our choice of dispersing rGO in o-DCB was two-fold: firstly to produce extremely stable dispersions of rGO and secondly it is a common solvent of conjugated polymers of the indaceno- type. Thus, we have prepared nine (9) different dispersions of hybrid materials using sonication for 6 h in each of the three different copolymers with 3, 5 and 10% wt rGO loading respectively as evident in Table 1.

Table 1. The oxidation and reduction potentials from the respective cyclic voltammograms for each system (recalibrated versus Fc/Fc+) and the corresponding HOMO, LUMO and band gaps.

Copolymer	$\overline{M}_n$ (kDa) *	E^{ox}_{onset} (V vs. Fc/Fc+)	E^{red}_{onset} (V vs. Fc/Fc+)	E_{HOMO} (eV)	E_{LUMO} (eV)	E_g^{ec}
1		0.05	−2.01	−5.15	−3.09	2.06
3 %rGO		−0.18	−1.73	−4.92	−3.37	1.55
5 %rGO	128	−0.13	−2.09	−4.88	−3.01	1.87
10 %rGO		0.11	−1.95	−5.21	−3.15	2.06
2		0.13	−2.35	−5.23	−2.75	2.48
3 %rGO		−0.16	−1.88	−4.94	−3.22	1.72
5 %rGO	153	−0.14	−1.79	−4.96	−3.31	1.65
10 %rGO		0.44	−2.07	−5.54	−3.03	2.51

Table 1. Cont.

Copolymer	$\overline{M}_n$ (kDa) *	E^{ox}_{onset} (V vs. Fc/Fc+)	E^{red}_{onset} (V vs. Fc/Fc+)	E_{HOMO} (eV)	E_{LUMO} (eV)	Eg^{ec}
3		0.55	−2.04	−5.65	−3.06	2.59
3 %rGO	150	0.03	−1.94	−5.13	−3.16	1.97
5 %rGO		0.01	−1.59	−5.11	−3.51	1.60
10 %rGO		0.17	−1.97	−5.27	−3.13	2.14

* Number average molecular weights were calculated from Gel Permeation Chromatography measurements. (For additional information, refer to Supporting Information).

3. Results and Discussion

3.1. Raman and X-ray Diffraction Analysis

Raman spectroscopy was performed to study the oxidation of pristine graphite to GO and its further reduction to rGO. In Figure S2 in the Supporting Information the Raman spectra of the two allotropic forms of graphene oxide obtained are illustrated. In the GO spectrum, the two major peaks D and G are 1310 cm^{-1} and 1590 cm^{-1} respectively possess quite large deviations and structural deficiencies in relation to graphene due to the existence of the oxygen groups that alter sp^2 hybridization and therefore its structure. The imperfections are represented by the I_D / I_G ratio, which equals to 1.99, a value that is justified by the oxidation process of graphite. Correspondingly for the case of rGO, D and G peaks at 1301 cm^{-1} and 1575 cm^{-1} respectively show the structural defects from the non-selective removal of the oxygen groups due to the reduction which induces gaps along the structure. In this case, the I_D/I_G ratio increases to 2.23, indicating the successful rGO reduction [29].

Concerning the XRD measurements conducted on GO and rGO a clearcut difference on the first reflection is evident in the XRD spectra presented in Figure S3. Taking into consideration the primary reflection (001) and by applying Bragg's relationship one concludes that the reduction of the original oxygen groups leads to a change in the arrangement of the graphene sheets. Specifically, $d_{(001)}$ GO equals to 7.4 Å while $d_{(001)}$ rGO to 9.1 Å.

The successful chemical modification is verified by the increase of the neighboring graphite sheets distance as indicated by the synergy of literature [30] and experimental results. In Figure S3 the comparative diffraction diagram for GO and rGO is presented showcasing a distinct difference on the first reflection which reveals the larger interstitial space between the rGO sheets. The specific properties are desirable for the preparation of nanocomposite materials due to the capability of the conjugated polymer to be accommodated between the interstitial spaces of the laminar material.

TGA experiments were performed to study the thermal behavior of both the GO and the rGO. A minimal weight loss (8%) at approximately 100 °C is evident and corresponds to the water molecules, while functional groups (-O(CO)-, -OH, -O-) are decomposed at approximately 230 °C, leading to a second significant weight loss (30%). The graphitic lattice decomposition (50% weight loss) took place at approximately 500 °C (Figure S4A in the Supporting Information). The TGA studies on the rGO revealed a weight loss at approximately 10% due to the removal of moisture. The remaining functional groups (-O(CO)-, -OH, -O-) after the chemical modification are removed at approximately 190 °C leading to 20% weight loss which is less than the one observed in the GO further verifying the successful chemical modification. The graphitic lattice decomposition occurred at approximately 530 °C indicating higher thermal stability of the rGO due to the chemical modification of the functional groups (Figure S4B in the Supporting Information). The thermograph of the pristine conjugated copolymer (C3) and hybrid materials consisting of the conjugated copolymer C3 with different weight percentages of rGO are also presented in Figure S4C–F. The original decomposition of the pure conjugated copolymer took place at approximately 238 °C (Figure S4C). It is evident that the decomposition of the conjugated copolymer with 3 wt% of rGO initiated at 242 °C (Figure S4D), while for the 5 wt% rGO at 250 °C (Figure S4E) and for the 10% rGO at 260 °C (Figure S4F) indicating that the higher the rGO ratio the greater the thermal stability of the final hybrid material. Coherent behavior was also observed in the remaining samples. Note that, the materials with the highest percentage of rGO not only showcased the highest stability but also less weight loss which is in accordance to previous studies [31].

3.2. Morphological Characterization

The morphological characterization of GO, rGO and the final hybrid materials was performed through SEM measurements. In Figure 1 the difference in the nanosheets of GO (Figure 1A) and rGO (Figure 1B) in the dispersion form (an aqueous dispersion for GO and an o-DCB dispersion for rGO which was left to dry at an elevated temperature overnight under vacuum) is depicted. The SEM micrographs presented in Figure 1C,D correspond to the solid form of the GO and the rGO respectively, indicating an apparent introduction of abnormalities in the rGO morphology compared to the defect-free GO, which are attributed to the reduction process as vacancies that occur in the lattice during the removal of the oxygen groups. From the SEM images it is evident that the conversion of the graphite morphology to graphene oxide was accomplished successfully maintaining its lamellar formation. Also, it is straightforward that the film and solid forms exhibit significant alternations due to the re-dispersion of GO and rGO in o-DCB for the case of film state.

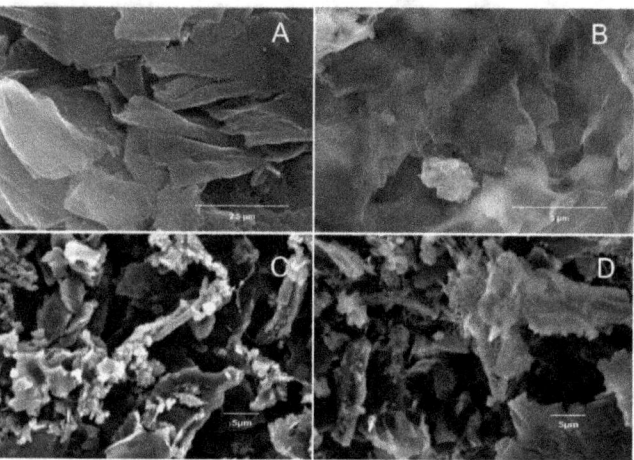

Figure 1. SEM images of the GO and rGO in film (**A**,**B**) and in solid (**C**,**D**) forms.

In Figure 2 the SEM micrograph in conjunction with the elemental mapping analysis of the hybrid material which is constituted by copolymer 2 (C2) with 5 wt% rGO is demonstrated. The interaction of the conjugated copolymer with the rGO due to π-π stacking led to the intercalation of the copolymer between the graphene nanosheets, a behavior that has been reported for similar layered systems [32] (Figure 2A). This is further supported by the EDX in which the elemental mapping analysis confirmed the existence of carbon (Figure 2B) (a common element of the hybrid system), sulfur (Figure 2C) which is derived from the thiophene groups in the copolymer structural unit and oxygen (Figure 2D), which is solely found in rGO and verifies the successful preparation of the hybrid material. The SEM/EDX data obtained from the different copolymers using 3, 5 and 10 wt% of rGO, showcased similar results, suggesting the non-covalent attachment of the rGO on the conjugated copolymers even in low concentration values.

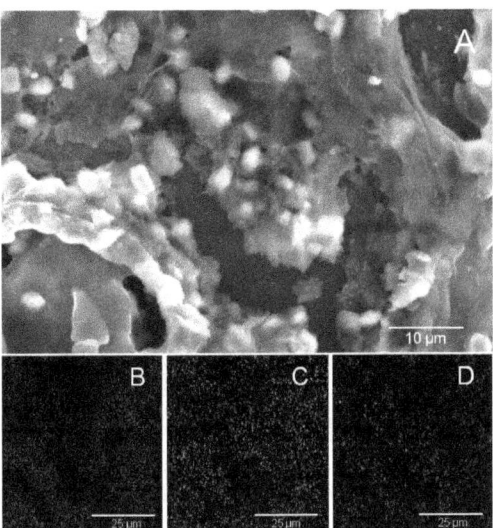

Figure 2. SEM images of a hybrid system with the Copolymer 2 and the rGO (**A**). There is an apparent intercalation that is supported with the aid of EDS, where the elemental mapping analysis confirms the existence of Carbon (**B**), Sulfur (**C**) and Oxygen (**D**).

Interestingly enough, organic cells-related applications are based on polymer films and therefore the morphological characterization of the hybrid material which is comprised by the C2 copolymer with either 5 wt% or 10 wt% rGO were conducted in thin film state. The spin coated samples were approximately ~100 nm thick and the SEM studies indicated similar structures to the ones obtained during bulk characterization. The results are promising for photovoltaic applications, but more comprehensive studies are required. Two representative SEM micrographs are presented in the Supporting Information (Figure S5a,b).

3.3. Optoelectronic Properties

In Figure 3 the UV-Vis spectra obtained from 300–900 nm for the three pristine copolymer systems and the hybrid materials with different rGO concentration are given. For copolymer 1 (C1) (Figure 3A) two distinct peaks at 423 nm and 723 nm can be clearly identified. With the addition of rGO, the first absorption peak was shifted from 423 nm to 429 nm while the second peak from 723 nm to 729 nm. This phenomenon indicates that even a relatively small (3% wt) addition of rGO is able to alter the absorption maxima of the copolymer and hence its energy gap. This phenomenon is also observed in samples with higher rGO concentration, which also exhibit similar displacements at a higher absorption intensity. Consistently, similar results were recorded for copolymer C2 (Figure 3B) which exhibits three absorption maxima (637 nm, 411 nm and 319 nm) but the shift only occurs for two maxima, since the peak at 637 nm shifts to 640 nm and the peak at 319 nm shifts to 322 nm. Finally, copolymer C3 (Figure 3C) demonstrates a similar shift in wavelength as C1 (from 713 nm to 719 nm and from 434 nm to 440 nm), thus reaffirming the impact of the rGO addition on the optical properties.

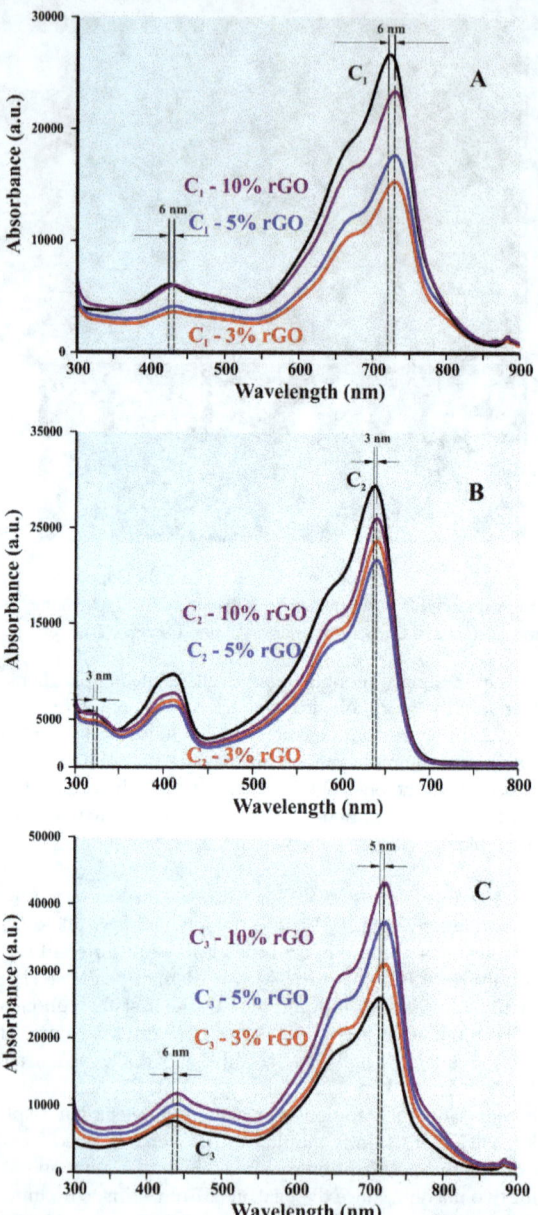

Figure 3. UV-Vis spectra of the: (**A**) C1 systems; (**B**) C2 systems; and (**C**) C3 systems. Distinct wavelength shifts can be observed in every system for even a minimum rGO addition, at (red line) C-3% rGO, (blue line) C-5% rGO, and (purple line) C-10% rGO.

Cyclic voltammetry was performed to correlate the oxidation and the reduction potential onsets with the respective HOMO (Highest Occupied Molecular Orbital) and LUMO (Lowest Unoccupied Molecular Orbital) levels as it has been previously described for such conjugated copolymers by our group [33]. The cyclic voltammograph of ferrocene in acetonitrile (ACN) was used as reference for the copolymers and is presented in the Supporting

Information (Figure S6). The oxidation and reduction potentials were recalibrated versus the Fc/Fc$^+$ redox couple and the corresponding HOMO and LUMO levels were calculated using the following equations:

$$E_{HOMO} = 5.1 + E^{ox}_{onset}, \qquad (1)$$

$$E_{LUMO} = 5.1 + E^{red}_{onset} \qquad (2)$$

The cyclic voltammographs (CVs) for the pristine C1 copolymer and the produced hybrid materials after the addition of 3, 5 and 10% wt rGO respectively are illustrated in Figure 4. It is apparent that for lower rGO concentrations (3 and 5 wt%) a significant alteration on the redox potentials is experienced and therefore to the HOMO and LUMO levels, while in higher concentration (10 wt%) the redox potentials are similar to those obtained for the pristine copolymer [34]. This effect is observed for all three hybrid systems since it is also evident in Figure 5 (C2) and in Figure 6 (C3). Table 1 summarizes the results provided from the CVs as well as the calculated HOMO, LUMO levels and the electrochemical band gap (E_g).

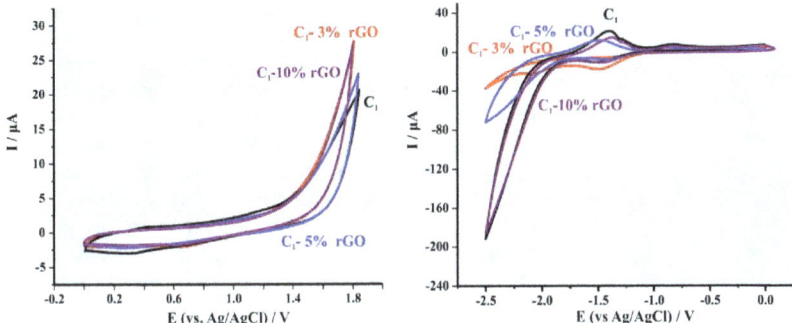

Figure 4. Cyclic voltammograms of the C1 systems for both the oxidation and the reduction process. in 0.1 mol L^{-1} Bu$_4$NBF$_4$. Scan rate 0.050 V s^{-1}. Signals presented as: (red line) C1-3% rGO, (blue line) C1-5% rGO, and (purple line) C1-10% rGO.

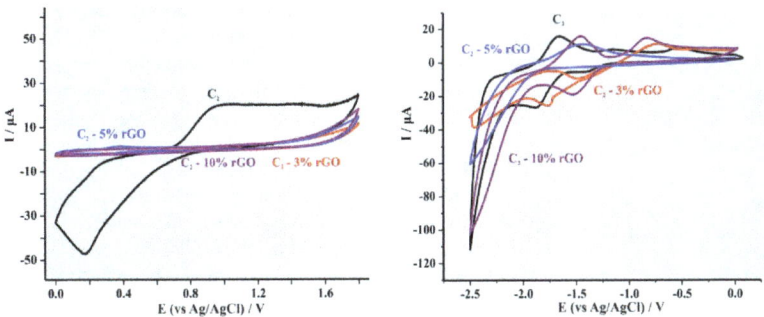

Figure 5. Cyclic voltammograms of the C2 systems for both the oxidation and the reduction process. in 0.1 mol L^{-1} Bu$_4$NBF$_4$. Scan rate 0.050 V s^{-1}. Signals presented as: (red line) C3-3% rGO, (blue line) C3-5% rGO, and (purple line) C3-10% rGO.

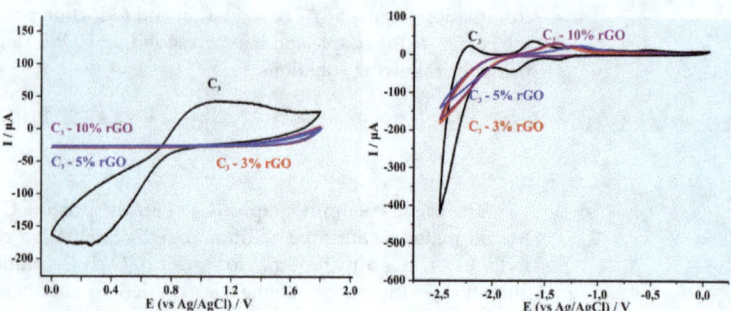

Figure 6. Cyclic voltammograms of the C3 systems for both the oxidation and the reduction process. in 0.1 mol L^{-1} Bu$_4$NBF$_4$. Scan rate 0.050 V s^{-1}.

4. Conclusions

Graphene oxide (GO) and reduced graphene oxide (rGO) were synthesized and successfully combined with indacenodithiophene- and indacenodithienothiophene-based conjugated copolymers for the preparation of hybrid materials in order to study the alteration on their optoelectronic properties due to non-covalent binding (π-π stacking) with rGO. The conjugated copolymers were synthesized through Stille aromatic coupling. Staudenmaier reaction was utilized for the synthesis of GO, which was further reduced for the preparation of rGO. The conjugated polymers were mixed using different ratios of rGO and were ultrasonicated in ortho-dichlorobenzene that constitutes an excellent dispersion medium for each system, to produce stable dispersions. To identify the successful synthesis of GO and rGO, morphological and structural characterization was performed using Scanning Electron Microscopy (SEM), Raman spectroscopy and X-Ray Diffraction analysis. UV-Vis absorption spectrophotometry was used to determine the optical properties of the pristine copolymers and the hybrid materials. The experiments indicated a significant wavelength shift in the absorption maxima for all rGOs of the order of 3–6 nm, which justifies the alteration and fine-tuning of the optical properties of the copolymers with the aid of rGO. Finally, electrochemical characterization was performed via Cyclic Voltammetry (CV) to determine the energy levels of copolymers and mixtures. From the determination of oxidation and reduction potentials and the calculated HOMO and LUMO levels it is concluded that small percentages of rGO (3–5 wt%) can alter the energy levels of the copolymers and ultimately decrease the electrochemical Energy gap (E_g), while larger concentrations of rGO can return these attributes to their original state. This suggests that the specific hybrid materials are quite promising for applications such as organic solar cells, where a tunable band gap and wavelength are of strategic importance.

Supplementary Materials: The following supporting information can be downloaded at: https://www.mdpi.com/article/10.3390/polym14235292/s1, Scheme S1: Chemical reactions and structures of: (**A**) indacenothiophene-alt -3,6-bis(5-bromothiophen-2-yl)-2,5-bis(2-octyldodecyl)pyrrolo [3,4-c]pyrrole (IDT and Dibromo-DPP) or C1; (**B**) indacenothiophene -alt-4,7-dibromo-5,6-difluorobenzo[1,2,5]thiadiazole (IDT and difluoro-BTD) or C2; and (**C**) Indacenodithienothiophene-alt-3,6-bis(5-bromothieno[3,2-b]thiophen-2-yl)-2,5 bis(2octyldodecyl)pyrrolo[3,4-c]pyrrole (IDTT and Dibromo-DPP) or C3 copolymers; Figure S1: Gel Permeation Chromatography (GPC) curves of the three different copolymers where the black line corresponds to the C1 copolymer, the red line to the C2 copolymer and the blue line to the C3; Table S1: Molecular characteristics of the three different copolymers as directly calculated from gel permission chromatography; Figure S2: Raman Spectra obtained for GO and rGO. Both spectra represent materials with a great number of defects which is evident by the respective I_G/I_D ratio. The red curve corresponds to the rGO while the black to the GO; Figure S3: XRD measurements for GO and rGO. The red curve corresponds to the rGO while the black to the GO; Figure S4: TGA thermograms under nitrogen atmosphere corresponding to: (**A**) GO; (**B**) rGO; (**C**) C3 copolymer; (**D**) C3 copolymer with 3 wt% rGO; (**E**) C3 copolymer with

5 wt% rGO and F) C3 copolymer with 10 wt% rGO; Figure S5: SEM images corresponding to the C2 copolymer with (**A**) 5 wt% rGO; and (**B**) 10 wt% rGO. The samples were prepared using the spin coating technique leading to thicknesses of approximately 100 nm; Figure S6: Cyclic voltammogram of ferrocene in ACN which was used as reference for the copolymers

Author Contributions: Conceptualization, C.L.C. and A.A.; methodology, C.L.C. and A.A.; validation, A.C.L., A.K., M.S., C.L.C., V.G.G. and A.A.; formal analysis, A.C.L., A.K., M.S., G.-M.M., I.M., D.V.V. and D.M.; investigation, A.C.L., A.K. and M.S.; resources, D.A.I., C.L.C., V.G.G. and A.A.; data curation, A.C.L., A.K., M.S., G.-M.M., I.M., D.M., C.G. and M.A.K.; writing—original draft preparation, A.C.L., C.L.C. and A.A.; writing—review and editing, A.C.L., A.K., G.-M.M., I.M., D.A.I., C.L.C. and A.A.; visualization, C.L.C. and A.A.; supervision, C.L.C., V.G.G. and A.A.; funding acquisition, D.A.I. and A.A. All authors have read and agreed to the published version of the manuscript.

Funding: This research was funded by [the Ministry of Science and Higher Education of the Russian Federation within State Contract], grant no. [075-15-2022-1105].

Data Availability Statement: The data presented in this study are available upon request from the corresponding author.

Conflicts of Interest: The authors declare no conflict of interest.

References

1. Novoselov, K.S.; Geim, A.K.; Morozov, S.V.; Jiang, D.; Zhang, Y.; Dubonos, S.V.; Grigorieva, I.V.; Firsov, A.A. Electric field in atomically thin carbon films. *Science* **2004**, *306*, 666–669. [CrossRef] [PubMed]
2. Staudenmaier, L. Method for the Preparation of the Graphite Acid. *Eur. J. Inorg. Chem.* **1898**, *31*, 1481–1487.
3. Stankovich, S.; Dikin, D.A.; Piner, R.D.; Kohlhaas, K.A.; Kleinhammes, A.; Jia, Y.; Wu, Y.; Nguyen, S.B.T.; Ruoff, R.S. Synthesis of graphene-based nanosheets via chemical reduction of exfoliated graphite oxide. *Carbon* **2007**, *45*, 1558–1565. [CrossRef]
4. Politakos, N.; Liontos, G.; Karanastasis, A.; Zapsas, G.; Moschovas, D.; Avgeropoulos, A. Surface Initiated Polymerization from Graphene Oxide. *Curr. Org. Chem.* **2015**, *19*, 1757–1772. [CrossRef]
5. Kern, K.; Gómez-Navarro, C.; Burghard, M.; Weitz, R.T.; Scolari, M.; Mews, A.; Bittner, A.M. Electronic Transport Properties of Individual Chemically Reduced Graphene Oxide Sheets. *Nano Lett.* **2007**, *7*, 3499–3503.
6. Wu, J.; Becerril, H.A.; Bao, Z.; Liu, Z.; Chen, Y.; Peumans, P. Organic solar cells with solution-processed graphene transparent electrodes. *Appl. Phys. Lett.* **2008**, *92*, 263302. [CrossRef]
7. Eda, G.; Lin, Y.-Y.; Miller, S.; Chen, C.-W.; Su, W.-F.; Chhowalla, M. Transparent and conducting electrodes for organic electronics from reduced graphene oxide. *Appl. Phys. Lett.* **2008**, *92*, 233305. [CrossRef]
8. Zhou, M.; Zhai, Y.; Dong, S. Electrochemical sensing and biosensing platform based on chemically reduced graphene oxide. *Anal. Chem.* **2009**, *81*, 5603–5613. [CrossRef]
9. Liu, Q.; Liu, Z.; Zhang, X.; Yang, L.; Zhang, N.; Pan, G.; Yin, S.; Chen, Y.; Wei, J. Polymer photovoltaic cells based on solytion-processable graphene and P3HT. *Adv. Funct. Mater.* **2009**, *19*, 894–904. [CrossRef]
10. Zhuang, X.-D.; Chen, Y.; Liu, G.; Li, P.-P.; Zhu, C.-X.; Kang, E.-T.; Noeh, K.-G.; Zhang, B.; Zhu, J.-H.; Li, Y.-X. Conjugated-polymer-functionalized graphene oxide: Synthesis and nonvolatile rewritable memory effect. *Adv. Mater.* **2010**, *22*, 1731–1735. [CrossRef]
11. Qi, X.; Tan, C.; Wei, J.; Zhang, H. Synthesis of graphene-conjugated polymer nanocomposites for electronic device applications. *Nanoscale* **2013**, *5*, 1440–1451. [CrossRef] [PubMed]
12. Liu, Q.; Liu, Z.; Zhang, X.; Zhang, N.; Yang, L.; Yin, S.; Chen, Y. Organic photovoltaic cells based on an acceptor of soluble graphene. *Appl. Phys. Lett.* **2008**, *92*, 223303. [CrossRef]
13. Liu, Z.; He, D.; Wang, Y.; Wu, H.; Wang, J. Solution-processable functionalized graphene in donor/acceptor-type organic photovoltaic cells. *Sol. Energy Mater. Sol. Cells* **2010**, *94*, 1196–1200. [CrossRef]
14. Gupta, O.; Roy, S.; Mitra, S. Nanocarbon-immobilized membranes for separation of tetrahydrofuran from water via membrane distillation. *ACS Appl. Nano Mater.* **2020**, *3*, 6344–6353. [CrossRef]
15. Sarapuu, A.; Kibena-Põldsepp, E.; Borghei, M.; Tammeveski, K. Electrocatalysis of oxygen reduction on heteroatom-doped nanocarbons and transition metal-nitrogen-carbon catalysts for alkaline membrane fuel cells. *J. Mater. Chem. A* **2018**, *6*, 776–804. [CrossRef]
16. Tang, C.; Titirici, M.; Zhang, Q. A review of nanocarbons in energy electrocatalysis: Multifunctional substrates and highly active sites. *J. Energy Chem.* **2017**, *26*, 1077–1093. [CrossRef]
17. Lee, H.; Lee, K.S. Interlayer Distance Controlled Graphene, Supercapacitor and Method of Producing the Same. U.S. Patent 10,214,422, 26 February 2019.
18. Panwar, N.; Soehartono, A.M.; Chan, K.K.; Zeng, S.; Xu, G.; Qu, J.; Coquet, P.; Yong, K.-T.; Chen, X. Nanocarbons for biology and medicine: Sensing, imaging, and drug delivery. *Chem. Rev.* **2019**, *119*, 9559–9656. [CrossRef]

19. Han, J.T.; Jang, J.I.; Cho, J.; Hwang, J.Y.; Woo, J.S.; Jeong, H.J.; Jeong, S.Y.; Seo, S.H.; Lee, G.-W. Synthesis of nanobelt-like 1-dimensional silver/nanocarbon hybrid materials for flexible and wearable electronics. *Sci. Rep.* **2017**, *7*, 4931. [CrossRef]
20. Meng, L.; Watson, W.B.; Qin, Y. Hybrid conjugated polymer/magnetic nanoparticle composite nanofibers through cooperative non-covalent interactions. *Nanoscale Adv.* **2020**, *2*, 2462–2470. [CrossRef]
21. Meng, L.; Fan, H.; Lane, J.M.D.; Qin, Y. Bottom-up approaches for precisely nanostructuring hybrid organic/inorganic multi-component composites for organic photovoltaics. *MRS Adv.* **2020**, *5*, 2055–2065. [CrossRef]
22. Liang, L.; Xie, W.; Fang, S.; He, F.; Yin, B.; Tlili, C.; Wang, D.; Qiu, S.; Li, Q. High-efficiency dispersion and sorting of single-walled carbon nanotubes via non-covalent interactions. *J. Mater. Chem. C* **2017**, *5*, 11339–11368. [CrossRef]
23. Girolamo, J.; Reiss, P.; Pron, A. Supramolecularly assembled hybrid materials via molecular recognition between diaminopyrimidine-functionalized poly(hexylthiophene) and thymine-capped CdSe nanocrystals. *J. Phys. Chem. C* **2007**, *111*, 14681–14688. [CrossRef]
24. Carsten, B.; He, F.; Son, H.J.; Xu, T.; Yu, L. Stille polycondensation for synthesis of functional materials. *Chem. Rev.* **2011**, *111*, 1493–1528. [CrossRef] [PubMed]
25. Zhang, W.; Smith, J.; Watkins, S.E.; Gysel, R.; McGehee, M.; Salleo, A.; Kirkpatrick, J.; Ashraf, S.; Anthopoulos, T.; Heeney, M.; et al. Indacenodithiophene Semiconducting Polymers for High-Performance, Air-Stable Transistors. *J. Am. Chem. Soc.* **2010**, *132*, 11437–11439. [CrossRef]
26. Chochos, C.L.; Drakopoulou, S.; Katsouras, A.; Squeo, B.M.; Sprau, C.; Colsmann, A.; Gregoriou, V.G.; Cando, A.P.; Allard, S.; Scherf, U.; et al. Beyond Donor–Acceptor (D–A) Approach: Structure–Optoelectronic Properties—Organic Photovoltaic Performance Correlation in New D–A_1–D–A_2 Low-Bandgap Conjugated Polymers. *Macromol. Rapid Commun.* **2017**, *38*, 1600720. [CrossRef] [PubMed]
27. Su, P.; Guo, H.-L.; Tian, L.; Ning, S.-K. An efficient method of producing stable graphene suspensions with less toxicity using dimethyl ketoxime. *Carbon* **2012**, *50*, 5351–5358. [CrossRef]
28. Konios, D.; Stylianakis, M.M.; Stratakis, E.; Kymakis, E. Dispersion behaviour of graphene oxide and reduced graphene oxide. *J. Colloid Interface Sci.* **2014**, *430*, 108–112. [CrossRef]
29. Němeček, D.; Thomas, G.J. Raman Spectroscopy of Viruses and Viral Proteins. In *Frontiers of Molecular Spectroscopy*; Elsevier: Amsterdam, The Netherlands, 2009; pp. 553–595.
30. Park, S.; An, J.; Potts, J.R.; Velamakanni, A.; Murali, S.; Ruoff, R.S. Hydrazine-reduction of graphite- and graphene oxide. *Carbon* **2011**, *49*, 3019–3023. [CrossRef]
31. Tuncel, D. Non-covalent interactions between carbon nanotubes and conjugated polymers. *Nanoscale* **2011**, *3*, 3545–3554. [CrossRef]
32. Alexandre, M.; Dubois, P. Polymer-layered silicate nanocomposites: Preparation, properties and uses of a new class of materials. *Mater. Sci. Eng. R Rep.* **2000**, *28*, 1–63. [CrossRef]
33. Gasparini, N.; Katsouras, A.; Prodromidis, M.I.; Avgeropoulos, A.; Baran, D.; Salvador, M.; Fladischer, S.; Spiecker, E.; Chochos, C.L.; Ameri, T.; et al. Photophysics of Molecular-Weight-Induced Losses in Indacenodithienothiophene-Based Solar Cells. *Adv. Funct. Mater.* **2015**, *25*, 4898–4907. [CrossRef]
34. Kausar, A. Conjugated Polymer/Graphene Oxide Nanocomposites. *J. Compos. Sci.* **2021**, *5*, 292. [CrossRef]

Article

Development of Double Hydrophilic Block Copolymer/Porphyrin Polyion Complex Micelles towards Photofunctional Nanoparticles

Maria Karayianni, Dimitra Koufi and Stergios Pispas *

Theoretical and Physical Chemistry Institute, National Hellenic Research Foundation, 48 Vassileos Constantinou Avenue, 116 35 Athens, Greece
* Correspondence: pispas@eie.gr

Abstract: The electrostatic complexation between double hydrophilic block copolymers (DHBCs) and a model porphyrin was explored as a means for the development of polyion complex micelles (PICs) that can be utilized as photosensitive porphyrin-loaded nanoparticles. Specifically, we employed a poly(2-(dimethylamino) ethyl methacrylate)-*b*-poly[(oligo ethylene glycol) methyl ether methacrylate] (PDMAEMA-*b*-POEGMA) diblock copolymer, along with its quaternized polyelectrolyte copolymer counterpart (QPDMAEMA-*b*-POEGMA) and 5,10,15,20-tetraphenyl-21H,23H-porphine-p,p',p'',p'''-tetrasulfonic acid tetrasodium hydrate (TPPS) porphyrin. The (Q)PDMAEMA blocks enable electrostatic binding with TPPS, thus forming the micellar core, while the POEGMA blocks act as the corona of the micelles and impart solubility, biocompatibility, and stealth properties to the formed nanoparticles. Different mixing charge ratios were examined aiming to produce stable nanocarriers. The mass, size, size distribution and effective charge of the resulting nanoparticles, as well as their response to changes in their environment (i.e., pH and temperature) were investigated by dynamic and electrophoretic light scattering (DLS and ELS). Moreover, the photophysical properties of the complexed porphyrin along with further structural insight were obtained through UV-vis (200-800 nm) and fluorescence spectroscopy measurements.

Keywords: double hydrophilic block copolymers; porphyrins; polyion complex micelles; photosensitizers

Citation: Karayianni, M.; Koufi, D.; Pispas, S. Development of Double Hydrophilic Block Copolymer/ Porphyrin Polyion Complex Micelles towards Photofunctional Nanoparticles. *Polymers* **2022**, *14*, 5186. https://doi.org/10.3390/polym14235186

Academic Editor: Nikolaos Politakos

Received: 31 October 2022
Accepted: 22 November 2022
Published: 29 November 2022

Publisher's Note: MDPI stays neutral with regard to jurisdictional claims in published maps and institutional affiliations.

Copyright: © 2022 by the authors. Licensee MDPI, Basel, Switzerland. This article is an open access article distributed under the terms and conditions of the Creative Commons Attribution (CC BY) license (https://creativecommons.org/licenses/by/4.0/).

1. Introduction

Successful cancer treatment is one of the most sought-after goals of modern-day medicine, since cancer remains a leading cause of death worldwide, accounting for nearly 10 million deaths in 2020 according to the World Health Organization [1]. Among the various proposed therapeutic strategies, photodynamic therapy (PDT) holds significant promise owing to its efficiency, site-specificity, and noninvasive characteristics. It is being used not only for the treatment of tumors related to various types of cancers (e.g., skin, esophageal, lung) but also a number of other diseases and medical conditions, such as atherosclerosis, rheumatoid arthritis, macular degeneration, psoriasis, and acne [2]. The working principle of PDT is based on the administration of photosensitizers that accumulate in pathological tissue and are subsequently exposed to a light source with appropriate wavelength so as to generate the production of reactive oxygen species (ROS), which in turn cause cell death [3–6]. Due to its selective action, it causes minimal damage to the surrounding healthy tissue, has no severe local or systemic side effects, is not painful, is well tolerated by patients, allows for outpatient use, and can even be applied in parallel with other therapeutic protocols [2,3]. All these advantages in combination with documented good therapeutic results render PDT a widespread contemporary method for the treatment of cancer and other infectious diseases.

Evidently, the role of the photosensitizer is of utmost importance for the effective application of PTD. Having entered the cell, the photoactive molecule is appropriately irradiated, leading to its excitation from the ground to the excited singlet state as a result of the photon absorption [3–5]. Next, the excited photosensitizers either relax back to the ground state via fluorescence photon emission or are transformed to the excited triplet state, whose energy under well-oxygenated conditions can be transferred to the surrounding tissue oxygen molecules, mainly producing singlet oxygen ($^1O_2$). These singlet oxygen particles are characterized by extremely strong oxidizing properties and can destroy tumor cells, either directly by inducing apoptotic and/or necrotic cell death or indirectly by provoking autophagy [3,4]. Over the years of PDT clinical practice, several characteristics have been established as prerequisites for ideal photosensitizers. These include high chemical purity, stability at room temperature, high photochemical reactivity, minimum dark cytotoxicity, preferential retention at the targeted tissue, facile solubility in bodily fluids, low cost, and wide availability, among others [2,3]. Of course, the most widely used photosensitizers consist of porphyrins (a class of heterocyclic macrocycle organic compounds) and their derivatives, since the historic discovery of hematoporphyrin and its photo-related medical properties in the beginning of the previous century [2].

Owing to their hydrophobic nature, porphyrins tend to self-assemble in aqueous media, forming aggregates of intriguing structures and unique features [7]. Nevertheless, aggregation reduces their photoactivity and thus therapeutic efficiency. In order to enhance porphyrin water solubility, as well as circulation lifetime and tumor specificity, and at the same time reduce aspecific tissue accumulation, the use of various types of polymeric nanocarriers has been proposed [4–6,8]. Among the plethora of available macromolecular architectures, the use of double hydrophilic block copolymers (DHBCs)—consisting of one charged and one hydrophilic block—as delivery vehicles is of particular interest due to their interesting self-assembly behavior [9–12]. This kind of copolymer is able to interact with oppositely charged bioactive species, forming polyion complex micelles (PICs) where the interacting charged species form the micellar core and the hydrophilic blocks the corona, thus providing solubility or even biocompatibility and stealth properties to the formed nanoparticles. This principle has been extensively exploited over the years as a means of developing porphyrin-loaded micelles for potential PDT applications. Starting with the work of Kataoka (a pioneer in PICs) and his coworkers on dendritic zinc porphyrins [13,14], followed by the extended studies of the Shi group on 5,10,15,20-tetrakis-(4-sulfonatophenyl)-porphyrin (TPPS) and its metal ions [15–20], as well as numerous other similar investigations involving not only DHBC [21–26] but also surfactants [27,28], polycations [29], nonionic triblock copolymers [30], or even polymeric membranes [31], one can only begin to fathom the importance of these systems considering their prospective therapeutic capability, and thus justify the ongoing considerable scientific interest they have attracted.

In this work, we report on the formation of photofunctional nanoparticles (owing to the intrinsic properties of the porphyrin) through the electrostatic complexation between the poly(2-(dimethylamino) ethyl methacrylate)-*b*-poly[(oligo ethylene glycol) methyl ether methacrylate] (PDMAEMA-*b*-POEGMA) double hydrophilic block copolymer or its quaternized strong polyelectrolyte counterpart (QPDMAEMA-*b*-POEGMA) and 5,10,15,20-tetraphenyl-21H,23H-porphine-p,p′,p″,p‴-tetrasulfonic acid tetrasodium hydrate (TPPS) porphyrin. The resulting PIC micelles comprise a mixed polyelectrolyte–porphyrin (Q)PDMAEMA/TPPS core and a hydrophilic biocompatible POEGMA corona/shell. Their solution properties in regard to their mass, size, size distribution and effective charge at varying mixing ratios with increasing porphyrin content were thoroughly investigated by means of dynamic and electrophoretic light-scattering techniques. In parallel, UV-vis and fluorescence detailed spectroscopic studies were conducted to probe the photophysical characteristics of the complexed porphyrin, extract additional information regarding its morphology and aggregation state, and of course validate the photosensitivity of the produced nanoparticles. The pH-dependent charge density of the PDMAEMA polyelectrolyte and aggregation behavior of TPPS porphyrin necessitated the investigation to be performed at different pH values, namely, pH 7 and 3. Moreover, the effect of temperature on the overall properties of the already formed PICs was examined.

2. Materials and Methods

2.1. Materials

The 5,10,15,20-tetraphenyl-21H,23H-porphine-p,p',p'',p'''-tetrasulfonic acid tetrasodium hydrate (TPPS) porphyrin (M_w = 1022.92 g/mol, anhydrous basis) (Scheme 1a) and all other reagents were purchased from Sigma-Aldrich (St. Louis, MO, USA) and used as received.

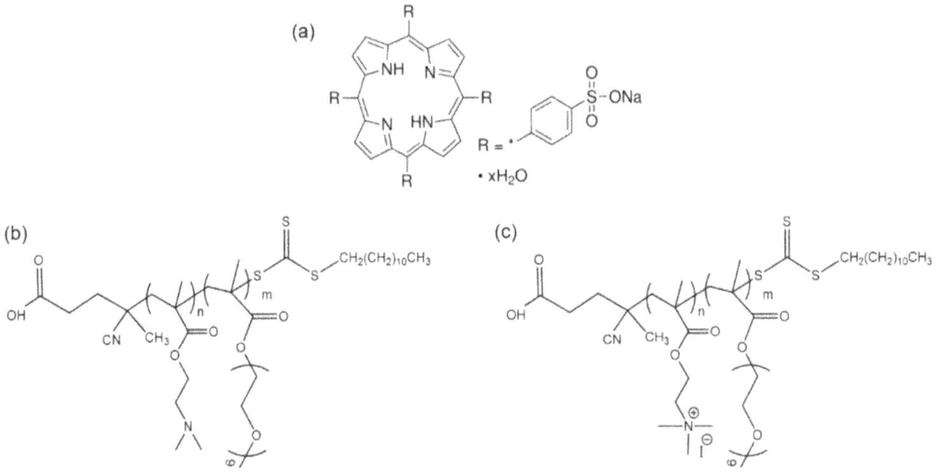

Scheme 1. Molecular structures of (**a**) TPPS porphyrin, (**b**) PDMAEMA-*b*-POEGMA copolymer and (**c**) its quaternized counterpart QPDMAEMA-*b*-POEGMA.

2.2. Block Copolymer Synthesis and Quaternization

The synthetic procedure of the PDMAEMA-*b*-POEGMA block copolymer by means of reversible addition fragmentation chain transfer (RAFT) polymerization, as well as the subsequent quaternization reaction, were performed in a similar manner to the one described in detail in a previous publication [32]. In brief, the synthetic route comprises two steps, starting with the initial polymerization of the DMAEMA monomer under appropriate conditions and the formation of the PDMAEMA homopolymer, which then acts as a macro-chain transfer agent (macro-CTA) for the polymerization of the OEGMA monomer (which bears 9 ethylene glycol units) and the formation of the POEGMA block. The molecular characteristics of the resulting copolymer as determined by SEC and ^{1}H NMR are summarized in Table 1, while its molecular structure is shown in Scheme 1b. The quaternization of the PDMAEMA-*b*-POEGMA precursor was performed by appropriate reaction with methyl iodide (CH$_3$I), thus transforming the tertiary amines of the PDMAEMA block to quaternary ammonium groups, as seen in Scheme 1c. The corresponding molecular characteristics are also presented in Table 1.

Table 1. Molecular characteristics of PDMAEMA-*b*-POEGMA and QPDMAEMA-*b*-POEGMA copolymers.

Copolymer	M_w (10^4 g/mol) [a]	M_w/M_n [a]	(Q)PDMAEMA Content (wt%)	POEGMA Content (wt%)	(Q)PDMAEMA Monomeric Units	POEGMA Monomeric Units
PDMAEMA-*b*-POEGMA	2.29	1.34	33 [b]	67 [b]	48 [c]	32 [c]
QPDMAEMA-*b*-POEGMA	2.97		48 [d]	52 [d]		

[a] Determined by SEC; [b] determined by ^{1}H NMR; [c] calculated based on the M_n of the corresponding monomeric units; [d] calculated assuming 100% quaternization and the composition of the precursor copolymer.

2.3. Preparation of the Polyion Complex Micelles (PICs)

Initially, stock solutions of the PDMAEMA-b-POEGMA copolymer or its quaternized counterpart QPDMAEMA-b-POEGMA at a concentration of 1 mg/mL and TPPS porphyrin at 0.53 mg/mL in water for injection (WFI) were prepared (magnetic stirring was usually employed in order to assist solubilization) and left overnight at ambient conditions to equilibrate. The pH of these solutions was around 6.5, henceforth called pH 7 in the following. For the formation of the complexes, appropriate aliquots, namely, 0.4, 1, 1.4, and 2 mL, of the porphyrin solution were added to 2 mL of the copolymer solution under stirring, mixing of the two constituent solutions was continued for 5 min, and at the final stage, dilution with WFI to a total volume of 10 mL was performed. In this way, the copolymer concentration is kept constant throughout the series of complex solutions, while porphyrin concentration increases thus changing the molar ratio of the two components. The resulting solutions were stable and no precipitation was observed. The stock solution concentration and mixing ratios were suitably chosen so as the molar ratio of the TPPS sulfonate groups to the corresponding tertiary amine groups of the DHBC or quaternary ammonium groups of its quaternized counterpart (QDHBC)—denoted as SO_3^-/NR_2 or NR_3^+, where R stands for the methyl group—ranged from 20% to 100%. The characteristics of the complex solutions in regard to the final concentration of the components and charged groups ratio are given in Table 2. Exactly the same procedure was followed for the preparation of the corresponding complexes at pH 3, starting with copolymer and porphyrin stock solutions whose pH was adjusted to 3 by appropriate addition of 0.1 M HCl, leading again to stable complex solutions.

Table 2. Characteristics of the complex solutions of the DHBC or the QDHBC and TPPS porphyrin.

Sample Name	$C_{(Q)DHBC}$ (mg/mL)	C_{TPPS} (mg/mL)	SO_3^-/NR_2 or NR_3^+ [a]
(Q)Comp(2 + 0.4)	0.2	0.021	20%
(Q)Comp(2 + 1)		0.053	50%
(Q)Comp(2 + 1.4)		0.074	70%
(Q)Comp(2 + 2)		0.105	100%

[a] $R = CH_3$.

2.4. Methods

Dynamic Light Scattering (DLS). DLS measurements were performed on an ALV/CGS-3 compact goniometer system (ALVGmbH, Hessen, Germany) equipped with an ALV-5000/EPP multi-tau digital correlator, a He-Ne laser (λ = 632.8 nm), and an avalanche photodiode detector. All sample solutions were filtered through 0.45 μm hydrophilic PVDF syringe filters (ALWSCI Group, Hangzhou, China) before measurement in order to remove any dust particles or large aggregates. The samples were loaded into standard 1 cm width soda-lime glass dust-free cylindrical cells and measurements were performed at a series of angles in the range of 45–135°. A circulating water bath was used to set the temperature at 25 °C, or at the desired value (25–60 °C) in the case of temperature-dependent measurements.

The measured normalized time autocorrelation functions of the scattered light intensity $g_2(t)$ were fitted with the aid of the CONTIN analysis, thus obtaining the distribution of relaxation times τ. Assuming that the observed fluctuations of the scattered intensity are caused by diffusive motions, the apparent diffusion coefficient D_{app} is related to the relaxation time τ as $D_{app} = 1/\tau q^2$, where q is the scattering vector defined as $4\pi n_0 \sin(\theta/2)/\lambda_0$ with n_0, θ and λ_0 the solvent refractive index, the scattering angle, and the wavelength of the laser in vacuum, respectively. From the apparent diffusion coefficient D_{app}, the hydrodynamic radius R_h can be obtained using the Stokes–Einstein relationship $R_h = k_B T / 6\pi \eta_0 D_{app}$, where k_B is the Boltzmann constant, T is the temperature, and η_0 is the viscosity of the solvent [33,34].

Electrophoretic Light Scattering (ELS). Zeta potential (ζ_P) measurements were conducted with a Zetasizer Nano-ZS (Malvern Panalytical Ltd., Malvern, United Kingdom) equipped with a He-Ne laser (λ = 633 nm) and an avalanche photodiode detector. The Henry equation in the Smoluchowski approximation was used for zeta potential calculation [35]. ζ_P values were determined using the Smoluchowski equation $\zeta_P = 4\pi\eta v/\varepsilon$, where η is the solvent viscosity, v the electrophoretic mobility, and ε the dielectric constant of the solvent, and are reported as averages of fifty repeated measurements at a 13° scattering angle and room temperature.

UV-vis spectroscopy. UV-vis absorption spectra of the complexes and neat porphyrin and copolymer solutions were recorded between 200 and 800 nm wavelength using a UV-vis NIR double-beam spectrophotometer (Lambda 19 by Perkin Elmer, Waltham, MA, USA) at room temperature. Appropriate dilutions of the samples were performed so as not to exceed maximum acceptable absorbance values. Specifically, the complex solutions were diluted with a factor 1:20 and the neat TPPS 1:100 (final concentration 5.3 μg/mL).

Fluorescence spectroscopy. Steady-state fluorescence spectra of the neat and complexed TPPS porphyrin were recorded with a double-grating excitation and a single-grating emission spectrofluorometer (Fluorolog-3, model FL3-21, Jobin Yvon-Spex, Horiba Ltd., Kyoto, Japan) at either room temperature or higher temperatures regulated by a circulating water bath. Excitation wavelength used was λ = 515 nm and emission spectra were recorded in the region 535–800 nm, with an increment of 1 nm, using an integration time of 0.1 s, and slit openings of 2 nm for both the excitation and the emitted beam. Under the employed experimental conditions, fluorescence from the TPPS was observed and utilized to extract information regarding its structure, while the neat copolymer solutions did not show any significant fluorescence.

3. Results and Discussion

3.1. PDMAEMA-b-POEGMA and TPPS Complexation

At first, we investigated the formation of PICs through the electrostatic complexation between the PDMAEMA-*b*-POEGMA DHBC and TPPS porphyrin at both pH 7 and 3. The pH value plays a significant role in the protonation state of both the tertiary amine groups of the DHBC, which become fully protonated at pH 3, thus increasing charge density, as well as the pyrrolic nitrogen atoms of the porphyrin molecule that are protonated at pH below 5, thus transforming the porphyrin from its free-base form (H_2TPPS^{4-}) to its diacid form (H_4TPPS^{2-}) [15–17]. The solution properties of the stable complex solutions were initially examined by means of DLS measurements, and Figure 1 shows the obtained scattering intensity values derived from measurements at 90° (I_{90}) as a function of the porphyrin concentration (C_{TPPS}) or equivalently charged groups ratio (SO_3^-/NR_2, R = CH_3) at both pH values. Note that the I_{90} values at C_{TPPS} = 0 correspond to those of the neat DHBC solutions at the same concentration as in the complexes (C_{DHBC} = 0.2 mg/mL).

The observed gradual increase in I_{90} values at both pH conditions serves as a proof of complexation and PIC formation, since the measured intensity is proportional to the mass of the scattering species in the solution. The apparent decrease at highest C_{TPPS} and pH 3 is most probably an undesired effect of the filtration of the samples (a standard practice for DLS measurements so as to remove any large dust particles or aggregates), which is always associated with some degree of solute retention and of course is more pronounced at higher concentrations. At low C_{TPPS} values (<0.06 mg/mL), the mass of the complexes formed at pH 3 is larger than the corresponding one at pH 7, a fact that indicates a higher degree of interaction and can be attributed to the increased charge density of the DHBC due to amine protonation. Nevertheless, as the concentration of the porphyrin increases (C_{TPPS} > 0.07 mg/mL), an abrupt increase in the mass of the formed PICs at pH 7 is observed and likely denotes some degree of aggregation.

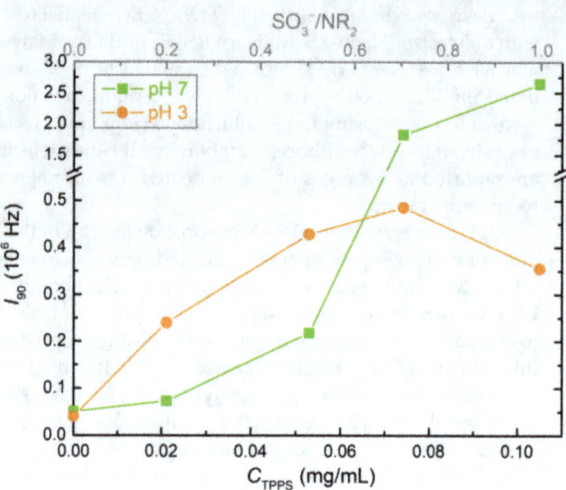

Figure 1. DLS intensity values at 90° I_{90} for the DHBC/TPPS complex solutions at pH 7 and 3, as a function of porphyrin concentration C_{TPPS} or charged groups ratio SO_3^-/NR_2 (R = CH$_3$).

Further insight regarding the properties of the complexes can be obtained from the equivalent size distribution functions (SDFs) extracted through the CONTIN regularization of the DLS measurements at 90° that are presented in Figure 2, along with the hydrodynamic radius R_h values of the individual peaks as a function of C_{TPPS} or the SO_3^-/NR_2 ratio at both pH values. The corresponding SDFs of the neat DHBC and TPPS solutions at relevant concentrations (C_{DHBC} = 0.2 mg/mL and C_{TPPS} = 0.053 mg/mL) are also included for comparison, with the extracted R_h values being noted in Figure 2c as the point at C_{TPPS} = 0 for the DHBC or the dashed line for TPPS.

With respect to the sizes of the species in solution, first of all we should point out that the DHBC exhibits two peaks, one small with an R_h about 9 nm, which possibly represents the molecularly dissolved copolymer single chains, and a second larger one about 90 or 70 nm depending on the pH that most likely indicates some degree of self-assembly and the formation of multichain aggregates. Although both blocks of the PDMAEMA-b-POEGMA copolymer are deemed hydrophilic, hydrophobic interactions stemming from the hydrophobic nature of the polymeric backbone can never be excluded and thus lead to the proposed formation of multichain domains, as also previously observed in analogue systems [36,37]. However, the solutions of the complexes exhibit only one peak in all cases, suggesting the presence of monodisperse PICs with R_h values ranging from 10 to 20 nm. As it seems upon interaction with the porphyrin the multichain domains of the DHBC breakup—likely because the electrostatic interactions in the system are stronger—and this way, only one type of complexes comprising of TPPS and single copolymer chains are formed. As the concentration of the porphyrin increases, so does the size of the PICs at both pH conditions, indicating further incorporation of TPPS. A more pronounced increase is evident at pH 3, leading to larger complexes, and is probably associated with the higher degree of interaction caused by protonation of the DHBC amine groups. Remarkably, the observed abrupt increase of the PICs mass at pH 7 and high C_{TPPS} is not accompanied by a similar increase in size; therefore, it must be correlated with some type of conformational change that leads to more compact structures. One final observation here is that the neat porphyrin has a hydrodynamic radius of about 10 nm, surely significantly larger than its actual molecular size, which is about 20 Å [38]. Hence, we conclude that porphyrin is also aggregated to some extent.

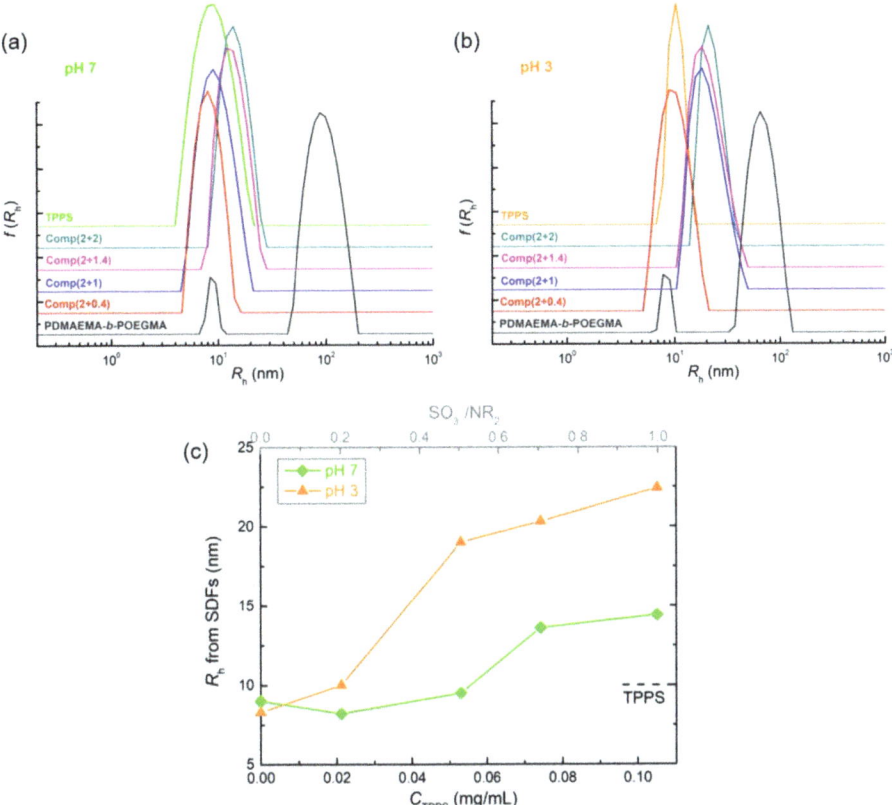

Figure 2. Size distribution functions (SDFs) derived from DLS measurements at 90° for the DHBC/TPPS complex solutions at pH (**a**) 7 and (**b**) 3, along with (**c**) the R_h values of the individual peaks as a function of C_{TPPS} or SO_3^-/NR_2 ratio. The dashed line denotes the corresponding R_h value of neat TPPS.

Another crucial characteristic of the formed PICs is their effective charge, which was acquired via electrophoretic light scattering (ELS) measurement, yielding the zeta potential ζ_P values presented in an identical manner to the previous DLS results in Figure 3. At both pH values, a decrease in ζ_P values of the complexes in regard to that of the neat DHBC (points at $C_{TPPS} = 0$) is seen, suggesting the neutralization of charges that takes place upon interaction of the two oppositely charged species. At pH 7, we observe a broader change that eventually leads to charge inversion from positive to negative values, or in other words the negative charge of TPPS ($\zeta_P \approx -20$ mV) prevails. On the contrary, at pH 3, though the initial decrease is more abrupt, the effective charge of the PICs remains positive for all complex solutions. This is apparently a consequence of the increased charge density of the positive DHBC (as also evident from the corresponding ζ_P values) and the parallel reduction of TPPS negative charges (note that at pH 3 TPPS has a $\zeta_P \approx -10$ mV) due to N protonation. Overall, the number of positive charges in the system is greater at pH 3, thus resulting in positively charged PICs.

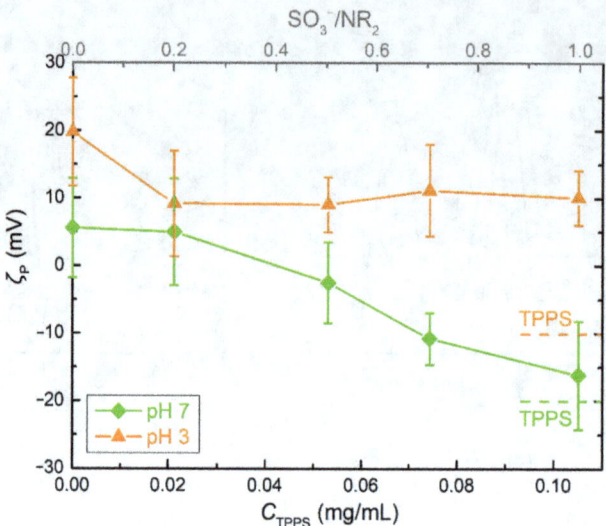

Figure 3. Zeta potential ζ_P values for the DHBC/TPPS complex solutions at pH 7 and 3, as a function of C_{TPPS} or SO_3^-/NR_2 ratio. The dashed lines denote the corresponding ζ_P values of neat TPPS.

Of equal importance to their solution characteristics are the optical properties of the formed PICs. For this reason, the complex solutions of the DHBC/TPPS system at both pH conditions were studied by UV-vis spectroscopy and the collected spectra are shown in Figure 4, where the corresponding spectra of the neat DHBC and TPPS solutions are also included for comparison. As expected, the DHBC does not exhibit any significant absorbance, so the observed peaks are attributed to the presence of the porphyrin. The spectroscopic features of TPPS are well known and can be directly correlated with the protonation and aggregation state of the porphyrin molecule [15–17,27,29,31,39]. As previously mentioned, at physiological pH values, TPPS exists in its free-base monomeric state (H_2TPPS^{4-}), which at high concentration or increased packing conditions can form face-to-face H-aggregates through π–π stacking of the molecules. On the other hand, at low pH values, the protonated diacid monomer (H_4TPPS^{2-}) is dominant and these zwitterionic molecules can interact electrostatically via their positive central and negative peripheral charges, forming side-by-side J-aggregates. Both types of TPPS monomers, as well as their aggregated counterparts, have distinctive absorption signatures.

In this case, TPPS solution at pH 7 shows a Soret band located at 413 nm accompanied by a Q-IV band at 516 nm, both indicative of the H_2TPPS^{4-} monomeric free-base form. In comparison, a gradual blue shift of about 8 nm is observed for the Soret band of the complex solutions, peaking at around 405 nm, while at the same time the Q-bands are slightly shifted to higher wavelengths (red shift). These two findings are related to the formation of H-type aggregates [15–17], an apparent consequence of the complexation with the DHBC, which increases the proximity of the porphyrin molecules and thus enhances their aggregation. Moreover, it is worth noting that for the complex solution with the highest C_{TPPS}, the Soret band shows also significant contribution from the 413 nm peak. This probably signifies the coexistence of H-aggregates and free-base monomers, meaning that some uncomplexed porphyrin molecules are still present in the solution. Therefore, it is possible that the maximum binding capacity of the DHBC has been exceeded.

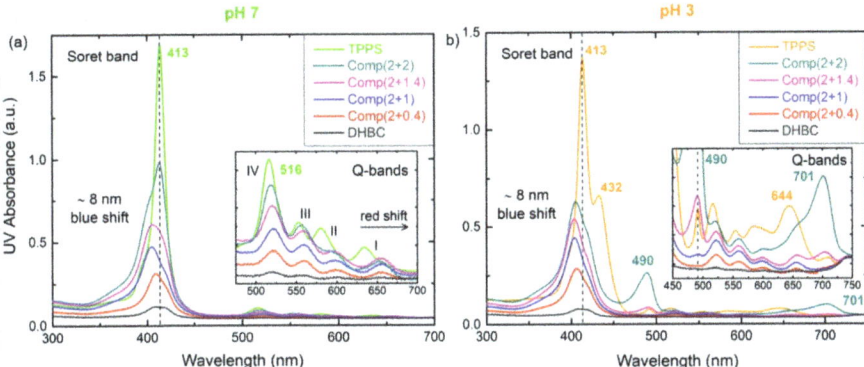

Figure 4. UV-vis spectra of the DHBC/TPPS complex solutions at pH (**a**) 7 and (**b**) 3. The insets show a scaling up of the Q-bands area. The corresponding spectra of neat TPPS and DHBC are also included for comparison.

At pH 3, the neat porphyrin shows a Soret band again peaking at 413 nm—the spectroscopic signature of the free-base monomer—but also a shoulder at 432 nm is observed. This shoulder, along with the distinctive Q-I peak at 644 nm, denote the presence of the diacid monomer [15–17]. Interestingly, it seems that both types of monomeric forms (H_2TPPS^{4-} and H_4TPPS^{2-}) are simultaneously present in the solution. A possible explanation for this is that the aggregation of the porphyrin molecules hinders their protonation to a full extent. It should be mentioned here that although no signs of H-aggregates can be seen in the spectra of neat TPPS at pH 7, these measurements were performed on highly diluted solutions (1:100) so as not to exceed maximum allowed absorbance values. Surely, the presence of aggregates in the initial stock solutions (C_{TPPS} = 0.53 mg/mL) that were used for the preparation of the PICs cannot be excluded, as was also evidenced by the corresponding hydrodynamic radii ($R_h \approx$ 10 nm). In regard to the complexes, the Soret band is once more shifted at 405 nm and the Q-bands are slightly red-shifted, denoting the existence of H-aggregates. Nevertheless, as porphyrin concentration increases the representative peaks of J-aggregates at 490 and 701 nm appear [15–17], as clearly obvious in the case of Comp (2 + 2) but also noticeable for the solutions of Comp (2 + 1.4) and neat TPPS, as seen in the inset of Figure 4b. Apparently, at higher C_{TPPS} values the number of TPPS molecules participating in the complexes and hence their proximity increases drastically, leading to J-type aggregation. Lastly, no signs of uncomplexed TPPS can be discerned in this pH value, one more indication of the increased interaction capability of the DHBC due to the protonation of its amine groups.

Fluorescence spectroscopy measurements of the same PIC solutions were also performed and the relative emission spectra are shown in Figure 5. Overall, the fluorescence spectral features corroborate the findings from the UV-vis measurements. At pH 7, the neat TPPS solution displays a double peak at 641 and 698 nm, which is typical of the free-base monomer [16,27,29]. A red shift of about 20 nm is observed for the complex solutions with the main emission peak appearing at 663 nm, thus signifying the presence of H-aggregates. Notably, as porphyrin concentration increases, contribution from the 641 nm peak becomes more and more prominent, culminating in the case of the Comp (2 + 2) solution, again indicating uncomplexed TPPS monomeric molecules. Respectively, at pH 3 a significant decrease in the fluorescence intensity of the neat TPPS is observed and the main peak is located around 677 nm, both spectral traits of the diacid H_4TPPS^{2-} form [16,27,29]. However, a small shoulder at 636 nm probably correlated with the free-base monomer is still discerned. The obtained emission spectra for the solutions of the PICs resemble the ones at pH 7, with the main peak about 667 nm (existence of H-aggregates). For the two complex solutions with higher porphyrin concentration, extended quenching is observed along with the appearance of the peak at 732 nm, transitions that can be attributed to the formation of J-aggregates [16,27,29].

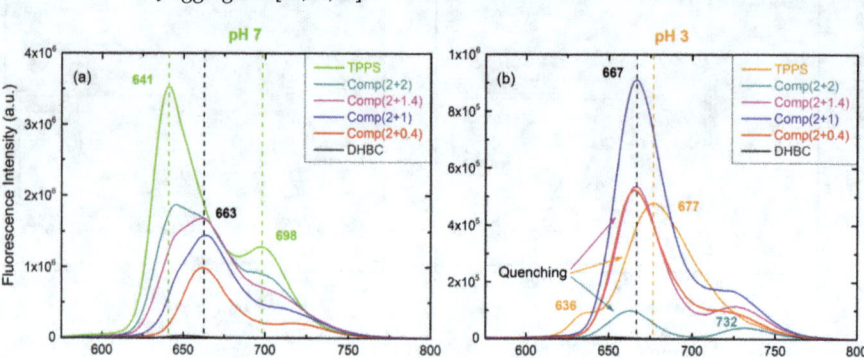

Figure 5. Fluorescence spectra of the DHBC/TPPS complex solutions at pH (**a**) 7 and (**b**) 3. The corresponding spectra of neat TPPS and DHBC are also included for comparison.

3.2. QPDMAEMA-b-POEGMA and TPPS complexation

Undoubtedly, electrostatic interactions play the most crucial role during the coupling of (macro)molecular species bearing opposite charges and essentially govern not only the complexation process itself but the properties of the resulting nanoparticles as well. Ergo, any change in the charged state of the individual components is expected to influence the final outcome and for this reason is worth investigating. On these grounds, we also studied the interaction of the quaternized homologue QPDMAEMA-*b*-POEGMA copolymer, which is a strong polyelectrolyte independently of pH with the TPPS porphyrin at both acidic and neutral pH conditions. Figure 6 presents the collective DLS and ELS results attained for the QDHBC/TPPS system in regard to scattering intensity, hydrodynamic radius, size distribution and effective charge, as a function of C_{TPPS} or equivalently the SO_3^-/NR_3^+ (R = CH_3) charged groups ratio at both pH 7 and 3.

As seen in Figure 6a, the mass of the complexes (proportional to the scattering intensity) increases with increasing C_{TPPS}, validating once again their formation. Somewhat larger I_{90} values are observed at pH 3 than at 7, although in this case the charge density of the QDHBC is the same at both pH values. Possibly, the higher mass of the complexes at pH 3 could be attributed to the differences in the protonation and/or aggregation state of the porphyrin molecule, as well as hydration state of the copolymer due to the additional CH_3 group. Moreover, generally higher scattering values are recorded in comparison to the previous DHBC/TPPS system (see Figure S1), indicating an analogous difference in the mass of the corresponding PICs. Quite interestingly, abrupt increase in I_{90} at pH 7 occurs only at the highest porphyrin concentration, demonstrating a more gradual and less intense transition to more compact conformations.

A better understanding of the observed changes in the mass of the complexes can be gained by examining the corresponding differences in their size and size distributions (Figure 6b–d). First of all, the QDHBC shows two peaks signifying the presence of single copolymer chains with a size of about 5 nm, along with multichain domains/aggregates giving peaks about 70 or 80 nm (at pH 7 or 3, respectively), as was the case for the precursor copolymer. The remarkable difference for this system is that the majority of the complex solutions also exhibit two peaks, which most likely denote the presence of two different types of complexes in regard to the morphology of the QDHBC. In other words, both the single chains and the multichain aggregates of the copolymer form complexes with the porphyrin. At pH 7 and high C_{TPPS}, the second peak is no longer discerned, suggesting that the multichain aggregates break up as the interaction with the TPPS becomes more prominent and only one type of PIC is preserved. On the contrary, at pH 3 both types of PICs coexist throughout the whole range of mixing ratios (i.e., component analogies), a fact that clearly justifies the observed increased mass. As it seems the different protonation/aggregation state of TPPS at pH 3 entails a different type of interaction with the quaternized copolymer, presumably weaker, that apparently is not able to cause the dissociation of the multichain domains. A small increase in most peaks (apart from R_{h2} at pH 7) can be seen, possibly indicating the incorporation of more porphyrin molecules.

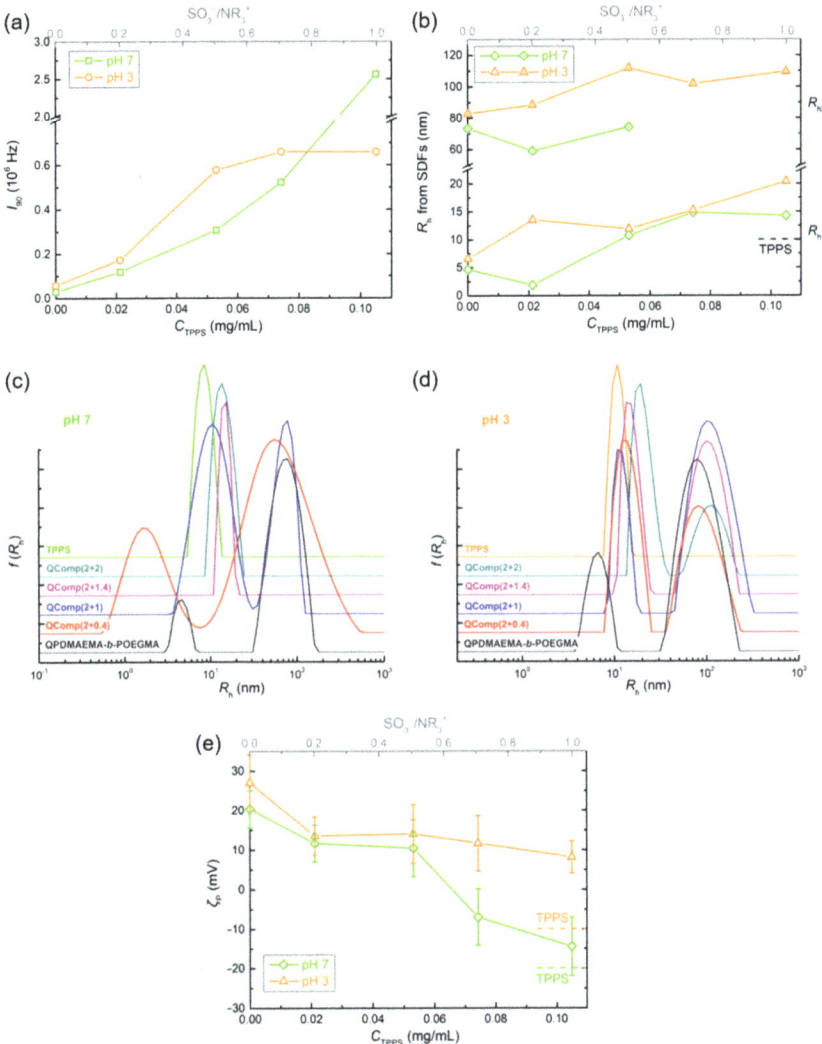

Figure 6. Collective DLS and ELS results in regard to (**a**) the scattering intensity at 90°, (**b**) the hydrodynamic radius derived from (**c**,**d**) the corresponding size distribution functions (SDFs) at 90°, and (**e**) zeta potential values for the QDHBC/TPPS complex solutions at pH 7 and 3, as a function of porphyrin concentration C_{TPPS} or charged groups ratio SO_3^-/NR_3^+ (R = CH$_3$).

As far as the effective charge of the complexes is concerned, the ζ_P values decrease as C_{TPPS} increases (Figure 6e), a direct and anticipated consequence of the charge neutralization that takes place upon interaction. Again, the overall change is greater at pH 7 than at 3 owing to the increased number of positive charges related to TPPS protonation and aggregation state.

Additional information about the state of the porphyrin molecules in the complexes can be derived from the UV-vis and fluorescence spectra of the QDHBC/TPPS system at both pH values shown in Figure 7. Absorbance spectra at pH 7 are almost identical to the ones measured for the previous system, characterized by a blue-shifted Soret band peaking about 404 nm and somewhat red-shifted Q-bands indicative of H-aggregation. Similarly, the same observations regarding the formation of H-type aggregates can be made at pH 3, while at high porphyrin concentrations the distinctive spectral features of J-aggregates are ascertained (peaks at 491 and 703 nm). Note that at both pH conditions and highest porphyrin content, the contribution of the 413 nm peak is clearly evident in the Soret band of the complexes, revealing the existence of uncomplexed TPPS and thus implying that the maximum binding capacity of the quaternized copolymer has been reached. Fluorescence spectroscopy measurements confirm the attained rationalization about H-aggregation by showing at both pH 7 and 3 a main peak located around 667 nm. Nevertheless, a more pronounced quenching of the emission signal occurs in this case, meaning that the aggregation of the porphyrin molecules is more extensive. Furthermore, at pH 3 and high C_{TPPS}, telltale signs of J-aggregates (extensive quenching and peak at 732 nm) can be seen.

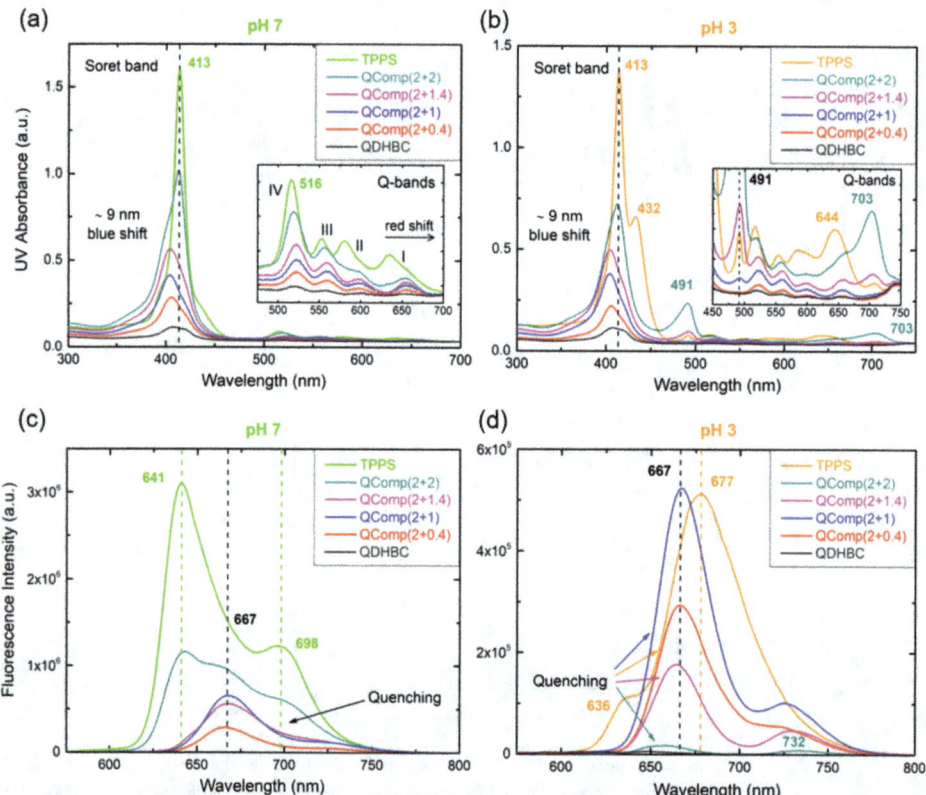

Figure 7. UV-vis (**a**,**b**) and fluorescence (**c**,**d**) spectra of the QDHBC/TPPS complex solutions at pH 7 (left) and 3 (right). The insets in the top row show a scaling up of the Q-bands area. The corresponding spectra of neat TPPS and QDHBC are also included for comparison.

3.3. Comparison of the Two Systems in Different Conditions

Certainly, the sum of the results presented thus far emphasizes the importance of self-assembly processes governed by electrostatic and/or hydrophobic interactions when one deals with such nanostructured hybrid systems. According to our observations, the nature of the macromolecular copolymer in regard to its protonation degree or equivalent charge density, as well as solvation state and latent hydrophobic nature that leads to the formation of multichain domains, along with the complementary features of the porphyrin molecule mainly involving its protonation and aggregation state, constitute the key factors influencing the structure and properties of the resulting PICs. Recapitulating, both the precursor PDMAEMA-b-POEGMA copolymer and its quaternized counterpart form multichain aggregates at either pH 7 or 3 (apparently stemming from hydrophobic backbone interactions) coexisting with single polymeric chains in solution. In an analogue manner, TPPS porphyrin exhibits two protonation states, that is, the free-base (H_2TPPS^{4-}) at pH 7 or diacid (H_4TPPS^{2-}) at pH 3 monomer, which in turn self-assemble when in close proximity into H- or J-type structures due to $\pi-\pi$ stacking or electrostatic binding, respectively. Upon mixing of these different species in the case of the DHBC/TPPS system at low porphyrin concentration and both pH conditions, monodisperse PICs comprising H-aggregated TPPS molecules and singe DHBC chains are formed. As the porphyrin content increases, so does its aggregation, leading to more compact complexed H-aggregates at pH 7, integrating more of TPPS molecules and eventually exceeding DHBC binding capacity. Accordingly, at pH 3 both H- and J-aggregates are incorporated in the resulting complexes, since both types are present in the corresponding neat TPPS solution, while the increased charge density of the DHBC causes a higher degree of interaction (no uncomplexed TPPS is discerned). When it comes to the quaternized DHBC, it seems that the consecutive change in the hydration state of the copolymer is associated with more stable multichain aggregates that do not easily dissociate during complexation. This way, two types of distinct PICs are formed at both pH values and low C_{TPPS} corresponding to single and multichain copolymer–porphyrin complexes. However, as the porphyrin concentration becomes higher at pH 7 only the single-chain type of PICs remains incorporating more TPPS molecules (suggesting that porphyrin aggregation is more prominent), while at pH 3 both types coincide and even complexes of J-aggregated TPPS are observed. Still, at both pH conditions spectroscopic signs of uncomplexed porphyrin are recorded, a fact that implies a reduced binding affinity for the quaternized copolymer in comparison to the precursor DHBC, a counterintuitive finding that points out the significance of the intrinsic polymeric solution properties. This proposed interaction scenario is schematically depicted in the following illustration (Scheme 2).

3.4. Temperature Effect on the Formed PICs

The ability of photosensitizers intended for clinical applications to respond to externally applied stimuli such as temperature is of particular interest, since it can open new treatment pathways and even enable combination of therapeutic protocols or methods, thus enhancing their efficiency potential [4,40]. To this end, we investigated the effect of temperature on the already formed PICs of both systems under study. Figure 8 shows the summary of DLS results regarding the values of the scattering intensity I_{90} and the hydrodynamic radius R_h derived from the corresponding size distribution functions (SDFs)-presented in Figures S2–S4 obtained via measurements performed at 90° and different temperatures ranging from 25 to 60 °C (with a 5 °C increment), for two representative solutions of complexes-Comp (2 + 1) and Comp (2 + 2)-of the DHBC or QDHBC/TPPS system at both pH 7 and 3. The corresponding values of the neat copolymers and porphyrin solutions are also included for comparison. In a similar manner, fluorescence spectra of the same representative solutions measured at temperatures of 25, 40, 60 °C, and back at 25 °C after heat treatment are displayed in Figure 9, along with the corresponding spectra of neat TPPS.

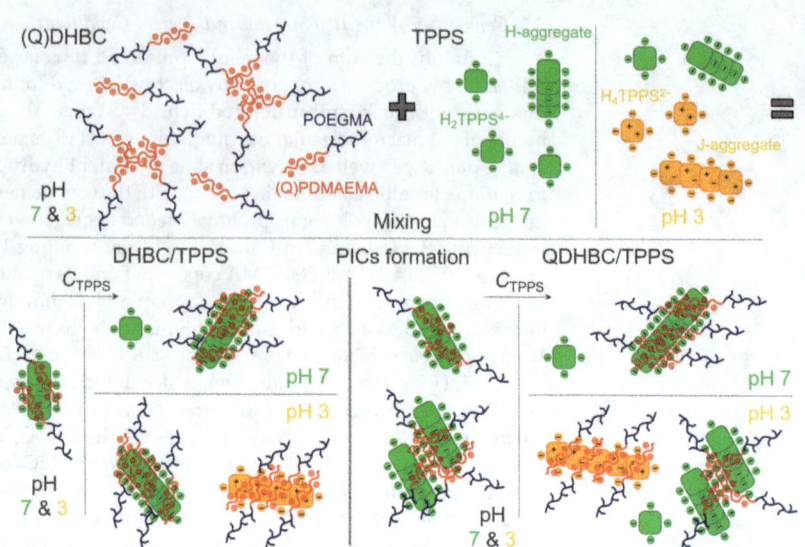

Scheme 2. Proposed interaction illustration for the formation of PICs between the DHBC or its quaternized counterpart QDHBC and TPPS porphyrin at different pH values and porphyrin contents.

At first, the effect of temperature on the pure constituents of the complex systems is presented. For both block copolymers, their mass (proportional to I_{90}) seems rather constant throughout the range of studied temperatures. In regard to their size, although the R_h of the first peak (single copolymer chains) shows small changes, an apparent decrease in the R_h of the second peak is observed (see also Figure S4). This fact indicates that the multichain aggregates shrink as temperature increases, most probably because hydrophobic interactions become stronger. When temperature drops back to ambient conditions, all sizes are restored. As far as porphyrin is concerned, its size is almost unaffected but a systematic decrease of its mass can be discerned, especially in the case of pH 3, where it is more pronounced. This decrease of mass is also accompanied by a significant transition in the respective fluorescence spectra, with the one at 60 °C showing clear signs of the free-base monomer (Figure 9e). Apparently, disaggregation (at least to some extent) of the porphyrin H-aggregates-present at both pH values, as previously discussed-takes place, which is eventually reestablished when the solution cools down.

For the complex solutions of the DHBC/TPPS system, only small changes in their mass are observed, apart from the noticeable decrease of the scattering intensity in the case of the Comp (2 + 2) solution at pH 7, which in combination with the parallel increase in emission intensity could denote a dissociation of the aggregated porphyrin molecules. At the same time, their sizes are either stable or suggest slight shrinking as a consequence of enhanced hydrophobic interactions. This seems to also be the case for the smaller population of complexes of the QDHBC/TPPS system, the one that corresponds to single copolymer chain–porphyrin complexes. Remarkably, the larger complexes that are correlated with the copolymer multichain aggregates either diminish or even disappear completely upon heating. This change is also expressed as an increase in the fluorescence intensity or in other words dequenching, denoting possible disaggregation of TPPS. Most likely at higher temperatures, the interaction between the copolymer and porphyrin molecules is more favorable, facilitated also by the apparent dissociation of porphyrin aggregates, thus leading to the dissociation of the multichain complexes. Overall, the increase in temperature showed that the formed PICs are susceptible to conformational rearrangements so as to compensate for the loss of hydration due to the stronger hydrophobic interactions.

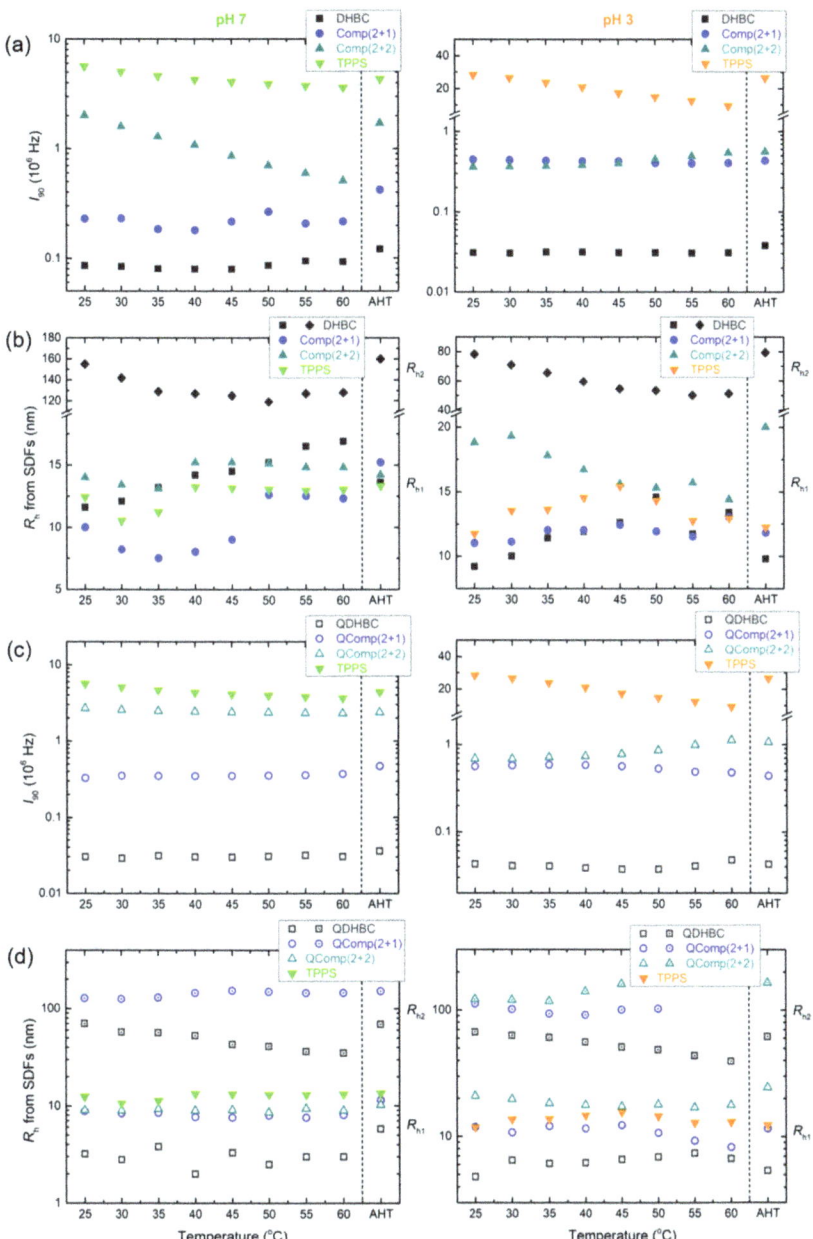

Figure 8. DLS results in regard to (**a,c**) the scattering intensity at 90° and (**b,d**) the hydrodynamic radius derived from the corresponding size distribution functions (SDFs) at 90°, for the complex solutions (Q)Comp (2 + 1) and (Q)Comp (2 + 2) of the (Q)DHBC/TPPS systems at pH 7 (left) and 3 (right) at different temperatures ranging from 25 to 60 °C (5 °C step), and cooled back at 25 °C after heat treatment (AHT). The corresponding values of neat (Q)DHBC and TPPS are also included for comparison.

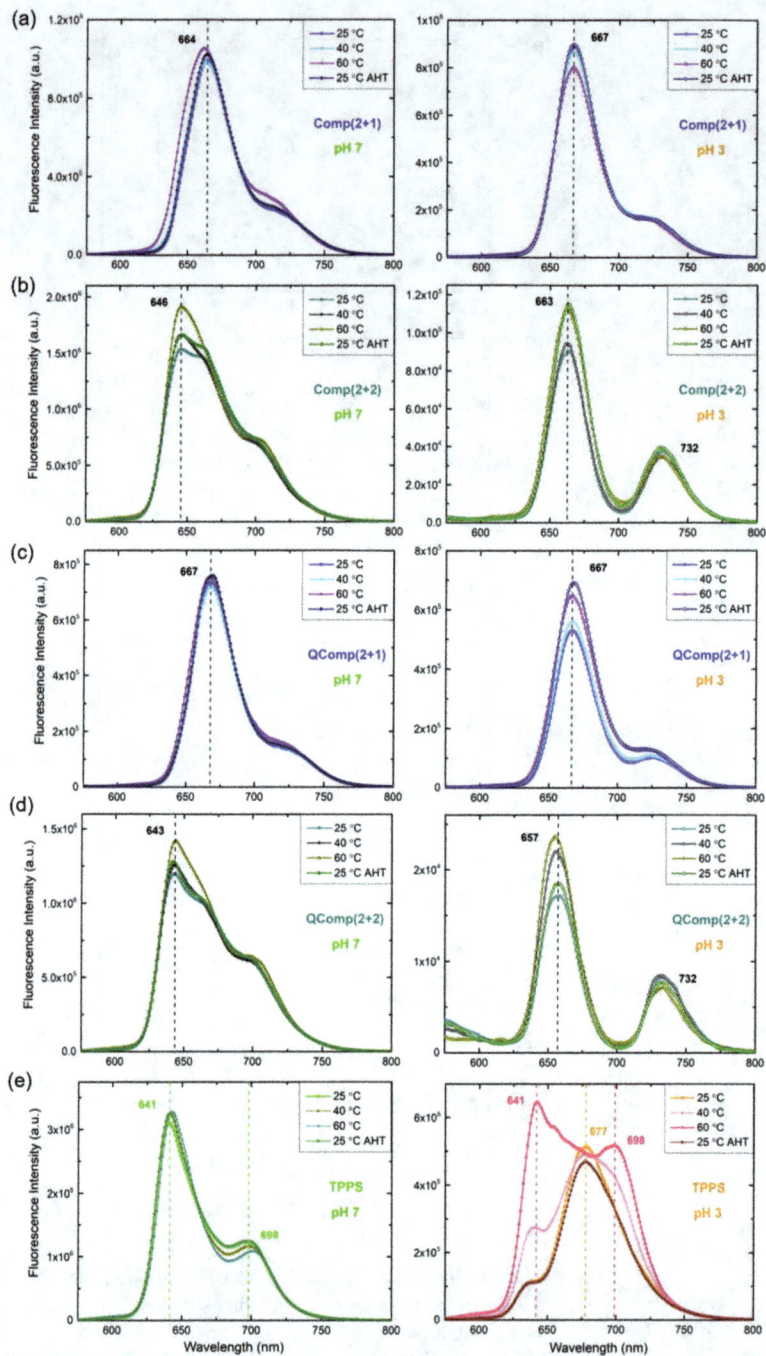

Figure 9. Fluorescence spectra of the complex solutions (**a,c**) (Q)Comp (2 + 1) and (**b,d**) (Q)Comp (2 + 2) of the (Q)DHBC/TPPS systems at pH 7 (left) and 3 (right) at different temperatures: 25, 40, 60 °C, and 25 °C after heat treatment (AHT). (**e**) The corresponding spectra of neat TPPS for comparison.

4. Conclusions

The electrostatic complexation between the DHBC PDMAEMA-*b*-POEGMA and the corresponding quaternized strong block polyelectrolyte QPDMAEMA-*b*-POEGMA with TPPS porphyrin led to the formation of nanostructured photoactive PICs. Both the solution and optical properties and the morphology and structure of these complexed hybrid species proved to be dependent on the protonation and aggregation state of their original constituents, the pH of the solution, and the analogy between the two components. Monodisperse small PICs were formed in the case of the DHBC/TPPS system at both pH 7 and 3, consisting of single copolymer chains and H- or J-type aggregates of TPPS depending on the pH. The increase in TPPS content resulted in more compact complexed porphyrin aggregates. Respectively, the analogous QDHBC/TPPS system was characterized by the presence of two different types of complexes, those resulting from the interaction of the single chains or the multichain aggregates of the quaternized copolymer and the corresponding H-aggregates of TPPS at both pH values and low porphyrin contents. At higher TPPS concentration, increased aggregation of the porphyrin occurred, subsequently affecting the structure of the complexes and thus leading to larger H- or even J-type complexed aggregates. The optical properties of the PICs for both systems were directly correlated with the intrinsic protonation and/or aggregation state of TPPS, which was evidently further enhanced upon interaction with the copolymers. Actually, the quaternized counterpart caused a greater degree of porphyrin aggregation compared to the precursor DHBC. Finally, the formed PICs responded to the increase in temperature by appropriate conformational rearrangements that led to disaggregation of the complexed porphyrin in the case of multichain–porphyrin complexes or more compact structures (shrinking) for the single-chain ones.

Supplementary Materials: The following supporting information can be downloaded at: https://www.mdpi.com/article/10.3390/polym14235186/s1, Figure S1: Comparison of DLS scattering intensity values at 90° I_{90} for the DHBC/TPPS (closed symbols) and the QDHBC/TPPS (open symbols) complex solutions at pH 7 and 3, as a function of porphyrin concentration C_{TPPS} or charged groups ratio SO_3^-/NR_2 or NR_3^+ (R = CH_3).; Figure S2: Size distribution functions (SDFs) derived from DLS measurement at 90° for the complex solutions (a, b) Comp (2 + 1) and (c, d) Comp (2 + 2) of the DHBC/TPPS system at pH 7 (left) and 3 (right) at different temperatures ranging from 25 to 60 °C (5 °C step), and cooled back to 25 °C after heat treatment (AHT); Figure S3: Size distribution functions (SDFs) derived from DLS measurement at 90° for the complex solutions (a,b) QComp (2 + 1) and (c, d) QComp (2 + 2) of the QDHBC/TPPS system at pH 7 (left) and 3 (right) at different temperatures ranging from 25 to 60 °C (5 °C step), and cooled back to 25 °C after heat treatment (AHT); Figure S4: Size distribution functions (SDFs) derived from DLS measurement at 90° for the neat (a,b) DHBC, (c,d) QDHBC and (e,f) TPPS solutions at pH 7 (left) and 3 (right) at different temperatures ranging from 25 to 60 °C (5 °C step), and cooled back to 25 °C after heat treatment (AHT).

Author Contributions: Conceptualization, S.P.; methodology, D.K and M.K.; formal analysis, D.K. and M.K.; data acquisition, D.K.; data curation, D.K. and M.K.; writing—original draft preparation, M.K.; writing—review and editing, S.P.; supervision, M.K. and S.P. All authors have read and agreed to the published version of the manuscript.

Funding: This research received no external funding.

Institutional Review Board Statement: Not applicable.

Informed Consent Statement: Not applicable.

Data Availability Statement: Data presented in this study are available on reasonable request from the corresponding author.

Acknowledgments: The authors would like to express their gratitude to Varvara Chrysostomou for the synthesis of the copolymers used in this study.

Conflicts of Interest: The authors declare no conflict of interest.

References

1. Sung, H.; Ferlay, J.; Siegel, R.L.; Laversanne, M.; Soerjomataram, I.; Jemal, A.; Bray, F. Global Cancer Statistics 2020: GLOBOCAN Estimates of Incidence and Mortality Worldwide for 36 Cancers in 185 Countries. *CA Cancer J. Clin.* **2021**, *71*, 209–249. [CrossRef] [PubMed]
2. Kou, J.; Dou, D.; Yang, L. Porphyrin Photosensitizers in Photodynamic Therapy and Its Applications. *Oncotarget* **2017**, *8*, 81591–81603. [CrossRef] [PubMed]
3. Kwiatkowski, S.; Knap, B.; Przystupski, D.; Saczko, J.; Kędzierska, E.; Knap-Czop, K.; Kotlińska, J.; Michel, O.; Kotowski, K.; Kulbacka, J. Photodynamic Therapy–Mechanisms, Photosensitizers and Combinations. *Biomed. Pharmacother.* **2018**, *106*, 1098–1107. [CrossRef] [PubMed]
4. Shi, X.; Zhang, C.Y.; Gao, J.; Wang, Z. Recent Advances in Photodynamic Therapy for Cancer and Infectious Diseases. *Wiley Interdiscip. Rev.: Nanomed. Nanobiotechnol.* **2019**, *11*, 1–23. [CrossRef] [PubMed]
5. Tian, J.; Huang, B.; Nawaz, M.H.; Zhang, W. Recent Advances of Multi-Dimensional Porphyrin-Based Functional Materials in Photodynamic Therapy. *Coord. Chem. Rev.* **2020**, *420*, 213410. [CrossRef]
6. Qindeel, M.; Sargazi, S.; Hosseinikhah, S.M.; Rahdar, A.; Barani, M.; Thakur, V.K.; Pandey, S.; Mirsafaei, R. Porphyrin-Based Nanostructures for Cancer Theranostics: Chemistry, Fundamentals and Recent Advances. *ChemistrySelect* **2021**, *6*, 14082–14099. [CrossRef]
7. Magna, G.; Monti, D.; Di Natale, C.; Paolesse, R.; Stefanelli, M. The Assembly of Porphyrin Systems in Well-Defined Nanostructures: An Update. *Molecules* **2019**, *24*, 4307. [CrossRef]
8. Dickerson, M.; Bae, Y. Block Copolymer Nanoassemblies for Photodynamic Therapy and Diagnosis. *Ther. Deliv.* **2013**, *4*, 1431–1441. [CrossRef]
9. Insua, I.; Wilkinson, A.; Fernandez-Trillo, F. Polyion Complex (PIC) Particles: Preparation and Biomedical Applications. *Eur. Polym. J.* **2016**, *81*, 198–215. [CrossRef]
10. Harada, A.; Kataoka, K. Polyion Complex Micelle Formation from Double-Hydrophilic Block Copolymers Composed of Charged and Non-Charged Segments in Aqueous Media. *Polym. J.* **2018**, *50*, 95–100. [CrossRef]
11. Cabral, H.; Miyata, K.; Osada, K.; Kataoka, K. Block Copolymer Micelles in Nanomedicine Applications. *Chem. Rev.* **2018**, *118*, 6844–6892. [CrossRef]
12. Magana, J.R.; Sproncken, C.C.M.; Voets, I.K. On Complex Coacervate Core Micelles: Structure-Function Perspectives. *Polymers* **2020**, *12*, 1953. [CrossRef] [PubMed]
13. Stapert, H.R.; Nishiyama, N.; Jiang, D.L.; Aida, T.; Kataoka, K. Polyion Complex Micelles Encapsulating Light-Harvesting Ionic Dendrimer Zinc Porphyrins. *Langmuir* **2000**, *16*, 8182–8188. [CrossRef]
14. Jang, W.-D.; Nishiyama, N.; Zhang, G.-D.; Harada, A.; Jiang, D.-L.; Kawauchi, S.; Morimoto, Y.; Kikuchi, M.; Koyama, H.; Aida, T.; et al. Supramolecular Nanocarrier of Anionic Dendrimer Porphyrins with Cationic Block Copolymers Modified with Polyethylene Glycol to Enhance Intracellular Photodynamic Efficacy. *Angew. Chemie Int. Ed.* **2005**, *44*, 419–423. [CrossRef] [PubMed]
15. Zhao, L.; Ma, R.; Li, J.; Li, Y.; An, Y.; Shi, L. J- and H-Aggregates of 5,10,15,20-Tetrakis-(4-Sulfonatophenyl)-Porphyrin and Interconversion in PEG-b-P4VP Micelles. *Biomacromolecules* **2008**, *9*, 2601–2608. [CrossRef] [PubMed]
16. Zhao, L.; Xiang, R.; Zhang, L.; Wu, C.; Ma, R.; An, Y.; Shi, L. Micellization of Copolymers via Noncovalent Interaction with TPPS and Aggregation of TPPS. *Sci. China Chem.* **2011**, *54*, 343–350. [CrossRef]
17. Zhao, L.; Xiang, R.; Ma, R.; Wang, X.; An, Y.; Shi, L. Chiral Conversion and Memory of TPPS J-Aggregates in Complex Micelles: PEG-b-PDMAEMA/TPPS. *Langmuir* **2011**, *27*, 11554–11559. [CrossRef] [PubMed]
18. Chai, Z.; Gao, H.; Ren, J.; An, Y.; Shi, L. MgTPPS/Block Copolymers Complexes for Enhanced Stability and Photoactivity. *RSC Adv.* **2013**, *3*, 18351–18358. [CrossRef]
19. Zhao, L.; Li, A.; Xiang, R.; Shen, L.; Shi, L. Interaction of Fe^{III}-Tetra-(4-Sulfonatophenyl)-Porphyrin with Copolymers and Aggregation in Complex Micelles. *Langmuir* **2013**, *29*, 8936–8943. [CrossRef]
20. Chai, Z.; Jing, C.; Liu, Y.; An, Y.; Shi, L. Spectroscopic Studies on the Photostability and Photoactivity of Metallo-Tetraphenylporphyrin in Micelles. *Colloid Polym. Sci.* **2014**, *292*, 1329–1337. [CrossRef]
21. Bo, Q.; Zhao, Y. Double-Hydrophilic Block Copolymer for Encapsulation and Two-Way pH Change-Induced Release of Metalloporphyrins. *J. Polym. Sci. Part A Polym. Chem.* **2006**, *44*, 1734–1744. [CrossRef]
22. Sorrells, J.L.; Shrestha, R.; Neumann, W.L.; Wooley, K.L. Porphyrin-Crosslinked Block Copolymer Assemblies as Photophysically-Active Nanoscopic Devices. *J. Mater. Chem.* **2011**, *21*, 8983–8986. [CrossRef]
23. Liu, S.; Hu, C.; Wei, Y.; Duan, M.; Chen, X.; Hu, Y. Transformation of H-Aggregates and J-Dimers of Water-Soluble Tetrakis (4-Carboxyphenyl) Porphyrin in Polyion Complex Micelles. *Polymers* **2018**, *10*, 494. [CrossRef] [PubMed]
24. Wang, C.; Zhao, P.; Jiang, D.; Yang, G.; Xue, Y.; Tang, Z.; Zhang, M.; Wang, H.; Jiang, X.; Wu, Y.; et al. In Situ Catalytic Reaction for Solving the Aggregation of Hydrophobic Photosensitizers in Tumor. *ACS Appl. Mater. Interfaces* **2020**, *12*, 5624–5632. [CrossRef] [PubMed]
25. Kubát, P.; Henke, P.; Raya, R.K.; Štěpánek, M.; Mosinger, J. Polystyrene and Poly(Ethylene Glycol)-b-Poly(ε-Caprolactone) Nanoparticles with Porphyrins: Structure, Size, and Photooxidation Properties. *Langmuir* **2020**, *36*, 302–310. [CrossRef] [PubMed]
26. Zhao, Y.; Zhu, Y.; Yang, G.; Xia, L.; Yu, F.; Chen, C.; Zhang, L.; Cao, H. A pH/H_2O_2 Dual Triggered Nanoplatform for Enhanced Photodynamic Antibacterial Efficiency. *J. Mater. Chem. B* **2021**, *9*, 5076–5082. [CrossRef] [PubMed]
27. Maiti, N.C.; Mazumdar, S.; Periasamy, N. J- and H-Aggregates of Porphyrin–Surfactant Complexes: Time-Resolved Fluorescence and Other Spectroscopic Studies. *J. Phys. Chem. B* **1998**, *102*, 1528–1538. [CrossRef]
28. Liu, Q.; Zhou, H.; Zhu, J.; Yang, Y.; Liu, X.; Wang, D.; Zhang, X.; Zhuo, L. Self-Assembly into Temperature Dependent Micro-/Nano-Aggregates of 5,10,15,20-Tetrakis(4-Carboxyl Phenyl)-Porphyrin. *Mater. Sci. Eng. C* **2013**, *33*, 4944–4951. [CrossRef]

29. Toncelli, C.; Pino-Pinto, J.P.; Sano, N.; Picchioni, F.; Broekhuis, A.A.; Nishide, H.; Moreno-Villoslada, I. Controlling the Aggregation of 5,10,15,20-Tetrakis-(4-Sulfonatophenyl)-Porphyrin by the Use of Polycations Derived from Polyketones Bearing Charged Aromatic Groups. *Dye. Pigment.* **2013**, *98*, 51–63. [CrossRef]
30. Steinbeck, C.A.; Hedin, N.; Chmelka, B.F. Interactions of Charged Porphyrins with Nonionic Triblock Copolymer Hosts in Aqueous Solutions. *Langmuir* **2004**, *20*, 10399–10412. [CrossRef]
31. Wang, M.; Yan, F.; Zhao, L.; Zhang, Y.; Sorci, M. Preparation and Characterization of a pH-Responsive Membrane Carrier for Meso-Tetraphenylsulfonato Porphyrin. *RSC Adv.* **2017**, *7*, 1687–1696. [CrossRef]
32. Haladjova, E.; Chrysostomou, V.; Petrova, M.; Ugrinova, I.; Pispas, S.; Rangelov, S. Physicochemical Properties and Biological Performance of Polymethacrylate Based Gene Delivery Vector Systems: Influence of Amino Functionalities. *Macromol. Biosci.* **2021**, *21*, 2000352. [CrossRef] [PubMed]
33. Pecora, R. *Dynamic Light Scattering*; Springer: Berlin/Heidelberg, Germany, 1985.
34. Chu, B. *Laser Light Scattering*; Elsevier: Amsterdam, The Netherlands, 1991.
35. Hunter, R.J. *Zeta Potential in Colloid Science*; Elsevier: Amsterdam, The Netherlands, 1981.
36. Karayianni, M.; Radeva, R.; Koseva, N.; Pispas, S. Electrostatic Complexation of a Double Hydrophilic Block Polyelectrolyte and Proteins of Different Molecular Shape. *J. Polym. Sci. Part B Polym. Phys.* **2016**, *54*, 1515–1529. [CrossRef]
37. Giaouzi, D.; Pispas, S. Synthesis and Self-assembly of Thermoresponsive Poly(N-isopropylacrylamide)-b-poly(Oligo Ethylene Glycol Methyl Ether Acrylate) Double Hydrophilic Block Copolymers. *J. Polym. Sci. Part A Polym. Chem.* **2019**, *57*, 1467–1477. [CrossRef]
38. Hollingsworth, J.V.; Richard, A.J.; Vicente, M.G.H.; Russo, P.S. Characterization of the Self-Assembly of Meso-Tetra(4-Sulfonatophenyl)Porphyrin ($H_2TPPS_4^-$) in Aqueous Solutions. *Biomacromolecules* **2012**, *13*, 60–72. [CrossRef] [PubMed]
39. Bolzonello, L.; Albertini, M.; Collini, E.; Di Valentin, M. Delocalized Triplet State in Porphyrin J-Aggregates Revealed by EPR Spectroscopy. *Phys. Chem. Chem. Phys.* **2017**, *19*, 27173–27177. [CrossRef] [PubMed]
40. Huang, B.; Tian, J.; Jiang, D.; Gao, Y.; Zhang, W. NIR-Activated "OFF/ON" Photodynamic Therapy by a Hybrid Nanoplatform with Upper Critical Solution Temperature Block Copolymers and Gold Nanorods. *Biomacromolecules* **2019**, *20*, 3873–3883. [CrossRef] [PubMed]

Article

Polyacrylonitrile-*b*-Polystyrene Block Copolymer-Derived Hierarchical Porous Carbon Materials for Supercapacitor

Ainhoa Álvarez-Gómez [1], Jiayin Yuan [2], Juan P. Fernández-Blázquez [3], Verónica San-Miguel [1,*] and María B. Serrano [1,*]

[1] Department of Materials Science and Engineering and Chemical Engineering (IAAB), University of Carlos III of Madrid, Av. Universidad, 30, 28911 Leganés, Spain
[2] Department of Materials and Environmental Chemistry, Stockholm University, 10691 Stockholm, Sweden
[3] IMDEA Materials Institute, C/Eric Kandel 2, 28906 Getafe, Spain
* Correspondence: vmiguel@ing.uc3m.es (V.S.-M.); berna@ing.uc3m.es (M.B.S.); Tel.: +34-91-624-6049 (V.S.-M.); +34-91-624-9469 (M.B.S.)

Abstract: The use of block copolymers as a sacrificial template has been demonstrated to be a powerful method for obtaining porous carbons as electrode materials in energy storage devices. In this work, a block copolymer of polystyrene and polyacrylonitrile (PS–*b*–PAN) has been used as a precursor to produce fibers by electrospinning and powdered carbons, showing high carbon yield (~50%) due to a low sacrificial block content ($f_{PS} \approx 0.16$). Both materials have been compared structurally (in addition to comparing their electrochemical behavior). The porous carbon fibers showed superior pore formation capability and exhibited a hierarchical porous structure, with small and large mesopores and a relatively high surface area (~492 m^2/g) with a considerable quantity of O/N surface content, which translates into outstanding electrochemical performance with excellent cycle stability (close to 100% capacitance retention after 10,000 cycles) and high capacitance value (254 F/g measured at 1 A/g).

Keywords: block copolymer template; porous carbon fibers; hierarchical pores; supercapacitor

Citation: Álvarez-Gómez, A.; Yuan, J.; Fernández-Blázquez, J.P.; San-Miguel, V.; Serrano, M.B. Polyacrylonitrile-*b*-polystyrene Block Copolymer-Derived Hierarchical Porous Carbon Materials for Supercapacitor. *Polymers* 2022, *14*, 5109. https://doi.org/10.3390/polym14235109

Academic Editors: Nikolaos Politakos and Apostolos Avgeropoulos

Received: 26 October 2022
Accepted: 19 November 2022
Published: 24 November 2022

Publisher's Note: MDPI stays neutral with regard to jurisdictional claims in published maps and institutional affiliations.

Copyright: © 2022 by the authors. Licensee MDPI, Basel, Switzerland. This article is an open access article distributed under the terms and conditions of the Creative Commons Attribution (CC BY) license (https://creativecommons.org/licenses/by/4.0/).

1. Introduction

Supercapacitors (SCs) are electrochemical devices that store energy by intercalating charges at the electrode−electrolyte interface. SCs development has increased in the last decade due to the higher capacitance and lower voltage limits compared to conventional capacitors [1]. From an energy storage point of view, SCs bridge the gap between electrolytic capacitors and batteries [2]. The main advantages of these devices over other systems include fast charge−discharge (on the level of seconds) without losing efficiency or degrading their internal structures, high power densities (>1 W/g), low heat generation, and long lifetime (>500,000 cycles) due to the storage mechanism, which does not involve irreversible reactions [3]. Nevertheless, supercapacitors present considerably lower energy density than conventional batteries, which limits their use in applications, such as electric vehicles, that demand high energy and power density.

Owing to their combination of outstanding physical and chemical properties such as high electrical conductivity, high surface area, good corrosion resistance, and thermal stability, activated carbons have been the most widely used and commercially available electrode materials for electrochemical double−layer capacitors (EDLCs) [4,5]. In particular, high conductivity and high surface area are considered key features in high−performance electrode materials [6]. However, active carbons present some limitations that must be overcome regarding their applications in supercapacitors. Although high surface areas (>1000 m^2/g) are achieved, high capacitance values are not necessarily obtained since not all of the micropores are accessible to the electrolyte ions [7]. The reason for this is that these materials do not usually show precise control over the porous structure due to the difficulty of optimizing the activation processes [8]. Pore size and pore size distributions have been found to strongly influence electrochemical performance (i.e., the accessibility of ionic species, capacitance, energy density, and power density) [9]. Therefore, the development of advanced carbon electrode materials is largely focused on designing and obtaining precise carbon nanostructures [10,11]. Hierarchical porosity comprising small and large mesopores (2–50 nm), which facilitate electrical double-layer formation (as well as macropores) has been shown to improve ion−accessibility and ion diffusion, allowing easy access to the micropores (<2 nm) [12].

Due to their high versatility and various synthetic routes that allow excellent control over molecular weight [13,14], block copolymers (BCPs) are ideal precursors for obtaining ordered nanostructured materials, as the controlled morphologies generated by the microphase−separation and self−assembly can be retained after pyrolysis [15,16]. The use of BCPs as templates (soft templating) has proven to be an effective method for obtaining porous carbon materials suitable for energy storage applications [17,18]. BCPs used as carbon precursors typically consist of a high carbon−yielding polymer as carbon matrix, usually polyacrylonitrile (PAN), and a thermally degradable sacrificial block. These block copolymers can be easily converted into porous carbon materials derived through subsequent thermal treatments [19]. A critical first step to ensure a high carbon yield is to stabilize PAN at temperatures around 300 °C in an oxidizing atmosphere, followed by carbonization under nitrogen. These thermal treatments allow one to retain the phase-separation morphologies and form the final porous carbon structures. Some of the most commonly used sacrificial blocks include poly(butyl acrylate) (PBA) [20], polystyrene (PS) [21], poly(ethylene oxide) [22], poly(acrylic acid) (PAA) [23], and poly(methyl methacrylate) (PMMA) [24].

Among different porous carbon structures, such as monoliths [25] or membranes [26], porous carbon fibers (PCFs) stand out for offering a combination of low density, free-standing nature, flexibility, good chemical stability, and excellent thermal and electrical conductivity [27,28]. PCFs have been successfully designed from diverse materials displaying high specific surface area (SSA) with excellent electrochemical performance [29]. Compared to powder carbons, fibers present two main advantages [30]: (i) the surface area accessible to the electrolyte is increased; and (ii) one−dimensional fibers provide continuous electron conduction pathways with small electrical resistance. In addition, to boost their electrochemical applications, ideal PCFs should possess hierarchical porous structures with mesopores and micropores interconnected. Although different polymer blends have been explored as template materials for obtaining PCFs as supercapacitor electrodes [31–33], BCPs have been recently proven to be a powerful new precursor for the fabrication of PCFs for electrochemical performance. G. Liu et al. reported one of the highest capacitance values found in the literature [34] using PMMA−b−PAN copolymer-based porous carbon fibers for supercapacitors. Therefore, exploring different copolymers to find optimal nanostructures and, consequently, improve their electrochemical performance elicits noteworthy attention. This article focuses on the structural and morphological comparative study of two different electrode materials for supercapacitors, constituted by PCFs based on PAN−b−PS copolymer with a hierarchical porous structure and the carbon powder as bulk material without any processing. (Scheme 1).

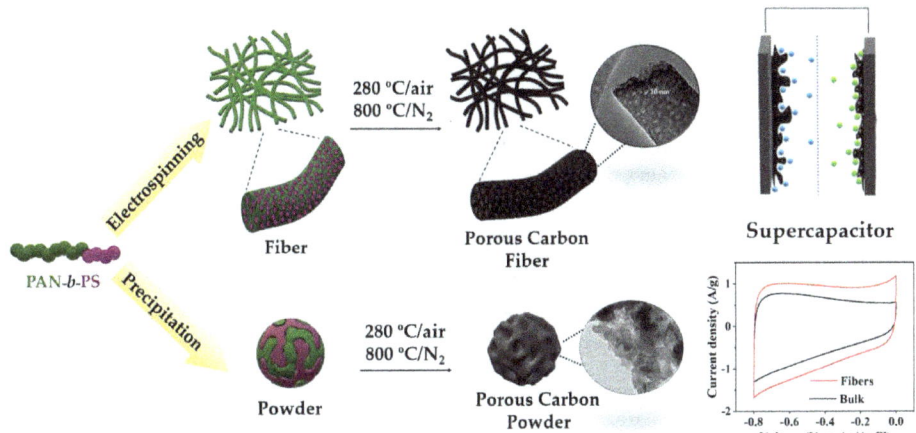

Scheme 1. Preparation scheme of porous carbon fibers and powders, from PAN−b−PS copolymer template for use as supercapacitor electrode materials.

2. Materials and Methods

2.1. Materials

Acrylonitrile (AN, ≥99%) and styrene (≥99%) were purchased from Sigma-Aldrich (Saint Louis, MO, USA) and purified by passing through a basic alumina column in order to remove the inhibitor before use. (2,2′−azobis(2−methylpropionitrile) (AIBN) (≥98%; Cymit), 2−cyano−2−propyl dodecyl trithiocarbonate (CPDT; ≥98%; Sigma-Aldrich), anhydrous N,N−dimethylformamide (DMF; ≥99.8%; Sigma-Aldrich), tetrahydrofuran (THF; ≥99%; Sigma-Aldrich), and methanol (MeOH; ≥99.8%; Sigma-Aldrich) were used as received without further purification.

2.2. Synthesis of PS−CPDT MacroCTA

PS−b−PAN copolymer was synthesized by reversible addition-fragmentation chain transfer polymerization (RAFT) using AIBN as initiator and CPDT as chain transfer agent (CTA), via two steps (Scheme S1). First, a macroCTA (PS−CPDT) was synthesized. Based on previously reported [35], a typical procedure was carried out as follows: styrene (2 mL, 17 mmol) and CPDT (30 µL, 0.085 mmol) were added to a Schlenk flask equipped with a magnetic stirrer and sealed with a rubber septum. The mixture was subjected to three freeze−pump−thaw cycles to remove oxygen. Afterwards, the Schlenk flask under N_2 was placed into a thermostatic bath at 140 °C for 30 min, followed by 48 h at 90 °C. Then, the reaction mixture was cooled down and diluted with a small amount of THF. The product was isolated by precipitation in a large amount of MeOH twice, filtered, and dried under vacuum for 24 h at 30 °C to give a yellow solid. Number average molecular weight (M_n) determined by size exclusion chromatography (SEC) was 18,615 g/mol.

2.3. Synthesis of PS−b−PAN Copolymer

In a second step, the purified PS−CPDT macro-agent was used to synthesize PS−b−PAN copolymer. A typical polymerization procedure is described as follows: PS−CPDT macroCTA (0.1 g, 0.0056 mmol), AIBN (0.6 mg, 0.0038 mmol), and AN (2.9 mL, 44 mmol) were placed in a Schlenk flask equipped with a magnetic bar and dissolved in anhydrous DMF (6 mL). The tube was sealed with a rubber septum and the mixture was subjected to three freeze-pump-thaw cycles to remove oxygen. Then, the flask was placed into a thermostatic bath at 70 °C for 24 h under a N_2 atmosphere. After this time, the reaction mixture was cooled down, precipitated in a large amount of MeOH, and filtered, repeated twice. Finally, the resulting block copolymer was dried under vacuum for 24 h at 30 °C as a white powder (M_n and polydispersity index (PDI) via SEC were 118,710 g/mol and 1.26, respectively).

2.4. Preparation of Porous Carbon Materials

Heat treatments were conducted in a horizontal tube furnace with a controlled atmosphere for both materials, block copolymer powder, and fibers. The as-obtained PS−*b*−PAN copolymer powder was heated up at a rate of 5 °C/min and isothermally stabilized at 280 °C for 1 h under air atmosphere. Then, the stabilized copolymer was carbonized at a rate of 5 °C/min at 800 °C for 1 h under N_2 flow. The final powder carbon product was named as bulk material. Porous carbon fibers were produced from a solution of PS−*b*−PAN copolymer in DMF at 20 wt% concentration. The solution was stirred at 50 °C for 20 h and electrospun under the following parameters: flow rate of 1.5 mL/h, 15 cm of working distance, and a high voltage power supply of 18 kV under humidity ~40% RH at 20−25 °C. The fiber mat was collected on a stationary plate, peeled off from the aluminum foil, and dried in a vacuum oven at 50 °C for 5 h. The fibers were stabilized and carbonized by the same heat treatments as the block copolymer powders; air oxidation at 280 °C for 1 h and pyrolysis under N_2 at 800 °C for 1 h.

2.5. Characterization

Molecular weights and blocks composition were measured by ^{1}H−NMR in deuterated DMF with a Bruker DPX 300 MHz (Bruker, Rheinstetten, Germany) equipment. Molecular weight and polydispersity (PDI) of macroCTA and block copolymer were analyzed by Size Exclusion Chromatography (SEC) using a Waters 515 HPLC pump instrument (Waters, Barcelona, Spain) equipped with a Waters 24214 Refractive Index Detector and Agilent PLgel columns (500, 100, and mixed C). DMF at a flow rate of 0.5 mL/min was used as eluent at 45 °C working temperature and polystyrene (7,500,000−4490 g/mol) was utilized as standards for calibration (PolyScience (PolyScience, Niles, IL, USA)). Thermal transitions were analyzed by Differential Scanning Calorimetry (DSC) using a Mettler Toledo DSC SC822 equipment (Mettler Toledo, Madrid, Spain). Samples were placed in sealed aluminum pans purged with N_2 flow using a heating and cooling rate of 10 °C min^{-1} from 25 to 350 °C. Glass transition temperature (T_g) of the macroinitiator was determined from the second heating cycle. Thermogravimetric analysis (TGA) was carried out using TA instruments Q50 (TA instruments, New Castle, DE, USA) equipment. Measurements were performed with 8 mg of sample under N_2 flow in a temperature range of 25−900 °C at a rate of 10 °C/min. According to Brunauer-Emmett-Teller (BET) method, surface areas were determined by N_2 adsorption/desorption isotherms measured at 77 K on a Micromeritics ASAP-2020 (Micromeritics, Norcross, GA, USA). Samples were degassed at 200 °C under N_2 atmosphere for 24 h prior to analysis. Pore size distributions were studied using Nonlocal Density Functional (NLDFT) method. The morphology and microstructure of the obtained porous carbon materials were imaged by Field Emission Scanning Electron Microscope (FESEM FEI TENEO LoVac (FEI, Hillsboro, OR, USA)) and Transmission Electron Microscope (TEM, Philips Tecnai 20 FEG (FEI, Hillsboro, OR, USA)). TEM samples were prepared by depositing the carbon material dispersion in ethanol into holey carbon copper grids. Raman spectroscopy (LabRAM HR800 spectrometer (Horiba, Kyoto, Japan) was performed with a 514.5 nm Ar laser excitation. X-ray photoelectron spectroscopy (SPECS GmbH with UHV system and energy analyzer PHOIBOS 150 9MCD (SPECS GmbH, Berlin, Germany) was carried out employing a non-monochromatic Al Mg X-ray source operated at 200 W.

Electrochemical characterization was evaluated using Biologic VSP-300 potentiostat (Biologic, Seyssinet-Pariset, France) working with a three electrodes cell configuration. Cyclic voltammetry, galvanic charge/discharge cycles, and electrochemical impedance spectroscopy measurements were carried out with Ag/AgCl as reference electrode, Pt as counter electrode, and aqueous KOH solution (6 M) as electrolyte. The carbon powder working electrode was prepared by pressing into a clean nickel foam a mixture of active material, black carbon, and PTFE 60% dispersion in water with 80:10:10 proportion, respectively. Fibers were tested as a self−standing electrode without the use of binder.

Capacitance values of the three−electrode cell were calculated from GCD and CV curve respectively using the following equations:

$$C_s = \frac{I \cdot \Delta t}{m \cdot \Delta V} \quad (1)$$

$$C_s = \frac{1}{m \cdot v \cdot \Delta V} \int_{V_0}^{V_t} |I(V) \cdot dV| \quad (2)$$

where, ΔV is the voltage window, m is the active material electrode mass, I is the current used in the measure, Δt is the time to take place in the discharge curve, and v is the sweep rate. For the symmetrical cell measurements, electrode capacitance value was calculated according to the equation below:

$$C_s = \frac{4 \cdot I \cdot \Delta t}{m \cdot \Delta V} \quad (3)$$

As for the three−electrode cell measurements, ΔV corresponded with the electrolyte voltage window, m is the sum of the two−electrode active material mass, I is the discharge current, and Δt is the discharge time.

Energy density (W h/Kg) and power density (W/Kg) were calculated according to the following equations:

$$E\,(Wh/Kg) = \left(\frac{1}{2} C_s V^2\right)/3.6 \quad (4)$$

$$P(W/Kg) = \frac{E \cdot 3600}{\Delta t} \quad (5)$$

where Δt (s) is the discharge time and V (V) is the discharge voltage range.

3. Results and Discussion

3.1. Synthesis of PS−b−PAN Copolymer

The copolymer PS−b−PAN, used as carbon templating material for fiber and bulk materials, was successfully synthesized by two-step RAFT polymerization (Scheme S1). Figure 1a shows the ^{1}H−NMR spectra of polystyrene macroCTA and block copolymer measured in d_7-DMF. Signals related to macroCTA and copolymer were fully identified in the spectra. Molecular weight and degree of polymerization (DP) were determined by comparison of the relative integration of the signals at 2.45 ppm assigned to the protons −CH$_2$−CH(CN)− of PAN and those at the range 6.7−7.4 ppm ascribed to the phenyl protons of PS. Calculations revealed the following composition: 0.16 volume fraction of polystyrene as sacrificial block ($M_{n,\,NMR}$ = 16,403 g/mol) and 0.84 of polyacrylonitrile ($M_{n,\,NMR}$ = 101,230 g/mol) as a carbon precursor block. SEC chromatograms presented unimodal narrow peaks for both macroCTA PS−CPDT and the BCP PS−b−PAN with low polydispersity index, 1.05 and 1.26, respectively (Figure 1b), indicating a successful control of the radical polymerization. Additionally, a large total molecular weight (sum of both blocks) has been obtained, which is demonstrated to be beneficial for strong block segregation [14,15] and facilitates fiber production by electrospinning.

3.2. Thermal Characterization

Bulk and fiber polymer materials were stabilized at 280 °C and carbonized at 800 °C under controlled conditions, after precipitation of the block copolymer and electrospinning, respectively. First, the morphology of the PAN phase is chemically fixed under air atmosphere through oxidative crosslinking reactions. In this step, side chain crosslinking and some cyclization occurred, converting C≡N into C=N bonds and turning them into thermally stables-triazine networks. Further cyclization followed by dehydrogenation (300–400 °C) and denitrogenation (>600 °C) leads to partially graphitic structures under an inert atmosphere [17,36]; whereas the sacrificial PS phase is thermally released as monomer in a gas phase, generating pores into the PAN carbon matrix [37]. PAN stabilization/oxidation processes were revealed in the DSC trace of block copolymer as a sharp exotherm at 260 °C (Figure 1c). The DSC trace of the macroCTA (PS−CPDT, inset in the Figure 1c) showed a T_g at around 100 °C, corresponding to the amorphous region of PS, which is slightly visible in the PS−b−PAN thermogram. According to DSC data, the stabilization process was fixed at the upper limit temperature of the exothermic peak (280 °C for 2 h) to avoid a low efficiency of cyclization reactions and ensure an effective stabilization before PS decomposition. No melting peak of the PAN block was detected, due to low heating rates (<30 °C/min) oxidation/cyclization reactions occurred before melting [38]. Endothermic peaks corresponding to the decomposition of PS did not appear up to 350 °C, therefore, stabilization of PAN and decomposition of PS are well separated, which is highly desirable to preserve the morphologies generated in the microphase separation. Weight loss of PS−b−PAN due to pyrolysis was evaluated by TGA (Figure 1d). The TGA profile of PS-CPDT displayed a single weight-loss stage at 400 °C, while PS−b−PAN profile showed three loss stages. The first stage (~220–290 °C) corresponded to partial dehydrogenation and crosslinking (conversion of −C≡N into s-triazine ring) [39]. The second loss stage (~310−460 °C) was attributed to the thermal decomposition of the sacrificial PS block. The third stage showed a slight weight loss between 460 and 870 °C corresponding to further fragmental process of the pre-stabilized PAN [12,40]. The PS−b−PAN showed a 5% of weight loss at 290 °C in a nitrogen atmosphere and a char yield of 50%. Carbonization temperature was fixed at 800 °C for both bulk and fibers. Higher pyrolysis temperatures (>900 °C) are related to the introduction of quaternary nitrogen into the basal plane of graphitic structures, which do not significantly participate in electrochemical reactions [17].

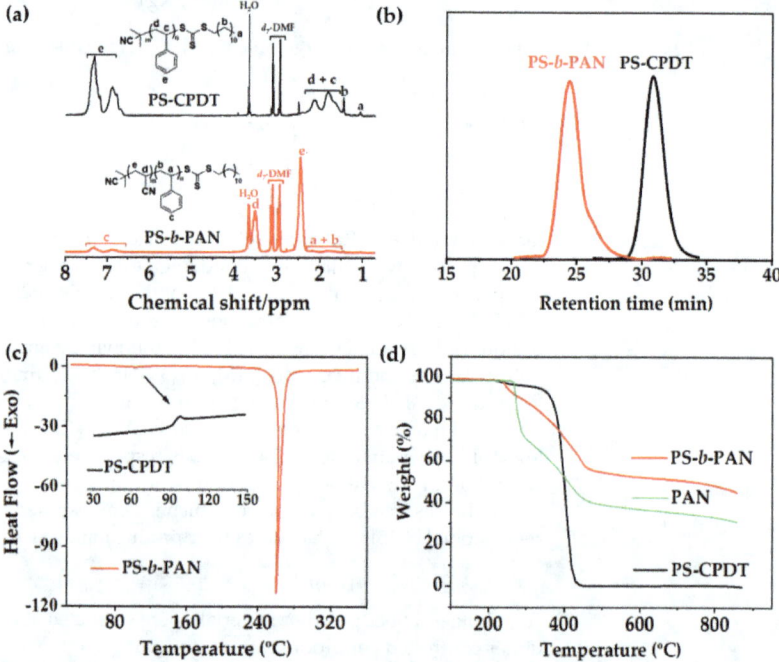

Figure 1. (a) ^{1}H−NMR spectra of PS−CPDT macroCTA (above) and PS−b−PAN polymer (below), (b) SEC chromatograms, (c) DSC thermograms of PS-CPDT (insert graph) and PS−b−PAN (d) TGA curves of PS−CPDT and PS−b−PAN.

3.3. Structural Characterization

The generated morphologies and microdomain sizes are affected by the phase separation between PS and PAN blocks [41]. Pore size is directly correlated to the domain size of the sacrificial block, which is related to molecular weight or degree of polymerization, Flory–Hugging's interaction parameter (χ), and volume fraction (f) [42,43]. The fraction of sacrificial block is especially significant for designing block copolymers to obtain suitable porous carbon materials for capacitive energy storage [8]. In this work, PS has been chosen as a thermally sacrificial block based on its high incompatibility with PAN (segmental interaction parameter, χ_{ij} = 0.83 of acrylonitrile and styrene monomers) [44]. Due to fast solvent vaporization during electrospinning, BCP does not self–assemble into any thermodynamically equilibrated morphologies and, independent of composition, PS–b–PAN forms disordered nanostructures assembled through kinetic pathways, preventing any classical block copolymer thermodynamic equilibrated morphology such as spherical, cylindrical, or lamellar structures [45]. Instead, disordered, and interconnected domains are formed into the electrospun PS–b–PAN fibers. Similarly, due to the co–precipitation process with a non-solvent (MeOH) followed to obtain the powder material, disorder domains were also formed. Although the powder or bulk material can form conventional morphologies derived from self–assembly.

SEM and TEM images (Figure 2b,c, respectively) of the carbon bulk material exhibited a continuous surface porous morphology along the inner material, confirming the interconnectivity of the pores in the bulk grains. This type of disordered carbon structure with percolated pores was also formed with other PS–BCPs [21,46,47]. SEM images of the as–electrospun fibers from the PS–b–PAN copolymer showed rough surfaces (Figure 3b), typical of a fast self-assembling of the BCP during electrospinning. The average diameter of fibers corresponded to 107 ± 4 nm. After carbonization, fibers maintained their shape, although diameters hardly decreased (105 ± 5 nm). Fiber diameter distribution before and after carbonization can be found in Figure S2. SEM and TEM images (Figure 3c–e) revealed abundant mesopores on the surface and cross–section of the fiber, with an average pore size of around 10 nm (Figure 3f). Additional SEM and TEM images of carbon bulk and carbon fibers can be seen in Figure S1.

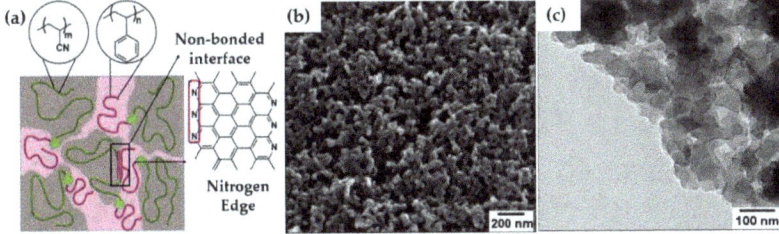

Figure 2. (a) Scheme of disordered morphology of PS–b–PAN upon uncontrolled phase-separation, (b) SEM image of bulk material after carbonization, (c) TEM image of bulk carbon material.

N–functionalities play an essential role in ensuring a rapid ion access into the pore structure. Pyridinic nitrogen, due to the combination of the lone pair and the π–electron system, leads to an enhancement of the ion conductivity and diffusion [48]. Thus, nitrogen exposed in the pore wall surfaces improves ion accessibility and, therefore, storage capability. In PAN–based BCPs, the interface between non–bonded PAN and PS domains contained pyridinic species that formed nitrogen–rich zigzag graphene edges (see schematic, Figure 2a) [17]. In disordered morphologies, such as those obtained here, non–bonded interfaces are predominant, guaranteeing a greater exposure of N during the electrochemical performance.

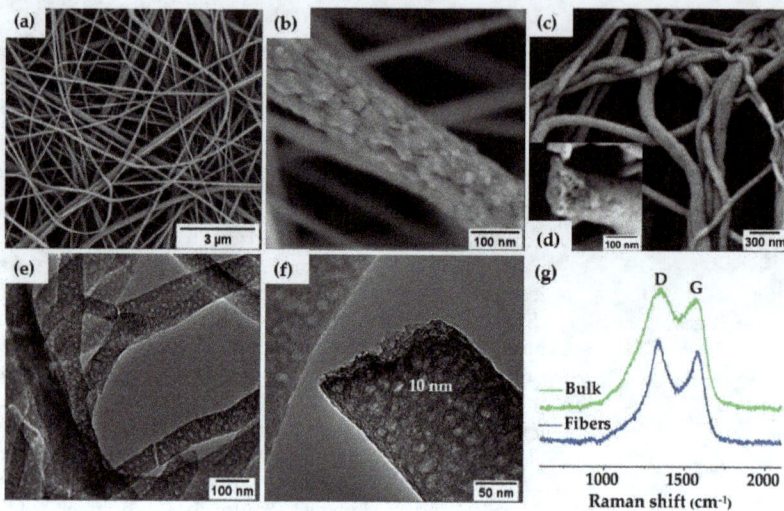

Figure 3. (a,b) electrospun PS−b−PAN fibers, (c) SEM image of fiber membrane after carbonization and (d) cross−section of carbon fibers inserted, (e,f) TEM images of the porous carbon fibers, (g) Raman spectra of both carbon materials.

Raman spectroscopy provides information about ordered and disordered carbon structures. In Figure 3g, Raman spectra of mesoporous carbon bulk materials and fibers revealed the characteristic bands for highly ordered graphitic structures ("G band") at ~1560−1600 cm^{-1} and disordered domains ("D band") at ~1320−1350 cm^{-1} [49]. The intensity ratio of the D and G bands (I_D/I_G) may help elucidate the extent of carbon-containing defects and, therefore, to obtain information about the degree of graphitization of the structures. The I_D/I_G ratio reached a value of 1.11 for the fibers and 1.10 for the bulk powders, indicating a similar formation of graphitic structure for both materials. Deconvolution of the Raman spectra can be found in Figure S3. In addition, X-ray photoelectron spectroscopy (XPS) revealed the surface chemical composition of the carbon fibers. XPS full spectrum (Figure S4a) confirmed the presence of three peaks corresponding to O 1s, N 1s, and C 1s at ~533.3, 401.3, and 284.6 eV, respectively. Sulfur peaks were not observed in the spectrum (S2s and S2p at ~240 and ~163 eV, respectively), which confirmed the elimination of sulfur−containing end−groups in the copolymer by pyrolysis. The carbon (Figure S4b), oxygen, and nitrogen 1s XPS spectra can be resolved by curve fitting into several peaks with attributable binding energies [50–53] to estimate the presence of different types of functional groups. The N 1s spectrum was deconvoluted into four peaks (Figure 4a), verifying the existence of pyridinic nitrogen (N−P) at 398.2 eV, amide nitrogen (O=C−N) at 399.5 eV, pyrrolic/pyridone nitrogen (N−X) at 400.8 eV, and pyridine−N-oxide (N−O) at 402.7 eV. For the oxygen 1s deconvoluted spectrum (Figure 4b), the binding energies at 530.6, 532.5, 534, and 535.5 eV corresponded to quinone type groups (C=O; O−I), phenol and/or ether groups (C−OH and/or C−O−C; O−II), ester groups (O−C=O; O−III) and amide groups (N−C=O; O−IV), respectively, demonstrating a highly oxygen-rich material. Nitrogen atoms were intrinsically present in PAN, whereas oxygen atoms were originated during crosslinking under air atmosphere. Table S1 presented the relative surface concentrations of nitrogen and oxygen species determined by fitting the N 1s and O 1s core level spectra. Furthermore, the atomic percentage of N, C, and O was determined from XPS. As discussed above, the presence of heteroatoms may lead to a potential enhancement of charge mobility and electrolyte wettability. The latter was confirmed by measuring the contact angles of the fiber mats, before and after carbonization, with KOH aqueous solution (6 M). Images in Figure S5 revealed that contact angle after pyrolysis was not noticeable due to the enhanced wettability of the material in presence of oxygen, which will improve the effective contact area between the electrode and the electrolyte [54].

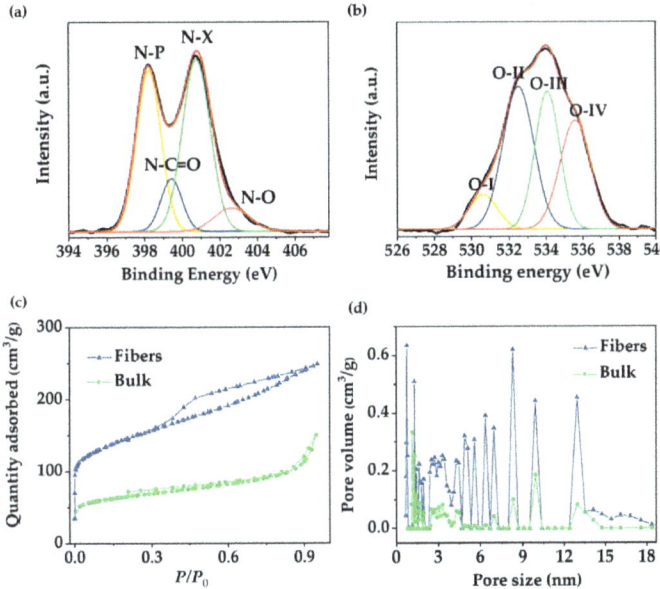

Figure 4. (**a**,**b**) XPS high resolution spectra of N and O, respectively, of the porous carbon fibers, (**c**) N_2 adsorption/desorption isotherms, and (**d**) NLDFT pore size distribution of fibers and bulk carbon material.

In order to confirm and further study the overall porous structure showed in SEM and TEM, N_2 adsorption/desorption analysis was conducted to test the specific surface area (SSA) and pore size distribution. Fiber material revealed a type IV isotherm with a large hysteresis loop (Figure 4c), which indicated the presence of mesopores, according to published IUPAC report [55]. Results were in good agreement with the pore structure previously observed in TEM images. Fast adsorption at the low relative pressure range ($P/P_0 < 0.1$) confirmed the presence of micropores, revealing a BET SSA of 491.6 m^2/g in which 170.4 m^2/g corresponded to micropore SSA. In contrast, the bulk materials reached a BET specific surface area of 242 m^2/g with micropore SSA of 93.9 m^2/g and an isotherm type II (Figure 4c), suggesting a low pore density in the material. For both materials, pore sizes are concentrated in the mesopore and micropore range (Figure 4d). In the case of the fibers, pore size distribution with predominant micropore sizes and mesopores between 5 and 18 nm is shown. The presence of different micropore/mesopore sizes suggests a hierarchical pore structure within the material, which has been reported to be beneficial, offering transport pathways for ions, reducing diffusion distances from the electrolyte to the micropores, and providing a large number of active sites which results in a high-rate capacity [26]. According to the isotherm data, the bulk material exhibited lower pore volume and reduced the number of peaks in the mesopore range, indicating that block copolymer−derived porous carbon fiber showed a considerable better control over the pore formation and double value of specific surface area with respect to the bulk material.

3.4. Electrochemical Characterization

In order to determine the influence of the pore structure on their electrochemical performance, fibers and bulk material were tested in a three−electrode electrochemical cell. Cyclic voltammetry (CV) was evaluated at different sweep rates between 5 and 500 mV/s (Figure S6). Figure 5a showed the comparison of the electrochemical behavior of both materials (voltammetry at 5 mV/s), showing a rectangular shape close to an electric double layer capacitance (EDLC) but slightly distorted from an ideal supercapacitor, which evidenced how charge is not stored through a purely capacitive mechanism. This deviation from the ideal behavior can be influenced by the participation of pseudocapacitive storage mechanism involving fast and reversible redox reactions due to the edge N functionalities inherited from PAN stabilization process [56]. Consistent with the CV data, galvanostatic charge–discharge (GCD) curves of the fiber material displayed a triangular shape (Figure 5b) and a gravimetric capacitance value of 254 F/g measured at a current density of 1 A/g and 308 F/g measured at 10 mV/s. By contrast, bulk material exhibited a considerable reduced capacitance value of 145 F/g calculated at the same current density (1 A/g). The superior SSA_{BET}, pore volume, and hierarchical porosity rich in micropores/mesopores presented in the fiber material resulted in an improvement in electrochemical storage. Electrochemical performance between various PCFs and carbon powder materials obtained through different templates using blends or copolymer as precursors is shown in Table 1.

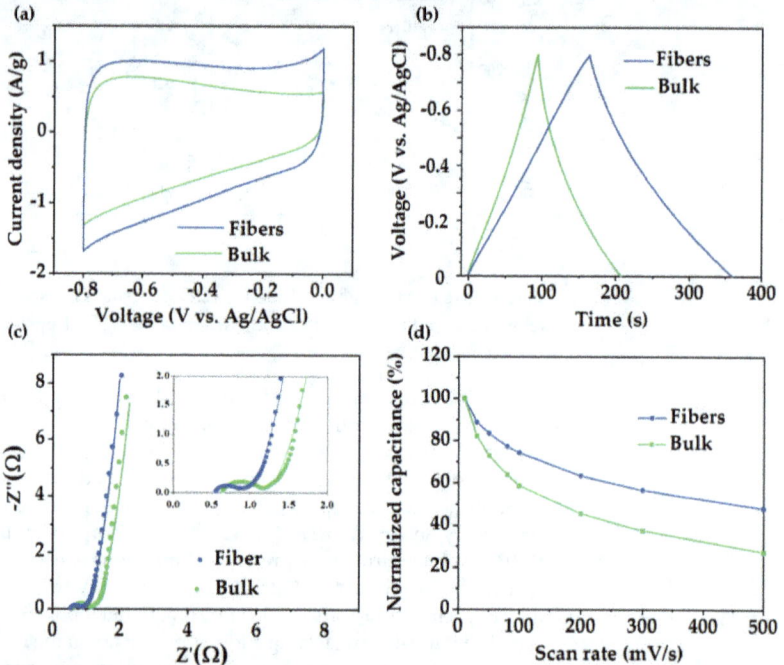

Figure 5. (a) CV curves at 5 mV/s scan rate, (b) GCD curves at 1 A/g current density, (c) EIS diagrams, and (d) capacitance retention of porous carbon fibers and carbon powders derived from PS–b–PAN.

Electrochemical impedance spectroscopy (EIS) was conducted from 0.1 to 100 KHz with a perturbance of 5 mV. Nyquist plots have been fitted with the equivalence circuit shown in Figure S6c. For both materials, Nyquist diagrams (Figure 5c) can be divided into two different regions [57], one at high frequencies displaying a semicircle shape, from which values of R_S (series resistance) and R_{CT} (charge–transfer resistance) can be obtained, and a low frequency region showing a line shape related to the Warburg impedance. R_S is mainly associated with the intrinsic resistances of the electrode, electrolyte, and current collector. Both materials exhibited R_S values less than 1 Ω, with a value of 0.5 Ω for the fiber and 0.6 Ω for the bulk material. Note that the lower value obtained for the PCFs is due to both, its continuous structure with minimum interface effects, providing intimate contact with the electrolyte, and the intrinsic channels (or large pores) between individual fibers, which allows the electrolyte to reach the electrode core faster compared to the powder carbon.

Table 1. Summary of specific surface area and specific capacitance values measured at different current densities (A/g) of various porous carbon fibers (PCFs) and porous carbon powder materials as electrodes for supercapacitors.

Material	Precursor	SSA (m^2/g)	Cs (F/g)	Electrolyte	Reference
Powder carbons					
S/N-doped porous carbons	PBA−b−PAN	478	236 (0.1 A/g)	KOH 6M	[58]
Open-ended hollow carbon spheres (HCS)	PS−b−P4VP/KOH activation	1583	249 (0.5 A/g)	KOH 6M	[59]
N-doped hierarchical porous carbons (NHPC)	PS−b−PAN/KOH activation	2105	230 (1 A/g)	KOH 6M	[47]
Mesoporous carbons	PAN−b−PS−b−PAN	954	185 (0.625 A/g)	KOH 2M	[21]
Fibers					
Linear-tube carbon nanofibers (LTCNF)	PAN/PS	212	188 (0.5 A/g)	LiOH 3M	[60]
Multichannel PCFs	PAN/PS	750	250 (1 A/g)	KOH 6M	[61]
N-doped multi-nano-channel PCFs	PAN/PS	840	461 (0.25 A/g)	H$_2$SO$_4$ 1M	[62]
Mesoporous carbon nanofibers (MCNFs)	PAN/PAA−b−PAN−b−PAA	250	256 (0.5 A/g)	KOH 4M	[23]
PCFs	PAN/PMMMA	683	140 (0.5 A/g)	KOH 6M	[63]
Interconnected meso-PCFs	PMMA−b−PAN	503	360 (1 A/g)	KOH 6M	[34]
Hierarchical PCFs	PS−b−PAN	492	254 (1 A/g)	KOH 6M	This work

Additionally, R$_{CT}$ mainly depends on the electronic and ionic resistances at the interfaces and in the whole system. Values of R$_{CT}$ in fiber and bulk material corresponded with 0.45 and 0.60 Ω, respectively. The lower value achieved by the fibers is due to the synergistic effect of higher pore density and continuous structure, which improved the ion diffusion. Fiber presented a 50% capacitance retention at high sweep rates (500 mV/s), which is significantly higher than the 30% retention of the bulk material (Figure 5d).

To further assess more precisely the behavior of the fiber material, electrodes were tested in a symmetrical Swagelok type cell. CV curves (Figure 6a) presented a rectangular shape even at high sweep rates. Charge−discharge curves (Figure 6b) also exhibited a triangular shape showing a specific capacitance value of 234 F/g at a current density of 1 A/g, slightly lower than the capacitance value obtained in the three−electrode cell. EIS diagram (inset in Figure 6c) showed a considerable decrease in the Rs with a value of 0.25 Ω; however, R$_{CT}$ value was the same as in the three−electrode cell measurements (0.5 Ω). Electrode stability was further evaluated. Capacitance remained stable without noticeable degradation along 10,000 cycles, demonstrating remarkable capacitance retention of 99.9% and electrode stability. Ragone plot of the fibers (Figure 6d) displayed a high energy density and power density of 20 W h/Kg and 12,300 W/Kg, respectively.

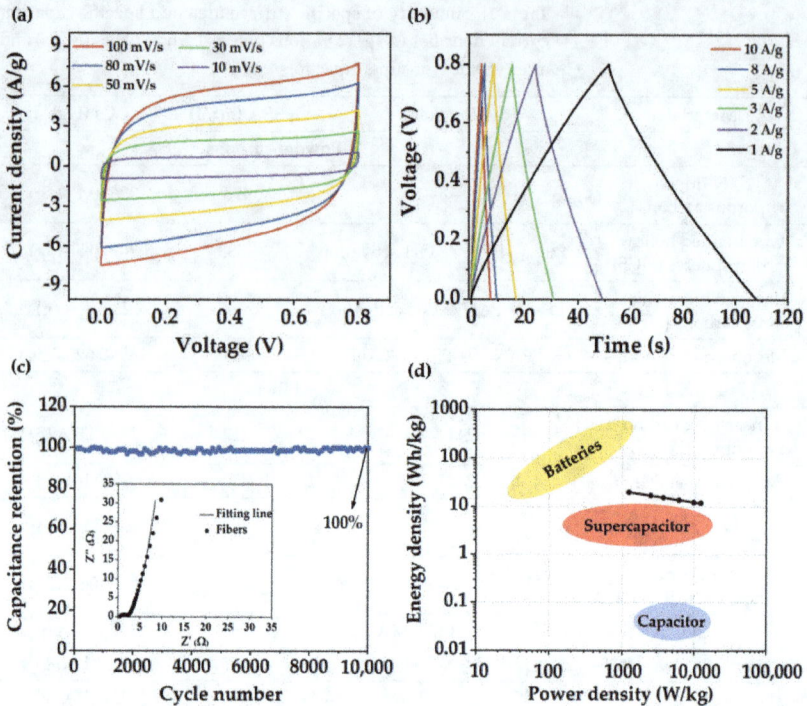

Figure 6. (a) CV curves at different sweep rates, (b) GCD curves at various current densities, (c) Electrode stability along 10,000 cycles at a current density of 100 A/g. Inset: EIS diagram, (d) Ragone plot of the porous carbon fibers.

4. Conclusions

Polystyrene and polyacrylonitrile copolymer (PS−b−PAN) of high molecular weight and low polydispersity has been successfully synthesized by RAFT polymerization. This copolymer had been proven to be a suitable precursor for the obtention of fibers and powder carbon materials. Due to the high incompatibility between PAN and PS blocks, a lower content of the sacrificial block is required ($f_{PS} \approx 0.16$) for obtaining highly porous structures with a considerably high carbon yield (~50%). Fibers of PS−b−PAN produced by electrospinning combined with a pyrolysis process showed hierarchical porosity with mainly micro and mesopores (2−20 nm) and relatively high SSA (~492 m^2/g), demonstrating a better control over pore formation than materials in bulk. The electrochemical behavior of both carbon materials was tested. As expected, fibers showed superior performance, almost doubling the capacitance value of the bulk material, 254 F/g and 145 F/g (measured at 1 A/g), respectively. Fibers also were tested as a free-standing electrode in a symmetrical cell showing excellent cycle stability. In short, this work provides valuable guidance for understanding the phase-separation behavior and the designing of block copolymers for their use as a template. Concretely, focusing on fibers−based carbon electrodes for energy storage devices and for other applications where high carbon-yield materials with good control over the porous structure are required.

Supplementary Materials: The following supporting information can be downloaded at: https://www.mdpi.com/article/10.3390/polym14235109/s1. Scheme S1: Synthetic route via RAFT polymerization for the obtention of PS−b−PAN block copolymer; Figure S1: (a,b) Supplemental SEM images of bulk material after carbonization, (c,d) electrospun PS−b−PAN fibers, (e,f) TEM images of the porous carbon fiber; Figure S2: Diameter distribution of fibers (a) before and (b) after carbonization; Figure S3: Deconvolution of RAMAN spectra of (a) fiber and (b) bulk material; Figure S4: (a) XPS full spectrum and (b) High-resolution XPS spectra showing C1s peaks in XPS of carbon porous fiber; Figure S5: Contact angle of the fiber mat (a) before and (b) after carbonization; Figure S6: (a) CVs of the fiber electrode at scan rates of 5−100 mV/s, (b) GCD curves at different current densities of the fiber electrode, and (c) Equivalent circuit used for the EIS fitting data; Table S1: Relative surface contents of nitrogen and oxygen species determined from N1s and O1s peaks in XPS spectra.

Author Contributions: Conceptualization, M.B.S. and V.S.-M.; design of the work, A.Á-G., V.S.-M. and M.B.S.; performance of the experiments and analysis, A.Á.-G. under the supervision of M.B.S. and V.S.-M. writing original draft preparation, A.Á.-G., V.S.-M. and M.B.S.; writing—review and editing A.Á.-G., J.Y., J.P.F.-B., V.S.-M. and M.B.S.; supervision, M.B.S. and V.S.-M. All authors contributed to the scientific discussion. All authors have read and agreed to the published version of the manuscript.

Funding: This research was financially supported by Spanish Ministry of Science and Innovation (PID2021-125302NB-I00). J.Y. is grateful for financial support from the Wallenberg Academy Fellow program (Grant KAW 2017.0166) from the Knut & Alice Wallenberg Foundation in Sweden.

Institutional Review Board Statement: Not applicable.

Informed Consent Statement: Not applicable.

Data Availability Statement: Not applicable.

Acknowledgments: The authors gratefully acknowledge J. Yuan's research group, the microscopy laboratory of IMDEA materials, and the microscopy laboratory of Carlos III University. Ainhoa Álvarez would like to acknowledge the mobility grant from Carlos III University.

Conflicts of Interest: The authors declare no conflict of interest.

References

1. Salanne, M.; Rotenberg, B.; Naoi, K.; Kaneko, K.; Taberna, P.L.; Grey, C.P.; Dunn, B.; Simon, P. Efficient Storage Mechanisms for Building Better Supercapacitors. *Nat. Energy* **2016**, *1*, 16070. [CrossRef]
2. Patrice, S.; Yuri, G.; Bruce, D. Where Do Batteries End and Supercapacitors Begin? *Science* **2014**, *343*, 1208–1210.
3. Pandolfo, A.G.; Hollenkamp, A.F. Carbon Properties and Their Role in Supercapacitors. *J. Power Sources* **2006**, *157*, 11–27. [CrossRef]
4. González, A.; Goikolea, E.; Barrena, J.A.; Mysyk, R. Review on Supercapacitors: Technologies and Materials. *Renew. Sustain. Energy Rev.* **2016**, *58*, 1189–1206. [CrossRef]
5. Liu, C.F.; Liu, Y.C.; Yi, T.Y.; Hu, C.C. Carbon Materials for High-Voltage Supercapacitors. *Carbon N. Y.* **2019**, *145*, 529–548. [CrossRef]
6. Simon, P.; Gogotsi, Y. Materials for Electrochemical Capacitors. *Nat. Mater.* **2008**, *7*, 845–854. [CrossRef]
7. Liu, N.; Shen, J.; Liu, D. Activated High Specific Surface Area Carbon Aerogels for EDLCs. *Microporous Mesoporous Mater.* **2013**, *167*, 176–181. [CrossRef]
8. Menéndez-Díaz, J.A.; Martín-Gullón, I. Chapter 1 Types of Carbon Adsorbents and Their Production. *Interface Sci. Technol.* **2006**, *7*, 1–47. [CrossRef]
9. Kim, Y.J.; Horie, Y.; Ozaki, S.; Matsuzawa, Y.; Suezaki, H.; Kim, C.; Miyashita, N.; Endo, M. Correlation between the Pore and Solvated Ion Size on Capacitance Uptake of PVDC-Based Carbons. *Carbon N. Y.* **2004**, *42*, 1491–1500. [CrossRef]
10. Vatamanu, J.; Bedrov, D. Capacitive Energy Storage: Current and Future Challenges. *J. Phys. Chem. Lett.* **2015**, *6*, 3594–3609. [CrossRef]
11. Wu, D.; Xu, F.; Sun, B.; Fu, R.; He, H.; Matyjaszewski, K. Design and Preparation of Porous Polymers. *Chem. Rev.* **2012**, *112*, 3959–4015. [CrossRef]
12. Serrano, J.M.; Liu, T.; Khan, A.U.; Botset, B.; Stovall, B.J.; Xu, Z.; Guo, D.; Cao, K.; Hao, X.; Cheng, S.; et al. Composition Design of Block Copolymers for Porous Carbon Fibers. *Chem. Mater.* **2019**, *31*, 8898–8907. [CrossRef]
13. Matyjaszewski, K. Atom Transfer Radical Polymerization (ATRP): Current Status and Future Perspectives. *Macromolecules* **2012**, *45*, 4015–4039. [CrossRef]
14. Keddie, D.J. A Guide to the Synthesis of Block Copolymers Using Reversible-Addition Fragmentation Chain Transfer (RAFT) Polymerization. *Chem. Soc. Rev.* **2014**, *43*, 496–505. [CrossRef] [PubMed]

15. Liu, Q.; Tang, Z.; Ou, B.; Liu, L.; Zhou, Z.; Shen, S.; Duan, Y. Design, Preparation, and Application of Ordered Porous Polymer Materials. *Mater. Chem. Phys.* **2014**, *144*, 213–225. [CrossRef]
16. Wang, H.; Shao, Y.; Mei, S.; Lu, Y.; Zhang, M.; Sun, J.K.; Matyjaszewski, K.; Antonietti, M.; Yuan, J. Polymer-Derived Heteroatom-Doped Porous Carbon Materials. *Chem. Rev.* **2020**, *120*, 9363–9419. [CrossRef]
17. McGann, J.P.; Zhong, M.; Kim, E.K.; Natesakhawat, S.; Jaroniec, M.; Whitacre, J.F.; Matyjaszewski, K.; Kowalewski, T. Block Copolymer Templating as a Path to Porous Nanostructured Carbons with Highly Accessible Nitrogens for Enhanced (Electro)Chemical Performance. *Macromol. Chem. Phys.* **2012**, *213*, 1078–1090. [CrossRef]
18. Liu, T.; Liu, G. Block Copolymer-Based Porous Carbons for Supercapacitors. *J. Mater. Chem. A* **2019**, *7*, 23476–23488. [CrossRef]
19. Yin, J.; Zhang, W.; Alhebshi, N.A.; Salah, N.; Alshareef, H.N. Synthesis Strategies of Porous Carbon for Supercapacitor Applications. *Small Methods* **2020**, *4*, 1900853. [CrossRef]
20. Kopeć, M.; Yuan, R.; Gottlieb, E.; Abreu, C.M.R.; Song, Y.; Wang, Z.; Coelho, J.F.J.; Matyjaszewski, K.; Kowalewski, T. Polyacrylonitrile-b-Poly(Butyl Acrylate) Block Copolymers as Precursors to Mesoporous Nitrogen-Doped Carbons: Synthesis and Nanostructure. *Macromolecules* **2017**, *50*, 2759–2767. [CrossRef]
21. Wang, Y.; Kong, L.B.; Li, X.M.; Ran, F.; Luo, Y.C.; Kang, L. Mesoporous Carbons for Supercapacitors Obtained by the Pyrolysis of Block Copolymers. *Xinxing Tan Cailiao/New Carbon Mater.* **2015**, *30*, 302–309. [CrossRef]
22. Kruk, M.; Dufour, B.; Celer, E.B.; Kowalewski, T.; Jaroniec, M.; Matyjaszewski, K. Well-Defined Poly(Ethylene Oxide)-Polyacrylonitrile Diblock Copolymers as Templates for Mesoporous Silicas and Precursors for Mesoporous Carbons. *Chem. Mater.* **2006**, *18*, 1417–1424. [CrossRef]
23. Dong, W.; Wang, Z.; Zhang, Q.; Ravi, M.; Yu, M.; Tan, Y.; Liu, Y.; Kong, L.; Kang, L.; Ran, F. Polymer/Block Copolymer Blending System as the Compatible Precursor System for Fabrication of Mesoporous Carbon Nanofibers for Supercapacitors. *J. Power Sources* **2019**, *419*, 137–147. [CrossRef]
24. Nguyen, C.T.; Kim, D.P. Direct Preparation of Mesoporous Carbon by Pyrolysis of Poly(Acrylonitrile-b-Methylmethacrylate) Diblock Copolymer. *J. Mater. Chem.* **2011**, *21*, 14226–14230. [CrossRef]
25. Carriazo, D.; Picó, F.; Gutiérrez, M.C.; Rubio, F.; Rojo, J.M.; Del Monte, F. Block-Copolymer Assisted Synthesis of Hierarchical Carbon Monoliths Suitable as Supercapacitor Electrodes. *J. Mater. Chem.* **2010**, *20*, 773–780. [CrossRef]
26. Ran, F.; Shen, K.; Tan, Y.; Peng, B.; Chen, S.; Zhang, W.; Niu, X.; Kong, L.; Kang, L. Activated Hierarchical Porous Carbon as Electrode Membrane Accommodated with Triblock Copolymer for Supercapacitors. *J. Memb. Sci.* **2016**, *514*, 366–375. [CrossRef]
27. Wen, Y.; Kok, M.D.R.; Tafoya, J.P.V.; Sobrido, A.B.J.; Bell, E.; Gostick, J.T.; Herou, S.; Schlee, P.; Titirici, M.M.; Brett, D.J.L.; et al. Electrospinning as a Route to Advanced Carbon Fibre Materials for Selected Low-Temperature Electrochemical Devices: A Review. *J. Energy Chem.* **2021**, *59*, 492–529. [CrossRef]
28. Newcomb, B.A. Processing, Structure, and Properties of Carbon Fibers. *Compos. Part A Appl. Sci. Manuf.* **2016**, *91*, 262–282. [CrossRef]
29. Kim, C.; Yang, K.S.; Lee, W.J. The Use of Carbon Nanofiber Electrodes Prepared by Electrospinning for Electrochemical Supercapacitors. *Electrochem. Solid-State Lett.* **2004**, *7*, 397–399. [CrossRef]
30. Josef, E.; Yan, R.; Guterman, R.; Oschatz, M. Electrospun Carbon Fibers Replace Metals as a Current Collector in Supercapacitors. *ACS Appl. Energy Mater.* **2019**, *2*, 5724–5733. [CrossRef]
31. Jo, E.; Yeo, J.G.; Kim, D.K.; Oh, J.S.; Hong, C.K. Preparation of Well-Controlled Porous Carbon Nanofiber Materials by Varying the Compatibility of Polymer Blends. *Polym. Int.* **2014**, *63*, 1471–1477. [CrossRef]
32. He, T.; Fu, Y.; Meng, X.; Yu, X.; Wang, X. A Novel Strategy for the High Performance Supercapacitor Based on Polyacrylonitrile-Derived Porous Nanofibers as Electrode and Separator in Ionic Liquid Electrolyte. *Electrochim. Acta* **2018**, *282*, 97–104. [CrossRef]
33. Wang, H.; Wang, W.; Wang, H.; Jin, X.; Niu, H.; Wang, H.; Zhou, H.; Lin, T. High Performance Supercapacitor Electrode Materials from Electrospun Carbon Nanofibers in Situ Activated by High Decomposition Temperature Polymer. *ACS Appl. Energy Mater.* **2018**, *1*, 431–439. [CrossRef]
34. Zhou, Z.; Liu, T.; Khan, A.U.; Liu, G. Block Copolymer–Based Porous Carbon Fibers. *Sci. Adv.* **2019**, *5*, eaau6852. [CrossRef]
35. Ponnusamy, K.; Babu, R.P.; Dhamodharan, R. Synthesis of Block and Graft Copolymers of Styrene by Raft Polymerization, Using Dodecyl-Based Trithiocarbonates as Initiators and Chain Transfer Agents. *J. Polym. Sci. Part A Polym. Chem.* **2013**, *51*, 1066–1078. [CrossRef]
36. Zhong, M.; Tang, C.; Kim, E.K.; Kruk, M.; Celer, E.B.; Jaroniec, M.; Matyjaszewski, K.; Kowalewski, T. Preparation of Porous Nanocarbons with Tunable Morphology and Pore Size from Copolymer Templated Precursors. *Mater. Horizons* **2014**, *1*, 121–124. [CrossRef]
37. Kopeć, M.; Lamson, M.; Yuan, R.; Tang, C.; Kruk, M.; Zhong, M.; Matyjaszewski, K.; Kowalewski, T. Polyacrylonitrile-Derived Nanostructured Carbon Materials. *Prog. Polym. Sci.* **2019**, *92*, 89–134. [CrossRef]
38. Gupta, A.K.; Paliwal, D.K.; Bajaj, P. Melting Behavior of Acrylonitrile Polymers. *J. Appl. Polym. Sci.* **1998**, *70*, 2703–2709. [CrossRef]
39. Dang, W.; Liu, J.; Wang, X.; Yan, K.; Zhang, A.; Yang, J.; Chen, L.; Liang, J. Structural Transformation of Polyacrylonitrile (PAN) Fibers during Rapid Thermal Pretreatment in Nitrogen Atmosphere. *Polymers* **2020**, *12*, 63. [CrossRef]
40. Zhou, Z.; Liu, G. Controlling the Pore Size of Mesoporous Carbon Thin Films through Thermal and Solvent Annealing. *Small* **2017**, *13*, 19–21. [CrossRef]
41. Cho, J. Analysis of Phase Separation in Compressible Polymer Blends and Block Copolymers. *Macromolecules* **2000**, *33*, 2228–2241. [CrossRef]

42. Majewski, P.W.; Yager, K.G. Rapid Ordering of Block Copolymer Thin Films. *J. Phys. Condens. Matter* **2016**, *28*, 403002. [CrossRef] [PubMed]
43. Matsen, M.W.; Bates, F.S. Unifying Weak- and Strong-Segregation Block Copolymer Theories. *Macromolecules* **1996**, *29*, 1091–1098. [CrossRef]
44. Cowie, J.M.; Lath, D. Miscibility mapping in some blends involving poly (styrene-co-acrylonitrile). In *Makromolekulare Chemie. Macromolecular Symposia*; Hüthig & Wepf Verlag: Basel, Switzerland, 1988; Volume 16, pp. 103–112.
45. Liu, T.; Liu, G. Block Copolymers for Supercapacitors, Dielectric Capacitors and Batteries. *J. Phys. Condens. Matter* **2019**, *31*, 233001. [CrossRef]
46. Seo, M.; Hillmyer, M.A. Reticulated Nanoporous Polymers by Controlled Polymerization-Induced Microphase Separation. *Science* **2012**, *336*, 1422–1425. [CrossRef]
47. Tong, Y.X.; Li, X.M.; Xie, L.J.; Su, F.Y.; Li, J.P.; Sun, G.H.; Gao, Y.D.; Zhang, N.; Wei, Q.; Chen, C.M. Nitrogen-Doped Hierarchical Porous Carbon Derived from Block Copolymer for Supercapacitor. *Energy Storage Mater.* **2016**, *3*, 140–148. [CrossRef]
48. Zhang, H.; Ling, Y.; Peng, Y.; Zhang, J.; Guan, S. Nitrogen-Doped Porous Carbon Materials Derived from Ionic Liquids as Electrode for Supercapacitor. *Inorg. Chem. Commun.* **2020**, *115*, 107856. [CrossRef]
49. Shimodaira, N.; Masui, A. Raman Spectroscopic Investigations of Activated Carbon Materials. *J. Appl. Phys.* **2002**, *92*, 902–909. [CrossRef]
50. Hulicova-Jurcakova, D.; Seredych, M.; Lu, G.Q.; Bandosz, T.J. Combined Effect of Nitrogen- and Oxygen-Containing Functional Groups of Microporous Activated Carbon on Its Electrochemical Performance in Supercapacitors. *Adv. Funct. Mater.* **2009**, *19*, 438–447. [CrossRef]
51. Lee, S.W.; Gallant, B.M.; Lee, Y.; Yoshida, N.; Kim, D.Y.; Yamada, Y.; Noda, S.; Yamada, A.; Yang, S.H. Self-Standing Positive Electrodes of Oxidized Few-Walled Carbon Nanotubes for Light-Weight and High-Power Lithium Batteries. *Energy Environ. Sci.* **2012**, *5*, 5437–5444. [CrossRef]
52. Rosenthal, D.; Ruta, M.; Schlögl, R.; Kiwi-Minsker, L. Combined XPS and TPD Study of Oxygen-Functionalized Carbon Nanofibers Grown on Sintered Metal Fibers. *Carbon N. Y.* **2010**, *48*, 1835–1843. [CrossRef]
53. He, H.; Hu, Y.; Chen, S.; Zhuang, L.; Ma, B.; Wu, Q. Preparation and Properties of A Hyperbranch-Structured Polyamine Adsorbent for Carbon Dioxide Capture. *Sci. Rep.* **2017**, *7*, 3913. [CrossRef] [PubMed]
54. Oda, H.; Yamashita, A.; Minoura, S.; Okamoto, M.; Morimoto, T. Modification of the Oxygen-Containing Functional Group on Activated Carbon Fiber in Electrodes of an Electric Double-Layer Capacitor. *J. Power Sources* **2006**, *158*, 1510–1516. [CrossRef]
55. Thommes, M.; Kaneko, K.; Neimark, A.V.; Olivier, J.P.; Rodriguez-Reinoso, F.; Rouquerol, J.; Sing, K.S.W. Physisorption of Gases, with Special Reference to the Evaluation of Surface Area and Pore Size Distribution (IUPAC Technical Report). *Pure Appl. Chem.* **2015**, *87*, 1051–1069. [CrossRef]
56. Frackowiak, E. Carbon Materials for Supercapacitor Application. *Phys. Chem. Chem. Phys.* **2007**, *9*, 1774–1785. [CrossRef] [PubMed]
57. Mei, B.A.; Munteshari, O.; Lau, J.; Dunn, B.; Pilon, L. Physical Interpretations of Nyquist Plots for EDLC Electrodes and Devices. *J. Phys. Chem. C* **2018**, *122*, 194–206. [CrossRef]
58. Yuan, R.; Wang, H.; Sun, M.; Damodaran, K.; Gottlieb, E.; Kopeć, M.; Eckhart, K.; Li, S.; Whitacre, J.; Matyjaszewski, K.; et al. Well-Defined N/S Co-Doped Nanocarbons from Sulfurized PAN- b-PBA Block Copolymers: Structure and Supercapacitor Performance. *ACS Appl. Nano Mater.* **2019**, *2*, 2467–2474. [CrossRef]
59. Cao, S.; Qu, T.; Zhang, A.; Zhao, Y.; Chen, A. N-Doped Hierarchical Porous Carbon with Open-Ended Structure for High-Performance Supercapacitors. *ChemElectroChem* **2019**, *6*, 1696–1703. [CrossRef]
60. Bhoyate, S.; Kahol, P.K.; Sapkota, B.; Mishra, S.R.; Perez, F.; Gupta, R.K. Polystyrene Activated Linear Tube Carbon Nanofiber for Durable and High-Performance Supercapacitors. *Surf. Coatings Technol.* **2018**, *345*, 113–122. [CrossRef]
61. Zhang, L.; Han, L.; Liu, S.; Zhang, C.; Liu, S. High-Performance Supercapacitors Based on Electrospun Multichannel Carbon Nanofibers. *RSC Adv.* **2015**, *5*, 107313–107317. [CrossRef]
62. Ramakrishnan, P.; Shanmugam, S. Nitrogen-Doped Porous Multi-Nano-Channel Nanocarbons for Use in High-Performance Supercapacitor Applications. *ACS Sustain. Chem. Eng.* **2016**, *4*, 2439–2448. [CrossRef]
63. He, G.; Song, Y.; Chen, S.; Wang, L. Porous Carbon Nanofiber Mats from Electrospun Polyacrylonitrile/Polymethylmethacrylate Composite Nanofibers for Supercapacitor Electrode Materials. *J. Mater. Sci.* **2018**, *53*, 9721–9730. [CrossRef]

Article

Graft Polymerization of Acrylamide in an Aqueous Dispersion of Collagen in the Presence of Tributylborane

Yulia L. Kuznetsova [1,2], Karina S. Sustaeva [2], Alexander V. Mitin [2], Evgeniy A. Zakharychev [2], Marfa N. Egorikhina [1], Victoria O. Chasova [1], Ekaterina A. Farafontova [1], Irina I. Kobyakova [1] and Lyudmila L. Semenycheva [1,*]

[1] Federal State Budgetary Educational Institution of Higher Education, Privolzhsky Research Medical University of the Ministry of Health of the Russian Federation, 603005 Nizhny Novgorod, Russia
[2] Department of Organic Chemistry, Faculty of Chemistry, National Research Lobachevsky State University of Nizhny Novgorod, 23, Gagarin Ave., 603022 Nizhny Novgorod, Russia
* Correspondence: llsem@yandex.ru

Abstract: Graft copolymers of collagen and polyacrylamide (PAA) were synthesized in a suspension of acetic acid dispersion of fish collagen and acrylamide (AA) in the presence of tributylborane (TBB). The characteristics of the copolymers were determined using infrared spectroscopy and gel permeation chromatography (GPC). Differences in synthesis temperature between 25 and 60 °C had no significant effect on either proportion of graft polyacrylamide generated or its molecular weight. However, photomicrographs taken with the aid of a scanning electron microscope showed a breakdown of the fibrillar structure of the collagen within the copolymer at synthesis temperatures greater than 25 °C. The mechanical properties of the films and the cytotoxicity of the obtained copolymer samples were studied. The sample of a hybrid copolymer of collagen and PAA obtained at 60 °C has stronger mechanical properties compared to other tested samples. Its low cytotoxicity, when the monomer is removed, makes materials based on it promising in scaffold technologies.

Keywords: collagen; polyacrylamide; tributylborane; hybrid copolymer; cytotoxicity

1. Introduction

Organoboron compounds occupy a special place both in the study of the polymerization of a wide range of monomers and in the production of finished goods. This results from the ability of alkylboranes to participate in all elementary stages of radical polymerization [1,2]. The alkylborane–oxygen system is the most promising [3–11]. On the one hand, it is a low-temperature radical initiator, suitable for use in graft polymerization, and on the other, a reversible inhibition agent. The combination of these properties determines the system's potentially wide application in (co)polymer synthesis and in macromolecular design. It should be noted that, frequently, alkylboranes themselves are not used directly to solve the relevant tasks, but instead, are substituted with oxidation-stable amine complexes that release alkylboranes during the polymerization. Papers [5,6] report on the pseudo vivo (reversible) radical polymerization of alkyl(meth)acrylates initiated by the binary trialkylborane–oxygen system. A method exists of modifying the surface of polypropylene to change its properties through graft polymerization of alkyl acrylates [8] or maleic aldehyde [9] using the alkylborane–oxygen system. Papers [10,11] report on a room temperature polymerization method under the action of the alkylborane amine complex, using acrylic monomers of methyl acrylate (MA) and copolymers of MA, and AA along with linear copolymers of polydimethylsiloxane with isopropylacrylamide [11], vinylpyrrolidone [10], etc. The alkylborane–oxygen system is already known as a low-temperature initiator for the radical polymerization of vinyl monomers for bonding thermoplastics and materials with low surface energy [12]. The approach of using trialkylborane amine complexes with an acrylate base originally attracted attention in adhesive development because of the speed of graft polymerization on the substrate surface, being a rapid curing material with a unique ability to promote adhesion to plastics with low surface energy (polypropylene, polyethylene, Teflon) due to the involvement of an extremely active boron centered radical [13]. Radical copolymerization using trialkylboranes is a method that can produce collagen-based hybrid materials, as described in recent literature [14]. TBB is used as a radical copolymerization initiator to obtain hybrid copolymers of methyl methacrylate and collagen [15], methyl methacrylate and gelatin [16], AA and gelatin [17], and butyl acrylate and collagen [18]. In the case of collagen, TBB enables polymerization even at 25 °C, avoiding collagen denaturation and thus maintaining its structure [15]. It should be noted that polymerization in the presence of an alkylborane–oxygen initiator system requires no heating or UV irradiation, this being a particular advantage for the encapsulation of heat-sensitive hydrophilic active substances into silicone copolymer particles [10,11]. This advantage also allows the use of the alkylborane oxygen system in the production of materials for 3D printing and in the synthesis of hybrid products incorporating natural polymers [15–18]. Given the uniqueness of organoboranes as important reagents in radical (co)polymerization, research in this field remains relevant and makes it possible for new aspects of their capabilities to be discovered. Such versatile participation of organoboranes in radical, including graft, polymerization contributed to the choice of these compounds for the synthesis of scaffold precursors based on fish collagen. The prevalence of fish collagen compared to animal collagen is currently very high [19–21]. This is due to the following factors: 96% identity to human collagen; hypoallergenic and transdermal properties; inertia towards viruses, as well as the absence of religious restrictions. The use of fish collagen in the production of dressings and coatings [20–27] is due to its biocompatibility, biodegradability, and low antigenicity.

Scaffold technologies require the creation of materials with the properties listed above, as well as certain performance characteristics. The solution to this problem is the creation of hybrid materials based on collagen and synthetic polymers [28–38] with the participation of specialists from various specialties: medicine, biology, physics, chemistry, etc. Since PAA is widely known as a component of hybrid copolymers used in medicine as hydrogels [39], biodegradable drug carriers [40], and also has a porous structure, which makes it a promising component in the creation of scaffolds, we considered PAA as a synthetic component of a hybrid copolymer.

The current research is aimed at investigating AA graft polymerization in aqueous dispersions of collagen in the presence of TBB as the initiating agent at various temperatures, the determination of molecular weight characteristics, structure, mechanical properties, and cytotoxicity of the obtained copolymers, and the evaluation of the prospects for using the obtained copolymers in scaffold technologies.

2. Materials and Methods

2.1. Materials

An acetic acid dispersion of collagen was obtained from cod skin using a patented technique [41]. Collagen characteristics: Mn = 170 kDa, Mw = 210 kDa, PDI = 1.20. A 1% collagen concentration in the dispersion was obtained by diluting with 3% CH_3COOH solution. The acetic acid dispersion was purified by recrystallization from benzene, chloroform was dried with heat-treated calcium chloride, then distilled and stored in a dark vessel [42].

The TBB was synthesized using the technique described in [17]. Mg shavings (19.46 g, 0.8 mol) were placed into a 2 L three-necked flask equipped with a mechanical stirrer and reflux condenser; the mixture was heated and cooled in an argon atmosphere. Then, BF_3-Et_2O (28.2 g, 0.2 mol), iodine crystals, and anhydrous diethyl ether (200 mL) were added to the reaction flask still under argon. The reaction was then initiated by adding 9.4 mL of 1-butyl bromide (1) dropwise while stirring the reaction mixture, and residue 1 (73.8 g, 0.6 mol) dissolved in ether (100 mL) was slowly added over the space of 1 h, the ether being gently boiled using a reflux condenser. Stirring was continued for another 1.5 h;

after completion of the addition of 1, water (3.6 mL) saturated with NH_4Cl was added. The reaction mixture was allowed to stand until the clear ether supernatant could be decanted into a distillation flask. The ether was then distilled off in an argon flow and the residual TBB was distilled off in a vacuum. ^{11}B NMR ($CDCl_3$, δ, m.d., J/Hz): 86.7, J = 128 MHz.

2.2. Polymerization

An amount of 30 mL samples of the 1% collagen acetic acid dispersion were placed into two-necked flasks filled with argon and heated in a water bath to 25, 45, or 60 °C, respectively. For each of these flasks, 0.08 g of TBB was placed in an ampoule, the mixture was degassed by freezing and thawing repeatedly under vacuum, then the ampoule was filled with argon. The contents of the ampoule containing the TBB were then poured into one of the argon-filled reaction flasks and incubated for 30 min at the appropriate temperature, with constant stirring. Then, a degassed solution containing 0.3 g AA in 3 mL water was added to the reaction flask. The reaction mixture was incubated for another 3 h.

2.3. Chromatography–Mass Spectrometry

The collagen was freeze-dried, placed in an ampoule, and vacuumized; then, a heptane solution of TBB was added, the mixture was incubated for 30 min, and the ampoule was filled with argon. The gas phase was sampled and analyzed on a QP-2010 (Shimadzu) quadrupole chromato-mass spectrometer, using 70 eV electron impact ionization. Separation was performed at 40 °C on a DB-1 column (30 m, 0.25 mm, 0.25 µm). The evaporator, transition line, and ion source temperatures were all 200 °C. Helium was used as the carrier gas, with a 50 kPa column inlet pressure, a 1:20 sample separation, and 0.5 mL sample injection volume. The chromatogram was recorded in positive ion mode, using a total ion current in the 40–200 Da range.

2.4. Fourier Transform Infrared Spectroscopy

For this, copolymer films were prepared on KBr plates. The IR absorption spectra were recorded on an "IRPrestige-21" FTIR-spectrophotometer (Shimadzu, Kyoto, Japan).

2.5. Gel Permeation Chromatography

The collagen and PAA copolymers' water dispersion was analyzed on an LC-20 HPLC system (Shimadzu, Japan) with an ELSD-LT II low-temperature light-scattering detector. Measurements were performed in the following conditions: the column was a Tosoh Bioscience TSKgel G3000SWxl (30.0 cm L, 7.8 mm i.d., 5.0 µm pore size), at a column temperature of 30 °C, the mobile phase was 0.5 M acetic acid in water, and the injection volume was 20 µL, with a flow rate of 0.8 mL/min. Calibration was performed using narrow dispersion dextran with a molecular weight (MW) range of 1–410,000 Da (Fluca). SEC data processing was performed with LC-Solutions-GPC software.

2.6. Determination and Removal of the Unreacted Monomer

Unreacted AA was measured by bromination according to the Knopp method. Bromine was generated by the reaction of a bromide–bromate solution (5.568 g $KBrO_3$, 40 g KBr in 1 L of water) and hydrochloric acid. An aqueous dispersion of copolymer ~2 g in 100 mL of water was placed into a flask and 25 mL of the bromide–bromate solution and 10 mL 10% hydrochloric acid were added, the mixture was stirred and left in a dark place for 2.5 h. Then, 15 mL of 10% KI was added, and the released iodine was titrated with a 0.1 N solution of $Na_2S_2O_3$. A control was conducted with distilled water. The AA concentration was determined by:

$X\% = ((a - b) \times M)/(200 \times m)$, where a and b are the volumes (in mL) of $Na_2S_2O_3$ used for the probe and control titrations, respectively; M is the molecular weight of the AA, g/mole; and m is the mass of the copolymer probe, g. Both the mass and percentage of the unreacted AA were calculated.

To remove the unreacted monomer the freeze-dried, weighed samples of copolymer were extracted with chloroform in a Soxhlet extractor for 40 h. Then, they were dried to reach stationary mass, weighed and the unreacted AA was calculated as described above.

2.7. Scanning Electron Microscopy

Nanometer resolution images of the surfaces of the collagen and copolymer films were taken with a JEOL JSM-IT300LV scanning electron microscope in low vacuum at 4.0 nm resolution, using a 30 kV accelerating voltage and 5 to 2000× magnifications to provide prints 10 cm × 12 cm in size. The samples were obtained by removing small pieces of the freeze-dried polymer films from the substrate.

2.8. Kinematic Viscosity Determination

The flow times of a 1% collagen solution and of further dilutions of solutions of the AA-collagen copolymers synthesized at 25, 45, and 60 °C initially diluted to 1% were determined using a d = 0.56 PZH-2 viscometer until the results converged. For the measurements, the viscometer was filled and placed in a thermostat heated to 25 °C. The kinematic viscosity was determined according to the formula:

$V = \frac{g}{9807} \times T \times K$, where:

K–viscometer constant (K = 0.011797);
V–fluid kinematic viscosity, mm^2/sec;
T–expiration time of the fluid, sec.

2.9. Measurement of the Mechanical Properties of Films Based on the Synthesized (co)polymers

The mechanical properties of films obtained by irrigation from the synthesized (co)polymer solutions were measured using an AG-Xplus 0.5 universal testing machine (Shimadzu, Japan). For testing, 5 mm × 40 mm rectangular samples of the 20 to 80 μm thick films were cut. The tensile test was carried out at a speed of 1 mm/min. The maximum stress (strength) and elongation at breakage were recorded. The average value of 5 tested samples was taken as the measurement result.

2.10. Cytotoxicity Assessment Using MTT Assay

2.10.1. Testing with Cultures

Human dermal fibroblasts (HDFs) were used as MTT assay samples. Active, morphologically homogeneous 5–6 passage cultures with cells that adhered well to the plastic were used. The culture cells' immunophenotype corresponded to that of mesenchymal cells and the culture viability was 95 to 98%. The culture of HDFs used in this study had previously been tested for sterility and infection.

2.10.2. MTT Assay

The MTT test is a colorimetric quantitative test used to measure cellular metabolic activity and viability. The method is based on the reaction of 3-(4,5-dimethylthiazol-2-yl)-2,5-tetrazolium bromide (MTT) being reduced to purple formazan inside living cells. The extent of MTT recovery depends on the metabolic activity of NADPH-dependent oxidoreductase enzymes; accordingly, living and actively dividing HDFs show a high degree of MTT recovery, while HDFs that are toxically damaged (low metabolic activity), or dead show a low degree of MTT recovery. Dimethyl sulfoxide (DMSO) is added as a solvent to produce colored solutions as it reduces the intracellular formazan crystals. The color intensity of the test samples after the addition of DMSO was quantitatively measured at a wavelength of 540 nm using a flatbed reader.

The samples tested had the following characteristics:

Sample I was an aqueous solution of acetic acid containing a dispersion of collagen and PAA copolymer resulting from synthesis at 60 °C. The viscous, turbid-white, translucent liquid, with an acidic pH, was neutralized to pH = 7.3 before examination.

Sample II was a freeze-dried copolymer sponge of collagen–PAA synthesized at 60 °C, with no further treatment. It was white, brittle, and paper-like. It dissolved completely and had a neutral pH.

Sample III was a freeze-dried copolymer sponge of collagen–PAA synthesized at 60 °C and then washed with chloroform in a Soxhlet extractor. It was white and brittle.

Samples II and III were placed in Petri dishes (60 mm diameter) filled with DMEM/F12 medium containing 1% antibiotics (penicillin-streptomycin 100x, PanEco) and 2% fetal calf serum, and were incubated for 1 day at 37 °C and 5% CO_2. After 24 h, complete dissolution of the samples was observed, and the extract obtained was used for the study. Sample I was a suspension (hereafter—"extract"). For the MTT assay, the extracts obtained were diluted and used in the following proportions: Control 0:1: Extract 1:0; with further dilutions 1:1; 1:2; 1:4, and 1:8 of extract: medium, respectively.

As a preliminary, one day before the MTT assay, HDFs at a concentration of 1×10^4 cells per square centimeter (well size 0.5 cm^2) were inoculated onto 96-well flatbeds in DMEM/F12 medium with 1% antibiotics (penicillin-streptomycin 100x, PanEco) and 10% fetal calf serum, 200 µL suspension per well. The flatbed with HDFs was then placed in an incubator at 37 °C and 5% CO_2.

After 1 day of cultivation, the growth medium in the flatbed wells was removed, discarded, and replaced with either the control medium or the above extract dilutions. The flatbeds were replaced in the incubator for 3 days. After 72 h, 20 µL of MTT solution was injected into each well and the flatbeds were replaced in the incubator for another 3 h. After 3 h, the liquid was gently removed from all the wells and replaced with 99% DMSO solution, 200 µL per well. The optical density (OD) was measured using a Sunrise analyzer (Austria).

The relative growth intensity (RGI) was calculated according to the formula:

$$RGI(\%) = \frac{\text{mean OD in testing culture}}{\text{mean OD in control}} \times 100$$

The results were evaluated using a cytotoxicity grading scale (Table 1).

Table 1. Cytotoxicity assessment scale.

Relative Growth Intensity (RGI)	Level (Rank) of Cytotoxicity
100	0
75–99	1
50–74	2
25–49	3
1–24	4
0	5

The 0–1–rank indicates the absence of cytotoxicity, 5 rank–maximum cytotoxicity.

3. Discussion of the Results

Collagen is a common, natural component used for biodegradable hybrid copolymers with synthetic fragments; but, at elevated temperatures, collagen undergoes denaturation, gradually transforming into gelatin, so it is important to study the effect of temperature on the structure of the copolymers in which it has been used. For collagen and polymethyl methacrylate (PMMA) copolymers obtained in the presence of TBB within the 25–60 °C temperature range, it was found that the collagen structure was only preserved at a 25 °C synthesis temperature [15]. Thus, although the molecular weight characteristics and composition of the copolymers have similar values, their properties depend on the synthesis temperature, e.g., the light transmission is higher for PMMA copolymer films and collagen films synthesized at 60 °C, where the collagen structure is already disrupted. It has previously been shown [43] that there are only insignificant differences between the properties of cod gelatin and cod collagen (CC) subjected to enzymatic protein hydrolysis, and in their functional properties in hybrid hydrogel scaffolds in combination with fibrinogen. This suggests that it is potentially valuable to use CC in hybrid compositions and at temperatures above room temperature. Furthermore, the selected temperature profile will allow researchers to vary some properties of the target product. We synthesized collagen and PAA copolymers at the previously suggested temperatures of 25, 45, and 60 °C [15]. The process was run under conditions minimizing the formation of active radicals through tributylboron oxidation. The component ratios of the polymerizing composition were chosen based on the results of [17] with copolymers of gelatin and PAA synthesized at different concentrations of the TBB initiator: the TBB was first injected into an acetic acid dispersion of collagen in an inert argon atmosphere (Scheme 1). As a result, boroxyl fragments formed at the protein surface due to a reaction with fragments of hydroxyproline units, promoting controlled radical graft polymerization of the synthetic monomer via a reversible inhibition mechanism (Scheme 2):

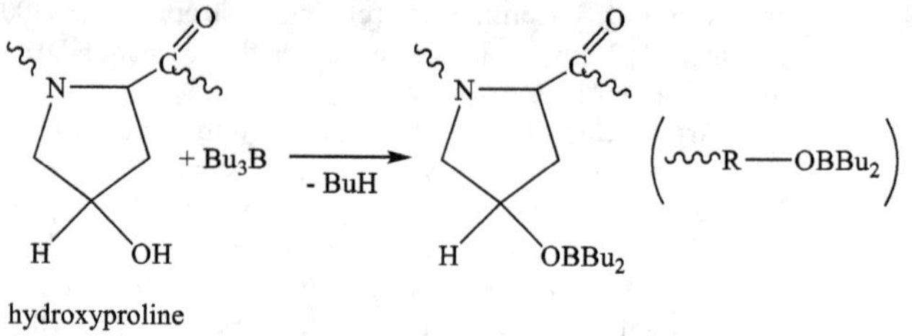

Scheme 1. Borylation of collagen at hydroxyl groups of hydroxyproline units.

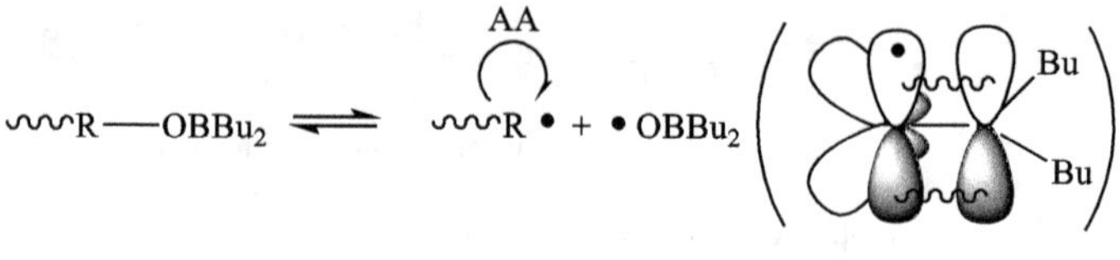

Scheme 2. Graft polymerization of AA by the stable free radical polymerization mechanism involving borylated collagen fragments.

At the end of the copolymerization, a clear colorless dispersion was obtained, which was examined by infrared spectroscopy (Figure 1) and gel permeation chromatography (GPS) (Figure 2).

In the infrared spectrum of the obtained copolymer dispersions (Figure 1, curves 1, 2, and 3), absorption bands related to those of collagen (Figure 1, curve 4) and PAA (Figure 1, curve 5) can be observed.

According to the IR spectra of the obtained copolymers, the intensities of the bands increase at 3275 cm^{-1} this being related to NH$_2$ group valence vibrations ν(N-H), while the absorption bands in the 1640 cm^{-1} region were related to valence vibrations ν(C = O) (amide I), and at 1540 cm^{-1} they were related to planar strain vibrations δ(N-H) (amide II) that occur when the synthesis temperature rises to 60 °C (Figure 1, curve 1). This is due to the freeing of amide groups when hydrogen bonds are broken inside the collagen triple helix on heating. Additionally, as the temperature of synthesis rises, intensity increases in the 1720 cm^{-1} range absorption band, this being related to the carboxyl group, perhaps occurring because of partial hydrolysis of the collagen molecules (Figure 1, curve 1). No noticeable changes corresponding to these were observed at 25 °C or 45 °C.

The dispersion contains a highly polymeric product with a molecular weight (MW) slightly higher than that of collagen, as shown by the GPC analysis (Table 2). The molecular weight distribution (MWD) of the obtained copolymers at 25 (Figure 2, curve 2), 45 (Figure 2, curve 3) and 60 °C (Figure 2, curve 4) show a slight shift to higher MWs compared to the original collagen MWD (Figure 2, curve 1). The trend of copolymer MW as a function of the synthesis temperature is consistent with the kinematic viscosity changes found for these samples (Table 2).

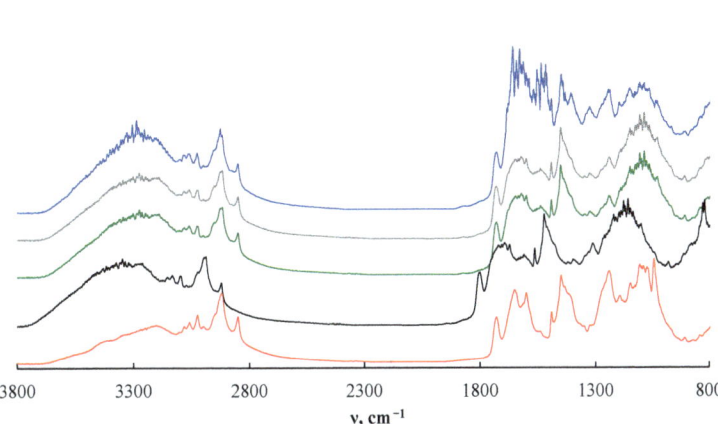

Figure 1. IR spectra of grafted copolymer obtained by polymerization of AA in the presence of TBB and collagen for 3.5 h at 60 (1), 45 (2), or 25 °C (3) and for collagen (4) and PAA (5).

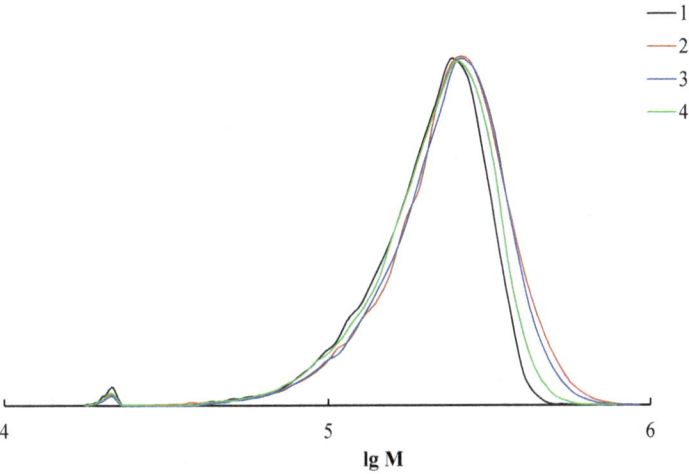

Figure 2. MWD of AA polymerization products in the presence of collagen and TBB synthesized for 3.5 h at 25 (2), 45 (3), or 60 °C (4), and of collagen (1).

Table 2. Molecular weight parameters of collagen and PAA polymerization products in the presence of collagen and TBB at different synthesis temperatures (3.5 h synthesis period).

Temperature, °C	$Mn \times 10^{-3}$	$Mw \times 10^{-3}$	Mw/Mn	Kinematic Viscosity, mm^2/s
Collagen	170	210	1.20	2.1
25	190	240	1.26	2.6
45	190	240	1.23	2.6
60	180	220	1.22	2.4

The acetic acid copolymer dispersions obtained after grafting contain unreacted AA amounting to 20.5–30.5% as the initial monomer (Table 3), its quantity being greater at higher synthesis temperatures.

Table 3. Unreacted monomer and PAA content in collagen copolymers formed at different synthesis temperatures.

Temperature, °C	Time, h	Residual Monomer, %	AA Conversion, %	Ratio of PAA in the Copolymer, %
25	3	20.5	79.5	44.2
45	3	28.8	71.2	41.6
60	3	30.5	69.5	41.0

The dispersions were air dried, brought to a constant weight, and, knowing the unreacted monomer content, the PAA content of the highly polymeric products was calculated (Table 3). As can be seen from Table 3, the content of PAA in the product slightly decreases with increasing synthesis temperature. Probably, at 25 °C, a significant part of the process proceeds under the action of the low-temperature initiating system alkylborane–oxygen, where borylated collagen acts as the alkylborane [3–6]. With an increase in temperature, the proportion of this process decreases, since the rate of oxidation increases, which leads to the formation of low molecular weight products. In this case, polymerization by the mechanism of reversible inhibition becomes predominant (Scheme 2), the rate of which is much lower than conventional radical polymerization. All of the above leads, on the one hand, to a decrease in the total conversion of AA and a decrease in MW.

To purify the copolymers by removing any PAA homopolymer and unreacted monomer, the copolymer dispersions were air-dried to remove water and then placed in a Soxhlet extractor, where they were extracted with chloroform for 40 h, the AA and PAA being soluble in chloroform, while collagen is not. No PAA was detected by IR spectroscopy after the extraction in chloroform, i.e., under these conditions, all the PAA had been grafted onto the collagen, this being consistent with the results of gelatin–PAA copolymer synthesis in the presence of TBB [17]. In addition, extraction of the copolymers with chloroform allowed the unreacted AA to be washed off, as confirmed by Knopp's method.

The resulting samples were examined using scanning electron microscopy (SEM). The freeze-dried copolymers, together with native collagen and PAA synthesized with a classical radical initiator were used to obtain photomicrographs.

Figure 3 shows significant differences between the copolymer photomicrographs (Figure 3c–e) and those of collagen (Figure 3a) and PAA (Figure 3b). The graft copolymer has denser contours of the collagen matrix due to the grafted synthetic fragments (Figure 3c–e). In particular, the porosity characteristic of the PAA can be observed. The pore size in the copolymers is about 50 μm. The photomicrographs obtained allow us to observe differences in the structure of the collagen fibers in copolymers synthesized at different temperatures. Collagen is characterized by a parallel arrangement of fibers (Figure 3a). With an increase in temperature, the SEM photographs show a noticeable transition from clear parallel collagen fibers (Figure 3a), which are preserved in the copolymer synthesized at 25 °C (Figure 3c), partially preserved in the copolymer synthesized at 45 °C (Figure 3d) and are almost completely absent in the copolymer synthesized at 60 °C (Figure 3e). Thus, collagen fiber integrity is impaired in the copolymer structure with increasing synthesis temperature; this has also been demonstrated for PMMA and collagen copolymers [15].

A wide range of applications for hybrid PAA copolymers occurs in their use as biomedical materials: hydrogels, scaffolds, etc. Pure collagen scaffolds have insufficient mechanical strength, thus limiting their use in tissue engineering. The inclusion of synthetic polymers into the structure provides for a significant increase in the strength of the final material. For all copolymers synthesized at 25, 45, and 60 °C, we measured the maximum force F at which the sample failed and the maximum stress (tensile strength) of sample films compared to the original copolymer constituents (Table 4).

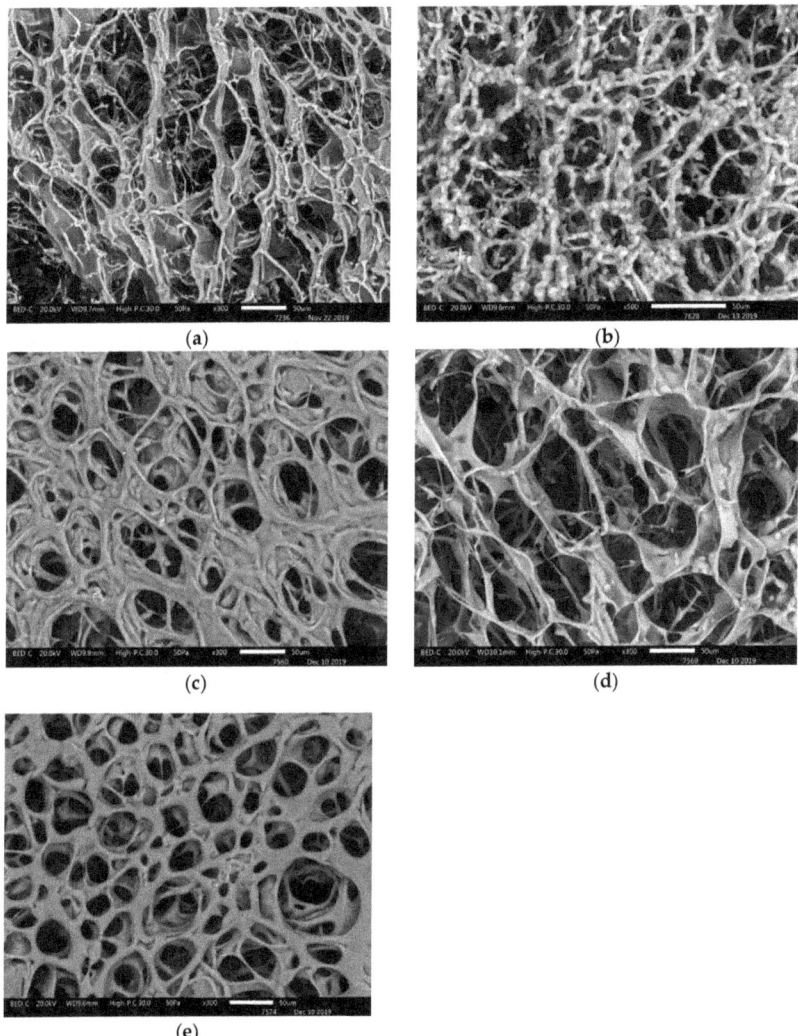

Figure 3. Structural SEM-photomicrographs of collagen (**a**), PAA (**b**), and copolymers of collagen and PAA synthesized at 25 (**c**), 45 (**d**), and 60 °C (**e**).

Table 4. Tensile mechanical properties of films based on PAA, and collagen copolymers synthesized at 25, 45, or 60 °C, and of pure PAA and collagen samples.

Co(Polymer)	Maximum Stress (MPa)	Elongation at Break, %
PAA-g-collagen (60 °C)	49.1	8.24
PAA-g-collagen (45 °C)	47.5	8.99
PAA-g-collagen (25 °C)	9.87	5.65
PAA	21.4	7.99
Collagen	33.8	4.44

According to the obtained data, grafted copolymers synthesized at 60 °C have stronger tensile properties than their individual constituents, PAA and collagen, respectively. According to the SEM data, such copolymer matrices have a denser structure with more cross-links, hence they are stiffer and can resist additional tensile force. This can be attributed to the fact that when the temperature increases, grafting is no longer carried out mainly on the collagen fibers, but on their denatured analog, i.e., gelatin. The identified differences are related to changes in the gelatin supramolecular structure during collagen denaturation: the collagen molecule is a left-handed helix of three α-chains of amino acid residues of known amino acids around a common axis. The gelatin molecule is a denatured helix with the bonds between the individual α-chains having broken. Unlike collagen fibers, as these parts are no longer bound together by hydrogen bonds, they are therefore more resistant to mechanical stress.

The cytotoxicity parameter is important when such copolymers are used as scaffolds; the cytotoxicity grade can be used to assess the likely extent of cell engraftment on a given material. For our cytotoxicity study using the MTT assay, a collagen and PAA copolymer synthesized at 60°C was selected. This copolymer was characterized by an even pore distribution (Figure 3e) due to the collagen structural breakdown, which also gave it better mechanical properties (Table 4), i.e., it is the most suitable for scaffolding techniques. The data are presented in Table 5 and Figure 4. The dispersion of the collagen and PAA copolymer obtained after synthesis (Sample I) exhibited moderate cytotoxicity—grade 3. It should be emphasized that this toxicity persisted even when the extract was diluted (1:1, 1:2, and 1:4). When examined microscopically, the field of view showed a large number of spherical HDF cells along with solitary flattened cells with regular morphology (Figure 4a). Only with a 1:8 dilution of the Sample I extract could we reduce the cytotoxic effect to cytotoxicity grade 2. However, both spherical cells and flattened HDFs could still be observed in the field of view under a microscope (Figure 4a′). Assuming that Sample I cytotoxicity was related to the presence of unreacted AA (Table 3), a lyophilic-dried copolymer of collagen and PAA synthesized at 60 °C (Sample II) was used for further studies, but, since Sample II also contained unreacted monomer, a lyophilic-dried and chloroform-washed copolymer of collagen and PAA was prepared in a Soxhlet extractor (Sample III) for cytotoxicity studies. The effectiveness of this method for purifying gelatin–PAA copolymer had already been demonstrated in [17]. Indeed, the highest cytotoxicity was observed in the extract and 1:1 dilution of Sample II—cytotoxicity grade 4 (Table 5). When examined microscopically, spherical HDFs with irregular cell membranes were observed, confirming the toxic effects on the culture (Figure 4b). These toxic effects were reduced when the Sample II extract was diluted. A 1:2 dilution demonstrated cytotoxicity grade 2. When examined microscopically, many spindle-shaped cells characteristic of HDFs adhering to plastic were observed, but some of the HDFs still took on a spherical shape. A 1:4 dilution of Sample II demonstrated cytotoxicity grade 1. Visually, the cells looked morphologically correct, with no spherical cells being observed in the field of view, but the cells were not growing as actively as in the control, as evidenced by the OD decrease. A 1:8 dilution (Figure 4b′) was neither visually nor quantitatively different from the control (Figure 4d′).

Table 5. Cytotoxicity assessment of collagen and PAA copolymer samples synthesized at 60 °C taken directly from synthesis (Sample I), freeze-dried (Sample II), and both freeze-dried and washed to remove unreacted AA (Sample III).

	Sample I			Sample II			Sample III		
	OD	RGI (%)	Rank	OD	RGI (%)	Rank	OD	RGI (%)	Rank
Control 0:1	0.664 ± 0.037	100	0	0.590 ± 0.037	100	0	0.857 ± 0.028	100	0
Extract 1:0	0.204 ± 0.006	31	3	0.100 ± 0.007	17	4	0.466 ± 0.036	54	2
Extract 1:1	0.186 ± 0.005	28	3	0.102 ± 0.002	17	4	0.617 ± 0.023	72	2
Extract 1:2	0.214 ± 0.009	32	3	0.3299 ± 0.03	51	2	0.772 ± 0.042	90	1
Extract 1:4	0.228 ± 0.008	34	3	0.550 ± 0.01	93	1	0.736 ± 0.039	86	1
Extract 1:8	0.364 ± 0.019	55	2	0.680 ± 0.033	100	0	0.713 ± 0.014	83	1

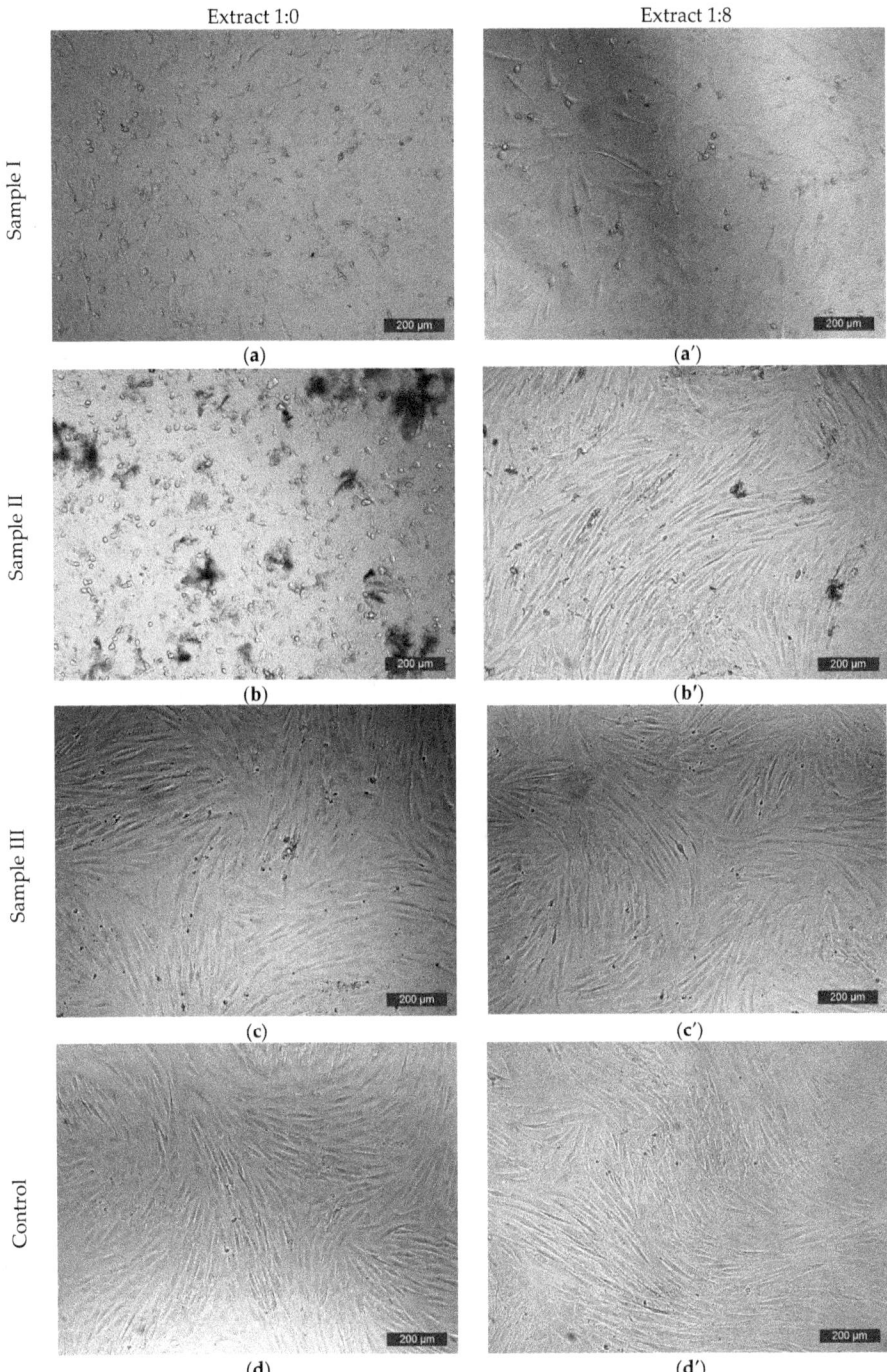

Figure 4. Representative photos, MTT test. (**a**)—the extract 1:0 of an aqueous solution of acetic acid containing a dispersion of collagen and PAA copolymer synthesized at 60 °C; (**a'**)—1:8 dilution of the (**a**); (**b**)—the extract 1:0 of lyophilic-dried copolymer of collagen and PAA synthesized at 60 °C;

(**b′**)—1:8 dilution of the (**b**); (**c**)—the extract 1:0 of lyophilic-dried and chloroform-washed copolymer of collagen and PAA synthesized at 60 °C; (**c′**)—1:8 dilution of the (**c**); (**d**)—control cells in the extract; (**d′**)—control cells in the 1:8 dilution.

Sample III showed moderate cytotoxicity with the extract and its 1:1 dilution corresponding to cytotoxicity grade 2. Morphologically correct HDFs like those adhering to plastic were visually observed in these dilutions, but spherical cells were also present in the field of view (Figure 4c). The MTT assay results of dilutions 1:2, 1:4, and 1:8 of Sample III were assessed at cytotoxicity grade 1. When examined microscopically, the condition of the cells at these dilutions (Figure 4c′) corresponded to that of the control (Figure 4d,d′). However, even at these dilutions, cell growth activity was markedly reduced, as evidenced by the optical density values (Table 5). A comparison of the MTT assay results for Sample II and Sample III indicated that the collagen copolymer and PAA cytotoxicity is caused by unreacted AA present as an impurity. Cytotoxicity is reduced by the removal of the unreacted monomer from the copolymer.

4. Conclusions

Thus, in the present work, grafted copolymers of PAA and collagen were synthesized in the presence of TBB at different temperatures. The synthesis temperature was found to have no appreciable effect on the composition and molecular weight characteristics of the copolymers. However, collagen denaturation means that the copolymer structure is significantly different when the synthesis temperature is increased. Such a structural change leads to a change in mechanical properties. When cytotoxicity was assessed, the collagen and PAA copolymers synthesized at 60 °C containing an AA impurity (Samples I and II) were found to have pronounced cytotoxicity. This manifestation of cytotoxicity is related to the presence of unreacted monomer in these samples and can be reduced via its removal by extraction with chloroform as in the case of Sample III. The low cytotoxicity of Sample III makes materials based on promising copolymers for further activities aimed at the development of new, artificial cytoskeletons for biomedical applications.

Author Contributions: Conceptualization, Y.L.K.; methodology, Y.L.K., L.L.S. and M.N.E.; validation, K.S.S., V.O.C., E.A.F. and I.I.K.; investigation, K.S.S., A.V.M., E.A.Z., V.O.C., E.A.F. and I.I.K.; data curation, Y.L.K., L.L.S. and M.N.E.; writing—original draft preparation, Y.L.K. and K.S.S.; writing—review and editing, A.V.M., M.N.E. and L.L.S.; visualization, K.S.S., E.A.F. and I.I.K.; supervision, M.N.E.; project administration, Y.L.K. All authors have read and agreed to the published version of the manuscript.

Funding: The work was carried out within the framework of the program "Priority-2030", by Minister of Science and Higher Education of the Russian Federation.

Institutional Review Board Statement: Not applicable.

Informed Consent Statement: Not applicable.

Data Availability Statement: Not applicable.

Conflicts of Interest: The authors declare no conflict of interest.

References

1. Grishin, D.F.; Semenycheva, L.L. Problems of managing the investigation of the macrorequest of radicals and the growth of polymer chains. *Adv. Chem.* **2001**, *70*, 425. [CrossRef]
2. Lv, C.; Du, Y.; Pan, X. Alkylboranes in conventional and controlled radical polymerization. *Polym. Chem.* **2020**, *58*, 14. [CrossRef]
3. Furukawa, J.; Tsuruta, T.; Nakayma, J. Socatalytical action of oxygen on vinyl polymerization initiated by metallorganic compounds. *Chem. Soc. Jpn.* **1960**, *63*, 876.
4. Kolesnikov, G.S.; Fedorova, I.S. Polymerization of acrylonitrile in presence of tributylborine. *Russ. Chem. Bull.* **1957**, *6*, 251–252. [CrossRef]
5. Zaremski, M.Y.; Garina, E.S.; Gurskii, M.E.; Bubnov, Y.N. Organoboranes-atmospheric oxygen systems as unconventional initiators of radical polymerization. *Polym. Sci. Ser. B* **2013**, *55*, 304–326. [CrossRef]

6. Zaremskii, M.Y.; Odintsova, V.V.; Plutalova, A.V.; Gurskii, M.E.; Bubnov, Y.N. Reactions of initiation and reinitiation in polymerization mediated by organoborane–oxygen systems. *Polym. Sci. Ser. B* **2018**, *60*, 162–171. [CrossRef]
7. Chung, T.C.; Janvikul, W.; Lu, H.L. A novel "stable" radical initiator based on the oxidation adducts of alkyl-9-BBN. *J. Am. Chem. Soc.* **1996**, *118*, 705. [CrossRef]
8. Okamura, H. Generation of radical species on polypropylene by alkylborane-oxygen system and its application to graft polymerization. *Polym. Sci.* **2009**, *47*, 6163–6167. [CrossRef]
9. Wang, Z.M.; Hong, H.; Chung, T.C. Synthesis of Maleic Anhydride Grafted Polypropylene with High Molecular Weight Using Borane/O_2 Radical Initiator and Commercial PP Polymers. *Macromolecules* **2005**, *38*, 8966–8970. [CrossRef]
10. Ahn, D.; Wier, K.A.; Mitchell, T.P.; Olney, P.A. Applications of fast, facile, radiation-free radical polymerization techniques enabled by room temperature alkylborane chemistry. *ACS Appl. Mater. Interfaces* **2015**, *7*, 23902–23911. [CrossRef]
11. Huber, R.O.; Beebe, J.M.; Smith, P.B.; Howell, B.A.; Ahn, D. Facile synthesis of thermoresponsive poly(NIPAAm-g-PDMS) copolymers using room temperature alkylborane chemistry. *Macromolecules* **2018**, *51*, 4259–4268. [CrossRef]
12. Dodonov, V.A.; Starostina, T.I. Radical bonding thermoplastics and materials with low surface energy using acrylate compositions. *Polym. Sci. Ser. D* **2018**, *11*, 60–66. [CrossRef]
13. Zharov, Y.V. An acrylate compound for adhesion of inert thermoplastics with metals. *Polym. Sci. Ser. D* **2015**, *8*, 33–36. [CrossRef]
14. Fujisawa, S.; Kadoma, Y. Tri-n-Butylborane/WaterComplex-Mediated copolymerization of methyl methacrylate with proteinaceous materials and proteins. *Polymers* **2010**, *2*, 575. [CrossRef]
15. Kuznetsova, Y.L.; Morozova, E.A.; Sustaeva, K.S.; Markin, A.V.; Mitin, A.V.; Batenkin, M.A.; Salomatina, E.V.; Shurygina, M.P.; Gushchina, K.S.; Pryaznikova, M.I.; et al. Tributylborane in the synthesis of graft copolymers of collagen and polymethyl methacrylate. *Russ. Chem. Bull.* **2022**, *71*, 389–398. [CrossRef]
16. Kuznetsova, Y.L.; Morozova, E.A.; Vavilova, A.S.; Markin, A.V.; Smirnova, O.N.; Zakharycheva, N.S.; Lyakaev, D.V.; Semenycheva, L.L. Synthesis of biodegradable graft copolymers of gelatin and polymethyl methacrylate. *Polym. Sci.* **2020**, *13*, 453. [CrossRef]
17. Kuznetsova, Y.L.; Sustaeva, K.S.; Vavilova, A.S.; Markin, A.V.; Lyakaev, D.V.; Mitin, A.V.; Semenycheva, L.L. Tributylborane in the synthesis of graft-copolymers of gelatin and acrylamide. *J. Organomet. Chem.* **2020**, *924*, 121431. [CrossRef]
18. Uromicheva, M.A.; Kuznetsova, Y.L.; Valetova, N.B.; Mitin, A.V.; Semenycheva, L.L.; Smirnova, O.N. Synthesis of grafted polybutyl acrylate copolymer on fish collagen. *Chem. Biotechnol.* **2021**, *11*, 16–25. [CrossRef]
19. Fatuma, F.F.; Chunlei, X.; Weiyan, Q.; Hanmei, X. Collagen from marine biological sources and medical applications. *Chem. Biodivers.* **2018**, *15*, 1700557.
20. Oliveira, V.M.; Assis, C.R.D.; de Aquino Marques Costa, B.; de Araújo Neri, R.C.; Monte, F.T.D.; da Costa Vasconcelos Freitas, H.M.S.; França, R.C.P.; Santos, J.F.; de Souza Bezerra, R.; Porto, A.L.F. Physical, biochemical, densitometric and spectroscopic techniques for characterization collagen from alternative sources: A review based on the sustainable valorization of aquatic by-products. *J. Mol. Struct.* **2021**, *1224*, 129203. [CrossRef]
21. Toledano, M.; Toledano-Osorio, M.; Carrasco-Carmona, Á.; Vallecillo, C.; Lynch, C.D.; Osorio, M.T.; Osorio, R. State of the Art on Biomaterials for Soft Tissue Augmentation in the Oral Cavity. Part I: Natural Polymers-Based Biomaterials. *Polymers* **2020**, *12*, 1850. [CrossRef] [PubMed]
22. Egorikhina, M.N.; Aleynik, D.Y.; Rubtsova, Y.P.; Levin, G.Y.; Charykova, I.N.; Semenycheva, L.L.; Bugrova, M.L.; Zakharychev, E.A. Hydrogel scaffolds based on blood plasma cryoprecipitate and collagen derived from various sources: Structural, mechanical and biological characteristics. *Bioact. Mater.* **2019**, *4*, 334–345. [CrossRef] [PubMed]
23. Egorikhina, M.N.; Aleynik, D.Y.; Rubtsova, Y.P.; Charykova, I.N. Quantitative analysis of cells encapsulated in a scaffold. *MethodsX* **2020**, *7*, 101146. [CrossRef]
24. Shanmugam, S.; Gopal, B. Antimicrobial and cytotoxicity evaluation of aliovalent substituted hydroxyapatite. *Appl. Surf. Sci.* **2014**, *303*, 277–281. [CrossRef]
25. Fan, C.; Wang, D.-A. Macroporous Hydrogel Scaffolds for Three-Dimensional Cell Culture and Tissue Engineering. *Tissue Eng. Part B Rev.* **2017**, *23*, 451–461. [CrossRef]
26. Caliari, S.R.; Burdick, J.A. A Practical Guide to Hydrogels for Cell Culture. *Nat. Methods* **2016**, *13*, 405–414. [CrossRef]
27. Egorikhina, M.N.; Rubtsova, Y.; Charykova, I.; Bugrova, M.; Bronnikova, I.; Mukhina, P.; Sosnina, L.; Aleynik, D. Biopolymer Hydrogel Scaffold as An Artificial Cell Niche for Mesenchymal Stem Cells. *Polymers* **2020**, *12*, 2550. [CrossRef]
28. Shpichka, A.I.; Koroleva, A.V.; Deiwick, A.; Timashev, P.S.; Semenova, E.F.; Moiseeva, I.Y.; Konoplyannikov, M.A.; Chichkov, B.N. Evaluation of the vasculogenic potential of hydrogels based on modified fibrin. *Cell Tissue Biol.* **2017**, *11*, 81–87. [CrossRef]
29. Galler, K.M.; Cavender, A.C.; Koeklue, U.; Suggs, L.J.; Schmalz, G.; D'Souza, R.N. Bioengineering of dental stem cells in a PEGylated fibrin gel. *Regen. Med.* **2011**, *6*, 191–200. [CrossRef]
30. Zhang, G.; Wang, X.; Wang, Z.; Zhang, J.; Suggs, L. A PEGylated fibrin patch for mesenchymal stem cell delivery. *Tissue Eng.* **2006**, *12*, 9–19. [CrossRef]
31. Ivanov, A.A.; Popova, O.P.; Danilova, T.I.; Kuznetsova, A.V. Strategies for selecting and use of scaffolds in bioengineering. *Biol. Bull. Rev.* **2019**, *139*, 196.
32. Zhang, D.; Wu, X.; Chen, J.; Lin, K. The development of collagen based composite scaffolds for bone regeneration. *Bioact. Mater.* **2018**, *3*, 129–138. [CrossRef]
33. Al Kayal, T.; Losi, P.; Pierozzi, S.; Soldani, G. A New Method for Fibrin-Based Electrospun/Sprayed Scaffold Fabrication. *Sci. Rep.* **2020**, *10*, 5111. [CrossRef]

34. Sousa, R.O.; Martins, E.; Carvalho, D.N.; Alves, A.L.; Oliveira, C.; Duarte, A.R.C.; Silva, T.H.; Reis, R.L. Collagen from Atlantic cod (*Gadus morhua*) skins extracted using CO_2 acidified water with potential application in healthcare. *J. Polym. Res.* **2020**, *27*, 73. [CrossRef]
35. Castilho, M.; Hochleitner, G.; Wilson, W.; Rietbergen, B.; Dalton, P.D.; Groll, J.; Malda, J.; Ito, K. Mechanical behavior of a soft hydrogel reinforced with three-dimensional printed microfibre scaffolds. *Sci. Rep.* **2018**, *8*, 1245. [CrossRef]
36. Jiang, H.-J.; Xu, J.; Qiu, Z.; Ma, X.-L.; Zhang, Z.-Q.; Tan, X.-X.; Cui, Y.; Cui, F.-Z. Mechanical properties and cytocompatibility improvement of vertebroplasty PMMA bone cements by incorporating mineralized collagen. *Materials* **2015**, *8*, 2616. [CrossRef]
37. Vedhanayagam, M.; Ananda, S.; Nair, B.U.; Sreeram, K.J. Polymethyl methacrylate (PMMA) grafted collagen scaffold re-inforced by $PdO-TiO_2$ nanocomposites. *Mater. Sci. Eng.* **2020**, *108*, 110378. [CrossRef]
38. Carrion, B.; Souzanchi, M.F.; Wang, V.T.; Tiruchinapally, G.; Shikanov, A.; Putnam, A.J.; Coleman, R.M. The synergistic effects of matrix stiffness and composition on the response of chondroprogenitor cells in a 3D precondensation microenvironment. *Adv. Healthc. Mater.* **2016**, *5*, 1192. [CrossRef]
39. Bai, Z.; Dan, W.; Yu, G.; Wang, Y.; Chen, Y.; Huang, Y.; Dan, N. Tough and tissue-adhesive polyacrylamide/collagen hydrogel with dopamine-grafted oxidized sodium alginate as crosslinker for cutaneous wound healing. *RSC Adv.* **2018**, *8*, 42123–42132. [CrossRef]
40. Luo, L.-J.; Lai, J.-Y. Epigallocatechin Gallate-Loaded Gelatin-g-Poly(N-Isopropylacrylamide) as a New Ophthalmic Pharmaceutical Formulation for Topical Use in the Treatment of Dry Eye Syndrome. *Sci. Rep.* **2017**, *7*, 9380. [CrossRef]
41. Semenycheva, L.L.; Astanina, M.V.; Kuznetsova, J.L.; Valetova, N.B.; Geras'kina, E.V.; Tarankova, O.A. Method for Production of Acetic Dispersion of High Molecular Fish Collagen. Patent RF No. 2,567,171, 10 November 2015.
42. Armarego, W.L.F.; Chai, C.C.L. *Purification of Laboratory Chemicals*, 7th ed.; Elsevier: Oxford, UK, 2012; p. 1024.
43. Chasova, V.; Semenycheva, L.; Egorikhina, M.; Charykova, I.; Linkova, D.; Rubtsova, Y.; Fukina, D.; Koryagin, A.; Valetova, N.; Suleimanov, E. Cod gelatin as an alternative to cod collagen in hybrid materials for regenerative medicine. *Macromol. Res.* **2022**, *30*, 212–221. [CrossRef]

Review

Well-Defined Nanostructures by Block Copolymers and Mass Transport Applications in Energy Conversion

Shuhui Ma, Yushuang Hou, Jinlin Hao, Cuncai Lin, Jiawei Zhao and Xin Sui *

College of Materials Science and Engineering, Qingdao University, Qingdao 266071, China
* Correspondence: suixin_1991@126.com

Abstract: With the speedy progress in the research of nanomaterials, self-assembly technology has captured the high-profile interest of researchers because of its simplicity and ease of spontaneous formation of a stable ordered aggregation system. The self-assembly of block copolymers can be precisely regulated at the nanoscale to overcome the physical limits of conventional processing techniques. This bottom-up assembly strategy is simple, easy to control, and associated with high density and high order, which is of great significance for mass transportation through membrane materials. In this review, to investigate the regulation of block copolymer self-assembly structures, we systematically explored the factors that affect the self-assembly nanostructure. After discussing the formation of nanostructures of diverse block copolymers, this review highlights block copolymer-based mass transport membranes, which play the role of "energy enhancers" in concentration cells, fuel cells, and rechargeable batteries. We firmly believe that the introduction of block copolymers can facilitate the novel energy conversion to an entirely new plateau, and the research can inform a new generation of block copolymers for more promotion and improvement in new energy applications.

Keywords: block copolymer; self-assembly; nanochannels; membranes; energy conversion

1. Introduction

Two or more thermodynamically incompatible polymer blocks with diverse physical and chemical properties are covalently bonded together to form so-called block copolymers (BCPs) [1]. In the 1950s, the discovery of the surfactant Pluronic (PEO-b-PPO-b-PEO) [2–4] first attracted the attention of scientists. The mass production of BCP received a boost in 1956 due to the invention of living anionic polymerization [5]. Subsequently, other controlling polymerization techniques can also be used to synthesize BCP, such as living cationic polymerization [6], atom transfer radical polymerization (ATRP) [7–11], and reversible addition-fragmentation chain transfer polymerization (RAFT) [12,13]. Edwards' self-consistent field theory (SCFT) [14] provided a premium theoretical tool for posterior simulations to explore the phase behavior of BCP. The foundations laid by these pioneers [15,16] have greatly facilitated the advancement of materials science, technology, and theory [17]. The chemical properties of various BCP blocks tend to vary with amenability to addition. To exploit the controllable properties of BCP, scientists have combined theoretical derivation with experimental practice to modify the structure of BCP for desired applications. This requires a thorough understanding of the factors that govern the nanostructure of BCP to regulate it according to the actual needs.

The BCP system can stand out and become the focus of research due to its incomparable self-assembly ability [18], and BCP can undergo accurate microphase separation in the range of 10–100 nm, which expands a new method to prepare nanodevices. The self-assembly of BCP forms rich nanostructures due to the difference in structure, block ratio, and sequences [19]. Surface structure and nanostructure are the keys to improving the properties in applications such as electronic lithography [20], organic photovoltaics [21], nanofiltration [22], and ultrafiltration [23], where electron transport or surface features dictate the performance of the devices. BCP has the potential to boost the properties of these nanodevices [24], which can result in their flourishing applications in different areas due to unique nanoscale customizable template nanostructures [25]. This review intends to highlight the effective exploitation and performance enhancement of BCP for novel energy sources instead of repeating the aforementioned applications. Our main focus will be on the construction and application of BCP-based nanostructured materials in mass transport for energy conversion. The massive exploitation and usage of nonrenewable resources, such as oil and coal, has become a burden on the environment [26]. In recent years, global warming has been responsible for frequent natural disasters [27,28], which have resulted in an irresistible surge to limit the traditional energy methods that incur huge carbon emissions and foster innovative and efficient clean energy sources. Concentration cells, fuel cells, and rechargeable batteries that can convert chemical energy into electricity without carbon emissions are new energy sources with the potential to replace traditional energy sources in the foreseeable future [29]. Using membranes for mass transportation is a key part to obtain high power efficiency from the cells of these devices related to new energy conversion processes [30], which makes the development of a high-performance battery diaphragm very important.

Because the so-called structure determines the properties, we devote a fairly lengthy section to the nanostructure of BCP before presenting their properties relevant for various practical applications. First, we briefly analyze the relevance of three vital parameters in the microphase separation of BCP. Second, various factors affect the self-assembly structure of BCP, such as the chemical structure and external conditions during the assembly process. We elaborate on the effect of various factors on the regulation of self-assembled structures, such as the chemical structure, solvent types, polymer solution concentration, nonsolvent, external field conditions, and additive aspects. Well-defined nanostructures with high order and high density can be obtained by this bottom-up self-assembly method of BCP, which can be used as nanochannel templates for mass transport. Finally, as the highlight of this paper, we investigate the application of BCP-based membranes in energy conversion, such as concentration cells, fuel cells, and rechargeable batteries. Finally, we conclude with futuristic perspectives. We hope that this review can provide a theoretical basis for how to regulate the nanostructure of polymers and offer ideas to further promote the application of block copolymers in new energy fields.

2. Microphase Separation of Block Copolymer

Spontaneous phase separation occurs in BCP due to repulsion between chemically and thermodynamically incompatible blocks. At the macroscopic scale, the phase separation between blocks cannot occur due to the presence of covalent bonds between the blocks of BCP [31]. Instead, in the macromolecular-length scale, phase separation can only occur at the microscopic scale, i.e., microphase separation [32]. Due to the microphase separation, there is structural arrangement of BCP at the macroscopic scale, which is known as the self-assembly phenomenon [33] and responsible for the generation of new boundaries between blocks and can result in various assembly structures.

Three significant parameters are responsible for the microphase separation between the blocks: (1) the relative volume fraction f of each block, which is used to characterize the microscopic composition of BCP; (2) the Flory–Huggins parameter χ, which is used to characterize the interaction between the blocks [34,35] and is inversely proportional to the temperature change; (3) the total degree of polymerization N of the polymer. Among them, χN, which represents the phase separation strength, plays a decisive function in the separation state of BCP [36]. Increasing the temperature or decreasing χN gradually decreased the incompatibility between the blocks. Molecular-level mixing changes polymeric molecular chain from the stretched chain state with obvious interfaces and ordered arrangement to the disordered Gaussian chain state, as shown in Figure 1a, which results in an order-to-disorder transition (ODT) [37] of the polymer. The critical χN value for the occurrence of ODT is approximately 10.5. Electron microscopy can be used to clearly observe the ODT transition process of some BCPs with lower molecular weights, as shown in Figure 1b, which shows TEM images of poly(styrene)-b-poly(imide) (PS-b-PI) before and after undergoing the ODT process.

Due to the introduction of new theories such as the thermal up-and-down effect and mean-field theory, the improved self-consistent field theory (SCFT) can more accurately calculate the BCP phase behavior by scientists [38]. The theoretical phase diagram of the AB diblock copolymer [39] based on self-consistent mean-field (SCMF) theory showing the changes in phase behavior with increasing or decreasing values of f_A and χN in a concrete manner is shown in Figure 1c (left). However, the experimental phase diagram deviates from the theoretical phase diagram because there are various practical and objective factors. Figure 1c (right) shows the experimental phase diagram of the PI-b-PS copolymer obtained by Bates et al. Both experimental and theoretical phase diagrams of PI-b-PS are qualitatively consistent with the variation occurring on the quantitative scale. First, the critical χN value for ODT is approximately 20, with the loss of symmetry for the phase diagram at approximately $f_A = 1/2$. Instead of the CPS phase, a mesostable porous lamellar PL phase appears between C and L regions (calculations confirmed the mesostability of this phase later on). The difference in the PI and PS blocks from the theoretically assumed AB blocks is the main reason behind these deviations. The molecular morphology, properties, and interactions between PI and PS are responsible for the difference in experimental and theoretical values. In the experimental phase diagram, the ODT transitions can directly proceed in both the disordered D region and ordered regions that only occur at the critical point in the theoretical phase diagram, and this phenomenon can be mainly attributed to the effect of the thermal up-and-down phenomenon. With an increasing number of blocks, the phase behavior of block copolymers becomes more complicated [40–42].

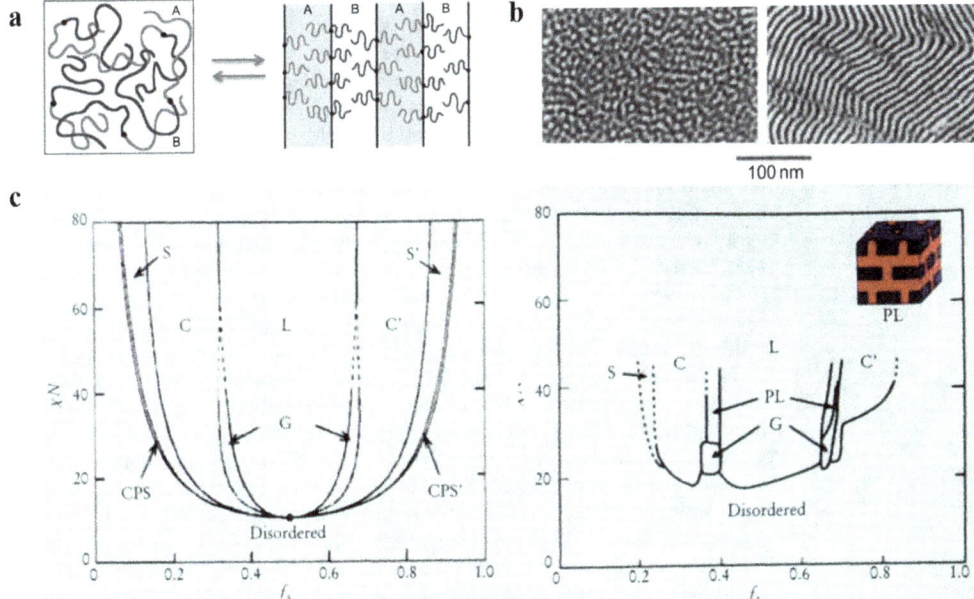

Figure 1. (**a**) Molecular chain distribution of AB-type diblock copolymer before and after the ODT [37]. (**b**) TEM images of PS-b-PI before and after the ODT [37]. (**c**) Theoretical phase diagrams predicted by the self-consistent field theory of the linear AB diblock copolymer and the actual phase diagram (spherical (S, S'), cylindrical (C, C'), bicontinuous gyroidal (G), and lamellar (L)) obtained from the PS-b-PI [39].

3. Regulation of the Self-Assembled Structure of the Block Copolymer

3.1. Influence of the Molecular Structure on the Self-Assembled Structure

The structural morphology of BCP films is our main interest. The self-assembly of BCP film is affected by many factors [43–45] which can be divided into molecular structure factors and assembly condition factors. The most fundamental strategy to regulate the nanoscale morphology of BCP is to adjust its molecular structure, since the BCP structure determines the self-assembly properties. Above the ODT transition curve, the thermodynamically stable nanostructures of BCP mainly consist of spherical S, cylindrical C, bicontinuous gyroidal G, and lamellar L geometries. As shown in Figure 2a, a change in volume fraction f of the AB block changes the morphology of the AB diblock copolymer [46]. Depending on whether $f_A < 0.5$ or $f_B < 0.5$, block A is dispersed within the continuous phase composed of block B or vice versa, and with increasing f_A, the transformation of S→C→G→L occurs, while the same S'→C'→G'→L transition occurs with increasing f_B. The size of each block in each form is determined by the molecular weight and compatibility of the blocks. With increasing molecular weight of the polymer, the blocks become segregated with increasing distance between the domains, and these factors make it challenging to undertake studies of high molecular weight BCPs [47].

The structure of BCP is also characterized by an essential parameter known as polydispersity index (PDI) [48] of molecular weight, which is calculated as PDI = M_n/M_w [49]. The BCPs obtained from the polymerization of reactive anions are monodisperse [50]. However, in recent years, with the emergence of novel methods for synthesizing BCPs, most of the obtained BCPs have been polydisperse [51]. It is necessary to study the effect of broadening the molecular-weight distribution on the phase behavior of BCPs. With the concept of maintaining thermodynamic equilibrium to the maximum possible intent, Ruzette et al. [52] investigated the self-assembly behavior of triblock copolymer ABA with varying molecular-weight distributions, with narrower dispersion of block B (poly (butyl acrylate) (PBA)) and wider dispersion of block A (poly (methyl methacrylate) (PMMA)). The resulting transparent BCP membranes are homogeneous. The TEM characterization reveals the possibility of forming classical S, C, G, and L structures. However, a highly curved interface between the blocks and the PMMA block was formed mainly due to the shift of the interface to release the stretching energy of the molecular chains.

Lynd et al. [53] investigated the effect of molecular-weight distribution on the occurrence of ODT for two different BCPs prepared of polydispersed poly (ethylene-acrylamide)-b-poly (DL-propyleneglycol) (PL) and polystyrene-b-polyisoprene (SI). Ultimately, during the occurrence of ODT, increasing polydispersity in blocks with fewer ($f < 0.5$) or more ($f > 0.5$) components decrease or increase in the critical χN value, respectively. Widin et al. [54] reported the structural effects of a low-dispersion block polystyrene (PS) and a high-dispersion block polybutylene (PB) on the triblock copolymer SBS. As shown in Figure 2b (left), the molecular chain of highly dispersive block PB is stabilized by the low dispersive PS block in the middle position. Finally, similar experimental results to those of Ruzette were obtained, where the phase separation interface was bent toward the polydisperse segment because the polydisperse block has a smaller filling volume than the monodisperse block of the same molecular weight (the phase interface bending process is shown in Figure 2b (right)). The polydisperse B-block also reduces the critical χN value for ODT and increases the position of the lamellar phase window. The investigations show that by playing with the polydispersity of a particular block in BCP, the microphase separation state of BCP and its thermodynamic stability can be regulated.

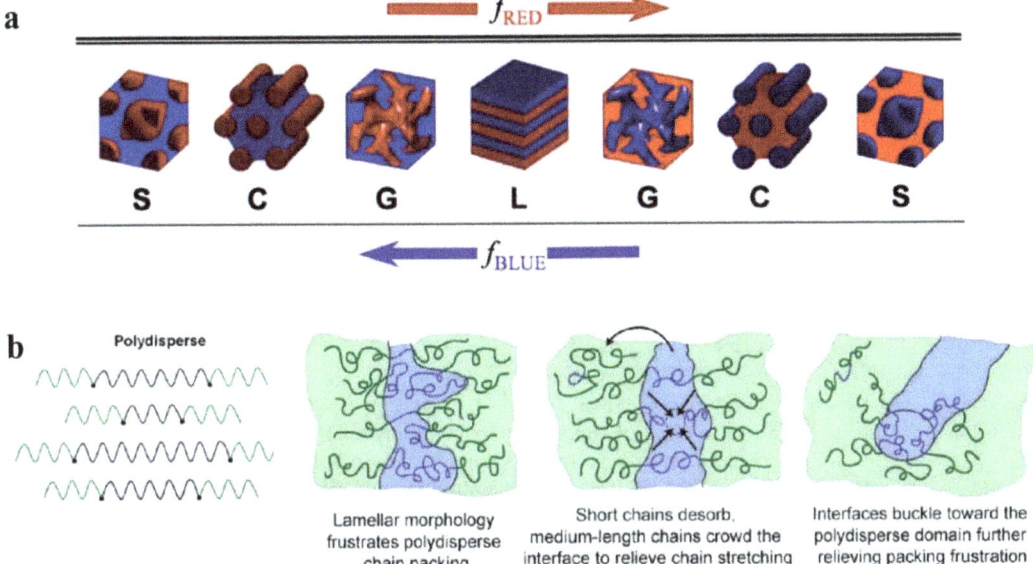

Figure 2. (a) Four equilibrium-ordered state transitions in diblock copolymers determined by relative volume fraction f. Spherical (S), cylindrical (C), bicontinuous gyroidal (G), and lamellar (L) [46]. (b) Schematic diagram of polydispersed B-blocks in ABA-type triblock copolymers leading to distortion of block microdomain interfaces [54].

3.2. Influence of Assembly Conditions on the Self-Assembled Structure

Normally, in addition to the interactions between polymer molecules, external film formation conditions can also affect the nanostructure. Hence, the type of solvent, polymer solution concentration, nonsolvent, additives, and external field conditions during the film formation process can affect the structure of BCP.

3.2.1. Solvents to Prepare the Membranes and Self-Assembly

The selectivity of solvents toward different blocks significantly affect the self-assembled structures. Hanley et al. [55] dissolved PS-b-PI in bis(2-ethylhexyl) phthalate (DOP), which is a nonselective solvent for PS and PI, di-n-butyl phthalate (DBP), and diethyl phthalate (DEP) solvents, which are sequentially more selective for the PS block, and tetradecane (C14), which is a selective solvent for the PI blocks, to observe their phase behavior by small-angle X-ray scattering (SAXS) (Figure 3a). The PS-b-PI membrane exhibited the transformation of G→hexagonal filled C→D, which, when dissolved in different solvents, exhibited different nanostructures. After dissolution in DBP, the phase transition process of PS-b-PI was L→G'→C'→S'. The phase behavior in DEP, which is more selective for PS, like DBP, was similar to that of DBP but with higher T_{ODT}. In contrast, in C14, for the PI selective solvent, the G'→C'→S' ordered phases were the only occurring transitions because enhancement of segregation between microdomains by the selective solvents improve the stability of the ordered phase, which results in the appearance of more ordered phase behavior. The formation and closure of pores in BCP nanostructures can also be regulated by selective solvents [56]. The spontaneously formed poly(styrene)-b-poly (2-vinyl pyridine) (PS-b-P2VP) membrane is a dense, nonporous film. However, when dissolved in P2VP selective solvent (e.g., ethanol), the P2VP blocks dissolve to form a membrane with round or elliptical pores surrounding the core of the PS blocks. With increasing time or temperature, these pores grow into a columnar network and finally form three-dimensional interconnected pores. This pore formation process is often reversible, and immersion of the porous membrane in a PS block selective solvent (e.g., cyclohexane) closes the pores to restore the dense, nonporous membrane.

In addition, dissolution of the self-assembled structure of BCPs in mixed solvents should be explored. Yi et al. [57] investigated the nanostructure of poly(styrene)-b-poly (4-vinyl pyridine) (PS-b-P4VP) by adding it to N, N-dimethylformamide, DMF, (selective for P4VP block, and slow volatilization), tetrahydrofuran, THF, (selective for PS, faster volatilization), and their mixtures. In pure DMF solvent, a nanoscale spherical structure was obtained for PS-b-P4VP, a dense and smooth surface was obtained in pure THF, while the mixed solvent yielded a rod-like aggregated structure where the diameter of the nanospheres was similar to that observed in DMF. When polymers are dissolved in a mixture of solvents with different solubilities and volatilities, the preferential evaporation of one solvent leads to a concentration gradient within the BCP. This concentration gradient leads to phase separation of the BCP, the so-called evaporation-induced phase separation (EIPS) [58] of the solvent.

In addition, the concentration of the polymer solution is an important factor that affects the nanostructure of BCP. Jiang et al. [59] was the first to simulate the membrane formation of a double hydrophobic poly(styrene)-b-poly (methyl methacrylate) (PS-b-PMMA). The effect of the polymer concentration on the morphology of polymer molecules was also investigated in the polymer concentration range of 10–55%. Figure 3b clearly shows the difference in morphologies corresponding to different polymer concentrations. For low polymer concentrations, microphase separation just begins to occur, and the formation of a continuous phase in the system under this condition is not possible because there is not a sufficient BCP concentration, which forms a spherical structure at lower concentrations. With increasing polymer concentration, a structure can be observed because a continuous phase forms by BCP. With a further increase in concentration, porous membranes with elliptical pores start to form. When the concentration increases to approximately 30%, the elliptical pores change to circular pores. From there on, a further increase in polymer concentration only decreases the pore size within the porous membrane and membrane wall thickening. When the polymer concentration reaches approximately 45%, the polymer shows an irregular sponge-like, cross-linked structure, and finally, a continued increase in polymer concentration causes the pores to plug and results in a continuous structure of the polymer. This experiment indicates that, within a certain concentration range, the polymer concentration can be adjusted to achieve different polymer morphologies and obtain porous membranes with ideal pore size and stability.

Formation of the BCP membrane by selective solvents requires three steps: solvent uptake, swelling, and solvent evaporation drying. The effects of the first two steps on the BCP have been described in the previous section, while the evaporation rate of the solvent has been found to affect the structural orientation of BCP. It has been proven that fast solvent evaporation leads to a vertical orientation of the internal structure [60], while slower solvent evaporation rates cause parallel structural orientation or structures with mixed orientation [61]. The evaporation rate determines the propagation of phase separation from the membrane surface to the interior. Similar conclusions were obtained by Phillip et al. [62] when comparing the morphology of poly(styrene)-b-poly (D, L-lactide) (PS-b-PLA) under rapid solvent evaporation and slow evaporation (Figure 3c). Because ultrafast solvent evaporation makes steep concentration gradients rapidly propagate the vertical direction of the membrane surface, the cylindrical structure orients along the surface's normal direction.

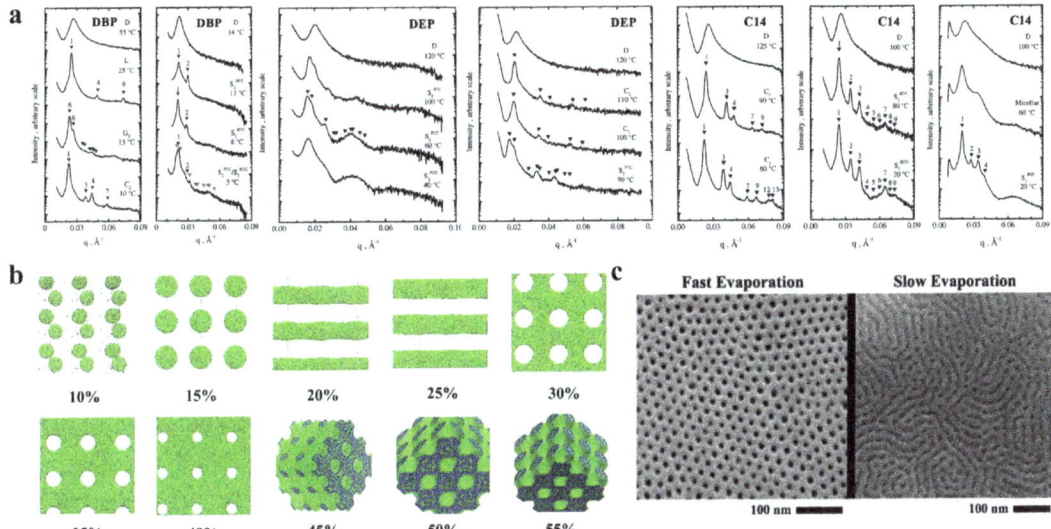

Figure 3. (a) Representative SAXS profiles as a function of temperature for DBP, DEP, and C14 solutions [55]. (b) Nanoscale morphology diagram of PS-b-PMMA porous membrane simulated by DPD when dissolved in tetrahydrofuran solvent at different polymer concentrations. PMMA is shown in green, and PS is shown in blue [59]. (c) SEM images of different morphologies formed by PS-b-PLA at varied solvent evaporation rates [62].

During the self-assembly process via solvent vapor annealing (SVA), the solvents are found to have a significant influence. For a given polymer system, the choice of different solvents can result in different nanostructures [63–65]. Peng et al. [66] investigated the effect of SVA on the morphology of poly(styrene)-b- poly (ethylene oxide) (PS-b-PEO). As shown in Figure 4a, the nanostructure of the PS-b-PEO membrane obtained by conventional casting at room temperature was a disordered and irregular worm-like morphology, while the PS-b-PEO membrane resulting from 5 min of steam annealing treatment with toluene (a selective solvent for PS) was transformed into a highly ordered arrangement of hexagonally filled perpendicular cylindrical microdomains.

However, due to preferential wetting of the surface, a thermodynamically preferred morphology should be formed with a cylindrical orientation parallel to the surface instead of perpendicular to the surface, as experimentally obtained in [67]. The formation mechanism of this morphology has not been understood, and various researchers have attempted to use computer simulations or experimental investigations to explain this phenomenon [68–70]. Phillip et al. [71] analyzed the evolution of the internal morphology of PS-b-PLA during SVA. Theoretical analysis and experimental verification found that the generation of such cylinders evolved from spherical intermediates nucleated at the vapor/polymer membrane interface due to the rapid evaporation of solvent, which created a concentration gradient in the BCP, and the perpendicular orientation of the intermediates with respect to the surface and epitaxial growth into cylinders. However, too-fast evaporation of the solvent does not allow sufficient time for nucleation. Lin et al. [72] used TEM analysis to demonstrate that the length of this ordered arrangement could reach the level of membrane thickness. In other words, SVA can achieve a directional ordered arrangement perpendicular to the membrane surface throughout the membrane, which is of great significance for the development of certain ion- or proton-conducting membranes or water-permeation membranes with controlled pore sizes. Kim et al. [73] demonstrated that SVA yielded a more ordered, structurally stable membrane than those obtained by conventional methods by imparting the molecular chain mobility, shielding the surface energy of the two blocks at the vapor/BCP interface and the interfacial energy between polymer and substrate, and rapidly eliminating defects in the BCP structure (Figure 4b).

When the blocks in BCP strongly interact, more variables in the SVA process will affect the structure [65,74,75]. For instance, Park et al. [76] proposed structural changes in a PS-b-PMMA membrane by choosing a selective or neutral substrate (different substrate treatments with the generated SEM images of varying morphologies are shown in Figure 4c). The membranes on neutral substrates exhibit a cylindrical structure perpendicular to the surface at the top (near the air) and bottom (near the substrate). For the SVA process on selective substrates, the top structure of the BCP membrane was unaffected, exhibiting a cylindrical structure perpendicular to the surface, but the bottom structure changed to a cylindrical structure parallel to the substrate. This result demonstrates the possibility of tailoring the pattern and structure of the BCP membrane by adjusting the membrane–substrate interaction. Cheng et al. [77] investigated various process parameters during the SVA process and ultimately demonstrated that the temperature of the substrate and vapor were also key factors in controlling the structural morphology of BCP membranes.

However, the ordered BCP membrane obtained by the selective solvent SVA method is only short-ranged. It has been shown that long-range ordering of BCP membrane in the lateral direction can be obtained via annealing with nonselective solvent (solvent with the same selectivity for each block and no specific interactions) vapors during SVA [78]. Bosworth et al. [79] obtained a long-range ordered BCP membrane by annealing poly(α-methylstyrene)-block-poly(4-hydroxystyrene) in acetone (selective solvent) vapor, which resulted in a spherical structure and tetrahydrofuran (nonselective solvent) vapor, forming a columnar phase parallel to the surface. This is due to the fact that the nonselective solvent is only uniformly distributed between the molecular chains and same selectivity for each block. It increases only their mobility without any other interactions, resulting in spontaneous formation of a cylindrical orientation parallel to the membrane plane with the lowest surface energy. SVA overcomes the shortcomings of conventional thermal annealing and enables highly ordered rearrangement of BCP in a relatively short time. By reducing the effective T_g, it weakens the interactions between chains and between chains and substrates to increase the mobility of polymer chain [80].

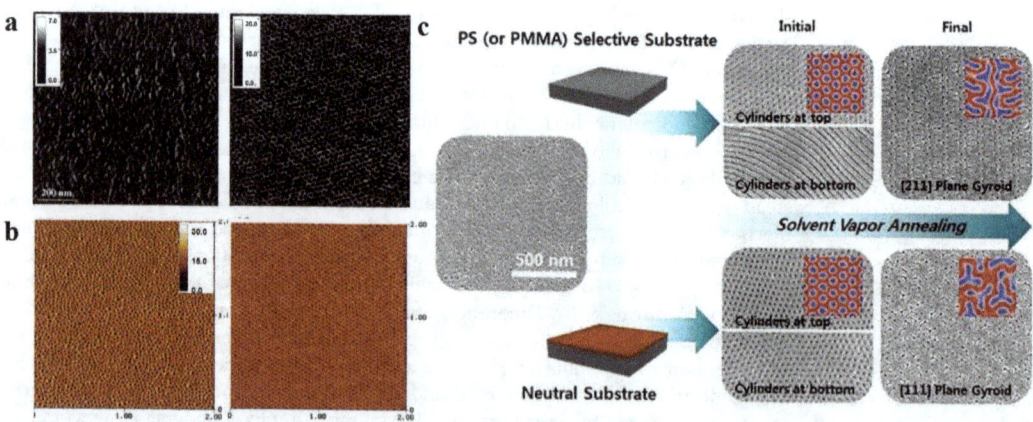

Figure 4. (**a**) AFM phase images (1×1 μm^2) of as-cast conventional PS-b-PEO film and PS-b-PEO film annealed in PS-selective toluene vapor for 5 min [66]. (**b**) AFM phase diagrams of PS-b-PEO films obtained by spin-coating method and annealed in neutral benzene vapor for 48 h [73]. (**c**) SEM images of the (111) and (211) planes produced by solvent vapor annealing of PS-b-PMMA films in neutral substrate and selective substrate [76].

3.2.2. Nonsolvent Approaches for Constructing Nanostructure

Researchers have also investigated the effect of nonsolvent-induced phase separation (NIPS) [81] of BCPs intensively [82–84]. The polymer is dissolved in solvent to form a homogeneous solution, followed by the slow addition of reagents which are more soluble in the solvent compared to the polymer (called extractants) to extract the solvent. The resulting two-phase structure with the polymer as continuous phase and solvent as the dispersed phase can lead to a polymer with a certain pore structure after the removal of solvent. Plisko et al. [85] modified poly (ethylene glycol)-b-poly (propylene glycol)-b-poly (ethylene glycol) (Pluronic F127) by the NIPS method employing polysulfone (PSF), and the polymer membrane obtained by the NIPS method was found to be porous in comparison to the dense polymer membrane prepared by the EIPS method (the surface structure of the porous film is shown in Figure 5a). Similarly, Gu et al. [86] fabricated porous PI-b-PS-b-P4VP membrane with sponge-like structure using the NIPS method, and Figure 5b shows the cross-sectional structure of the sponge-like porous film. NIPS is the most commonly used method for the preparation of membranes for separation [87,88], but the pore size and distribution of pores on the surface of NIPS-formed membranes are irregular and not well-controlled [89,90], and presence of some large pores can lead to material defects resulting in reduced mechanical strength of the membrane. This feature is responsible for limiting the application of BCP membranes made by the NIPS method in a great way.

In order to overcome the shortcoming mentioned above, a new NIPS strategy has emerged in the recent years: a short self-assembly is formed via short solvent evaporation after pouring the BCP into a solvent or mixed solvent, and then it was immersed in a nonsolvent for NIPS [91,92]. By this method, a monolithic, asymmetrically structured membrane, as shown in Figure 5c,d, can be formed [93]. Self-assembly results in the formation of highly ordered, dense pores in the upper thin selective layer which are uniformly perpendicular to the membrane surface [94]; pores of the lower layer which exhibit a sparse, disordered sponge-like interconnected pore channel are formed by NIPS [95]. This method of combining self-assembly (S) of BCP with nonsolvent-induced phase separation (NIPS) is named SNIPS [96,97]. As the first step of membrane formation via SNIPS is dissolution of BCP for self-assembly, different solvent parameters that affect the nanostructure (different selective solvents, mixed solvents, and solvent volatilization rate) discussed in the previous section also affect the membrane structure formed by the SNIPS method [98–100].

3.2.3. The Influence of External Field on Self-Assembled Structure of BCP

Thermal annealing has proved to be one of the efficient methods for preparing long-range ordered BCP membranes [101–104], and temperature is an effective and adjustable experimental condition during the process of self-assembly. Thermal annealing refers to the heating of BCP above its glass transition temperature T_g [105] for allowing the molecular chains within the polymer to self-assemble to the state closest to thermodynamic equilibrium [106–108]. Stehlin et al. [109] used AFM to observe the morphological variation of PS-b-PMMA at 180 °C and 200 °C (well above its T_g). Their resultant AFM images are shown in Figure 6a. It illustrates that thermal annealing results in ordered arrangement of the molecular chains, and the orderliness of PS-b-PMMA structure at 200 °C is higher than that of 180 °C for the same time, which indicates that the higher the annealing temperature, the faster the order of the arrangement. However, the BCP membrane produced by conventional thermal annealing techniques often shows structural defects, and the range of order regulation is limited, and for a thick membrane, ordering of the molecular chains over a large area cannot be obtained [110]. Moreover, for some polymers, the spacing between T_g and the viscous flow temperature T_f is small, which may not be sufficient for the rearrangement of molecular chains to take place within a short time.

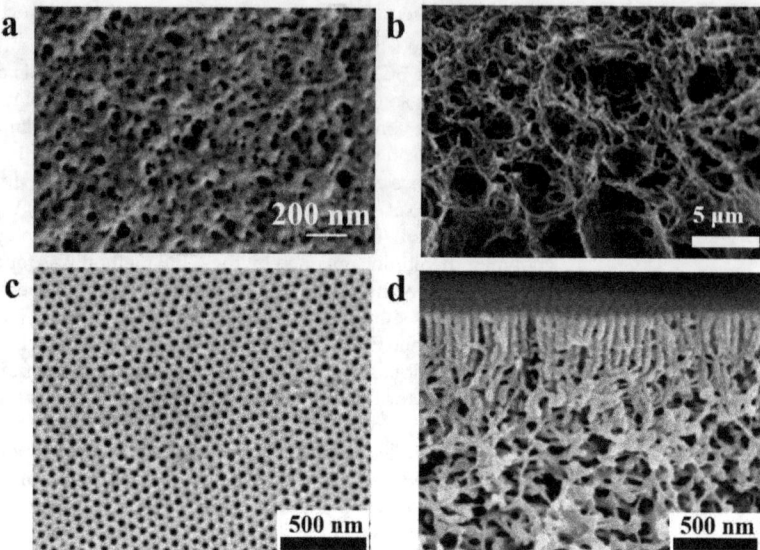

Figure 5. (**a**) SEM image of the surface of porous ultrafiltration membrane formed by the NIPS method for Pluronic F127 [85]. (**b**) SEM image of cross-section of multiporous sponge-like structure PI-b-PS-b-P4VP formed by NIPS method [86]. SEM images of the surface (**c**) and cross section (**d**) of PS-b-P4VP membrane obtained by SNIPS [93].

In some systems, a combined method of the thermal annealing and solvent annealing methods can be used [111]. Kim et al. [112] experimentally investigated PS-b-P4VP membrane in chloroform vapor (the schematic diagram is shown in Figure 6b). Since chloroform is selective for the PS block, the PS block is pulled toward the outside of the membrane, and the P4VP block forms multilayer spherical micelles. The solvent vapor increases the mobility of the polymer molecular chains by swelling the membrane, and consequently, some of the P4VP blocks fuse. A single SVA process is insufficient to drive the fusion of P4VP spherical structures and subsequently convert them into columnar structures perpendicular to the membrane surface as a result of strong repulsion between the PS and P4VP blocks (B). During the cooling process, chloroform vapor escapes from the inside along the direction perpendicular to the membrane as the membrane shrinks, which allows the spherical structures to aggregate along the vertical direction (C). During further SVA with programmed heating, the aggregation of spherical microdomains can propagate to the inner center of the film (E), resulting in membrane expansion (D), and shrink when cooled. After several cycles of operation, vertically oriented P4VP cylindrical microdomains are obtained (F). This experiment provides an effective orientation method for the orderly arrangement, with strong interactions between the blocks or thick membrane. Furthermore, combining thermal annealing with SVA is enabled to surmount the structural uncontrollability difficulties of high molecular weight BCP, which is due to the low mobility of molecular chains caused by chain entanglement [113].

As early as 1991, Gurovich [114] made a theoretical derivation of the force and phase behavior of BCP in an electric field, but application of electric fields in the research of BCP has stopped at the theoretical level for a while since then. This delay ended when A. Bo¨ker [115] observed the morphological changes of due to the application of DC electric field on the PS-b-PI prepared via reactive anion polymerization. In the absence of an applied electric field, the PS-b-PI self-assembled as ordered lamellar structures in the control group, while under the application of an electric field, lamellar structures of the samples underwent macroscopic orientation along the electric field lines. The application of electric field achieves long-range ordered nanostructures. Schmidt [116] verified the changes in the lamellar structure of PS-b-PI with increasing strength of the electric field. The phase space parallel to the electric field direction decreases, while the phase space perpendicular to the electric field direction increases, mainly resulting from the stretching of polymer molecular chains along the electric field lines (the mechanism is shown above in the Figure 6c). Schoberth et al. [117] used quasi in situ scanning force microscopy (SFM) capable of solvent vapor treatment of BCP despite high electric fields to characterize the phase behavior. Figure 6c below shows the SFM image of the disordered Polystyrene-b-poly (2-hydroxyethyl methacrylate)-b-poly (methyl methacrylate) $S_{47}H_{10}M_{43}$ oriented along the electric field direction after the application of an electric field [118]. Subsequently, it was experimentally demonstrated that application of an electric field decreases the T_{ODT} of PS-b-PI [119].

BCP orientation can also be regulated by the application of shear fields [120]. Unidirectional shear force is applied on the BCP membrane by contacting with suitably selected moving solid surface to induce the internal structure transformation of BCP. The shear modification of polystyrene-b-polybutadiene-b-polystyrene (SBS) and the study of its rheological behavior were carried out by Morrison et al. [121]. The structure of SBS can become oriented along the direction of shear force. Angelescu et al. [122] demonstrated that the rate of shearing plays a decisive role for obtaining the molecular alignment. Low shearing rates allow the molecular chains to arrange in an ordered manner along the shear direction, while too-high shear rates may cause a disordered arrangement of the intramolecular structures. It was then shown that for similar temperature, shear orientation requires less time compared to annealing and generates structures with other defects such as disruption of grain boundaries and absence of dislocations almost in the shear range [123]. Pujari et al. [124] observed a lamellar structure perpendicular to the substrate when the self-assembly of BCP was obtained through the application of shear stress to PS-b-PMMA during annealing. Different structures of BCP such as S, C, and L structures were oriented through the application of shearing. Davis et al. [125] investigated the morphological changes of poly(styrene)-b-poly (n-hexyl methacrylate) (PS-b-PHMA) due to application of shearing and also probed the effect of shearing on different polymer composition (see Figure 6d for a schematic diagram of the apparatus for applying shear and its results). It was finally concluded that shear causes orientation along the shear direction, and that for membranes with varying PS content, the dislocation density is maximum for the membrane containing the highest amount of PS. The ordering of shear field presents a fresh approach toward nanostructural adjustment of BCP membrane, which can achieve unprecedented structural ordering at lower temperatures and under simpler conditions [126,127].

3.2.4. Additives for Regulating Self-Assembly Structure

Researchers also explored various types of substances that can act as electrostatic additives [128] to regulate the nanomorphology of BCP [129], especially, to obtain some morphology in narrow interval [130]. Ionic liquid (IL) is a liquid composed of only anions and cations, with high ionic conductivity, chemical stability, and high thermal stability. Ionic liquids, when doped into BCP, can selectively complex with the charged blocks and induce lyotropic phase transitions via strong electrostatic cohesion [131]. Kim et al. [132] impregnated BCP into different ionic liquids to observe the variation in morphologies. It was demonstrated that the IL, 1-ethyl-3-methylimidazolium p-toluenesulfonate, could induce the structural transformation of the BCP from L to C morphology at a particular concentration, and different ILs result in the induction of different morphologies. Later, Bennett [133] added the IL 1-ethyl-3-methylimidazolium bis(trifluoromethanesulfonyl)imide (EMIM) to PS-b-PMMA to obtain a phase transition in the following sequence: D→L→G→C→D→S. IL-induced lyotropic phase transitions of BCP correspond to the single ion-induced phase transitions, while the role of multicharged ions on the phase transitions have rarely been reported, although it is easier to obtain strong electrostatic cohesion by multicharged ions. Polyoxometalate (POM) is a class of polymetallic oxygen cluster compounds formed by pretransition metal ions linked by oxygen, which has recently been applied for controlling the BCP nanostructures [134,135]. Zhang et al. [136] observed a change of an originally ordered columnar phase structure of the block copolymer into a disordered bicontinuous phase structure via incorporating Keggin-type POM $H_4SiW_{12}O_{40}$ (SiW) into PS-b-P4VP, which was due to strong electrostatic cohesion between the POM and P4VP chains.

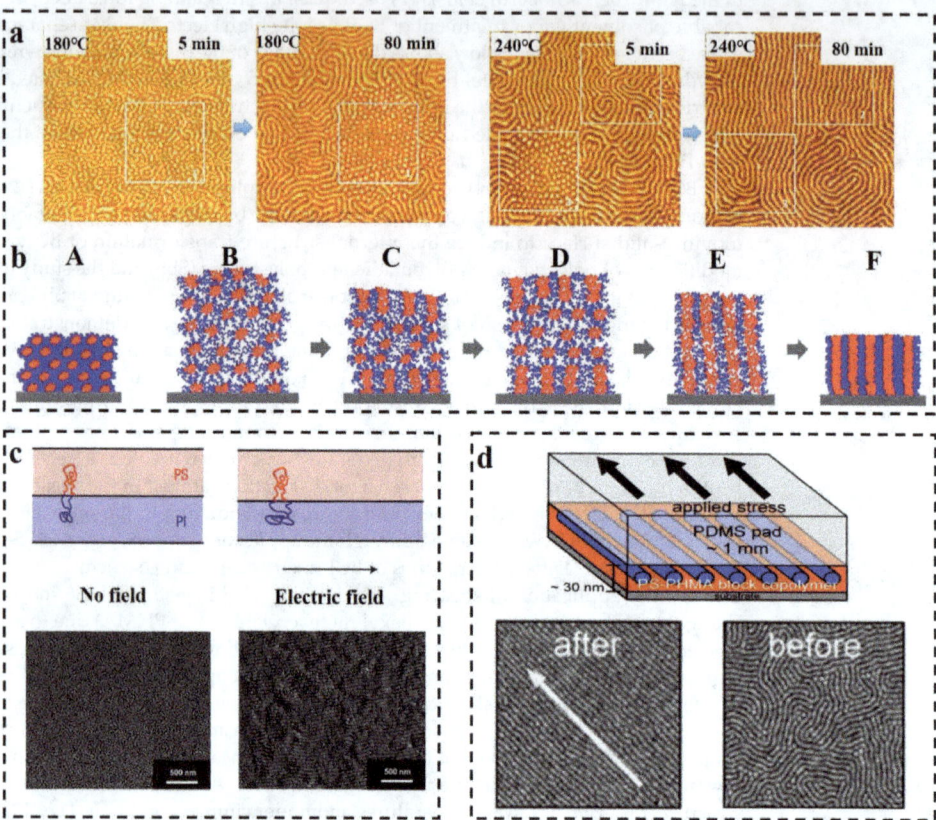

Figure 6. (**a**) In situ AFM images (1.25 × 1.25 μm^2) of PS-b-PMMA obtained by thermal annealing at 180 °C and 240 °C for 5 min and 80 min, respectively [109]. (**b**) The cylindrical microdomain structure of PS-b-P4VP perpendicular to the surface is illustrated by the combination of programmed temperature rise and fall and SVA [112]. (**c**) The impact of electric field on the stretching of BCP molecular chains is illustrated and the microdomains are macroscopically oriented along the electric field direction in SEM images [116,118]. (**d**) Schematic diagram of the unidirectional shear force applied to PS-b-PHMA and TM-AFM phase images of the block nanoscale before and after shear force application. Scale bar: 500 nm [125].

4. Mass Transport Applications in Energy Conversion of Block Copolymers

BCPs are one of the most important materials for the preparation of nanostructures by the bottom-up self-assembly method. Self-assembly of BCPs results in well-defined nanostructures with suitable volume fractions as well as Flory–Huggins interaction parameters. These structures can be used as nanochannels' template for mass transport. Regulation of the functional groups and structures of the channels is possible via molecular structure designing. The channel size can be tuned between 10 to 100 nm and channel density can reach up to 10^{11} cm^{-2}. High-resolution nanopatterns can be widely used in the energy field, such as sensors, memory, battery separator, etc. We concentrate our focus on mass transport applications of BCP in the energy conversion.

4.1. Ion Selective Membranes for Concentration Cell

Use of ion-selective membranes for converting osmotic energy into electrical energy is known as reverse electrodialysis (RED) [137–140]. Concentration cells use this energy conversion system based on salinity gradient [141–143]. Concentration cells are expected to solve the current energy crisis [144,145] by converting the salinity-gradient energy between seawater and river water into electricity [146]. However, the performance of conventional commercial ion-selective membranes has not been satisfactory due to their high resistance and low electrical power density. The key in developing salinity-gradient-dependent energy conversion is the preparation of high-performance ion-selective membranes [147,148]. In living organisms, ion channels are mainly composed of asymmetric membrane proteins embedded in lipid layers; these channels are capable of controlling the transport of specific ions or molecules through the membrane and also participate in various complex life activities of the human body.

Researchers have been working to imitate nature in order to find high-performance membrane materials [149]. Different shapes of nanopores comparable to lipid layers on membranes have been etched by methods known as ion-path etching [150]. Then, asymmetric nanochannels can be obtained by modifying the charge on the surface of the nanochannels [151]. Surface charge inside the inner wall of channels is the major controlling factor of ion transport [152,153]. The surface charge shows ion selectivity via adsorption of counter ions and repulsion of the co-ions [154], which can be used for salinity-gradient-driven energy conversion. Due to distinctive and controllable self-assembly properties, ion-selective membranes based on BCP have been used as separators in concentration cells [155]. Researchers carried out controlled synthesis of BCP membranes with different self-assembled structure for ion transportation. The key to improving the energy conversion efficiency of membranes mainly lies in the construction of ordered [156], high-density nanochannels having asymmetric chemical composition, pore size, and surface charges [157].

The common approach toward construction of asymmetric nanochannels is to combine BCP with some pre-existing and well-studied nanochannels. For example, based on their excellent physical and mechanical properties and dimensional stability, polyethylene terephthalate (PET) membranes are often fabricated into nanopores of various shapes by ion-path etching [158]. Zhang et al. [159] added self-assembled positively charged PS_{48400}-b-$P4VP_{21300}$ onto the surface of negatively charged PET membrane. The structure of the hybridized membrane is shown in Figure 7a (left). PS_{48400}-b-$P4VP_{21300}$ self-assemble into highly ordered, hexagonally packed pores, and when the *pH* is below the *pKa*, the P4VP chains exhibit a swollen, positively charged, and hydrophilic state. Ion-path-etched PET membrane contains conical channels. In the hybrid membrane, nanochannels with asymmetric chemical structure, channel shape and size, and surface charge polarity were formed, and the resulting membrane showed anion selectivity and ultrahigh ion rectification (rectification ratio f_{rec} up to 1075). The asymmetric structure of the hybrid membrane increases the energy conversion efficiency by eliminating the concentration polarization phenomenon to obtain a power density of 0.35 W m^{-2} in a 50-fold salinity gradient (Figure 7a (right)).

Wang's group [160] introduced the strategy of layer-by-layer (LBL) self-assembly to regulate the channel size and improve the ion transport capacity of the channels. The PET membrane with bullet-shaped pores was immersed in a mixture of PS-b-P4VP and homopolymerized polystyrene (h-PS). Then, the BCP and homopolymer was bonded onto the surface of the PET membrane and the inner surface of the pores by employing the solvent annealing-induced nanowetting in the template (SAINT) method. The self-assembly of LBL on the surface of PET membrane and the inner surface of the channel improves the energy conversion efficiency by reducing the effective pore size. Furthermore, as shown in Figure 7b, pH value affected the stretching state of the P4VP chains introduced within the bullet PET channel, leading to pH-responsive behavior of the wettability, ion rectification, ionic conductivity and ion selectivity of this hybrid membrane. Previous studies have shown that the performance of the membranes in concentration cell systems is usually pH-responsive, while few systems have maintained good ion rectification even in multiple pH environments. Yang's group [161] incorporated poly (tert-butyl methacrylate) (PtBuMA) into the PS-b-P2VP/PET hybrid nanochannels (the structure of the hybrid membrane is shown in Figure 7c (left)), which hydrolyzes in alkaline solution to produce negative charges, while the pyridine group in the P2VP block provide positive charges in acidic conditions. The membrane maintained a high f_{rec} (up to 200) with changing pH values (its rectification properties are shown in Figure 7c (right)). Similarly, Wu's group [162] employed the LBL method to graft thermally responsive PBOB-b-PNIPAM (with negative charge) into PET conical pores (schematic diagram of the synthesis process is shown in Figure 7d). The f_{rec} increased from 1.64 to 10.66 by introduction of the polymer into the PET hole, while increasing the number of layers resulted in decreasing conductivity and f_{rec}. As wettability and molecular conformation of PBOB-b-PNIPAM in channels depend on temperature, selective change of ion rectification and conductivity of the channels can be obtained by changing the temperature.

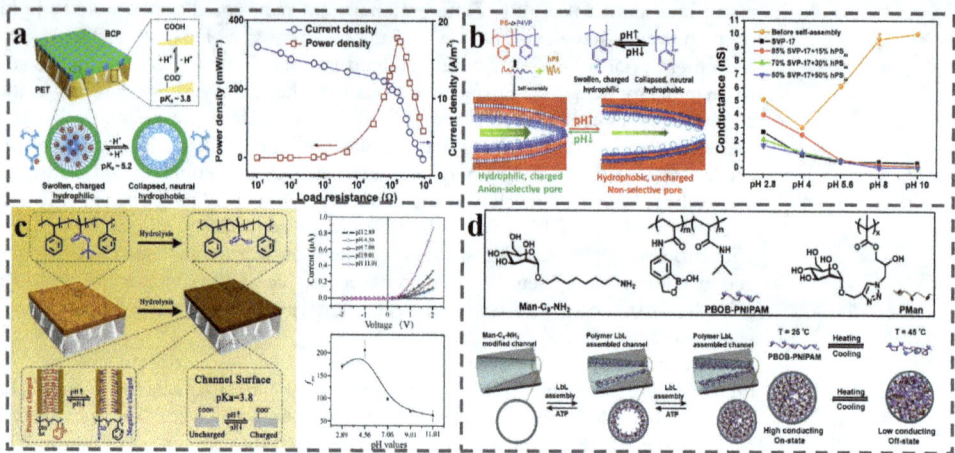

Figure 7. (**a**) Schematic diagram of a hybrid membrane consisting of a pH-responsive BCP membrane and a PET membrane with tapered pores obtained by ion-path etching and its permeation energy conversion efficiency [159]. (**b**) The pH-sensitive ionic conductivity of PS-b-P4VP with h-PS formed by self-assembly of molecular chain morphology with h-PS within bullet-shaped PET nanopores [160]. (**c**) The pH-responsive carboxyl groups, pyridine groups on BCP, and carboxyl groups in the PET pores enable pH-responsive, unidirectional rectification of ions in BCP-PET-laminated membrane [161]. (**d**) Molecular structures of feedstock polymers and synthetic routes to temperature-sensitive hybrid films [162].

In addition, one-dimensional (1D) and two-dimensional (2D) hybrid membranes can also be obtained when BCP is composited with 2D layered nanochannels. 2D nanomaterials have attracted the interest of researchers due to their simplicity of preparation, ease of controlling their structure and components, and possibility to extend in a large scale [163]. The unique layered stacking structure of 2D nanomaterials possesses large specific surface area and can be easily modified with functional groups. For example, MXene, with intrinsic functional groups between its layers, has received wide interest in the application of concentration cell [164]. Lin et al. [165], in an innovative method, hybridized negatively charged MXene membrane with positively charged PS-b-P2VP membrane into inhomogeneous membrane containing asymmetric structure, components, and charges. The ion selectivity experiment indicated dominant anionic selectivity for the BCP membrane with respect to MXene. This hybridization resulted in a power density of 6.74 W m^{-2} at pH = 11.

Beside these commercial BCPs, a newly designed functional BCP has also been developed. Our group [166] synthesized a UV-sensitive PEO-hv-PChal by ATRP. Introduction of o-nitrobenzyl ester as a degradable group between the two blocks resulted in breakage of this group, leading to the in situ formation of carboxylic acid groups under ultraviolet light. Self-supporting ordered nanochannels can be prepared by one-step degradation and cross-linking under the same UV light by employing PEO with good solubility as the degradation removal phase and the liquid crystal molecule containing chalone group as the continuous phase (Figure 8a). The size, density, and degree of order of the channels can be controlled by molecular weight and dispersion of the BCP. Then, we constructed asymmetric nanochannels by compounding the ordered nanochannels with AAO membrane. AAO is used by researchers in various fields of research because of its precise, stable, and un-deformed honeycomb structure with uniform pore size distribution and adjustable aperture. This asymmetric organic–inorganic hybrid membrane owned different channels and groups. Solution pH controls the ionization states of both the carboxyl group in the PChal and the hydroxyl group in the AAO pore (the pH-influenced groups within the hybridized pore channel are shown in Figure 8b). High energy conversion efficiency can be achieved over a wide pH range due to the synergistic effect of the hybrid nanochannels [167].

Although the above hybridization methods of combining BCP with other organic or inorganic membranes are able to achieve high rectification ratios or high power densities, further improvement of their performance is limited by the high internal resistance of the membrane (membrane thicknesses are often only in the micron range) and weak adhesion between the heterogeneous membranes. This motivated researchers to search for a strategy that can achieve high conversion efficiency with low internal resistance of the membrane. Zhang et al. [168] achieved innovative hybridization of two BCP membranes composed of different structural units (PEO-b-PChal and PS-b-P4VP) into an ultrathin ion-selective Janus membrane having only 500 nm thickness (the structure is shown in Figure 8c). This Janus membrane achieved excellent ion rectification due to asymmetric chemical structure, geometric aperture structure, and opposite surface charges. The results show that the P4VP chain in the channel of PS-b-P4VP membrane plays a dominant role in ion selectivity, making the hybrid membrane anion-selective. However, the operating conditions of the Janus membrane depend on the state of P4VP chains, as the performance of Janus membrane is mainly affected by the P4VP chains. Finally, this hybrid membrane was able to achieve a power density of 2.04 W m^{-2} in a seawater–river water environment at pH = 4.3.

However, the ion transport of BCP-based nanochannels is hindered due the resulting disorder and embedding of functional group distribution originating from disorder due to chaotic nature of the random coil conformation of P4VP flexible chain in the range of 1–10 nm, which makes them easy to unwind. This is a limiting factor for the further improvement of BCP-based nanochannels' performance. Our group [169] further proposed hybridization of rigid rod-like-structured PS-b-PPLG onto the prepared PEO-b-PChal porous membrane. Poly (γ-benzyl-L-glutamate) (PBLG) is a poly(polypeptide) with stable α-helical rigid structure, and its branched chain makes it easier to modify with ionic liquids as functional groups. After the introduction of ionic liquid imidazole bromide on the α-helical rigid-structured PBLG to provide positive charges, this hybrid membrane possesses nanochannels with asymmetric surface charge and nanostructure (the structure of the asymmetric nanochannel membrane is shown in Figure 8d). The rigid PPLG segments exist in a straight chain conformation in the channel, and the arrangement is ordered in the size range of 1–10 nm, forming hierarchical ordered structures with different sizes. This method provides a strategy to improve the orderliness of the channel and reduce the internal resistance of nanochannels.

To further improve the power density of energy conversion, Li et al. [170] demonstrated a robust mushroom-shaped (with stem and cap) nanochannel array membrane with an ultrathin selective layer and ultrahigh pore density. The stem parts, negative-charged 1D channels, are prepared from the previous PEO-b-PChal self-assembly with a density of ~10^{11} cm^{-2}, while the cap parts, positive-charged 3D channels, are formed by chemically grafted hyperbranched polyethyleneimine. As shown in Figure 8e, the hyperbranched polyethyleneimine cap parts as the selective layer are equivalent to tens of 1D nanochannels per stem. These robust mushroom-shaped nanochannels achieve f_{rec} of 17.3 and power density of 15.4 W m^{-2} at a KCl solution with 500-fold salinity gradient.

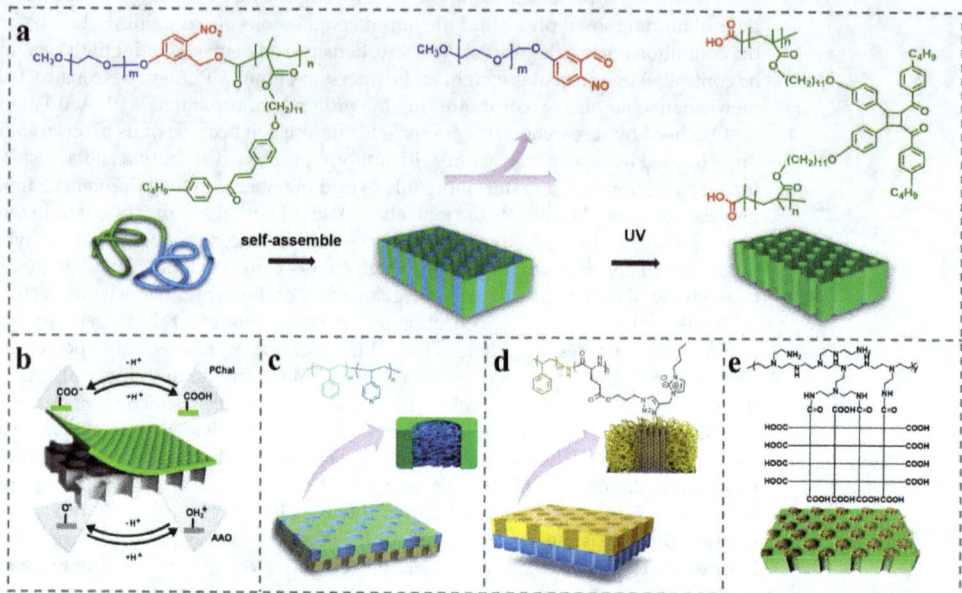

Figure 8. (a) Structural formula of PEO-hv-PChal molecule which can be degraded and cross-linked simultaneously under UV light and its self-assembly process [166]. (b) Schematic diagram of BCP-AAO asymmetric nanochannel membrane structure [166]. (c) Schematic diagram of PS-b-P4VP nanopore channel and schematic diagram of asymmetrical ultrathin Janus film hybridized with PEO-b-PChal porous membrane [168]. (d) Schematic diagram of the nanopore channels of PS-b-PPLG and its hybridization with PEO-b-PChal porous membrane to form a membrane with asymmetric nanopore channels [169]. (e) Schematic diagram of the molecular chain structure of PEO-b-PChal and polyethyleneimine and their hybridization to form mushroom-like asymmetric nanopore channels [170].

The SNIPS method also gives a new perspective for the preparation of monolayer asymmetric membrane. Koo et al. [171] combined SNIPS with a thermodynamic method for controlling the growth process of nanochannels in the membrane to achieve an asymmetric porous PS-b-P2VP membrane. The porous BCP membrane formed by the SNIPS method can spontaneously form asymmetric nanopore channels with dense nanopores on the surface and spongy and looser micropores inside. Parameters such as membrane thickness can also be controlled by thermodynamic methods. In addition, the P2VP channels can be made positively charged by quaternization, which imparts anion selectivity into the membrane and reduces the pore dimensions. Connecting a large number of salinity-gradient cells with PS-P2VP membranes can be used for the operation of small electronic devices.

Xie et al. [172] exploited a new method for the utilization of BCP in the construction of nanochannels: the nanochannels can be regulated by utilizing and modifying the different chemical properties of each block of BCP as a template. Different hydrophilic properties of PEO and PPO in triblock copolymer Pluronic F127 were used as soft templates to synthesize 1D anisotropic carbon-ordered mesoporous nanowires (CMWs). Then, dense CMWs membranes (with negative charge) were synthesized on porous AAO membranes (with positive charge) by the vacuum filtration method. In the CMWs membrane, ions can pass through the 3D interconnected channels formed by the gaps of these CMWs.

Therefore, the prepared asymmetric hybrid membranes possess asymmetric chemical components, nanostructures, and surface charges, and the membrane exhibited excellent cation selectivity.

4.2. Ion Exchange Membranes for Fuel Cell

A fuel cell is a device capable of converting chemical energy from fuels (such as methanol, ethanol, pure hydrogen, natural gas, and gasoline) and directly oxidizing it into electrical energy through an electrochemical reaction [173,174]. It has the advantages of high efficiency, no toxic gas emission, and no pollution, and is regarded as the most promising way for generation of electricity [175,176]. For example, in hydrogen fuel cell, the basic principle is the reverse reaction of water electrolysis, where hydrogen and oxygen are supplied to the anode and cathode, respectively. After outward diffusion of hydrogen through the anode, it reacts with the electrolyte to release electrons which then reach the cathode through external load. Electrolyte diaphragm is an important part of fuel cells [177]. Its main function is to conduct ions while keeping the oxidizer and reducing agent separate [178]. The ion transport occurs through the groups present in the membrane-forming material, through the combination and separation of ions, and via forming a strip of ion channels [179]. For example, proton-exchange membranes usually have some strong electrolyte groups such as sulfonic acid radical that can easily bond with the protons [180]. Protons can easily bond with the groups and release, allowing the protons to pass through the membrane to form a current without direct contact between the positive and negative oxidizers and the fuel. Similarly, the function of cation-exchange membrane [181] and anion-exchange membrane is to allow the cations or anions to pass through, forming a current, while blocking positive and negative oxidizer and fuel. The principle is based on the selective permeability of ion-exchange membranes [182].

A proton-exchange membrane fuel cell is a new type of fuel cell with high efficiency and low pollution [183,184]. Proton-exchange membrane (PEM) [185], also known as polymer electrolyte membrane, is a very thin, rigid, plastic-like polymer material with proton conductivity that can be used to replace conventional liquid electrolytes [186,187]. The biggest challenge of PEM is to improve ionic conductivity and thermal stability without compromising the mechanical strength [188–190]. Inspired by the common BCP structures having rigid blocks as the skeletal support and flexible blocks as ion conductors, BCPs are usually used as precursors for nanostructured polymer materials [191].

The bicontinuous phase of self-assembled BCP has attracted much attention. There are two separate and mixed structures, the domain is throughout the material structure, it has a very high interfacial area and connectivity, and it is advantageous to the ion transport in the membrane. However, bicontinuous phase in the phase diagram has a relatively narrow interval, and it is difficult to achieve by controlling the molecular weight and proportion. Doping of other molecules to regulate the interaction forces is a good strategy to regulate the self-assembled structure of BCP. Electrostatic interaction between IL [192,193] and BCP is a promising pathway to induce BCP to form a bicontinuous phase. Morgan et al. [194] used 1-butyl-3-methylimidazolium bis (trifluoromethylsulfonyl) imide (BMITFSI) to induce the self-assembly of AB-type diblock copolymer PS-b-PEO. BMITFSI is immiscible with PS, and it enters the PEO block to induce the separation of PS-b-PEO assembly into a bicontinuous phase (the bicontinuous phase structure is shown in Figure 9a). Additionally, as shown in Figure 9a, IL also reduces the intermolecular forces between the polymer chains to enhance the fluidity of polymer chain, improves the ionic transport property of the membrane, and affects the free volume and flexibility of the membrane, improving the conductivity and electrochemical stability of the membrane.

It has also been found that in spite of IL, POM is also used to induce BCP to establish a bicontinuous phase structure. Zhang et al. [195] induced the transition of an AB-type diblock copolymer PS-b-P2VP from layered structure to bicontinuous phase structure using $H_4SiW_{12}O_{40}$ (SiW) (the structural transformation process is shown in the top diagram of Figure 9b). POM in BCP plays the role of electrostatic morphology control system, improves the plasma conductivity, and also plays the role of nanointensifier for improving the modulus. Ultimately, as shown in the lower graph of Figure 9b, POM increased the proton conductivity σ and Young's modulus of the bicontinuous structured material at room temperature by 0.1 mS cm^{-1} and by 7.4 G Pa, respectively. Zhai et al. [196] introduced $H_3PW_{12}O_{40}$ (PW) into the synthesized ABA-type triblock copolymer PVP-b-PS-b-PVP. As shown in Figure 9c, weak electrostatic interactions between PW and PVP units induced the BCP to form a three-dimensional continuous-charged domain (for ionic conduction) and an inverse cylindrical phase structure with embedded neutral domains (for mechanical support). The nanocomposite corresponding to this configuration was shown to be a novel PEM with proton conductivity σ of 1.32 mS cm^{-1}.

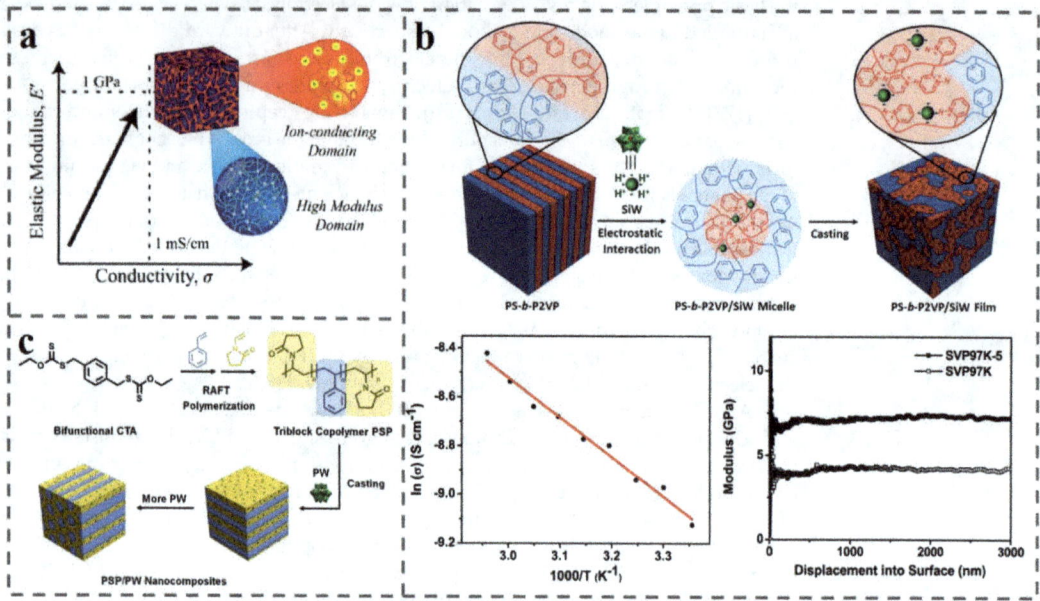

Figure 9. (a) BMITFSI induces PS-b-PEO to form a bicontinuous phase microdomain structure and enhances the elastic modulus E' and conductivity σ of PS-b-PEO [194]. (b) The electrostatic interaction between SiW and PS-b-P2VP induces the formation of a bicontinuous phase favorable to proton conduction, leading to an increase in proton conductivity and Young's modulus [195]. (c) PW induces PVP-b-PS-b-PVP by electrostatic interaction to form a proton-conducting continuous phase with cylindrical, mechanically enhanced phase with great proton-conduction advantage [196].

The alkaline fuel cell composed of anion exchange membrane (AEM) is similar to the PEM fuel cell in terms of mechanism and complete reaction equilibrium [197], but the electrode reaction is different from that of the PEM fuel cell, as the reaction takes place under alkaline conditions. Alkaline fuel cells have distinct advantages: the cost of fuel cell production can be reduced by some inexpensive catalysts such as iron and nickel [198,199]; liquid fuels such as methanol and ethanol that can be stored and transported easily are used [200]; and the corrosion of metal catalysts is less than that of acidic environment, prolonging the fuel cell life [201]. AEMs are the critical components of alkaline fuel cells. Their ionic conductivity and stability under alkaline environments are critical to the performance of alkaline fuel cells [202,203]. Microstructure of the polymer determines the material performance, and the design of a customized AEM with a rational structure is essential for the production of high-performance alkaline fuel cells [204,205].

Sulfonated poly (styrene-ethylene-butylene-styrene) copolymer (SEBS) is a promising diaphragm material, having high thermal and chemical stability, adjustable mechanical properties, and good proton conductivity and cost-effectiveness [206–208]. S-SEBS-g-MA AEM was prepared by grafting sulfonic acid and maleic anhydride to form ion channel for improvement of ion conductivity [209] (Figure 10a). From electrochemical analysis, as shown in Figure 10a (right), the modified membranes showed increased ionic conductivity, ionic exchange capacity (*IEC*), and water absorption, all of which were higher than those of the conventional commercial Nafion 117 membrane. These results suggest that modified AEMs based on BCPs such as SEBS can have comparable potential as AEMs with respect to the commercial Nafion 117.

Polyolefin-based AEMs exhibit good alkali resistance and have great potential for mass production due to their easy processability and low cost [210,211]. In the anionic polymerization, the chemical composition, molecular weight, and conformational distribution of the products can be controlled during the process of polymer synthesis. Polyisoprene-b-poly(4-methylstyrene) (SCP) AEMs with a star topology and improved properties were prepared (see Figure 10b left diagram for SCP-AEM structural formula) through anionic polymerization by Pan et al [212]. This star-shaped structure is a head-to-head cross-linked structure. The conductivity and low water absorption and alkaline stability of AEM can be significantly improved through functionalization such as bromination and quaternary ammonium near the "star-shaped", nucleus resulting in the construction of continuous ion transport channels (AFM phase diagram is shown in Figure 10b (middle)). The *IEC* and hydroxide conductivity of the functionalized AEM can reach 2.15 mmol g^{-1} and 68.1 mS cm^{-1}, respectively. With increasing current density, maximum power density of the fuel cell reaches 120.2 mW cm^{-2}. This method provides a new approach to enhance the performance of AEM by taking advantage of the customizable structural properties of BCP. The conductivity and basic stability of AEM can be improved by placing the functional groups that can build continuous ion channels and the chain segments that are easily attacked by hydroxide in the inside structure of BCP by structural design.

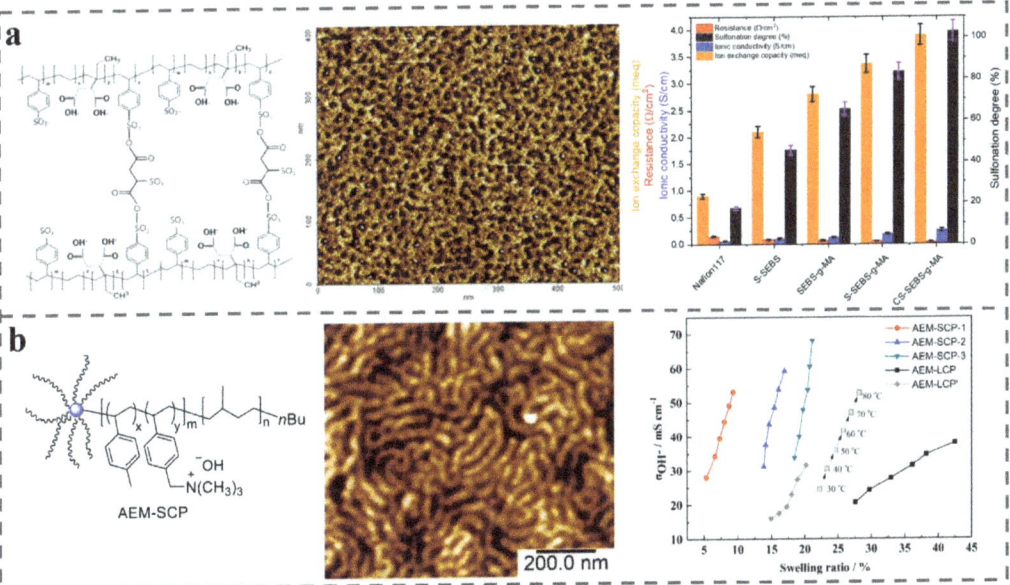

Figure 10. (a) Molecular structure formula of S-SEBS-g-MA AEM and its AFM phase image. Compared with AEM formed by grafting different molecular chains and Nafion 117, S-SEBS-g-MA AEM has the superior ionic conductivity and *IEC* performance [209]. (b) Molecular structure and AFM image of AEM-SCP membrane and its anion conductivity versus swelling ratio [212].

4.3. Battery Separator for Rechargeable Batteries (e.g., Lithium-Ion Batteries)

Among commercially implemented rechargeable batteries, high specific energy and stable cycling performance have resulted in the broad applications of lithium-ion batteries (LIBs) [213,214] in cell phones, notebook computers, electric vehicles, and grid energy storage, with prospects for further development. However, the performance and safety of lithium-ion batteries still need improvement [215,216]. As a key component, lithium ions transfer between the positive and negative electrodes during the charging and discharging process and are controlled by a battery separator, which in turn controls the efficiency of the lithium-ion battery [217,218]. An important parameter characterizing the migration of lithium ions is $t+$, which is defined as the ratio of Li^+ migration rate to that of the migration of all ions in the electrolyte for a lithium-ion battery [219,220]. In liquid electrolytes, $t+$ is normally below 1, with the general value being between 0.3 and 0.4 [221]. Traditional LIBs commonly used organic solvent-resistant, high-strength porous polyolefin membrane as a battery separator, which has good chemical stability and mechanical stability, but the thermal stability is poor. During heat build-up and rise of temperature inside the battery, the diaphragm can melt, which results in a short-circuit between the positive and negative terminals, causing an accident [222]. In recent years, there have been frequent safety accidents concerning LIBs, which seriously threaten the safety of human life and limit the application prospects of LIBs. Researchers are committed to finding ways to improve the safety of LIBs [223,224], especially by enhancing the thermal stability of the battery diaphragm to prevent accidents.

Researchers have employed polymer membranes with higher melting points and good thermal stability, such as polyimide, cellulose, and PSF, to replace traditional polyolefin materials, which has shown primary results regarding enhancement of thermal stability of LIBs. Yang et al. [225] studied PSF-b-PEG, a BCP composed of PSF and hydrophilic polyethylene glycol PEG, as a diaphragm for LIBs (diaphragm structure and lithium-ion transport process are shown in Figure 11a). This BCP has the following advantages: high thermal stability of PSF, strong affinity for the liquid organic electrolytes, and good complexation ability of PEG toward lithium salts, allowing for transport of lithium ions, which greatly improves the lithium-ion conductivity. In addition, presence of benzene ring and ether bond on the PSF chain of PSF-b-PEG provides rigidity and flexibility, performing good mechanical strength. A strong interaction force between the two blocks results in good mechanical stability for PSF-b-PEG during the formation of membrane and channels, resulting in no ion channel blockage. As is shown at Figure 11b,c, PSF-b-PEG has been tested to be comparable to commercial polypropylene Celgard 2400 membranes, both with respect to thermal stability and porosity at 380 °C, but the PSF-b-PEG membranes have additional advantages of much higher wettability, mechanical strength, and electrolyte absorption than Celgard 2400. During the tests in the temperature range of 75 °C to 150 °C, PSF-b-PEG membranes consistently maintained relatively low thermal shrinkage, ensuring the safety of LIBs. Additionally, at temperatures higher than 125 °C, the channels of PSF-b-PEG membrane will automatically close due to thermal annealing, switching "off" the working state of the battery diaphragm, and during the process of channels closure, there will not be any change of membrane size, ensuring the safety of LIBs.

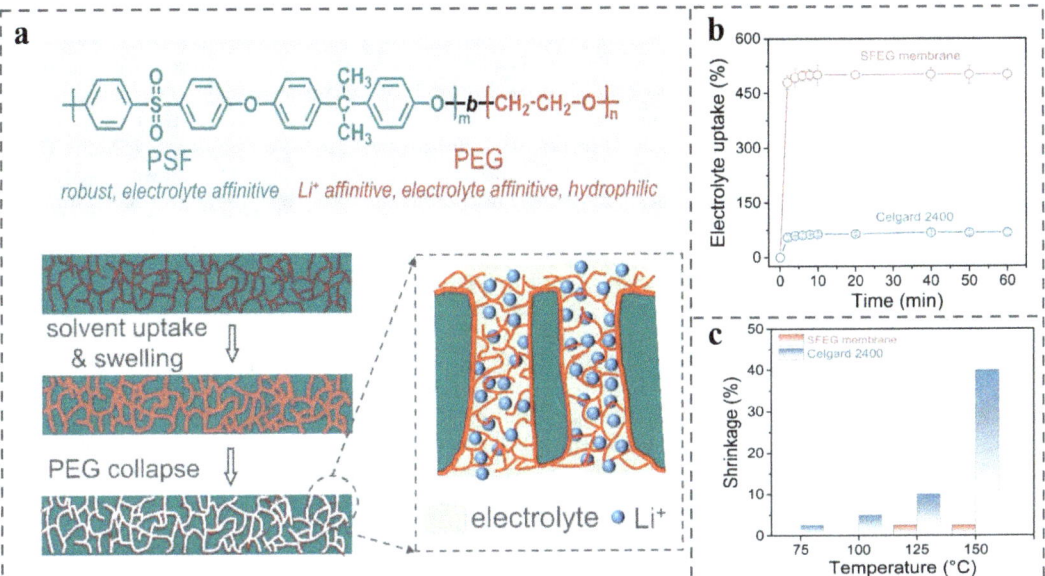

Figure 11. (a) Molecular structure of PSF-b-PEG membrane and its selective swelling application in lithium-ion transport. Compared to the Celgard 2400 diaphragm, PSF-b-PEG membrane exhibits exceptional electrolyte uptake performance (b) and thermal stability (c) [225].

Even after using thermally stable diaphragms, safety accidents cannot be eliminated at the root, and liquid organic electrolytes still have safety hazards [226]. Solid polymer lithium-ion batteries, which use solid polymer electrolytes (SPE) instead of traditional liquid electrolytes [227], are safer than traditional liquid LIBs and will gradually become the mainstream of LIBs by replacing the traditional liquid LIBs [228,229]. The ion channels of SPE should be perpendicular to the electrode surface to facilitate ion transport [230]. BCP electrolytes (BCPEs) are one of the most attractive alternatives to traditional liquid electrolytes in LIBs, as they can improve thermal and mechanical stability while maintaining ionic conductivity.

In general, rigid blocks are used as the skeleton support to provide mechanical strength, whereas flexible blocks are used to enhance ionic conductivity. PEG has been widely used in LIBs but has associated problems such as low ionic conductivity, poor mechanical strength, and narrow electrochemical window, which can be improved by reintroducing rigid blocks. Lin et al. [231] utilized cross-linking copolymerization reactions of flexible PEG blocks and rigid hexamethylene di-isocyanate trimer (HDIt) blocks in different ratios to synthesize a series of new BCPEs, named PH-BCPE, with 3D networks (see Figure 12a (left)). It has been experimentally demonstrated that the ratio of two blocks R (n_{PEG}/n_{HDIt}) can be a controlling factor for the performance of PH-BCPE. By increasing the proportion of flexible PEG blocks (elevation of R value), the ionic conductivity of PH-BCPE can be increased (as shown in Figure 12a (middle)) and the interfacial resistance with the electrode can be decreased, whereas increasing the proportion of rigid HDIt blocks (decrease in R value) can improve the electrochemical window (as shown in Figure 12a (right)) and mechanical strength of PH-BCPE. Finally, the ionic conductivity of obtained PH-BCPE resulted in an ionic conductivity up to $5.7 \cdot 10^{-4}$ S cm^{-1}, $t+$ value of 0.49, and wide electrochemical window up to 4.65 V (vs. Li$^+$/Li).

To achieve a BCPE with high mechanical stability and ionic conductivity, He et al. [232] employed RAFT polymerization to synthesize a difunctional P(DBEA-co-MA)-b-PEG which consists of the UV-cross-linkable block P(DBEA-co-MA) and suspended PEG chains (the network structure is shown in Figure 12b). The mechanical strength can be improved due to the presence of tethered double bonds in P(DBEA-co-MA) which can form a cross-linked network, while the suspended PEG molecular chains with low-crystallinity can increase molecular mobility and reduce chain entanglement to improve ionic conductivity. The interfacial resistance between the electrode and P(DBEA-co-MA)-b-PEG-SPE can be reduced from 5500 Ω cm^2 to 100 Ω cm^2 compared to the conventional PEO-SPE. The cycle stability of the battery is doubled, and the battery can be cycled for more than 700 h at 22 °C, showing significant improvement of durability. Using bis (trifluoromethane) sulfonimide lithium salt (LITFSI) as the ion carrier, the final value of $t+$ was calculated to be 0.35.

Among the various forms of BCP, cylindrical (C) pore channels have been applied, though the lamellar (L) structures possess higher f values and theoretically higher ionic conductivity than C, but L has been explored very little. Liu et al. [233] employed an intermediate layer of azobenzene to synthesize a series of tablet-b-bottlebrush (TB) BCP electrolytes (Figure 12c). The liquid crystal polymer segments can form nanopores in a directional and rapid manner after solvent thermal annealing and the presence of polyethylene oxide (PEO) side chains leads to increasing electrical conductivity. The performance of electrolyte membrane with a thickness of 200 mm or more prepared by this method is superior to other spin-coated nanoscale membrane. The final synthesized SPE also possesses an ionic conductivity of 2.19 mS cm^{-1} at 200 °C, along with superior thermal stability.

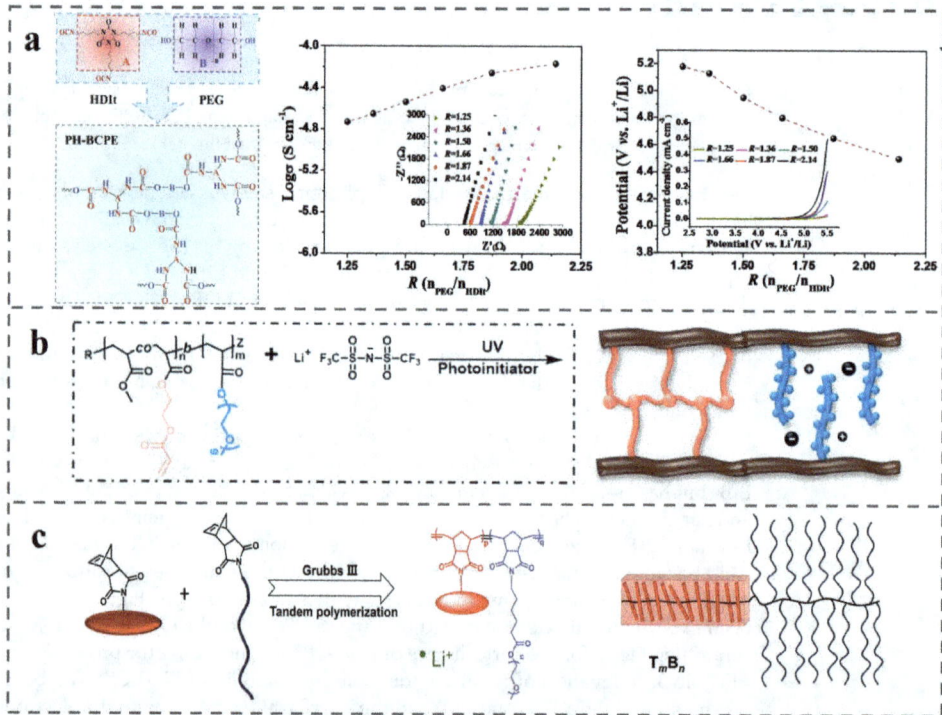

Figure 12. (**a**) PEG and HDIt molecules were copolymerized to obtain PH-BCPE with 3D network structure. R has a decisive effect on the ionic conductivity and electrochemical window of PH-BCPE [231]. (**b**) Synthesis of P(DBEA-co-MA)-b-PEG with 3D cross-linked network structure [232]. (**c**) TB-BCPEs were synthesized by tandem polymerization and have a lamellar structure favorable for ion transport [233].

5. Conclusions

The self-assembly behaviors of BCP can be regulated by mastering the molecular structure, solvent type, concentration of polymer solution, nonsolvent, external environmental conditions, and additives. Additionally, the self-assembly behaviors of BCP can precisely control the structure of BCP at the nanoscale. Moreover, due to their unparalleled surface-controllable and nanostructure-controllable properties, BCP can be introduced into a broad spectrum of applications. This review covers advancements in block copolymers for the optimization of energy conversion in novel batteries. The integration of BCP into new energy sources such as concentration cells, fuel cells, and lithium-ion batteries can improve energy conversion efficiency while sustaining high performance, which is unachievable with classical materials.

In spite of the attractive options provided by BCP in enhancing energy conversion efficiency, for new energy devices, there is still considerable work to be done to make them competitive with existing technologies in terms of performance and cost. The first challenge is to utilize novel materials for the synthesis of block copolymers with new and superior structures and properties. The materials currently used to synthesize block copolymers are conventional polymeric materials, and it is critical to explore new materials as well as new structures for BCPs. This becomes potentially changing as more applications are discovered. The second challenge is how to precisely tune the interface during self-assembly to tailor BCPs with absolutely precise nanostructures on demand. In many cases, different disciplines are not well-integrated, and perhaps the collision of different sciences will bring chances for precise tuning of BCPs. Finally, due to the constraints of the established technology platform and cost issues, the research on BCP-based composites is still at the primary stage of laboratory research rather than large-scale commercial application. This requires not only a change in technology, but also a long process of financial investment and training of personnel. Certainly, all of these bottlenecks are common barriers to the implementation of new technologies, and the future is still bright with the endless possibilities of BCP. We are convinced that block copolymers can enable a new era of energy conversion; the question is not if, but when.

Author Contributions: The original writing framework was constructed by S.M. and X.S.; S.M. completed the first draft; X.S. completed the revision of the article. Y.H., J.H., C.L. and J.Z. provided many suggestions during the writing process. All authors have read and agreed to the published version of the manuscript.

Funding: The work was supported by the National Natural Science Foundation of China (No. 22005162), the Natural Science Foundation of Shandong Province (No. ZR2020QE093), and the Special Financial Aid to Postdoctor Research Fellow (No. 2020T130330).

Data Availability Statement: Not applicable.

Conflicts of Interest: The authors declare no conflict of interest.

References

1. Schacher, F.H.; Rupar, P.A.; Manners, I. Functional Block Copolymers: Nanostructured Materials with Emerging Applications. *Angew. Chem. Int. Ed.* **2012**, *51*, 7898–7921. [CrossRef] [PubMed]
2. Bates, C.M.; Bates, F.S. 50th Anniversary Perspective: Block Polymers-Pure Potential. *Macromolecules* **2017**, *50*, 3–22. [CrossRef]
3. Mankowich, A.M. Micellar Molecular Weights of Selected Surface Active Agents. *J. Phys. Chem.* **1954**, *58*, 1027–1030. [CrossRef]
4. Vaughn, T.H.; Suter, H.R.; Lundsted, L.G.; Kramer, M.G. Properties of some newly developed nonionic detergents. *J. Am. Oil. Chem. Soc.* **1951**, *28*, 294–299. [CrossRef]
5. Szwarc, M.; Levy, M.; Milkovich, R. Polymerization initiated by electron transfer to monomer. A new method of formation of block polymers. *J. Am. Chem. Soc.* **1956**, *78*, 2656–2657. [CrossRef]
6. Lanson, D.; Ariura, F.; Schappacher, M.; Borsali, R.; Deffieux, A. Application of living ionic polymerizations to the design of AB-type comb-like copolymers of various topologies and organizations. *Macromol. Res.* **2007**, *15*, 173–177. [CrossRef]
7. Charleux, B.; Delaittre, G.; Rieger, J.; D'Agosto, F. Polymerization-Induced Self-Assembly: From Soluble Macromolecules to Block Copolymer Nano-Objects in One Step. *Macromolecules* **2012**, *45*, 6753–6765. [CrossRef]

8. Nasrullah, M.J.; Vora, A.; Webster, D.C. Block Copolymer Synthesis via a Combination of ATRP and RAFT Using Click Chemistry. *Macromol. Chem. Phys.* **2011**, *212*, 539–549. [CrossRef]
9. Siegwart, D.J.; Oh, J.K.; Matyjaszewski, K. ATRP in the design of functional materials for biomedical applications. *Prog. Polym. Sci.* **2012**, *37*, 18–37. [CrossRef] [PubMed]
10. Matyjaszewski, K. Advanced Materials by Atom Transfer Radical Polymerization. *Adv. Mater.* **2018**, *30*, 1706441. [CrossRef] [PubMed]
11. Keddie, D.J. A guide to the synthesis of block copolymers using reversible-addition fragmentation chain transfer (RAFT) polymerization. *Chem. Soc. Rev.* **2014**, *43*, 496–505. [CrossRef]
12. Eggers, S.; Eckert, T.; Abetz, V. Double thermoresponsive block-random copolymers with adjustable phase transition temperatures: From block-like to gradient-like behavior. *J. Polym. Sci. Pol. Chem.* **2018**, *56*, 399–411. [CrossRef]
13. Gyorgy, C.; Hunter, S.J.; Girou, C.; Derry, M.J.; Armes, S.P. Synthesis of poly(stearyl methacrylate)-poly(2-hydroxypropyl methacrylate) diblock copolymer nanoparticles via RAFT dispersion polymerization of 2-hydroxypropyl methacrylate in mineral oil. *Polym. Chem.* **2020**, *11*, 4579–4590. [CrossRef]
14. Edwards, S.F. The statistical mechanics of polymers with excluded volume. *Proc. Phys. Soc.* **1965**, *85*, 613–624. [CrossRef]
15. Matsen, M.W.; Schick, M. Stable and unstable phases of a diblock copolymer melt. *Phys. Rev. Lett.* **1994**, *72*, 2660–2663. [CrossRef] [PubMed]
16. Matsen, M.W.; Bates, F.S. Unifying weak- and strong-segregation block copolymer theories. *Macromolecules* **1996**, *29*, 1091–1098. [CrossRef]
17. Spencer, R.K.W.; Matsen, M.W. Fluctuation effects in blends of A plus B homopolymers with AB diblock copolymer. *J. Chem. Phys.* **2018**, *148*, 204907. [CrossRef] [PubMed]
18. Kang, S.; Kim, G.-H.; Park, S.-J. Conjugated Block Copolymers for Functional Nanostructures. *Acc. Chem. Res.* **2022**, *55*, 2224–2234. [CrossRef]
19. Rosler, A.; Vandermeulen, G.W.M.; Klok, H.A. Advanced drug delivery devices via self-assembly of amphiphilic block copolymers. *Adv. Drug Deliv. Rev.* **2012**, *64*, 270–279. [CrossRef]
20. Cummins, C.; Lundy, R.; Walsh, J.J.; Ponsinet, V.; Fleury, G.; Morris, M.A. Enabling future nanomanufacturing through block copolymer self-assembly: A review. *Nano Today* **2020**, *35*, 100936. [CrossRef]
21. Meng, L.; Fan, H.; Lane, J.M.D.; Qin, Y. Bottom-Up Approaches for Precisely Nanostructuring Hybrid Organic/Inorganic Multi-Component Composites for Organic Photovoltaics. *MRS Adv.* **2020**, *5*, 2055–2065. [CrossRef]
22. Yu, H.Z.; Qiu, X.Y.; Moreno, N.; Ma, Z.W.; Calo, V.M.; Nunes, S.P.; Peinemann, K.V. Self-Assembled Asymmetric Block Copolymer Membranes: Bridging the Gap from Ultra- to Nanofiltration. *Angew. Chem. Int. Ed.* **2015**, *54*, 13937–13941. [CrossRef] [PubMed]
23. Radjabian, M.; Abetz, V. Advanced porous polymer membranes from self-assembling block copolymers. *Prog. Polym. Sci.* **2020**, *102*, 101219. [CrossRef]
24. Cheng, X.Q.; Wang, Z.X.; Jiang, X.; Li, T.X.; Lau, C.H.; Guo, Z.H.; Ma, J.; Shao, L. Towards sustainable ultrafast molecular-separation membranes: From conventional polymers to emerging materials. *Prog. Mater. Sci.* **2018**, *92*, 258–283. [CrossRef]
25. Park, H.B.; Kamcev, J.; Robeson, L.M.; Elimelech, M.; Freeman, B.D. Maximizing the right stuff: The trade-off between membrane permeability and selectivity. *Science* **2017**, *356*, eaab0530. [CrossRef] [PubMed]
26. Khan, I.; Hou, F.J.; Le, H.P. The impact of natural resources, energy consumption, and population growth on environmental quality: Fresh evidence from the United States of America. *Sci. Total Environ.* **2021**, *754*, 142222. [CrossRef]
27. Schiermeier, Q. Increased flood risk linked to global warming. *Nature* **2011**, *470*, 316. [CrossRef] [PubMed]
28. Zandalinas, S.I.; Fritschi, F.B.; Mittler, R. Global Warming, Climate Change, and Environmental Pollution: Recipe for a Multifactorial Stress Combination Disaster. *Trends Plant Sci.* **2021**, *26*, 588–599. [CrossRef]
29. Fan, E.S.; Li, L.; Wang, Z.P.; Lin, J.; Huang, Y.X.; Yao, Y.; Chen, R.J.; Wu, F. Sustainable Recycling Technology for Li-Ion Batteries and Beyond: Challenges and Future Prospects. *Chem. Rev.* **2020**, *120*, 7020–7063. [CrossRef] [PubMed]
30. Orilall, M.C.; Wiesner, U. Block copolymer based composition and morphology control in nanostructured hybrid materials for energy conversion and storage: Solar cells, batteries, and fuel cells. *Chem. Soc. Rev.* **2011**, *40*, 520–535. [CrossRef]
31. Bates, F.S. Polymer-polymer phase behavior. *Sci.-New York* **1991**, *251*, 898–905. [CrossRef] [PubMed]
32. Wong, C.K.; Qiang, X.L.; Muller, A.H.E.; Groschel, A.H. Self-Assembly of block copolymers into internally ordered microparticles. *Prog. Polym. Sci.* **2020**, *102*, 101211. [CrossRef]
33. Feng, H.B.; Lu, X.Y.; Wang, W.Y.; Kang, N.G.; Mays, J.W. Block Copolymers: Synthesis, Self-Assembly, and Applications. *Polymers* **2017**, *9*, 494. [CrossRef] [PubMed]
34. Farrell, R.A.; Fitzgerald, T.G.; Borah, D.; Holmes, J.D.; Morris, M.A. Chemical Interactions and Their Role in the Microphase Separation of Block Copolymer Thin Films. *Int. J. Mol. Sci.* **2009**, *10*, 3671–3712. [CrossRef]
35. Bates, C.M.; Seshimo, T.; Maher, M.J.; Durand, W.J.; Cushen, J.D.; Dean, L.M.; Blachut, G.; Ellison, C.J.; Willson, C.G. Polarity-Switching Top Coats Enable Orientation of Sub-10-nm Block Copolymer Domains. *Science* **2012**, *338*, 775–779. [CrossRef]
36. Hagita, K.; Aoyagi, T.; Abe, Y.; Genda, S.; Honda, T. Deep learning-based estimation of Flory–Huggins parameter of A–B block copolymers from cross-sectional images of phase-separated structures. *Sci. Rep.* **2021**, *11*, 12322. [CrossRef]
37. Yoshida, H.; Takenaka, M. 1-Physics of block copolymers from bulk to thin films. In *Directed Self-Assembly of Block Co-Polymers for Nano-Manufacturing*; Gronheid, R., Nealey, P., Eds.; Woodhead Publishing: Sawston, UK, 2015; pp. 3–26.

38. Liu, M.J.; Qiang, Y.C.; Li, W.H.; Qiu, F.; Shi, A.C. Stabilizing the Frank-Kasper Phases via Binary Blends of AB Diblock Copolymers. *Acs Macro Lett.* **2016**, *5*, 1167–1171. [CrossRef]
39. Bates, F.S.; Fredrickson, G.H. Block copolymers-Designer soft materials. *Phys. Today* **1999**, *52*, 32–38. [CrossRef]
40. Groschel, A.H.; Muller, A.H.E. Self-assembly concepts for multicompartment nanostructures. *Nanoscale* **2015**, *7*, 11841–11876. [CrossRef]
41. Wanka, G.; Hoffmann, H.; Ulbricht, W. Phase-Diagrams and aggregation behavior of poly(oxyethylene)-poly(oxypropylene)-poly(oxyethylene) triblock copolymers in aqueous-solutions. *Macromolecules* **1994**, *27*, 4145–4159. [CrossRef]
42. Bates, F.S.; Hillmyer, M.A.; Lodge, T.P.; Bates, C.M.; Delaney, K.T.; Fredrickson, G.H. Multiblock Polymers: Panacea or Pandora's Box? *Science* **2012**, *336*, 434–440. [CrossRef] [PubMed]
43. Majewski, P.W.; Yager, K.G. Rapid ordering of block copolymer thin films. *J. Phys.-Condes. Matter* **2016**, *28*, 403002. [CrossRef] [PubMed]
44. Abetz, V.; Kremer, K.; Muller, M.; Reiter, G. Functional Macromolecular Systems: Kinetic Pathways to Obtain Tailored Structures. *Macromol. Chem. Phys.* **2019**, *220*, 1800334. [CrossRef]
45. Cui, H.G.; Chen, Z.Y.; Zhong, S.; Wooley, K.L.; Pochan, D.J. Block copolymer assembly via kinetic control. *Science* **2007**, *317*, 647–650. [CrossRef]
46. Lynd, N.A.; Meuler, A.J.; Hillmyer, M.A. Polydispersity and block copolymer self-assembly. *Prog. Polym. Sci.* **2008**, *33*, 875–893. [CrossRef]
47. Hadziioannou, G.; Skoulios, A. Molecular weight dependence of lamellar structure in styrene isoprene two- and three-block copolymers. *Macromolecules* **1982**, *15*, 258–262. [CrossRef]
48. Carrot, C.; Guillet, J. From dynamic moduli to molecular weight distribution: A study of various polydisperse linear polymers. *J. Rheol.* **1997**, *41*, 1203–1220. [CrossRef]
49. Llorens, J.; Rude, E.; Marcos, R.M. Polydispersity index from linear viscoelastic data: Unimodal and bimodal linear polymer melts. *Polymer* **2003**, *44*, 1741–1750. [CrossRef]
50. Levy, M. The impact of the concept of "Living Polymers" on material science. *Polym. Adv. Technol.* **2007**, *18*, 681–684. [CrossRef]
51. Nguyen, H.T.; Tran, T.T.; Nguyen-Thai, N.U. Preparation of polydisperse polystyrene-block-poly(4-vinyl pyridine) synthesized by TEMPO-mediated radical polymerization and the facile nanostructure formation by self-assembly. *J. Nanostruct. Chem.* **2018**, *8*, 61–69. [CrossRef]
52. Ruzette, A.V.; Tence-Girault, S.; Leibler, L.; Chauvin, F.; Bertin, D.; Guerret, O.; Gerard, P. Molecular disorder and mesoscopic order in polydisperse acrylic block copolymers prepared by controlled radical polymerization. *Macromolecules* **2006**, *39*, 5804–5814. [CrossRef]
53. Lynd, N.A.; Hillmyer, M.A. Effects of polydispersity on the order-disorder transition in block copolymer melts. *Macromolecules* **2007**, *40*, 8050–8055. [CrossRef]
54. Widin, J.M.; Schmitt, A.K.; Schmitt, A.L.; Im, K.; Mahanthappa, M.K. Unexpected Consequences of Block Polydispersity on the Self-Assembly of ABA Triblock Copolymers. *J. Am. Chem. Soc.* **2012**, *134*, 3834–3844. [CrossRef] [PubMed]
55. Hanley, K.J.; Lodge, T.P.; Huang, C.-I. Phase Behavior of a Block Copolymer in Solvents of Varying Selectivity. *Macromolecules* **2000**, *33*, 5918–5931. [CrossRef]
56. Wang, Y. Nondestructive Creation of Ordered Nanopores by Selective Swelling of Block Copolymers: Toward Homoporous Membranes. *Acc. Chem. Res.* **2016**, *49*, 1401–1408. [CrossRef]
57. Yi, Z.; Zhang, P.-B.; Liu, C.-J.; Zhu, L.-P. Symmetrical Permeable Membranes Consisting of Overlapped Block Copolymer Cylindrical Micelles for Nanoparticle Size Fractionation. *Macromolecules* **2016**, *49*, 3343–3351. [CrossRef]
58. Gohil, J.M.; Choudhury, R.R. Chapter 2-Introduction to Nanostructured and Nano-enhanced Polymeric Membranes: Preparation, Function, and Application for Water Purification. In *Nanoscale Materials in Water Purification*; Thomas, S., Pasquini, D., Leu, S.-Y., Gopakumar, D.A., Eds.; Elsevier: Amsterdam, The Netherlands, 2019; pp. 25–57.
59. Jiang, H.; Chen, T.; Chen, Z.; Huo, J.; Zhang, L.; Zhou, J. Computer simulations on double hydrophobic PS-b-PMMA porous membrane by non-solvent induced phase separation. *Fluid Phase Equilib.* **2020**, *523*, 112784. [CrossRef]
60. Paradiso, S.P.; Delaney, K.T.; Garcia-Cervera, C.J.; Ceniceros, H.D.; Fredrickson, G.H. Block Copolymer Self Assembly during Rapid Solvent Evaporation: Insights into Cylinder Growth and Stability. *Acs Macro Lett.* **2014**, *3*, 16–20. [CrossRef]
61. Russell, T.P.; Coulon, G.; Deline, V.R.; Miller, D.C. Characteristics of the surface-induced orientation for symmetric diblock PS/PMMA copolymers. *Macromolecules* **1989**, *22*, 4600–4606. [CrossRef]
62. Phillip, W.A.; O'Neill, B.; Rodwogin, M.; Hillmyer, M.A.; Cussler, E.L. Self-Assembled Block Copolymer Thin Films as Water Filtration Membranes. *ACS Appl. Mater. Interfaces* **2010**, *2*, 847–853. [CrossRef]
63. Tseng, Y.H.; Lin, Y.L.; Ho, J.H.; Chang, C.T.; Fan, Y.C.; Shen, M.H.; Chen, J.T. Reversible and tunable morphologies of amphiphilic block copolymer nanorods confined in nanopores: Roles of annealing solvents. *Polymer* **2021**, *228*, 123859. [CrossRef]
64. Lee, S.; Cheng, L.-C.; Gadelrab, K.R.; Ntetsikas, K.; Moschovas, D.; Yager, K.G.; Avgeropoulos, A.; Alexander-Katz, A.; Ross, C.A. Double-Layer Morphologies from a Silicon-Containing ABA Triblock Copolymer. *ACS Nano* **2018**, *12*, 6193–6202. [CrossRef] [PubMed]
65. Chavis, M.A.; Smilgies, D.M.; Wiesner, U.B.; Ober, C.K. Widely Tunable Morphologies in Block Copolymer Thin Films Through Solvent Vapor Annealing Using Mixtures of Selective Solvents. *Adv. Funct. Mater.* **2015**, *25*, 3057–3065. [CrossRef]

66. Peng, J.; Han, Y.C.; Knoll, W.; Kim, D.H. Development of nanodomain and fractal morphologies in solvent annealed block copolymer thin films. *Macromol. Rapid Commun.* **2007**, *28*, 1422–1428. [CrossRef]
67. Stenbock-Fermor, A.; Rudov, A.A.; Gumerov, R.A.; Tsarkova, L.A.; Boker, A.; Moller, M.; Potemkin, I.I. Morphology-Controlled Kinetics of Solvent Uptake by Block Copolymer Films in Nonselective Solvent Vapors. *Acs Macro Lett.* **2014**, *3*, 803–807. [CrossRef] [PubMed]
68. Tsarkova, L. Distortion of a Unit Cell versus Phase Transition to Nonbulk Morphology in Frustrated Films of Cylinder-Forming Polystyrene-b-polybutadiene Diblock Copolymers. *Macromolecules* **2012**, *45*, 7985–7994. [CrossRef]
69. Cavicchi, K.A.; Berthiaume, K.J.; Russell, T.P. Solvent annealing thin films of poly(isoprene-b-lactide). *Polymer* **2005**, *46*, 11635–11639. [CrossRef]
70. Xuan, Y.; Peng, J.; Cui, L.; Wang, H.; Li, B.; Han, Y. Morphology Development of Ultrathin Symmetric Diblock Copolymer Film via Solvent Vapor Treatment. *Macromolecules* **2004**, *37*, 7301–7307. [CrossRef]
71. Phillip, W.A.; Hillmyer, M.A.; Cussler, E.L. Cylinder Orientation Mechanism in Block Copolymer Thin Films Upon Solvent Evaporation. *Macromolecules* **2010**, *43*, 7763–7770. [CrossRef]
72. Lin, Z.Q.; Kim, D.H.; Wu, X.D.; Boosahda, L.; Stone, D.; LaRose, L.; Russell, T.P. A rapid route to arrays of nanostructures in thin films. *Adv. Mater.* **2002**, *14*, 1373–1376. [CrossRef]
73. Kim, S.H.; Misner, M.J.; Xu, T.; Kimura, M.; Russell, T.P. Highly oriented and ordered arrays from block copolymers via solvent evaporation. *Adv. Mater.* **2004**, *16*, 226–231. [CrossRef]
74. Kim, K.; Park, S.; Kim, Y.; Bang, J.; Park, C.; Ryu, D.Y. Optimized Solvent Vapor Annealing for Long-Range Perpendicular Lamellae in PS-b-PMMA Films. *Macromolecules* **2016**, *49*, 1722–1730. [CrossRef]
75. Li, X.; Peng, J.; Wen, Y.; Kim, D.H.; Knoll, W. Morphology change of asymmetric diblock copolymer micellar films during solvent annealing. *Polymer* **2007**, *48*, 2434–2443. [CrossRef]
76. Park, S.; Kim, Y.; Lee, W.; Hur, S.-M.; Ryu, D.Y. Gyroid Structures in Solvent Annealed PS-b-PMMA Films: Controlled Orientation by Substrate Interactions. *Macromolecules* **2017**, *50*, 5033–5041. [CrossRef]
77. Cheng, X.; Boker, A.; Tsarkova, L. Temperature-Controlled Solvent Vapor Annealing of Thin Block Copolymer Films. *Polymers* **2019**, *11*, 1312. [CrossRef] [PubMed]
78. Gowd, E.B.; Koga, T.; Endoh, M.K.; Kumar, K.; Stamm, M. Pathways of cylindrical orientations in PS-b-P4VP diblock copolymer thin films upon solvent vapor annealing. *Soft Matter* **2014**, *10*, 7753–7761. [CrossRef] [PubMed]
79. Bosworth, J.K.; Paik, M.Y.; Ruiz, R.; Schwartz, E.L.; Huang, J.Q.; Ko, A.W.; Smilgies, D.M.; Black, C.T.; Ober, C.K. Control of self-assembly of lithographically patternable block copolymer films. *Acs Nano* **2008**, *2*, 1396–1402. [CrossRef] [PubMed]
80. Arias-Zapata, J.; Bohme, S.; Garnier, J.; Girardot, C.; Legrain, A.; Zelsmann, M. Ultrafast Assembly of PS-PDMS Block Copolymers on 300 mm Wafers by Blending with Plasticizers. *Adv. Funct. Mater.* **2016**, *26*, 5690–5700. [CrossRef]
81. Guillen, G.R.; Ramon, G.Z.; Kavehpour, H.P.; Kaner, R.B.; Hoek, E.M.V. Direct microscopic observation of membrane formation by nonsolvent induced phase separation. *J. Membr. Sci.* **2013**, *431*, 212–220. [CrossRef]
82. Wang, D.-M.; Lai, J.-Y. Recent advances in preparation and morphology control of polymeric membranes formed by nonsolvent induced phase separation. *Curr. Opin. Chem. Eng.* **2013**, *2*, 229–237. [CrossRef]
83. Guillen, G.R.; Pan, Y.; Li, M.; Hoek, E.M.V. Preparation and Characterization of Membranes Formed by Nonsolvent Induced Phase Separation: A Review. *Ind. Eng. Chem. Res.* **2011**, *50*, 3798–3817. [CrossRef]
84. Tan, X.; Li, K. Inorganic hollow fibre membranes in catalytic processing. *Curr. Opin. Chem. Eng.* **2011**, *1*, 69–76. [CrossRef]
85. Plisko, T.V.; Penkova, A.V.; Burts, K.S.; Bildyukevich, A.V.; Dmitrenko, M.E.; Melnikova, G.B.; Atta, R.R.; Mazur, A.S.; Zolotarev, A.A.; Missyul, A.B. Effect of Pluronic F127 on porous and dense membrane structure formation via non-solvent induced and evaporation induced phase separation. *J. Membr. Sci.* **2019**, *580*, 336–349. [CrossRef]
86. Gu, Y.; Wiesner, U. Tailoring Pore Size of Graded Mesoporous Block Copolymer Membranes: Moving from Ultrafiltration toward Nanofiltration. *Macromolecules* **2015**, *48*, 6153–6159. [CrossRef]
87. Abed, M.R.M.; Kumbharkar, S.C.; Groth, A.M.; Li, K. Ultrafiltration PVDF hollow fibre membranes with interconnected bicontinuous structures produced via a single-step phase inversion technique. *J. Membr. Sci.* **2012**, *407–408*, 145–154. [CrossRef]
88. Gin, D.L.; Noble, R.D. Designing the Next Generation of Chemical Separation Membranes. *Science* **2011**, *332*, 674–676. [CrossRef]
89. Li, D.F.; Chung, T.S.; Ren, J.Z.; Wang, R. Thickness dependence of macrovoid evolution in wet phase-inversion asymmetric membranes. *Ind. Eng. Chem. Res.* **2004**, *43*, 1553–1556. [CrossRef]
90. Widjojo, N.; Chung, T.S. Thickness and air gap dependence of macrovoid evolution in phase-inversion asymmetric hollow fiber membranes. *Ind. Eng. Chem. Res.* **2006**, *45*, 7618–7626. [CrossRef]
91. Abetz, V. Isoporous Block Copolymer Membranes. *Macromol. Rapid Commun.* **2015**, *36*, 10–22. [CrossRef] [PubMed]
92. Foroutani, K.; Ghasemi, S.M.; Pourabbas, B. Molecular tailoring of polystyrene-block-poly (acrylic acid) block copolymer toward additive-free asymmetric isoporous membranes via SNIPS. *J. Membr. Sci.* **2021**, *623*, 119099. [CrossRef]
93. Peinemann, K.V.; Abetz, V.; Simon, P.F.W. Asymmetric superstructure formed in a block copolymer via phase separation. *Nat. Mater.* **2007**, *6*, 992–996. [CrossRef]
94. Karunakaran, M.; Nunes, S.P.; Qiu, X.Y.; Yu, H.Z.; Peinemann, K.V. Isoporous PS-b-PEO ultrafiltration membranes via self-assembly and water-induced phase separation. *J. Membr. Sci.* **2014**, *453*, 471–477. [CrossRef]
95. Stegelmeier, C.; Filiz, V.; Abetz, V.; Perlich, J.; Fery, A.; Ruckdeschel, P.; Rosenfeldt, S.; Forster, S. Topological Paths and Transient Morphologies during Formation of Mesoporous Block Copolymer Membranes. *Macromolecules* **2014**, *47*, 5566–5577. [CrossRef]

96. Jung, A.; Rangou, S.; Abetz, C.; Filiz, V.; Abetz, V. Structure Formation of Integral Asymmetric Composite Membranes of Polystyrene-block-Poly(2-vinylpyridine) on a Nonwoven. *Macromol. Mater. Eng.* **2012**, *297*, 790–798. [CrossRef]
97. Rahman, M.M. Selective Swelling and Functionalization of Integral Asymmetric Isoporous Block Copolymer Membranes. *Macromol. Rapid Commun.* **2021**, *42*, 2100235. [CrossRef]
98. Dorin, R.M.; Phillip, W.A.; Sai, H.; Werner, J.; Elimelech, M.; Wiesner, U. Designing block copolymer architectures for targeted membrane performance. *Polymer* **2014**, *55*, 347–353. [CrossRef]
99. Rangou, S.; Buhr, K.; Filiz, V.; Clodt, J.I.; Lademann, B.; Hahn, J.; Jung, A.; Abetz, V. Self-organized isoporous membranes with tailored pore sizes. *J. Membr. Sci.* **2014**, *451*, 266–275. [CrossRef]
100. Radjabian, M.; Abetz, C.; Fischer, B.; Meyer, A.; Abetz, V. Influence of Solvent on the Structure of an Amphiphilic Block Copolymer in Solution and in Formation of an Integral Asymmetric Membrane. *Acs Appl. Mater. Interfaces* **2017**, *9*, 31224–31234. [CrossRef]
101. Tan, K.W.; Jung, B.; Werner, J.G.; Rhoades, E.R.; Thompson, M.O.; Wiesner, U. Transient laser heating induced hierarchical porous structures from block copolymer-directed self-assembly. *Science* **2015**, *349*, 54–58. [CrossRef]
102. Seshimo, T.; Maeda, R.; Odashima, R.; Takenaka, Y.; Kawana, D.; Ohmori, K.; Hayakawa, T. Perpendicularly oriented sub-10-nm block copolymer lamellae by atmospheric thermal annealing for one minute. *Sci. Rep.* **2016**, *6*, 19481. [CrossRef]
103. Albalak, R.J.; Thomas, E.L.; Capel, M.S. Thermal annealing of roll-cast triblock copolymer films. *Polymer* **1997**, *38*, 3819–3825. [CrossRef]
104. Tong, Q.Q.; Zheng, Q.; Sibener, S.J. Alignment and Structural Evolution of Cylinder-Forming Diblock Copolymer Thin Films in Patterned Tapered-Width Nanochannels. *Macromolecules* **2014**, *47*, 4236–4242. [CrossRef]
105. Sepe, A.; Hoppe, E.T.; Jaksch, S.; Magerl, D.; Zhong, Q.; Perlich, J.; Posselt, D.; Smilgies, D.M.; Papadakis, C.M. The effect of heat treatment on the internal structure of nanostructured block copolymer films. *J. Phys.-Condes. Matter* **2011**, *23*, 254213. [CrossRef]
106. Shi, L.-Y.; Yin, C.; Zhou, B.; Xia, W.; Weng, L.; Ross, C.A. Annealing Process Dependence of the Self-Assembly of Rod–Coil Block Copolymer Thin Films. *Macromolecules* **2021**, *54*, 1657–1664. [CrossRef]
107. Majewski, P.W.; Yager, K.G. Latent Alignment in Pathway-Dependent Ordering of Block Copolymer Thin Films. *Nano Lett.* **2015**, *15*, 5221–5228. [CrossRef]
108. Wang, H.S.; Kim, K.H.; Bang, J. Thermal Approaches to Perpendicular Block Copolymer Microdomains in Thin Films: A Review and Appraisal. *Macromol. Rapid Commun.* **2019**, *40*, 1800728. [CrossRef] [PubMed]
109. Stehlin, F.; Diot, F.; Gwiazda, A.; Dirani, A.; Salaun, M.; Zelsmann, M.; Soppera, O. Local Reorganization of Diblock Copolymer Domains in Directed Self-Assembly Monitored by in Situ High-Temperature AFM. *Langmuir* **2013**, *29*, 12796–12803. [CrossRef]
110. Majewski, P.W.; Yager, K.G. Reordering transitions during annealing of block copolymer cylinder phases. *Soft Matter* **2016**, *12*, 281–294. [CrossRef]
111. Zhou, Z.P.; Liu, G.L. Controlling the Pore Size of Mesoporous Carbon Thin Films through Thermal and Solvent Annealing. *Small* **2017**, *13*, 1603107. [CrossRef]
112. Kim, S.; Jeon, G.; Heo, S.W.; Kim, H.J.; Kim, S.B.; Chang, T.; Kim, J.K. High aspect ratio cylindrical microdomains oriented vertically on the substrate using block copolymer micelles and temperature-programmed solvent vapor annealing. *Soft Matter* **2013**, *9*, 5550–5556. [CrossRef]
113. Kim, E.; Ahn, H.; Park, S.; Lee, H.; Lee, M.; Lee, S.; Kim, T.; Kwak, E.-A.; Lee, J.H.; Lei, X.; et al. Directed Assembly of High Molecular Weight Block Copolymers: Highly Ordered Line Patterns of Perpendicularly Oriented Lamellae with Large Periods. *ACS Nano* **2013**, *7*, 1952–1960. [CrossRef] [PubMed]
114. Gurovich, E. On Microphase Separation of Block Copolymers In an Electric Field: Four Universal Classes. *Macromolecules* **1994**, *27*, 7339–7362. [CrossRef]
115. Boker, A.; Elbs, H.; Hansel, H.; Knoll, A.; Ludwigs, S.; Zettl, H.; Urban, V.; Abetz, V.; Muller, A.H.E.; Krausch, G. Microscopic mechanisms of electric-field-induced alignment of block copolymer microdomains. *Phys. Rev. Lett.* **2002**, *89*, 135502. [CrossRef]
116. Schmidt, K.; Schoberth, H.G.; Ruppel, M.; Zettl, H.; Hansel, H.; Weiss, T.M.; Urban, V.; Krausch, G.; Boker, A. Reversible tuning of a block-copolymer nanostructure via electric fields. *Nat. Mater.* **2008**, *7*, 142–145. [CrossRef] [PubMed]
117. Schoberth, H.G.; Olszowka, V.; Schmidt, K.; Boker, A. Effects of Electric Fields on Block Copolymer Nanostructures. In *Complex Macromolecular Systems I*; Muller, A.H.E., Schmidt, H.W., Eds.; Springer: Amsterdam, The Netherlands, 2010; Volume 227, pp. 1–31.
118. Olszowka, V.; Hund, M.; Kuntermann, V.; Scherdel, S.; Tsarkova, L.; Boker, A. Electric Field Alignment of a Block Copolymer Nanopattern: Direct Observation of the Microscopic Mechanism. *Acs Nano* **2009**, *3*, 1091–1096. [CrossRef] [PubMed]
119. Schoberth, H.G.; Pester, C.W.; Ruppe, M.; Urban, V.S.; Boker, A. Orientation-Dependent Order-Disorder Transition of Block Copolymer Lamellae in Electric Fields. *Acs Macro Lett.* **2013**, *2*, 469–473. [CrossRef]
120. Marencic, A.P.; Chaikin, P.M.; Register, R.A. Orientational order in cylinder-forming block copolymer thin films. *Phys. Rev. E* **2012**, *86*, 021507. [CrossRef]
121. Morrison, F.; Le Bourvellec, G.; Winter, H.H. Flow-induced structure and rheology of a triblock copolymer. *J. Appl. Polym. Sci.* **1987**, *33*, 1585–1600. [CrossRef]
122. Angelescu, D.E.; Waller, J.H.; Adamson, D.H.; Deshpande, P.; Chou, S.Y.; Register, R.A.; Chaikin, P.M. Macroscopic orientation of block copolymer cylinders in single-layer films by shearing. *Adv. Mater.* **2004**, *16*, 1736–1740. [CrossRef]
123. Angelescu, D.E.; Waller, J.H.; Register, R.A.; Chaikin, P.M. Shear-induced alignment in thin films of spherical nanodomains. *Adv. Mater.* **2005**, *17*, 1878–1881. [CrossRef]

124. Pujari, S.; Keaton, M.A.; Chaikin, P.M.; Register, R.A. Alignment of perpendicular lamellae in block copolymer thin films by shearing. *Soft Matter* **2012**, *8*, 5358–5363. [CrossRef]
125. Davis, R.L.; Chaikin, P.M.; Register, R.A. Cylinder Orientation and Shear Alignment in Thin Films of Polystyrene-Poly(n-hexyl methacrylate) Diblock Copolymers. *Macromolecules* **2014**, *47*, 5277–5285. [CrossRef]
126. Marencic, A.P.; Adamson, D.H.; Chaikin, P.M.; Register, R.A. Shear alignment and realignment of sphere-forming and cylinder-forming block-copolymer thin films. *Phys. Rev. E* **2010**, *81*, 011503. [CrossRef]
127. Kim, S.Y.; Nunns, A.; Gwyther, J.; Davis, R.L.; Manners, I.; Chaikin, P.M.; Register, R.A. Large-Area Nanosquare Arrays from Shear-Aligned Block Copolymer Thin Films. *Nano Lett.* **2014**, *14*, 5698–5705. [CrossRef] [PubMed]
128. Wang, X.J.; Goswami, M.; Kumar, R.; Sumpter, B.G.; Mays, J. Morphologies of block copolymers composed of charged and neutral blocks. *Soft Matter* **2012**, *8*, 3036–3052. [CrossRef]
129. Sing, C.E.; Zwanikken, J.W.; de la Cruz, M.O. Electrostatic control of block copolymer morphology. *Nat. Mater.* **2014**, *13*, 694–698. [CrossRef]
130. Goswami, M.; Sumpter, B.G.; Mays, J. Controllable stacked disk morphologies of charged diblock copolymers. *Chem. Phys. Lett.* **2010**, *487*, 272–278. [CrossRef]
131. Zhou, H.; Liu, C.G.; Gao, C.Q.; Qu, Y.Q.; Shi, K.Y.; Zhang, W.Q. Polymerization-Induced Self-Assembly of Block Copolymer Through Dispersion RAFT Polymerization in Ionic Liquid. *J. Polym. Sci. Pol. Chem.* **2016**, *54*, 1517–1525. [CrossRef]
132. Kim, S.Y.; Yoon, E.; Joo, T.; Park, M.J. Morphology and Conductivity in Ionic Liquid Incorporated Sulfonated Block Copolymers. *Macromolecules* **2011**, *44*, 5289–5298. [CrossRef]
133. Bennett, T.M.; Jack, K.S.; Thurecht, K.J.; Blakey, I. Perturbation of the Experimental Phase Diagram of a Diblock Copolymer by Blending with an Ionic Liquid. *Macromolecules* **2016**, *49*, 205–214. [CrossRef]
134. Cui, T.; Li, X.; Dong, B.; Li, X.; Guo, M.; Wu, L.; Li, B.; Li, H. Janus onions of block copolymers via confined self-assembly. *Polymer* **2019**, *174*, 70–76. [CrossRef]
135. He, Q.B.; Zhang, Y.J.; Li, H.L.; Chen, Q. Rheological Properties of ABA-Type Copolymers Physically End-Cross-Linked by Polyoxometalate. *Macromolecules* **2020**, *53*, 10927–10941. [CrossRef]
136. Zhang, L.Y.; Liu, C.; Shang, H.Y.; Cao, X.; Chai, S.C.; Chen, Q.; Wu, L.X.; Li, H.L. Electrostatic tuning of block copolymer morphologies by inorganic macroions. *Polymer* **2016**, *106*, 53–61. [CrossRef]
137. Guo, W.; Cao, L.X.; Xia, J.C.; Nie, F.Q.; Ma, W.; Xue, J.M.; Song, Y.L.; Zhu, D.B.; Wang, Y.G.; Jiang, L. Energy Harvesting with Single-Ion-Selective Nanopores: A Concentration-Gradient-Driven Nanofluidic Power Source. *Adv. Funct. Mater.* **2010**, *20*, 1339–1344. [CrossRef]
138. Logan, B.E.; Elimelech, M. Membrane-based processes for sustainable power generation using water. *Nature* **2012**, *488*, 313–319. [CrossRef] [PubMed]
139. Wang, W.; Hao, J.; Sun, Q.; Zhao, M.; Liu, H.; Li, C.; Sui, X. Carbon nanofibers membrane bridged with graphene nanosheet and hyperbranched polymer for high-performance osmotic energy harvesting. *Nano Res.* **2022**. [CrossRef]
140. Siria, A.; Bocquet, M.L.; Bocquet, L. New avenues for the large-scale harvesting of blue energy. *Nat. Rev. Chem.* **2017**, *1*, 0091. [CrossRef]
141. Mei, Y.; Tang, C.Y.Y. Recent developments and future perspectives of reverse electrodialysis technology: A review. *Desalination* **2018**, *425*, 156–174. [CrossRef]
142. Kingsbury, R.S.; Chu, K.; Coronell, O. Energy storage by reversible electrodialysis: The concentration battery. *J. Membr. Sci.* **2015**, *495*, 502–516. [CrossRef]
143. van Egmond, W.J.; Saakes, M.; Porada, S.; Meuwissen, T.; Buisman, C.J.N.; Hamelers, H.V.M. The concentration gradient flow battery as electricity storage system: Technology potential and energy dissipation. *J. Power Sources* **2016**, *325*, 129–139. [CrossRef]
144. Zaffora, A.; Culcasi, A.; Gurreri, L.; Cosenza, A.; Tamburini, A.; Santamaria, M.; Micale, G. Energy Harvesting by Waste Acid/Base Neutralization via Bipolar Membrane Reverse Electrodialysis. *Energies* **2020**, *13*, 5510. [CrossRef]
145. Tedesco, M.; Cipollina, A.; Tamburini, A.; Micale, G. Towards 1 kW power production in a reverse electrodialysis pilot plant with saline waters and concentrated brines. *J. Membr. Sci.* **2017**, *522*, 226–236. [CrossRef]
146. Siria, A.; Poncharal, P.; Biance, A.L.; Fulcrand, R.; Blase, X.; Purcell, S.T.; Bocquet, L. Giant osmotic energy conversion measured in a single transmembrane boron nitride nanotube. *Nature* **2013**, *494*, 455–458. [CrossRef] [PubMed]
147. Ma, H.; Wang, S.; Yu, B.; Sui, X.; Shen, Y.; Cong, H. Bioinspired nanochannels based on polymeric membranes. *Sci. China Mater.* **2021**, *64*, 1320–1342. [CrossRef]
148. Li, C.; Jiang, H.; Liu, P.; Zhai, Y.; Yang, X.; Gao, L.; Jiang, L. One Porphyrin Per Chain Self-Assembled Helical Ion-Exchange Channels for Ultrahigh Osmotic Energy Conversion. *J. Am. Chem. Soc.* **2022**, *144*, 9472–9478. [CrossRef]
149. Xiao, K.; Wen, L.P.; Jiang, L. Biomimetic Solid-State Nanochannels: From Fundamental Research to Practical Applications. *Small* **2016**, *12*, 2810–2831. [CrossRef]
150. Laucirica, G.; Albesa, A.G.; Toimil-Molares, M.E.; Trautmann, C.; Marmisolle, W.A.; Azzaroni, O. Shape matters: Enhanced osmotic energy harvesting in bullet-shaped nanochannels. *Nano Energy* **2020**, *71*, 104621. [CrossRef]
151. Chen, W.; Xiang, Y.; Kong, X.-Y.; Wen, L. Polymer-based membranes for promoting osmotic energy conversion. *Giant* **2022**, *10*, 100094. [CrossRef]
152. Emmerich, T.; Vasu, K.S.; Nigues, A.; Keerthi, A.; Radha, B.; Siria, A.; Bocquet, L. Enhanced nanofluidic transport in activated carbon nanoconduits. *Nat. Mater.* **2022**, *26*, 696–702. [CrossRef]

153. Lin, C.Y.; Combs, C.; Su, Y.S.; Yeh, L.H.; Siwy, Z.S. Rectification of Concentration Polarization in Mesopores Leads To High Conductance Ionic Diodes and High Performance Osmotic Power. *J. Am. Chem. Soc.* **2019**, *141*, 3691–3698. [CrossRef]
154. Wu, Y.D.; Qian, Y.C.; Niu, B.; Chen, J.J.; He, X.F.; Yang, L.S.; Kong, X.Y.; Zhao, Y.F.; Lin, X.B.; Zhou, T.; et al. Surface Charge Regulated Asymmetric Ion Transport in Nanoconfined Space. *Small* **2021**, *17*, 2101199. [CrossRef] [PubMed]
155. Sharon, D.; Bennington, P.; Dolejsi, M.; Webb, M.A.; Dong, B.X.; de Pablo, J.J.; Nealey, P.F.; Patel, S.N. Intrinsic Ion Transport Properties of Block Copolymer Electrolytes. *Acs Nano* **2020**, *14*, 8902–8914. [CrossRef] [PubMed]
156. DuChanois, R.M.; Porter, C.J.; Violet, C.; Verduzco, R.; Elimelech, M. Membrane Materials for Selective Ion Separations at the Water-Energy Nexus. *Adv. Mater.* **2021**, *33*, 2101312. [CrossRef] [PubMed]
157. Wang, J.; Hou, J.; Zhang, H.C.; Tian, Y.; Jiang, L. Single Nanochannel-Aptamer-Based Biosensor for Ultrasensitive and Selective Cocaine Detection. *Acs Appl. Mater. Interfaces* **2018**, *10*, 2033–2039. [CrossRef]
158. Apel, P.Y. Fabrication of functional micro- and nanoporous materials from polymers modified by swift heavy ions. *Radiat. Phys. Chem.* **2019**, *159*, 25–34. [CrossRef]
159. Zhang, Z.; Kong, X.Y.; Xiao, K.; Liu, Q.; Xie, G.H.; Li, P.; Ma, J.; Tian, Y.; Wen, L.P.; Jiang, L. Engineered Asymmetric Heterogeneous Membrane: A Concentration-Gradient-Driven Energy Harvesting Device. *J. Am. Chem. Soc.* **2015**, *137*, 14765–14772. [CrossRef]
160. Wang, J.; Liu, L.; Yan, G.L.; Li, Y.C.; Gao, Y.; Tian, Y.; Jiang, L. Ionic Transport and Robust Switching Properties of the Confined Self-Assembled Block Copolymer/Homopolymer in Asymmetric Nanochannels. *ACS Appl. Mater. Interfaces* **2021**, *13*, 14520–14530. [CrossRef]
161. Yang, L.S.; Liu, P.; Zhu, C.C.; Zhao, Y.Y.; Yuan, M.M.; Kong, X.Y.; Wen, L.P.; Jiang, L. Ion transport regulation through triblock copolymer/PET asymmetric nanochannel membrane: Model system establishment and rectification mapping. *Chin. Chem. Lett.* **2021**, *32*, 822–825. [CrossRef]
162. Wu, Y.F.; Yang, G.; Lin, M.C.; Kong, X.Y.; Mi, L.; Liu, S.Q.; Chen, G.S.; Tian, Y.; Jiang, L. Continuously Tunable Ion Rectification and Conductance in Submicrochannels Stemming from Thermoresponsive Polymer Self-Assembly. *Angew. Chem. Int. Ed.* **2019**, *58*, 12481–12485. [CrossRef]
163. Macha, M.; Marion, S.; Nandigana, V.V.R.; Radenovic, A. 2D materials as an emerging platform for nanopore-based power generation. *Nat. Rev. Mater.* **2019**, *4*, 588–605. [CrossRef]
164. Hao, J.; Wang, W.; Zhao, J.; Che, H.; Chen, L.; Sui, X. Construction and application of bioinspired nanochannels based on two-dimensional materials. *Chin. Chem. Lett.* **2022**, *33*, 2291–2300. [CrossRef]
165. Lin, X.B.; Liu, P.; Xin, W.W.; Teng, Y.F.; Chen, J.J.; Wu, Y.D.; Zhao, Y.F.; Kong, X.Y.; Jiang, L.; Wen, L.P. Heterogeneous MXene/PS-b-P2VP Nanofluidic Membranes with Controllable Ion Transport for Osmotic Energy Conversion. *Adv. Funct. Mater.* **2021**, *31*, 2105013. [CrossRef]
166. Sui, X.; Zhang, Z.; Zhang, Z.Y.; Wang, Z.W.; Li, C.; Yuan, H.; Gao, L.C.; Wen, L.P.; Fan, X.; Yang, L.J.; et al. Biomimetic Nanofluidic Diode Composed of Dual Amphoteric Channels Maintains Rectification Direction over a Wide pH Range. *Angew. Chem. Int. Ed.* **2016**, *55*, 13056–13060. [CrossRef] [PubMed]
167. Sui, X.; Zhang, Z.; Li, C.; Gao, L.C.; Zhao, Y.; Yang, L.J.; Wen, L.P.; Jiang, L. Engineered Nanochannel Membranes with Diode-like Behavior for Energy Conversion over a Wide pH Range. *Acs Appl. Mater. Interfaces* **2019**, *11*, 23815–23821. [CrossRef] [PubMed]
168. Zhang, Z.; Sui, X.; Li, P.; Xie, G.H.; Kong, X.Y.; Xiao, K.; Gao, L.C.; Wen, L.P.; Jiang, L. Ultrathin and Ion-Selective Janus Membranes for High-Performance Osmotic Energy Conversion. *J. Am. Chem. Soc.* **2017**, *139*, 8905–8914. [CrossRef]
169. Hao, J.L.; Yang, T.; He, X.L.; Tang, H.Y.; Sui, X. Hierarchical nanochannels based on rod-coil block copolymer for ion transport and energy conversion. *Giant* **2021**, *5*, 100049. [CrossRef]
170. Li, C.; Wen, L.P.; Sui, X.; Cheng, Y.R.; Gao, L.C.; Jiang, L. Large-scale, robust mushroom-shaped nanochannel array membrane for ultrahigh osmotic energy conversion. *Sci. Adv.* **2021**, *7*, eabg2183. [CrossRef] [PubMed]
171. Koo, J.M.; Park, C.H.; Yoo, S.; Lee, G.W.; Yang, S.Y.; Kim, J.H.; Yoo, S.I. Selective ion transport through three-dimensionally interconnected nanopores of quaternized block copolymer membranes for energy harvesting application. *Soft Matter* **2021**, *17*, 3700–3708. [CrossRef]
172. Xie, L.; Zhou, S.; Liu, J.R.; Qiu, B.L.; Liu, T.Y.; Liang, Q.R.; Zheng, X.Z.; Li, B.; Zeng, J.; Yan, M.; et al. Sequential Superassembly of Nanofiber Arrays to Carbonaceous Ordered Mesoporous Nanowires and Their Heterostructure Membranes for Osmotic Energy Conversion. *J. Am. Chem. Soc.* **2021**, *143*, 6922–6932. [CrossRef]
173. Sazali, N.; Salleh, W.N.W.; Jamaludin, A.S.; Razali, M.N.M. New Perspectives on Fuel Cell Technology: A Brief Review. *Membranes* **2020**, *10*, 99. [CrossRef]
174. Sharaf, O.Z.; Orhan, M.F. An overview of fuel cell technology: Fundamentals and applications. *Renew. Sust. Energ. Rev.* **2014**, *32*, 810–853. [CrossRef]
175. Zaman, S.; Huang, L.; Douka, A.I.; Yang, H.; You, B.; Xia, B.Y. Oxygen Reduction Electrocatalysts toward Practical Fuel Cells: Progress and Perspectives. *Angew. Chem. Int. Ed.* **2021**, *60*, 17832–17852. [CrossRef] [PubMed]
176. Cullen, D.A.; Neyerlin, K.C.; Ahluwalia, R.K.; Mukundan, R.; More, K.L.; Borup, R.L.; Weber, A.Z.; Myers, D.J.; Kusoglu, A. New roads and challenges for fuel cells in heavy-duty transportation. *Nat. Energy* **2021**, *6*, 462–474. [CrossRef]
177. He, G.W.; Li, Z.; Zhao, J.; Wang, S.F.; Wu, H.; Guiver, M.D.; Jiang, Z.Y. Nanostructured Ion-Exchange Membranes for Fuel Cells: Recent Advances and Perspectives. *Adv. Mater.* **2015**, *27*, 5280–5295. [CrossRef] [PubMed]
178. Li, W.; Liu, J.; Zhao, D.Y. Mesoporous materials for energy conversion and storage devices. *Nat. Rev. Mater.* **2016**, *1*, 16023. [CrossRef]

179. Ogungbemi, E.; Ijaodola, O.; Khatib, F.N.; Wilberforce, T.; El Hassan, Z.; Thompson, J.; Ramadan, M.; Olabi, A.G. Fuel cell membranes-Pros and cons. *Energy* **2019**, *172*, 155–172. [CrossRef]
180. Adamski, M.; Skalski, T.J.G.; Britton, B.; Peckham, T.J.; Metzler, L.; Holdcroft, S. Highly Stable, Low Gas Crossover, Proton-Conducting Phenylated Polyphenylenes. *Angew. Chem. Int. Ed.* **2017**, *56*, 9058–9061. [CrossRef]
181. Cheng, J.; He, G.H.; Zhang, F.X. A mini-review on anion exchange membranes for fuel cell applications: Stability issue and addressing strategies. *Int. J. Hydrog. Energy* **2015**, *40*, 7348–7360. [CrossRef]
182. Luo, T.; Abdu, S.; Wessling, M. Selectivity of ion exchange membranes: A review. *J. Membr. Sci.* **2018**, *555*, 429–454. [CrossRef]
183. Wang, Y.; Ruiz Diaz, D.F.; Chen, K.S.; Wang, Z.; Adroher, X.C. Materials, technological status, and fundamentals of PEM fuel cells—A review. *Mater. Today* **2020**, *32*, 178–203. [CrossRef]
184. Scofield, M.E.; Liu, H.Q.; Wong, S.S. A concise guide to sustainable PEMFCs: Recent advances in improving both oxygen reduction catalysts and proton exchange membranes. *Chem. Soc. Rev.* **2015**, *44*, 5836–5860. [CrossRef]
185. Xing, L.; Shi, W.; Su, H.; Xu, Q.; Das, P.K.; Mao, B.; Scott, K. Membrane electrode assemblies for PEM fuel cells: A review of functional graded design and optimization. *Energy* **2019**, *177*, 445–464. [CrossRef]
186. Kim, D.J.; Jo, M.J.; Nam, S.Y. A review of polymer–nanocomposite electrolyte membranes for fuel cell application. *J. Ind. Eng. Chem.* **2015**, *21*, 36–52. [CrossRef]
187. Tellez-Cruz, M.M.; Escorihuela, J.; Solorza-Feria, O.; Compan, V. Proton Exchange Membrane Fuel Cells (PEMFCs): Advances and Challenges. *Polymers* **2021**, *13*, 3064. [CrossRef] [PubMed]
188. Wang, Y.; Chen, K.S.; Mishler, J.; Cho, S.C.; Adroher, X.C. A review of polymer electrolyte membrane fuel cells: Technology, applications, and needs on fundamental research. *Appl. Energy* **2011**, *88*, 981–1007. [CrossRef]
189. Zhang, J.; Liu, Y.; Lv, Z.; Zhao, T.; Li, P.; Sun, Y.; Wang, J. Sulfonated Ti3C2Tx to construct proton transfer pathways in polymer electrolyte membrane for enhanced conduction. *Solid State Ion.* **2017**, *310*, 100–111. [CrossRef]
190. Yu, L.; Yue, B.; Yan, L.; Zhao, H.; Zhang, J. Proton conducting composite membranes based on sulfonated polysulfone and polysulfone-g-(phosphonated polystyrene) via controlled atom-transfer radical polymerization for fuel cell applications. *Solid State Ion.* **2019**, *338*, 103–112. [CrossRef]
191. Ciftcioglu, G.A.; Frank, C.W. Effect of Increased Ionic Liquid Uptake via Thermal Annealing on Mechanical Properties of Polyimide-Poly(ethylene glycol) Segmented Block Copolymer Membranes. *Molecules* **2021**, *26*, 2143. [CrossRef]
192. Watanabe, M.; Thomas, M.L.; Zhang, S.G.; Ueno, K.; Yasuda, T.; Dokko, K. Application of Ionic Liquids to Energy Storage and Conversion Materials and Devices. *Chem. Rev.* **2017**, *117*, 7190–7239. [CrossRef]
193. Elwan, H.A.; Mamlouk, M.; Scott, K. A review of proton exchange membranes based on protic ionic liquid/polymer blends for polymer electrolyte membrane fuel cells. *J. Power Sources* **2021**, *484*, 229197. [CrossRef]
194. Schulze, M.W.; McIntosh, L.D.; Hillmyer, M.A.; Lodge, T.P. High-Modulus, High-Conductivity Nanostructured Polymer Electrolyte Membranes via Polymerization-Induced Phase Separation. *Nano Lett.* **2014**, *14*, 122–126. [CrossRef] [PubMed]
195. Zhang, L.Y.; Cui, T.T.; Cao, X.; Zhao, C.J.; Chen, Q.; Wu, L.X.; Li, H.L. Inorganic-Macroion-Induced Formation of Bicontinuous Block Co-polymer Nanocomposites with Enhanced Conductivity and Modulus. *Angew. Chem. Int. Ed.* **2017**, *56*, 9013–9017. [CrossRef] [PubMed]
196. Zhai, L.; Chai, S.C.; Wang, G.; Zhang, W.; He, H.B.; Li, H.L. Triblock Copolymer/Polyoxometalate Nanocomposite Electrolytes with Inverse Hexagonal Cylindrical Nanostructures. *Macromol. Rapid Commun.* **2020**, *41*, 2000438. [CrossRef] [PubMed]
197. Xiao, F.; Wang, Y.C.; Wu, Z.P.; Chen, G.Y.; Yang, F.; Zhu, S.Q.; Siddharth, K.; Kong, Z.J.; Lu, A.L.; Li, J.C.; et al. Recent Advances in Electrocatalysts for Proton Exchange Membrane Fuel Cells and Alkaline Membrane Fuel Cells. *Adv. Mater.* **2021**, *33*, 2006292. [CrossRef]
198. Guerrero Moreno, N.; Cisneros Molina, M.; Gervasio, D.; Pérez Robles, J.F. Approaches to polymer electrolyte membrane fuel cells (PEMFCs) and their cost. *Renew. Sust. Energ. Rev.* **2015**, *52*, 897–906. [CrossRef]
199. Postole, G.; Auroux, A. The poisoning level of Pt/C catalysts used in PEM fuel cells by the hydrogen feed gas impurities: The bonding strength. *Int. J. Hydrog. Energy* **2011**, *36*, 6817–6825. [CrossRef]
200. Wang, J. System integration, durability and reliability of fuel cells: Challenges and solutions. *Appl. Energy* **2017**, *189*, 460–479. [CrossRef]
201. Garzon, F.; Uribe, F.A.; Rockward, T.; Urdampilleta, I.G.; Brosha, E.L. The Impact of Hydrogen Fuel Contaminates on Long-Term PMFC Performance. *ECS Trans.* **2006**, *3*, 695–703. [CrossRef]
202. Ramaswamy, N.; Mukerjee, S. Alkaline Anion-Exchange Membrane Fuel Cells: Challenges in Electrocatalysis and Interfacial Charge Transfer. *Chem. Rev.* **2019**, *119*, 11945–11979. [CrossRef]
203. Varcoe, J.R.; Atanassov, P.; Dekel, D.R.; Herring, A.M.; Hickner, M.A.; Kohl, P.A.; Kucernak, A.R.; Mustain, W.E.; Nijmeijer, K.; Scott, K.; et al. Anion-exchange membranes in electrochemical energy systems. *Energy Environ. Sci.* **2014**, *7*, 3135–3191. [CrossRef]
204. Chen, N.J.; Wang, H.H.; Kim, S.P.; Kim, H.M.; Lee, W.H.; Hu, C.; Bae, J.Y.; Sim, E.S.; Chung, Y.C.; Jang, J.H.; et al. Poly(fluorenyl aryl piperidinium) membranes and ionomers for anion exchange membrane fuel cells. *Nat. Commun.* **2021**, *12*, 2367. [CrossRef] [PubMed]
205. You, W.; Noonan, K.J.T.; Coates, G.W. Alkaline-stable anion exchange membranes: A review of synthetic approaches. *Prog. Polym. Sci.* **2020**, *100*, 101177. [CrossRef]
206. Gao, J.F.; Wang, L.; Guo, Z.; Li, B.; Wang, H.; Luo, J.C.; Huang, X.W.; Xue, H.G. Flexible, superhydrophobic, and electrically conductive polymer nanofiber composite for multifunctional sensing applications. *Chem. Eng. J.* **2020**, *381*, 122778. [CrossRef]

207. Zhu, J.Q.; Birgisson, B.; Kringos, N. Polymer modification of bitumen: Advances and challenges. *Eur. Polym. J.* **2014**, *54*, 18–38. [CrossRef]
208. Zhu, S.; So, J.H.; Mays, R.; Desai, S.; Barnes, W.R.; Pourdeyhimi, B.; Dickey, M.D. Ultrastretchable Fibers with Metallic Conductivity Using a Liquid Metal Alloy Core. *Adv. Funct. Mater.* **2013**, *23*, 2308–2314. [CrossRef]
209. Park, H.S.; Hong, C.K. Anion Exchange Membrane Based on Sulfonated Poly (Styrene-Ethylene-Butylene-Styrene) Copolymers. *Polymers* **2021**, *13*, 1669. [CrossRef] [PubMed]
210. Zhu, M.; Su, Y.X.; Wu, Y.B.; Zhang, M.; Wang, Y.G.; Chen, Q.; Li, N.W. Synthesis and properties of quaternized polyolefins with bulky poly(4-phenyl-1-butene) moieties as anion exchange membranes. *J. Membr. Sci.* **2017**, *541*, 244–252. [CrossRef]
211. Zhang, X.J.; Chu, X.M.; Zhang, M.; Zhu, M.; Huang, Y.D.; Wang, Y.G.; Liu, L.; Li, N.W. Molecularly designed, solvent processable tetraalkylammonium-functionalized fluoropolyolefin for durable anion exchange membrane fuel cells. *J. Membr. Sci.* **2019**, *574*, 212–221. [CrossRef]
212. Pan, Y.; Jiang, K.; Sun, X.R.; Ma, S.Y.; So, Y.M.; Ma, H.W.; Yan, X.M.; Zhang, N.; He, G.H. Facilitating ionic conduction for anion exchange membrane via employing star-shaped block copolymer. *J. Membr. Sci.* **2021**, *630*, 119290. [CrossRef]
213. Tarascon, J.M.; Armand, M. Issues and challenges facing rechargeable lithium batteries. *Nature* **2001**, *414*, 359–367. [CrossRef]
214. Sun, Y.M.; Liu, N.A.; Cui, Y. Promises and challenges of nanomaterials for lithium-based rechargeable batteries. *Nat. Energy* **2016**, *1*, 16071. [CrossRef]
215. Ye, Y.S.; Chou, L.Y.; Liu, Y.Y.; Wang, H.S.; Lee, H.K.; Huang, W.X.; Wan, J.Y.; Liu, K.; Zhou, G.M.; Yang, Y.F.; et al. Ultralight and fire-extinguishing current collectors for high-energy and high-safety lithium-ion batteries. *Nat. Energy* **2020**, *5*, 786–793. [CrossRef]
216. Liu, K.; Liu, Y.; Lin, D.; Pei, A.; Cui, Y. Materials for lithium-ion battery safety. *Sci. Adv.* **2018**, *4*, eaas9820. [CrossRef] [PubMed]
217. Lee, H.; Yanilmaz, M.; Toprakci, O.; Fu, K.; Zhang, X.W. A review of recent developments in membrane separators for rechargeable lithium-ion batteries. *Energy Environ. Sci.* **2014**, *7*, 3857–3886. [CrossRef]
218. Zhang, L.P.; Li, X.L.; Yang, M.R.; Chen, W.H. High-safety separators for lithium-ion batteries and sodium-ion batteries: Advances and perspective. *Energy Storage Mater.* **2021**, *41*, 522–545. [CrossRef]
219. Zhang, Y.; Yuan, J.J.; Song, Y.Z.; Yin, X.; Sun, C.C.; Zhu, L.P.; Zhu, B.K. Tannic acid/polyethyleneimine-decorated polypropylene separators for Li-Ion batteries and the role of the interfaces between separator and electrolyte. *Electrochim. Acta* **2018**, *275*, 25–31. [CrossRef]
220. Wang, X.; Peng, L.Q.; Hua, H.M.; Liu, Y.Z.; Zhang, P.; Zhao, J.B. Magnesium Borate Fiber Coating Separators with High Lithium-Ion Transference Number for Lithium-Ion Batteries. *Chemelectrochem* **2020**, *7*, 1187–1192. [CrossRef]
221. Zahn, R.; Lagadec, M.F.; Hess, M.; Wood, V. Improving Ionic Conductivity and Lithium-Ion Transference Number in Lithium-Ion Battery Separators. *ACS Appl. Mater. Interfaces* **2016**, *8*, 32637–32642. [CrossRef]
222. Ren, D.S.; Feng, X.N.; Liu, L.S.; Hsu, H.J.; Lu, L.G.; Wang, L.; He, X.M.; Ouyang, M.G. Investigating the relationship between internal short circuit and thermal runaway of lithium-ion batteries under thermal abuse condition. *Energy Storage Mater.* **2021**, *34*, 563–573. [CrossRef]
223. Zheng, Q.F.; Yamada, Y.; Shang, R.; Ko, S.; Lee, Y.Y.; Kim, K.; Nakamura, E.; Yamada, A. A cyclic phosphate-based battery electrolyte for high voltage and safe operation. *Nat. Energy* **2020**, *5*, 291–298. [CrossRef]
224. Zhang, X.Q.; Chen, X.; Cheng, X.B.; Li, B.Q.; Shen, X.; Yan, C.; Huang, J.Q.; Zhang, Q. Highly Stable Lithium Metal Batteries Enabled by Regulating the Solvation of Lithium Ions in Nonaqueous Electrolytes. *Angew. Chem. Int. Ed.* **2018**, *57*, 5301–5305. [CrossRef] [PubMed]
225. Yang, H.; Shi, X.S.; Chu, S.Y.; Shao, Z.P.; Wang, Y. Design of Block-Copolymer Nanoporous Membranes for Robust and Safer Lithium-Ion Battery Separators. *Adv. Sci.* **2021**, *8*, 2003096. [CrossRef]
226. Wang, Y.; Zhong, W.H. Development of Electrolytes towards Achieving Safe and High-Performance Energy-Storage Devices: A Review. *Chemelectrochem* **2015**, *2*, 22–36. [CrossRef]
227. Xiao, Y.R.; Turcheniuk, K.; Narla, A.; Song, A.Y.; Ren, X.L.; Magasinski, A.; Jain, A.; Huang, S.; Lee, H.; Yushin, G. Electrolyte melt infiltration for scalable manufacturing of inorganic all-solid-state lithium-ion batteries. *Nat. Mater.* **2021**, *20*, 984–990. [CrossRef]
228. Schnell, J.; Gunther, T.; Knoche, T.; Vieider, C.; Kohler, L.; Just, A.; Keller, M.; Passerini, S.; Reinhart, G. All-solid-state lithium-ion and lithium metal batteries-paving the way to large-scale production. *J. Power Sources* **2018**, *382*, 160–175. [CrossRef]
229. Wang, Q.S.; Jiang, L.H.; Yu, Y.; Sun, J.H. Progress of enhancing the safety of lithium ion battery from the electrolyte aspect. *Nano Energy* **2019**, *55*, 93–114. [CrossRef]
230. Manthiram, A.; Yu, X.W.; Wang, S.F. Lithium battery chemistries enabled by solid-state electrolytes. *Nat. Rev. Mater.* **2017**, *2*, 16103. [CrossRef]
231. Lin, Z.Y.; Guo, X.W.; Yang, Y.B.; Tang, M.X.; Wei, Q.; Yu, H.J. Block copolymer electrolyte with adjustable functional units for solid polymer lithium metal battery. *J. Energy Chem.* **2021**, *52*, 67–74. [CrossRef]
232. He, Y.B.; Liu, N.; Kohl, P.A. Difunctional block copolymer with ion solvating and crosslinking sites as solid polymer electrolyte for lithium batteries. *J. Power Sources* **2021**, *481*, 228832. [CrossRef]
233. Liu, D.; Yang, W.L.; Liu, Y.; Yang, S.C.; Shen, Z.H.; Fan, X.H.; Yang, H.; Zhou, Q.F. Enhancing ionic conductivity in tablet-bottlebrush block copolymer electrolytes with well-aligned nanostructures via solvent vapor annealing. *J. Mater. Chem. C* **2022**, *10*, 4247–4256. [CrossRef]

Article

Crystallization Behavior of Isotactic Propene-Octene Random Copolymers

Miriam Scoti *, Fabio De Stefano, Angelo Giordano, Giovanni Talarico and Claudio De Rosa

Dipartimento di Scienze Chimiche, Università Degli Studi di Napoli Federico II, Complesso Monte S. Angelo, Via Cintia, 80126 Napoli, Italy
* Correspondence: miriam.scoti@unina.it

Abstract: The crystallization behavior of random propene-octene isotactic copolymers (iPPC8) prepared with a homogeneous metallocene catalyst has been studied. Samples of iPPC8 with low octene content up to about 7 mol% were isothermally crystallized from the melt at various crystallization temperatures. The samples crystallize in mixtures of the α and γ forms of isotactic polypropylene (iPP). The relative amount of γ form increases with increasing crystallization temperature, and a maximum amount of γ form ($f_γ$(max)) is achieved for each sample. The crystallization behavior of iPPC8 copolymers is compared with the crystallization from the melt of propene–ethylene, propene–butene, propene–pentene, and propene–hexene copolymers. The results show that the behavior of iPPC8 copolymers is completely different from those described in the literature for the other copolymers of iPP. In fact, the maximum amount of γ form achieved in samples of different copolymers of iPP generally increases with increasing comonomer content, while in iPPC8 copolymers the maximum amount of γ form decreases with increasing octene content. The different behaviors are discussed based on the inclusion of co-monomeric units in the crystals of α and γ forms of iPP or their exclusion from the crystals. In iPPC8 copolymers, octene units are excluded from the crystals giving only the interruption effect that shortens the length of regular propene sequences, inducing crystallization of the γ form at low octene concentrations, lower than 2 mol%. At higher octene concentration, the crystallization of the kinetically favored α form prevails.

Keywords: isotactic polypropylene; copolymers; metallocene catalysts; role of defects excluded from crystals

1. Introduction

Copolymerization of propene with other α-olefins of different chain lengths to isotactic random copolymer is a well know strategy to modify the molecular structure and physical properties and mechanical behavior of isotactic polypropylene (iPP) and expand the possible applications of iPP [1–3]. The introduction of constitutional defects as comonomers of different sizes into polypropylene chains produces, generally, a decrease of melting temperature and density, resulting in higher clarity and an improvement of flexibility and ductility of iPP [1,4]. The efficient change of the mechanical properties depends on the size and concentration of the comonomer, and, in general, on the type of the incorporated defect. Therefore, understanding the effect of different defects on the crystallization and mechanical properties of iPP allows for a controlled modification of properties [4,5].

The effect of comonomers on the crystallization behavior of iPP and crystallization of the different polymorphic forms has been extensively investigated in samples prepared either with heterogeneous Ziegler–Natta catalysts [6–33] or homogeneous metallocene catalysts [34–93]. These studies have shown that modification of mechanical properties of iPP, as the deformation behavior, depends mostly on the crystallization of α and γ forms.

The α form is the stable form of iPP and generally crystallizes in the common conditions of crystallization from solution or from the melt and in fibers [5,94].

In standard highly stereoregular samples of iPP synthesized with heterogeneous Ziegler–Natta catalysts, the γ form has been observed only in low molecular mass samples [95], in copolymers of iPP with various comonomers [6], and by crystallization at high pressures [96–100].

The discovery of single-center homogeneous metallocene catalysts [101–103] allowed the synthesis of homogeneous samples of iPP that crystallize easily into the γ form in common conditions of crystallization at atmospheric pressure in samples of high molecular mass [104–115] and in random copolymers of iPP [34–93]. In fact, the γ form that crystallizes in iPP samples is characterized by chains containing defects of stereoregularity (as *rr* triads) or regioregularity (for instance 2,1 secondary propene units) [104–115], and constitutional defects, such as comonomers [34–93]. In these defective samples of iPP and in its copolymers with various comonomers, the γ form generally crystallizes in mixture with α form, and the relative amount of γ form increases with increasing the concentration of stereo-defects [104–112], regio-defects [115], and of comonomeric units [34–93]. In these conditions, and in the special case of copolymers, the crystallization of the γ form significantly modifies the mechanical behavior of iPP [1,4,75,78,86–92,116–120].

The easy crystallization of γ form in metallocene-based iPP and copolymers is related to the shortening of the length of the regular isotactic propene sequences due to the presence of defects randomly distributed along the chains generated by the catalyst. In fact, it is known that the crystallization of γ form is induced by short regular propene sequences, that is, when iPP chains contain any type of defect that interrupts the regular propene sequences [52,72,74,107,110]. The crystallization of α form is instead preferred when the regular propene sequences are very long, which are generated when the content of defects is low or when defects are segregated in blocks of the macromolecules [110,111,113,114].

In general, for iPP samples and iPP-based copolymers synthesized with heterogeneous multi-site Ziegler–Natta catalysts, the distribution of defects (and comonomeric units) along the macromolecules is not random, but they are segregated in blocks of the macromolecules [113,114,121,122]. Moreover, copolymer chains are characterized by mixtures of different macromolecules that may have different composition and type of distribution of the comonomers along the chains [123,124]. Therefore, in these systems, the regular propene sequences are generally much longer, giving crystallization of α form even for high concentration of defects. The nonrandom distribution of comonomers along the chains of Ziegler–Natta copolymers and the presence of other types of microstructural defects have so far prevented the study of the effect of a single comonomeric unit on the crystallization behavior and physical properties of iPP.

Copolymers of iPP and iPP homopolymer prepared with single-center metallocene catalysts are instead characterized by a more homogeneous molecular structure with a perfectly random and uniform distribution of molecular defects along the chain. Therefore, even a low content of defects decreases the length of the regular propene sequences, inducing the crystallization of the γ form [107,110,111,115].

The crystallization of α and γ forms depending on the different molecular structure and architecture has a great influence on the physical and mechanical properties of iPP [4,5,75,88–92,110,116–119]. Therefore, in general, the molecular architecture and topology of copolymers, from standard random to block or multiblock copolymers, greatly affects the crystallization behavior and properties of polymers because of the different length of crystallizable sequences [123–133].

The random distribution of defects and comonomers along the macromolecules allows considering iPPs and its copolymers synthesized with metallocene catalysts as model systems for the crystallization behavior of iPP, which is defined by the average length of the regular propene sequences that is directly related (inversely) to the concentration of defects [110]. Since the crystallization behavior of iPP depends on the length of regular propene sequences, the inclusion of defects in the crystals or their exclusion from the crystals plays a fundamental role. The inclusion of defects in the crystals gives longer crystallizable propene sequences, even for high concentrations of defects, whereas exclusion produces a shortening of the regular propene sequences [72,74,93,110]. This aspect is particularly relevant in the case of iPP-based copolymers with different α-olefins, as different comonomeric units may or may not be incorporated in the crystals of α or γ forms, depending on the size of comonomers and their compatibility with the crystal structures of the α and γ forms. In this context, detailed studies of the crystallization behavior and properties of copolymers of iPP with ethylene, butene, pentene, and hexene prepared with metallocene catalysts have been reported in the literature [34–93]. However, all these comonomers (ethylene, butene, pentene, and hexene) are in part included in the crystals and in part excluded from the crystals of α and γ forms, and their degree of inclusion and the partitioning of defects between the crystalline and amorphous phases have been also evaluated by solution and solid-state ^{13}C NMR and ab initio calculations [51,61,108,109]. Therefore, only the effect of comonomers included in the crystals or only the effect of comonomers excluded from the crystals cannot be understood yet.

This paper reports an analysis of the crystallization behavior of isotactic propene-octene copolymers (iPPC8) synthesized with a metallocene catalyst. Octene units are completely excluded from the crystals of both α and γ forms, therefore, we aim at clarifying the specific effect of defects excluded from the crystals on the crystallization behavior of iPP. The crystallization behavior of iPPC8 copolymers is compared with those of copolymers of iPP with ethylene, butene, pentene, and hexene. The effect of the octene units excluded from the crystals on the crystallization of α and γ forms of iPP is analyzed and compared to those of the different mentioned comonomers that are partially included in the crystals of α and γ forms with a different degree of incorporation.

2. Materials and Methods

Propene–octene isotactic copolymers (iPPC8) were prepared with the metallocene catalyst of Scheme 1 [134] in combination with methylalumoxane (MAO) (from Lanxess, Cologne, Germany), as reported in ref. [79]. The used catalyst is highly isospecific in both homo- and copolymerizations, and the iPPC8 copolymers result in being highly isotactic and contain very low amounts of defects of stereoregularity and regioregularity, with only about 0.2 mol% of 2,1-eryhro propene units [134] (Table 1). Moreover, octene incorporation does not affect the molecular mass, and all samples present high values of molecular mass (Table 1).

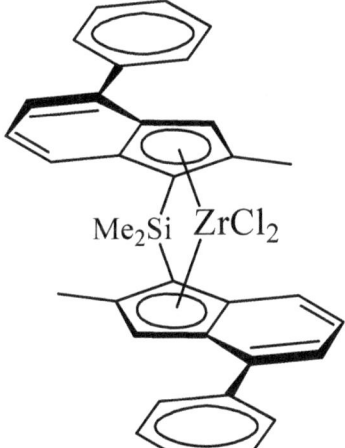

Scheme 1. Structure of metallocene complex dimethylsilyl (2,2'-dimethyl-4,4'-diphenylindenyl)ZrCl$_2$ used as catalyst for the synthesis of iPPC8 copolymers.

Table 1. Feed (mL octene) and copolymer composition (mol% octene), melting temperature of as-prepared samples (T_m), molecular mass (M_w), and dispersity (M_w/M_n) of iPPC8 copolymers [79].

Sample	Feed (mL Octene)	Composition (mol% Octene)	T_m (°C) [a]	M_w [b]	M_w/M_n [b]
iPPC8-1	1	1.9	132.5	398,576	2.1
iPPC8-2	4	4.3	114.4	257,274	2.0
iPPC8-3	6	7.1	91.8	302,205	2.1
iPPC8-4	7	10.3	72.6/53.2	252,380	2.1
iPPC8-5	8	12.8	47.3	230,985	2.0
iPPC8-6	10	15.9	44.9	225,750	2.0

[a] Determined from DSC heating curves recorded at 10 °C/min. [b] Determined by GPC.

The composition and comonomer distribution were determined by ^{13}C NMR analysis [135,136] (Table 1). All spectra were obtained using a Bruker DPX-400 spectrometer operating in the Fourier transform mode at 120 °C at 100.61 MHz (Bruker Company, Billerica, Massachusetts, USA). The samples were dissolved with 8% wt/v concentration in 1,1,2,2-tetrachloroethane-d$_2$ at 120 °C. The carbon spectra were acquired with a 90° pulse and 12 s of delay between pulses and CPD (WALTZ 16) to remove ^{1}H-^{13}C coupling. About 1500–3000 transients were stored in 32 K data points using a spectral window of 6000 Hz. For all copolymer samples, the peak of the propene methine carbon atoms was used as an internal reference at 28.83 ppm. The resonances were assigned according to ref. [135], and the 1-octene concentrations in the copolymers were evaluated from the constitutional diads PP, PO, and OO concentration (P = propene, O = octene). The NMR analysis showed that all the copolymers present a statistical distribution of comonomers ($r_1 \times r_2 \approx 1$) and homogeneous intermolecular composition.

The molecular masses and the dispersity were determined by gel permeation chromatography (GPC), using a Polymer Laboratories GPC220 apparatus equipped with a differential refractive index (RI) detector and a Viscotek 220R viscometer (Agilent Company, Santa Clara, CA, USA), on polymer solutions in 1,2,4-trichlorobenzene at 135 °C.

The calorimetry measurements were performed with differential scanning calorimeter (DSC) Mettler Toledo DSC-822 (Columbus, OH, USA) performing scans in a flowing N$_2$ atmosphere and a scanning rate of 10 °C/min.

X-ray powder diffraction profiles were recorded with Ni filtered Cu Kα radiation by using an Empyrean diffractometer (Malvern Panalytical, Worcestershire, UK), performing continuous scans of the 2θ Bragg angle from 2θ = 5° to 2θ = 40°.

All samples of iPPC8 copolymers were isothermally crystallized from the melt at different crystallization temperatures (T_c). Powder samples were melted at 200 °C and kept for 5 min at this temperature in a N$_2$ atmosphere. They were then rapidly cooled to the crystallization temperature, T_c, and kept at this temperature, still in a N$_2$ atmosphere, for a time long enough to allow complete crystallization at T_c. After the complete crystallization, the samples were quenched to room temperature and analyzed by X-ray diffraction and DSC. In the various isothermal crystallization experiments, the crystallization time is different depending on the crystallization temperature. The crystallization time necessary to have complete crystallization was evaluated by recording the crystallization exotherms in DSC and evaluating the crystallization kinetics. The crystallization time is about 2 h for low crystallization temperatures and about 2 weeks for the highest crystallization temperatures.

The degrees of crystallinity (x_c) were determined from the powder diffraction profiles by the ratio between the crystalline diffraction area (A_c) and the area of the whole diffraction profiles ($A_t = A_c + A_{am}$, $x_c = (A_c/A_t) \times 100$. The area of the crystalline phase (A_c) has been determined subtracting a baseline and the scattering halo of the amorphous phase (A_{am}) from the whole diffraction profile. For iPPC8 copolymer samples with low comonomer concentration, the amorphous halo has been obtained from the X-ray diffraction profile of a sample of atactic polypropylene. For iPPC8 copolymer samples with high octene concentration, the amorphous halo has been obtained from the X-ray diffraction profile of the amorphous sample iPPC8-6 with the highest octene concentration (15.9 mol%).

In samples that crystallize in mixtures of α and γ forms, the weight fraction of crystals of γ form f_γ, with respect to that of the α form, was evaluated from the intensity of the $(117)_\gamma$ reflection at $2\theta = 20.1°$ of the γ form, with respect to that of the $(130)_\alpha$ reflection at $2\theta = 18.6°$ of the α form, as the ratio: $f_\gamma = 100 \times (I(117)_\gamma / [I(117)_\gamma + I(130)_\alpha])$. The intensities of $(117)_\gamma$ and $(130)_\alpha$ reflections were measured from the area of the corresponding diffraction peaks above the diffuse amorphous halo in the X-ray powder diffraction profiles. The amorphous halo was obtained as described above. Next, it was scaled and subtracted to the X-ray diffraction profiles of the melt-crystallized samples. This method was applied to samples of iPPC8 copolymers of low octene concentrations that present X-ray diffraction profiles with well-defined and separated $(117)_\gamma$ reflection at $2\theta = 20.1°$ of the γ form and $(130)_\alpha$ reflection at $2\theta = 18.6°$ of the α form. For samples with high octene concentrations that crystallize from the melt in disordered modifications of γ form intermediate between the α and γ forms [112], the structural disorder reduces the intensities of both $(130)_\alpha$ and $(117)_\gamma$ reflections, and in limit of high degree of disorder in the γ crystals, both $(130)_\alpha$ and $(117)_\gamma$ reflections disappear. In these cases, the absence of the $(130)_\alpha$ and $(117)_\gamma$ reflections prevents the application of the method based on the intensities of the $(130)_\alpha$ and $(117)_\gamma$ reflections, and the amount of γ form has been evaluated using the method described in ref. [112], based on the simulation of the diffraction profiles by calculating the X-ray diffraction as the sum of the contributions of the diffraction of crystals of α form and of the diffraction of disordered crystals of γ form. The amount of γ form in the mixture of α crystals and disordered crystals of γ form corresponds to that which gives the best agreement between experimental and calculated diffraction profiles.

3. Results and Discussion

The X-ray powder diffraction profiles of as-prepared (precipitated from the polymerization solution) samples of iPPC8 copolymers are shown in Figure 1. The values of the degree of crystallinity evaluated from the diffraction profiles are also shown in Figure 1. The corresponding DSC heating curves are reported in Figure 2A. These data indicate that the presence of octene produces a decrease of crystallinity and melting temperature with increasing octene concentration from $x_c = 41\%$ and $T_m = 132\,°C$ of the sample with 1.9 mol% of octene down to $x_c = 5\%$ and $T_m = 45\,°C$ of the sample with 15.9 mol% of octene. For octene concentrations higher than 16 mol%, the copolymers do not crystallize any more.

It is apparent that samples with octene content up to about 12 mol% crystallize in the α form, as indicated by the presence in the diffraction profiles a–d of Figure 1 of the $(110)_\alpha$, $(040)_\alpha$, and $(130)_\alpha$ reflections at $2\theta = 14.1, 17.0,$ and $18.6°$, respectively, of the α form of iPP, and the absence of the $(117)_\gamma$ reflection at $2\theta = 20.1°$ of the γ form. The diffraction profiles of samples with octene concentration higher than nearly 12 mol% present a diffuse halo with some shoulders or very weak and broad diffraction peaks (profiles e, f of Figure 1). This indicates that these samples are basically amorphous. In the sample iPPC8-5 with 12.8 mol% of octene (profile e of Figure 1), very small and broad reflections are indeed still observed at $2\theta \approx 14, 17,$ and $21°$, corresponding to the $(110)_\alpha$, $(040)_\alpha$, and $(111)_\alpha$ reflections of the α form, and the amorphous scattering of the sample iPPC8-6 with 15.9 mol% of octene is clearly asymmetric with a broad peak at $2\theta = 14°$ (profile f of Figure 1). This indicates that in the samples iPPC8-5 and iPPC8-6, a very small residual crystallinity ($x_c = 10$ and 5%, respectively), which can be attributed to disordered crystals of the α form, is still present [79]. This is also demonstrated by the DSC heating curves e and f of Figure 2A, which still present endothermic peaks at about 47 and 45 °C [79].

The DSC curves of the iPPC8 copolymers recorded during cooling from the melt are reported in Figure 2B. It is apparent that the crystallization temperature and enthalpy decrease with increasing octene concentration, and the samples iPPC8-5 and iPPC8-6 with 12 mol% and 15.9 mol% of octene do not crystallize by cooling from the melt. In the curves e and f of Figure 2B, only very small and broad exothermic signals are visible at very low temperature of nearly −10 °C. The DSC cooling curves recorded at low cooling rates of 2.5 °C/min (data not shown) are very similar to those of Figure 2B. The low crystallinity that these two samples show in the as-prepared specimens (profiles e and f of Figures 1 and 2A) is due to further crystallization upon aging at room temperature.

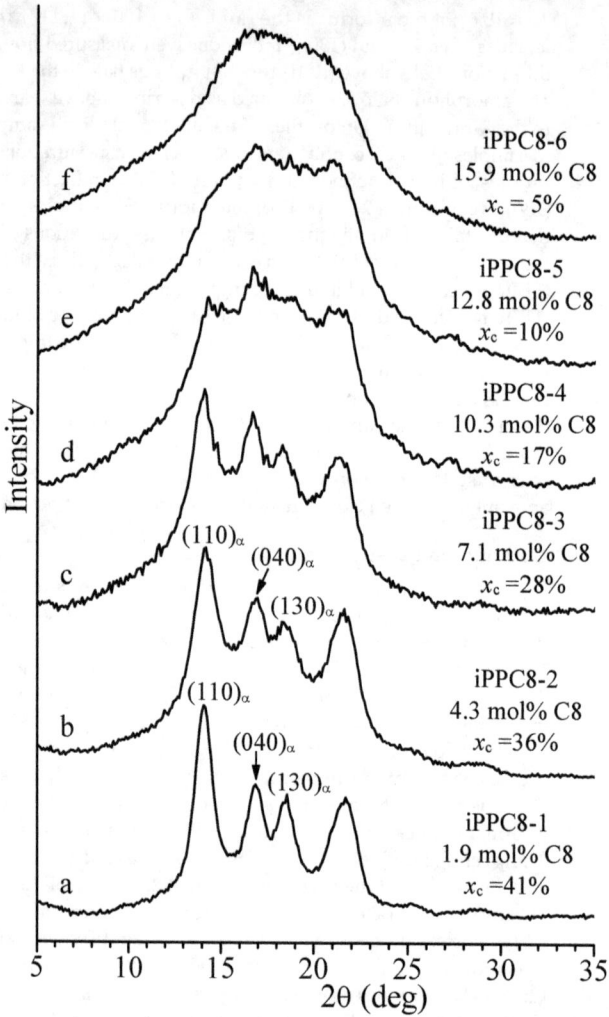

Figure 1. X-ray diffraction profiles of as-prepared samples of iPPC8 copolymers of the indicated octene concentration (**a–f**). The $(110)_\alpha$, $(040)_\alpha$, and $(130)_\alpha$ reflections at $2\theta = 14°$, $17°$, and $18.6°$, respectively, of the α form of iPP and the values of the degree of crystallinity x_c are indicated.

These data indicate that the γ form of iPP does not crystallize in these iPPC8 copolymers, even at high octene concentrations. Instead, this occurs in copolymers of iPP with ethylene (iPPC2) [72], butene (iPPC4) [72,81,92], pentene (iPPC5) [82,90,91,93], and hexene (iPPC6) [68,69,74,77]. Moreover, Figure 1 also indicates that the trigonal δ form does not crystallize in these iPPC8 copolymers, as instead occurs in iPPC5 [73,82,90,91,93] and iPPC6 [68–70,74,77] copolymers.

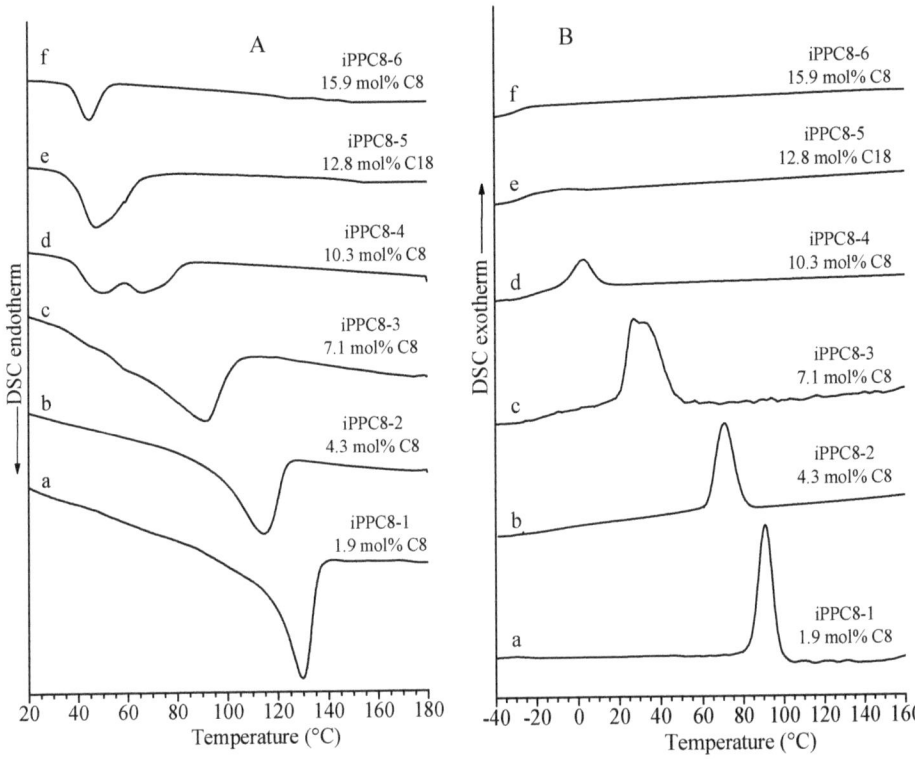

Figure 2. DSC heating (**A**) and cooling (**B**) curves of as-prepared samples of iPPC8 copolymers of the indicated 1-octene concentration (**a**–**f**).

The analysis of the 2θ positions of the $(110)_\alpha$ and $(040)_\alpha$ reflections of α form in the diffraction profiles of Figure 1 indicates that the Bragg distances of the diffracting planes are the same as those of the homopolymer, indicating that there is not expansion of the unit cell dimensions and that, therefore, octene co-units are excluded from the crystals of the α form. This behavior is different from those of iPPC4 [72,81,92], iPPC5 [73,82,90,91,93], and iPPC6 [68–70,74,75,77] copolymers, in which the comonomers are included, at least in part, in the crystals of α form with different degree of inclusion depending on the size and type of the comonomer unit. For iPPC6 and iPPC5 copolymers with high comonomer concentration, a huge amount of hexene and pentene comonomeric units is included in the unit cell of α form, inducing increase of crystal density that, in turn, induces crystallization of the δ form [68–70,73–75,77,82,90,91,93].

The exclusion of octene from the crystals of α and γ forms and the consequent absence of the crystallization of the trigonal δ form explains the fact that iPPC8 copolymers crystallize only up to 10–12 mol% of octene, whereas iPPC5 and iPPC6 crystallize up to very high pentene and hexene concentrations of about 55 mol% of pentene [73,82,90,91,93] and 25–30 mol% of hexene [68–70,74,75,77].

To study the effect of the octene comonomeric units excluded from the crystals on the crystallization of α and γ forms, samples of iPPC8 copolymers have been isothermally crystallized from the melt at high crystallization temperatures in conditions close to the thermodynamic conditions. The diffraction profiles of samples of iPPC8 copolymers isothermally crystallized from the melt at different temperatures are reported in Figure 3. The diffraction profiles of the as-prepared samples, (already presented in Figure 1) are also reported in Figure 3 (profiles a) for comparison. The isothermal crystallizations have been performed only for the three samples of iPPC8 copolymers with low octene concentrations of 1.9, 4.3, and 7.1 mol% that crystallize from the melt and develop a significant degree of crystallinity. As discussed above, samples with higher octene concentration do not crystallize from the melt or develop very low crystallinity and crystallize by cold crystallization.

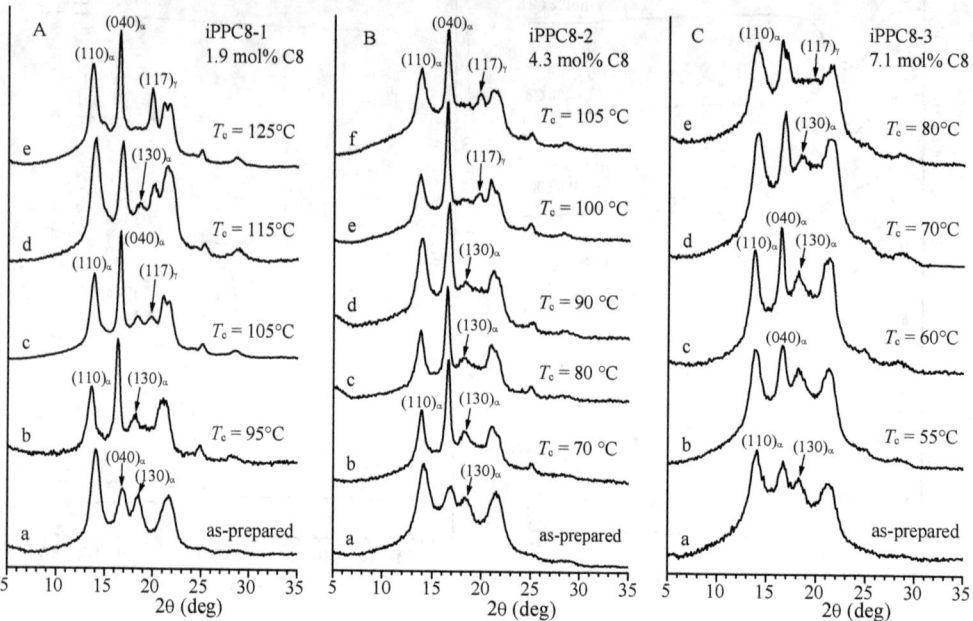

Figure 3. X-ray powder diffraction profiles of samples isothermally crystallized from the melt at the indicated crystallization temperatures T_c (**b–f**) of the samples iPPC8-1 with 1.9 mol% of octene (**A**), iPPC8-2 with 4.3 mol% of octene (**B**), and iPPC8-3 with 7.1 mol% of octene (**C**). The diffraction profiles of the as-prepared samples are also reported (profiles (**a**)). The $(110)_\alpha$, $(040)_\alpha$, and $(130)_\alpha$ reflections at $2\theta = 14°$, $17°$, and $18.6°$, respectively, of the α form of iPP and the $(117)_\gamma$ reflection of the γ form at $2\theta = 20.1°$, are indicated.

Samples of iPPC8 copolymers with low octene concentrations of 1.9 and 4.3 mol% crystallize at any crystallization temperature in mixtures of α and γ forms, as indicated by the presence of the $(130)_\alpha$ and $(117)_\gamma$ reflections at $2\theta \approx 18.6°$ and $20.1°$ of the α and γ forms, respectively, in all the diffraction profiles of Figure 3A,B. In these two samples, the intensity of the $(117)_\gamma$ reflection at $2\theta = 20.1°$ of the γ form increases up to achieve a maximum, whereas the intensity of the $(130)_\alpha$ reflection at $2\theta = 18.6°$ of the α form decreases with increasing crystallization temperature. This indicates that the relative amount of γ form increases, and that of the α form decreases, with increasing crystallization temperature up to achieve a maximum (Figure 3A,B).

The sample iPPC8-3 with higher octene concentration of 7.1 mol% instead crystallizes at all crystallization temperatures only in the α form, as indicated by the presence of only the $(130)_\alpha$ reflection at $2\theta = 18.6°$ of the α form and the absence of the $(117)_\gamma$ reflection at $2\theta = 20.1°$ of the γ form in all the diffraction profiles of Figure 3C. Only at the highest crystallization temperature of 80 °C does a very small broad peak at about $2\theta = 20.1°$ appear, while the $(130)_\alpha$ reflection at $2\theta = 18.6°$ disappears (profile e of Figure 3C). This indicates development at high crystallization temperature of a small amount of γ form. However, the lack of well-defined $(130)_\alpha$ and $(117)_\gamma$ reflections at $2\theta = 18.6°$ and $20.1°$ of α and γ forms in the diffraction profile e of Figure 3C indicates that the sample iPPC8-3 crystallizes in a disordered modification of the γ form intermediate between the ordered α and γ forms [110–112]. Contrary to propene–pentene and propene–hexene copolymers that, for high comonomer concentration, crystallize in the trigonal δ form [74,93], no traces of the trigonal δ form are observed in all three samples of iPPC8 copolymers isothermally crystallized from the melt (Figure 3).

The values of the amount of γ form (f_γ), with respect to the α form, that crystallizes in the isothermal crystallizations, determined from the intensities of $(117)_\gamma$ and $(130)_\alpha$ reflections in the diffraction profiles of Figure 3, are reported in Figure 4 as a function of the crystallization temperature. For all samples, the amount of γ form increases with increasing crystallization temperature up to achieve a maximum $f_\gamma(\max)$. In samples of different octene concentrations, a maximum amount of γ form is achieved at different crystallization temperatures (Figure 4). The values of the maximum amount of γ form ($f_\gamma(\max)$) that are obtained in each sample are reported in Figure 5 as a function of the octene concentration. It is apparent that the maximum amount of γ form achieves the highest value of 95% at the lowest octene concentration of 1.9 mol% and then decreases with increasing octene concentration down to a value of about 50% for the sample iPPC8-3 with 7.1 mol%. This sample always crystallizes in the α form and crystallizes in the disordered modification intermediate between the ordered α and γ forms only at the highest crystallization temperature (Figure 3C). For this sample (diffraction profile e of Figure 3C), because of the absence of both $(130)_\alpha$ and $(117)_\gamma$ reflections at $2\theta = 18.6°$ and $20.1°$ of α and γ forms, the amount of γ form has been calculated from the simulation of the diffraction profile by calculating the X-ray diffraction as the sum of the contributions of the diffraction of crystals of α form and of the diffraction of disordered crystals of γ form [112]. A value of $f_\gamma = 50\%$ has been obtained.

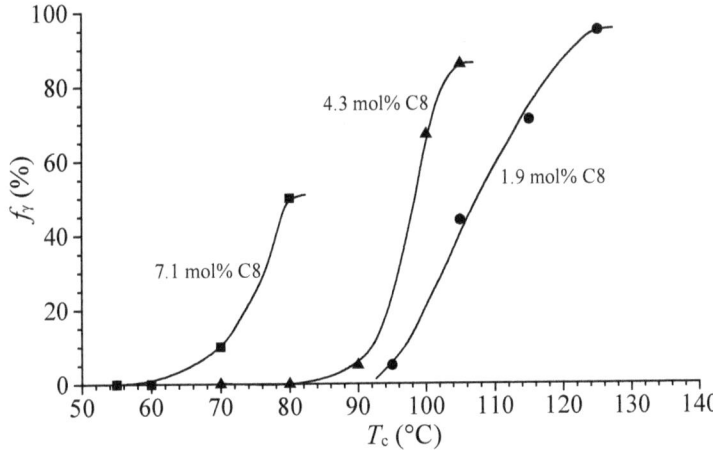

Figure 4. Relative amount of γ form (f_γ) that crystallizes from the melt in samples of iPPC8 copolymers isothermally crystallized from the melt at the crystallization temperatures, T_c, as a function of the crystallization temperature T_c. (●) sample iPPC8-1 with 1.9 mol% of octene, (▲) sample iPPC8-2 with 4.3 mol% of octene, (■) sample iPPC8-3 with 7.1 mol% of octene.

The behavior of iPPC8 copolymers of Figure 5 is completely different from the behaviors observed in iPPC2 [72], iPPC4 [72], iPPC5 [93], and iPPC6 [74] copolymers. In fact, the maximum amount of γ form achieved in each sample of different copolymers depends on the comonomer concentration and generally increases with increasing comonomer concentration, while in iPPC8 copolymers, the maximum amount of γ form decreases with increasing octene concentration (Figure 5).

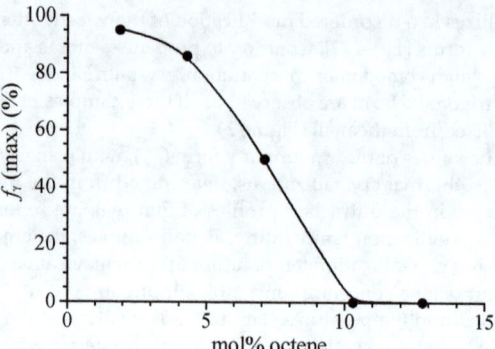

Figure 5. Maximum amount of γ form (f_γ(max)) obtained in isothermal crystallization from the melt of iPPC8 copolymers (the highest values of Figure 4) as a function of octene concentration.

The data of the maximum amount of γ form of iPPC8 copolymers of Figure 5 are also plotted in Figure 6 as a function of the total concentration of defects, defined as the sum of steric defects (*rr* diads), 2,1 regiodefects, and octene units, $\varepsilon = [rr] + [1,2] + [octene]$, and compared with the values of f_γ(max) observed and found in the literature for highly stereoregular iPP homopolymer [72,74,93,115], iPPC2 [72], iPPC4 [72], iPPC5 [93], and iPPC6 [74] copolymers. The literature values refer to copolymers synthesized with similar isospecific catalysts [72,74,93,115] that produce copolymers with isotacticity as high as that of the iPPC8 copolymers analyzed in this paper. Therefore, the differences among the different copolymers observed in Figure 6 are due to the different effects of the different comonomers on the crystallization of the α and γ forms. In the plot of Figure 6, the data of maximum amount of γ form produced by melt crystallizations of stereodefective iPPs containing different amounts of defects of stereoregularity (only *rr* diad defects) synthesized with various metallocene catalysts are also reported [110]. Figure 6 shows that, compared to the homopolymer sample of similar high isotacticity synthesized with the same or similar catalyst, for all copolymers the maximum amount of γ form rapidly increases with increasing comonomers concentration. For iPPC8 copolymers, the increase of f_γ(max) from that of the homopolymer is very fast, and the highest value of f_γ(max) is achieved at a very low octene concentration of 1.9 mol%. Then, for octene contents higher than 1.9 mol%, f_γ(max) decreases (Figures 5 and 6).

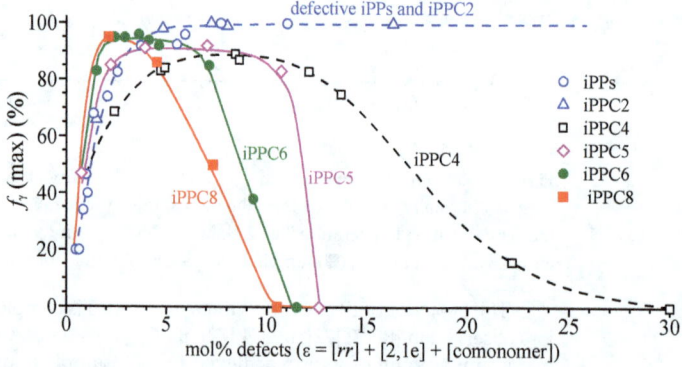

Figure 6. Maximum amount of γ form (f_γ(max)) obtained in samples of iPPC8 copolymers isothermally crystallized from the melt (■) as a function of the total concentration of defects ε, compared to literature values found for stereoirregular iPPs (○) [110], and for copolymers iPPC2 (△) [72], iPPC4 (□) [72], iPPC5 (◇) [93], and iPPC6 (●) [74].

The compared data of Figure 6 indicate that for iPPC2 copolymers, f_γ(max) increases with increasing ethylene content to achieve the maximum value of f_γ(max) = 100%, corresponding to the crystallization of the pure γ form [72]. A similar effect is visible in the stereodefective samples of iPP homopolymer containing variable content of stereodefects (only rr diad defects) (Figure 6) [110]. In fact, for these iPP samples the values of f_γ(max) increase with increasing concentration of rr defects and achieve the highest maximum value f_γ(max) = 100% for rr defects content higher than 5–7 mol% (Figure 6). Both iPPC2 copolymers and defective iPPs crystallize from the melt in the pure γ form for ethylene and rr defect concentrations higher than 5–7 mol% [72,110]. This suggests that rr diad defects and ethene units provide a similar effect in inducing crystallization of γ form, and the values of f_γ(max) of defective iPPs and iPPC2 copolymers in Figure 6 are interpolated by the same curve [72,110].

Different behaviors are instead observed in Figure 6 for the other copolymers. In fact, in copolymers iPPC8, iPPC6, iPPC5, and iPPC4, the maximum amount of γ form f_γ(max) first increases, achieves the highest value of nearly 90–95% and then decreases with further increase of comonomer concentration (Figure 6). The rates of increase and then of decrease of f_γ(max) with the comonomer concentration are, however, different in these copolymers. For low defects concentration (lower than 2–3 mol%), the increase of f_γ(max) observed in iPPC8 copolymers is faster than that in iPPC6, iPPC5, and iPPC4 copolymers, whereas the increase of f_γ(max) observed in iPPC6 copolymers is faster than that in iPPC5 and iPPC4 copolymers, and, finally, the increase of f_γ(max) in iPPC5 copolymers is, in turn, faster than that in iPPC4 copolymers (Figure 6). The fastest increase of f_γ(max) is, indeed, observed for iPPC8 copolymers characterized by the largest size comonomer. Moreover, the highest values of f_γ(max) in iPPC5 and iPPC6 copolymers are obtained soon for low pentene and hexene concentrations (2–3 mol%), whereas in iPPC8 copolymers is achieved at the lowest comonomer concentration of 1.9 mol% (Figure 6). Furthermore, the highest value of the maximum amount of γ form in iPPC8 copolymers (95%) is higher than those in copolymers iPPC6, iPPC5, and iPPC4, whereas iPPC6 copolymers produce a maximum amount of γ form (90%) higher than those in copolymers iPPC5 and iPPC4. Finally, iPPC5 copolymers give a highest maximum amount of γ form higher than that developed in iPPC4 copolymers [72,74,93].

At high concentrations of comonomers, f_γ(max) decreases in the three copolymers iPPC4, iPPC5, and iPPC6 down to f_γ(max) = 0, corresponding to crystallization of the pure α form. The decrease of f_γ(max) in iPPC6 copolymers is faster than in iPPC5 copolymers, which is, in turn, faster than in iPPC4 copolymers. In the case of iPPC8 copolymers, the decrease of f_γ(max) from the highest value is faster than in iPPC6, iPPC5 and iPPC4 copolymers, because the value of f_γ(max) = 50% is obtained already for the sample with 7.1 mol% of octene, which gives only a disordered modification of γ form with structure intermediate between those of the α and γ forms (Figure 3C). The drop off of f_γ(max) starts at concentrations of comonomers of 2 mol% for iPPC8 copolymers, nearly 4–5 mol% in iPPC6 copolymers, 10–11 mol% in iPPC5 copolymers, and 14–15 mol% in iPPC4 copolymers (Figure 6) [72,74,93]. Moreover, for concentrations of pentene and hexene higher than 10–11 mol%, iPPC5 and iPPC6 copolymers do not crystallize any more in the γ form but crystallize in the α form or in mixtures of α and δ forms [74,93]. Instead, at these high concentrations of butene, iPPC4 copolymers continue to crystallize in a mixture of α and γ forms, and only for butene concentration higher than 30 mol% they crystallize in the pure α form [72]. In iPPC8 copolymers, crystallization of the γ form is no longer observed at octene concentrations higher than 7 mol%, and for this composition, iPPC8 copolymers do not crystallize or develop only very low crystallinity of the α form (Figure 5).

Since in the analyzed copolymers there is no effect of stereo- or regio-defects, the observed different behavior is due to the different effect of different comonomers on the crystallization of iPP, related to the inclusion of different comonomers into crystals of α and γ forms or their exclusion from the crystals. Defects as comonomers [53,72,74,90–93], as well as defects of stereoregularity [105–111] and regioregularity [108,115], give the same effect of interruption the propene sequences, shortening the average length of regular propene sequences <L_{iPP}> and favoring the crystallization of the γ form. When the defects are incorporated in part or totally in the crystals of α and γ forms, the length of the crystallizable sequences increases, favoring the crystallization of the form that better accommodates the defect into crystals [72,74,90–93].

Ethylene, butene, pentene, hexene, and octene comonomers show different degrees of inclusion in crystals of iPP. A small amount of ethene is included in crystals of α and γ forms, and iPPC2 copolymers do not crystallize at high ethylene concentrations [51,52,72]. A high amount of butene is instead easily incorporated in the crystals of α form and, as a consequence, iPPC4 copolymers crystallize in the whole range of composition [61,72,92]. The degree of inclusion of pentene and hexene comonomeric units is very low in copolymers with low comonomer content but is very high in samples of high comonomer concentration [64,65,67–70,73–75,77,82,90,91,93]. Therefore, at low concentrations, pentene and hexene comonomeric units act as defects interrupting the regular propene sequences and inducing crystallization of the γ form [74,93]. At high comonomer concentrations, instead, the high fraction of co-units included in the crystals of the α form induces crystallization of the α form [68,69,73,74,77,93] and produces an increase of crystal density that then induces crystallization of the δ form for comonomer concentrations higher than about 15–16 mol% [68–70,73,74,77,82,90,91,93]. As discussed above, octene units are instead excluded from the crystals and are mainly segregated in the amorphous phase.

The two competing effects of interruption of the regular propene sequences of defects excluded from the crystals and the inclusion of defects into crystals define the crystallization behavior of iPP. The interruption effect induces crystallization of γ form [72,74,93,110], whereas the inclusion effect induces crystallization of α and δ forms [68–70,72,73,77,82,90–93]. For random copolymers and, generally, for iPP chains characterized by random distribution of defects along the macromolecules, the average length of the regular propene sequences is inversely proportional to the total concentration of defects ε and can be evaluated as <L_{iPP}> ≈ 1/ε [110]. The data of the maximum amount of γ form f_γ(max) of Figure 6 of all copolymers and of defective iPPs [72,74,93,110] are reported in Figure 7 as a function of the average length of the regular propene sequences <L_{iPP}>. For the different copolymers, different relationships between <L_{iPP}> and f_γ(max) have been obtained [72,74,93,110], and iPPC8 copolymers give a new different behavior.

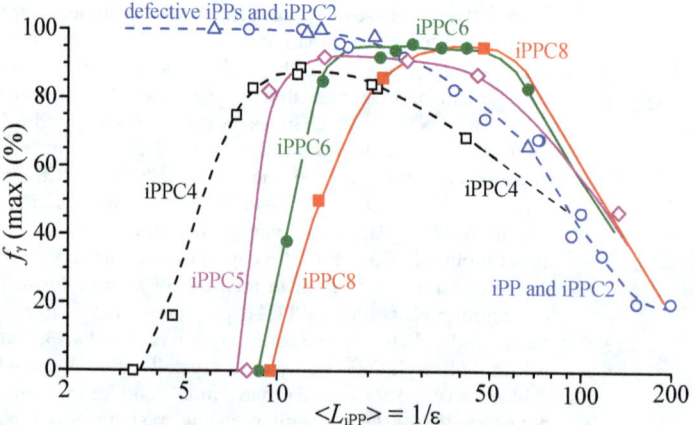

Figure 7. Maximum amount of γ form (f_γ(max)) obtained in iPPC8 copolymers isothermally crystallized from the melt (■) as a function of the average length of regular propylene sequences <L_{iPP}> compared to literature data of f_γ(max) found for stereoirregular iPPs (○) [110], iPPC2 (△) [72], iPPC4 (□) [72], iPPC5 (◇) [93], and iPPC6 (●) [74] copolymers.

The plot of Figure 7 indicates that the data of iPPC2 copolymers and of defective iPPs follow the same relation between f_γ(max) and $<L_{iPP}>$, which essentially corresponds to the interruption effect [72,110]. Ethylene co-units and rr stereodefects are mainly excluded from crystals (or partially included), and the effect of interruption and of shortening the length of the regular propene sequences prevails [72,110]. For the iPPC8 copolymers, octene defects are also mainly excluded from the crystals, and therefore, the interruption effect predominates and at octene concentrations lower than 2 mol% iPPC8 copolymers give the highest maximum amount of γ form, similar to that of stereodefective iPPs and iPPC2 copolymers (Figure 7). At higher octene concentration, the behavior of iPPC8 copolymers deviates from the master curve of iPPC2 and stereodefective iPPs, and the maximum amount of γ form does not depend anymore on the average length of regular propene sequences $<L_{iPP}>$ and decreases with increasing octene concentration and decreasing $<L_{iPP}>$. This is due to the fact that the long octene units makes the crystallization of the γ form too much slow even for short values of $<L_{iPP}>$, inducing crystallization of the kinetically favored α form, and prevents complete crystallization of both α and γ forms at octene concentrations higher than 12–13 mol%.

The three copolymers iPPC4, iPPC5, and iPPC6 give three different relationships between f_γ(max) and $<L_{iPP}>$ (Figure 7), because different amounts of the three comonomers are included in the crystals of the α form [72,74,93]. At low concentrations of comonomer (lower than 5–6 mol% and $<L_{iPP}>$ of 200-30 monomeric units), the included amounts of pentene and hexene are low, as in the case of ethylene, and the effect of interruption prevails inducing crystallization of the γ form. The bigger the comonomer, the more efficient the interruption effect, and the higher the amount of γ form. In fact, for this composition, iPPC5 and iPPC6 copolymers give an amount of γ form higher than those of iPPC2 copolymers and defective iPPs but lower than that of iPPC8 copolymers (Figure 7) [74,93]. High amounts of butene co-units are, instead, included in the crystals of α form, even at low concentrations, and iPPC4 copolymers give a maximum amount of γ form lower than those of iPPC5, iPPC6, and iPPC8 copolymers, and lower than those of iPPC2 copolymers and streoirregular iPPs (Figure 7) [72].

For comonomer concentrations higher than 5 mol% and $<L_{iPP}>$ lower that 20–30 monomeric units, hexene, pentene, and butene units are incorporated in the α form to very high extents with corresponding increase of crystalline density, and the inclusion effect with the stabilization of the α form prevails over the interruption effect, inducing the crystallization of the α and δ forms, with a fast decrease of the maximum amount of γ form down to zero for iPPC6, iPPC5, and iPPC4 copolymers (Figure 7) [72,74,93]. A fast or slow decrease of f_γ(max) depends on the fast or slow increase of crystal density, which depends on the size of the included comonomer. Therefore, the decrease of f_γ(max) is faster in iPPC6 copolymers because of the bigger hexene units, and the decrease of f_γ(max) in iPPC5 copolymers is faster than that of iPPC4 copolymers. It is worth noting that the fastest decrease of the maximum amount of γ form in iPPC8 copolymers is not due to incorporation of octene units in the crystals of α form, but, as mentioned above, is rather correlated to the fact that the excluded octene units make the crystallization of γ form too slow, inducing the faster crystallization of the kinetically favored α form, and then, at high concentrations, completely prevents crystallization of both α and γ forms (Figure 7).

These data on iPPC8 copolymers and the comparison with the literature demonstrate that the crystallization behavior of iPP may be described in terms of a model that defines a double role exerted by defects, the interruption of the regular propene sequences, and the inclusion effect. The crystallization of α, γ, and δ forms of iPP depends on which effect prevails, which in turn depends on the size and type of the defect.

4. Conclusions

Random isotactic propene–octene copolymers with octene concentrations ranging from 1.9 to 16 mol% have been synthesized with a homogeneous single center metallocene catalyst. The as-polymerized samples with octene content up to 10–12 mol% crystallize in the α form of iPP, whereas for higher octene concentration, the samples are basically amorphous or show very low crystallinity. The melting temperatures decrease with increasing octene content from 132 °C of the sample with 1.9 mol% of octene down to 47–45 °C for the samples with 12.8 and 15.9 mol% of octene.

Three samples of iPPC8 copolymers with low octene concentration of 1.9, 4.3, and 7.1 mol% have been isothermally crystallized from the melt at different crystallization temperatures. The samples crystallize at any crystallization temperature in mixtures of α and γ forms and the relative amount of γ form increases with increasing crystallization temperature and achieves a maximum value, which depends on the octene concentration. Contrary to propene–pentene and propene–hexene copolymers that, for high concentrations of pentene and hexene, crystallize in the δ form [74,93], the crystallization of the trigonal δ form has not been observed in all three samples of iPPC8 copolymers crystallized from the melt.

The behavior of iPPC8 copolymers is completely different from the behaviors observed in iPPC2 [72], iPPC4 [72], iPPC5 [93], and iPPC6 [74] copolymers. In fact, the maximum amount of γ form achieved in each sample of different copolymers depends on the concentration of comonomer and generally increases with increasing comonomer content, while in iPPC8 copolymers, the maximum amount of γ form decreases with increasing octene concentration. This different behavior is due to the fact that in iPPC8 copolymers octene units are excluded from the crystals, giving only the interruption effect that shortens the length of the regular propene sequences, inducing crystallization of the γ form. A maximum amount of γ form is achieved at low octene concentrations of nearly 1.9 mol%. At higher octene concentration, the amount of γ form crystallized from the melt rapidly decreases.

In copolymers of iPP with butene, pentene, and hexene, the comonomer units are, instead, incorporated in the crystals of α form to a very high extent. At a high concentration of comonomers, the inclusion effect that favors crystallization of the α form prevails over the interruption effect. The efficiency of the incorporation effect depends on the size of comonomers, as a rapid or slow increase of density is obviously correlated to the size of the incorporated comonomer.

For iPPC8 copolymers, the observed fastest decrease of the maximum amount of γ form and the consequent crystallization of the α form for high octene concentrations is not due to incorporation of octene units in the crystals of α form, but is due to the fact that the excluded octene units make the crystallization of γ form too slow, inducing the faster crystallization of the kinetically favored α form and, then, at high concentration, prevent the crystallization of both α and γ forms.

The reported results on crystallization of iPPC8 copolymers and the comparison with the literature demonstrate that the crystallization behavior of iPP may be described in terms of a model that defines two effects exerted by defects: the interruption of the regular propene sequences and the inclusion effect. The crystallization of α, γ, and δ forms of iPP depends on which effect prevails, which in turn depends on the size and type of the defect.

Author Contributions: M.S. and C.D.R. conceived the experiments, G.T. synthesized the samples, F.D.S., A.G. and M.S. performed the experiments. All authors have read and agreed to the published version of the manuscript.

Funding: This research received no external funding.

Institutional Review Board Statement: Not applicable.

Informed Consent Statement: Not applicable.

Data Availability Statement: The data in this study are available on reasonable request from the corresponding author.

Acknowledgments: The task force "Polymers and biopolymers" of the University of Napoli Federico II is acknowledged. The Department of Chemical Science of the University of Napoli Federico II is greatefully acknowledged for the funding of the article processing charge.

Conflicts of Interest: The authors declare no conflict of interest.

References

1. Gahleitner, M.; Tranninger, C.; Doshev, P. Polypropylene Copolymers. In *Polypropylene Handbook, Morphology, Blends and Composites*; Karger-Kocsis, J., Bárány, T., Eds.; Springer Nature: Cham, Switzerland, 2019; Chapter 6; p. 295.
2. Galli, P.; Haylock, J.C.; Simonazzi, T. Manufacturing and properties of polypropylene copolymers. In *Polypropylene: Structure, Blends and Composites. Vol. 2 Copolymers and Blends*; Karger-Kocsis, J., Ed.; Chapman & Hall: London, UK, 1995; pp. 1–24.
3. Pasquini, N. (Ed.) *Polypropylene Handbook*; Hanser Publishers: Munich, Germany, 2005.
4. De Rosa, C.; Auriemma, F.; Ruiz de Ballesteros, O.; Resconi, L.; Camurati, I. Tailoring the Physical Properties of Isotactic Polypropylene through Incorporation of Comonomers and the Precise Control of Stereo- and Regio-Regularity by Metallocene Catalysts. *Chem. Mater.* **2007**, *19*, 5122. [CrossRef]
5. De Rosa, C.; Scoti, M.; Di Girolamo, R.; Ruiz de Ballesteros, O.; Auriemma, F.; Malafronte, A. Polymorphism in polymers: A tool to tailor material's properties. *Polym. Cryst.* **2020**, *3*, e10101. [CrossRef]
6. Turner-Jones, A. Development of the γ-crystal form in random copolymers of propylene and their analysis by DSC and X-ray methods. *Polymer* **1971**, *12*, 487–508. [CrossRef]
7. Cimmino, S.; Martuscelli, E.; Nicolais, L.; Silvestre, C. Thermal and mechanical properties of isotactic random propylene-butene-1 copolymers. *Polymer* **1978**, *19*, 1222. [CrossRef]
8. Crispino, L.; Martuscelli, E.; Pracella, M. Influence of composition on the melt crystallization of isotactic random propylene/1-butene copolymers. *Makromol. Chem.* **1980**, *181*, 1747. [CrossRef]
9. Cavallo, P.; Martuscelli, E.; Pracella, M. Effect of thermal treatment on solution grown crystals of isotactic propylene/butene-1 copolymers. *Polymer* **1997**, *18*, 891–896. [CrossRef]
10. Starkweather, H.W., Jr.; Van-Catledge, F.A.; MacDonald, R.N. Crystalline order in copolymers of ethylene and propylene. *Macromolecules* **1982**, *15*, 1600–1604. [CrossRef]
11. Guidetti, G.P.; Busi, P.; Giulianetti, I.; Zanetti, R. Structure-properties relationships in some random copolymers of propylene. *Eur. Polym. J.* **1983**, *19*, 757–759. [CrossRef]
12. Busico, V.; Corradini, P.; De Rosa, C.; Di Benedetto, E. Physico-chemical and structural characterization of ethylene-propene copolymers with low ethylene content from isotactic-specific Ziegler-Natta catalysts. *Eur. Polym. J.* **1985**, *21*, 239–244. [CrossRef]
13. Avella, M.; Martuscelli, E.; Della Volpe, G.; Segre, A.; Rossi, E.; Simonazzi, T. Composition-properties relationships in propene-ethene random copolymers obtained with high-yield Ziegler-Natta supported catalysts. *Makromol. Chem.* **1986**, *187*, 1927. [CrossRef]
14. Marigo, A.; Marega, C.; Zanetti, R.; Paganetto, G.; Canossa, E.; Coletta, F.; Gottardi, F. Crystallization of the γ-form of isotactic poly(propylene). *Makromol. Chem.* **1989**, *190*, 2805. [CrossRef]
15. Monasse, B.; Haudin, J.M. Effect of random copolymerization on growth transition and morphology change in polypropylene. *Colloid Polym. Sci.* **1988**, *266*, 679–687. [CrossRef]
16. Xu, Z.K.; Feng, L.X.; Wang, D.; Yang, S.L. Copolymerization of propene with 1-alkenes using a $MgCl_2/TiCl_4$ catalyst. *Makromol. Chem.* **1991**, *192*, 1835–1840. [CrossRef]
17. Yang, S.L.; Xu, Z.K.; Feng, L.X. Copolymerization of propene with high-1-olefin using a MgCl2/TiCl4 catalyst. *Makromol. Chem. Macromol. Symp.* **1992**, *63*, 233–243. [CrossRef]
18. Zimmermann, H.J. Structural analysis of random propylene-ethylene copolymers. *J. Macromol. Sci. Phys.* **1993**, *32*, 141–161. [CrossRef]
19. Sugano, T.; Gotoh, Y.; Fujita, T. Effect of catalyst isospecificity on the copolymerization of propene with 1-hexene. *Makromol. Chem.* **1992**, *193*, 43. [CrossRef]
20. Hingmann, R.; Rieger, J.; Kersting, M. Rheological Properties of a Partially Molten Polypropylene Random Copolymer during Annealing. *Macromolecules* **1995**, *28*, 3801–3806. [CrossRef]
21. Morini, G.; Albizzati, E.; Balbontin, G.; Mingozzi, I.; Sacchi, M.C.; Forlini, F.; Tritto, I. Microstructure Distribution of Polypropylenes Obtained in the Presence of Traditional Phthalate/Silane and Novel Diether Donors: A Tool for Understanding the Role of Electron Donors in MgCl2-Supported Ziegler–Natta Catalysts. *Macromolecules* **1996**, *29*, 5770. [CrossRef]
22. Pérez, E.; Benavente, R.; Bello, A.; Pereña, J.M.; Zucchi, D.; Sacchi, M.C. Crystallization behavior of fractions of a copolymer of propene and 1-hexene. *Polymer* **1997**, *38*, 5411. [CrossRef]
23. Laihonen, S.; Gedde, U.W.; Werner, P.-E.; Martinez-Salazar, J. Crystallization kinetics and morphology of poly(propylene-*stat*-ethylene) fractions. *Polymer* **1997**, *38*, 361–369. [CrossRef]
24. Laihomen, S.; Gedde, U.W.; Werner, P.E.; Westdahl, M.; Jääskeläinen, P.; Martinez-Salazar, J. Crystal structure and morphology of melt-crystallized poly(propylene-*stat*-ethylene) fractions. *Polymer* **1997**, *38*, 371–377. [CrossRef]
25. Abiru, T.; Mizuno, A.; Weigand, F. Microstructural characterization of propylene–butene-1 copolymer using temperature rising elution fractionation. *J. Appl. Polym. Sci.* **1998**, *68*, 1493–1501. [CrossRef]
26. Feng, Y.; Jin, X.; Hay, J.N. Crystalline structure of propylene–ethylene copolymer fractions. *J. Appl. Polym. Sci.* **1998**, *68*, 381–386. [CrossRef]
27. Feng, Y.; Hay, J.N. The characterization of random propylene–ethylene copolymer. *Polymer* **1998**, *39*, 6589–6596. [CrossRef]
28. Xu, J.; Feng, Y. Application of temperature rising elution fractionation in polyolefins. *Eur. Polym. J.* **2000**, *36*, 867. [CrossRef]

29. Zhao, Y.; Vaughan, A.S.; Sutton, S.J.; Swingler, S.G. On nucleation and the evolution of morphology in a propylene/ethylene copolymer. *Polymer* **2001**, *42*, 6599–6608. [CrossRef]
30. Foresta, T.; Piccarolo, S.; Goldbeck-Wood, G. Competition between α and γ phases in isotactic polypropylene: Effects of ethylene content and nucleating agents at different cooling rates. *Polymer* **2001**, *42*, 1167–1176. [CrossRef]
31. Marega, C.; Marigo, A.; Saini, R.; Ferrari, P. The influence of thermal treatment and processing on the structure and morphology of poly(propylene-ran-1-butene) copolymers. *Polym. Int.* **2001**, *50*, 442. [CrossRef]
32. Marigo, A.; Causin, V.; Marega, C.; Ferrari, P. Crystallization of the γ form in random propylene-ethylene copolymers. *Polym. Int.* **2004**, *53*, 2001–2008. [CrossRef]
33. Dimenska, A.; Phillips, P.J. High pressure crystallization of random propylene–ethylene copolymers: α–γ Phase diagram. *Polymer* **2006**, *47*, 5445–5456. [CrossRef]
34. Arnold, M.; Henschke, O.; Knorr, J. Copolymerization of propene and higher α-olefins with the metallocene catalyst Et[Ind]2HfCl$_2$/methylaluminoxane. *Macromol. Chem. Phys.* **1996**, *197*, 563–573. [CrossRef]
35. Arnold, M.; Bornemann, S.; Köller, F.; Menke, T.J.; Kressler, J. Synthesis and characterization of branched polypropenes obtained by metallocene catalysis. *Macromol. Chem. Phys.* **1998**, *199*, 2647–2653. [CrossRef]
36. Galimberti, M.; Destro, M.; Fusco, O.; Piemontesi, F.; Camurati, I. Ethene/Propene Copolymerization from Metallocene-Based Catalytic Systems: Role of the Alumoxane. *Macromolecules* **1999**, *32*, 258–263. [CrossRef]
37. Busse, K.; Kressler, J.; Maier, R.D.; Scherble, J. Tailoring of the α-, β-, and γ-Modification in Isotactic Polypropene and Propene/Ethene Random Copolymers. *Macromolecules* **2000**, *33*, 8775–8780. [CrossRef]
38. Forlini, F.; Fan, Z.-Q.; Tritto, I.; Locatelli, P.; Sacchi, M.C. Metallocene-catalyzed propene 1-hexene copolymerization-influence of amount and bulkiness of cocatalyst and of solvent polarity. *Macromol. Chem. Phys.* **1997**, *198*, 2397. [CrossRef]
39. Tritto, I.; Donetti, R.; Sacchi, M.C.; Locatelli, P.; Zannoni, G. Evidence of Zircononium−Polymeryl Ion Pairs from 13C NMR in Situ ^{13}C$_2$H$_4$ Polymerization with Cp$_2$Zr(^{13}CH$_3$)$_2$-Based Catalysts. *Macromolecules* **1999**, *32*, 264–269. [CrossRef]
40. Pérez, E.; Zucchi, D.; Sacchi, M.C.; Forlini, F.; Bello, A. Obtaining the γ phase in isotactic polypropylene: Effect of catalyst system and crystallization conditions. *Polymer* **1999**, *40*, 675–681. [CrossRef]
41. Forlini, F.; Tritto, I.; Locatelli, P.; Sacchi, M.C.; Piemontesi, F. ^{13}C NMR studies of zirconocene-catalyzed propylene/1-hexene copolymers: In-depth investigation of the effect of solvent polarity. *Macromol. Chem. Phys.* **2000**, *201*, 401–408. [CrossRef]
42. Sacchi, M.C.; Forlini, F.; Losio, S.; Tritto, I.; Wahner, U.M.; Tincul, I.; Joubert, D.J.; Sadiku, E.R. Microstructure of Metallocene-Catalyzed Propene/1-Pentene Copolymers. *Macromol. Chem. Phys.* **2003**, *204*, 1643. [CrossRef]
43. Wahner, U.M.; Tincul, I.; Joubert, D.J.; Sadiku, E.R.; Forlini, F.; Losio, S.; Tritto, I.; Sacchi, M.C. 13C NMR Study of Copolymers of Propene with Higher 1-Olefins with New Microstructures by ansa-Zirconocene Catalysts. *Macromol. Chem. Phys.* **2003**, *204*, 1738. [CrossRef]
44. Sacchi, M.C.; Forlini, F.; Tritto, I.; Stagnaro, P. Unexpected Formation of Atactic Blocks in Propylene/1-Pentene Copolymers from rac-Me$_2$Si(2-MeBenz[e]Ind)$_2$ZrCl$_2$. *Macromol. Chem. Phys.* **2004**, *205*, 1804. [CrossRef]
45. Sacchi, M.C.; Forlini, F.; Losio, S.; Tritto, I.; Costa, G.; Stagnaro, P.; Tincul, I.; Wahner, U.M. Microstructural characteristics and thermal properties of ansa-zirconocene catalyzed copolymers of propene with higher α-olefins. *Macromol. Symp.* **2004**, *213*, 57. [CrossRef]
46. Costa, G.; Stagnaro, P.; Trefiletti, V.; Sacchi, M.C.; Forlini, F.; Alfonso, G.C.; Tincul, I.; Wahner, U.M. Thermal Behavior and Structural Features of Propene/1-Pentene Copolymers by Metallocene Catalysts. *Macromol. Chem. Phys.* **2004**, *205*, 383–389. [CrossRef]
47. Stagnaro, P.; Costa, G.; Trefiletti, V.; Canetti, M.; Forlini, F.; Alfonso, G.C. Thermal Behavior, Structure and Morphology of Propene/Higher 1-Olefin Copolymers. *Macromol. Chem. Phys.* **2006**, *207*, 2128–2141. [CrossRef]
48. Stagnaro, P.; Boragno, L.; Canetti, M.; Forlini, F.; Azzurri, F.; Alfonso, G.C. Crystallization and morphology of the trigonal form in random propene/1-pentene copolymers. *Polymer* **2009**, *50*, 5242–5249. [CrossRef]
49. Kim, I. Copolymerization of propene and 1-hexene using metallocene amide compounds. *Macromol. Rapid Commun.* **1998**, *19*, 299. [CrossRef]
50. Kim, I.; Kim, Y.J. Copolymerization of propene and 1-hexene with isospecific and syndiospecific metallocene catalysts. *Polym. Bull.* **1998**, *40*, 415–421. [CrossRef]
51. Alamo, R.G.; VanderHart, D.L.; Nyden, M.R.; Mandelkern, L. Morphological Partitioning of Ethylene Defects in Random Propylene−Ethylene Copolymers. *Macromolecules* **2000**, *33*, 6094–6105. [CrossRef]
52. Hosier, I.L.; Alamo, R.G.; Esteso, P.; Isasi, G.R.; Mandelkern, L. Formation of the α and γ Polymorphs in Random Metallocene-Propylene Copolymers. Effect of Concentration and Type of Comonomer. *Macromolecules* **2003**, *36*, 5623–5636. [CrossRef]
53. Hosier, I.L.; Alamo, R.G.; Lin, J.S. Lamellar morphology of random metallocene propylene copolymers studied by atomic force microscopy. *Polymer* **2004**, *45*, 3441–3455. [CrossRef]
54. Alamo, R.G.; Ghosal, A.; Chatterjee, J.; Thompson, K.L. Linear growth rates of random propylene ethylene copolymers. The change over from γ dominated growth to mixed (α+γ) polymorphic growth. *Polymer* **2005**, *46*, 8774–8789. [CrossRef]
55. Fan, Z.; Yasin, T.; Feng, L. Copolymerization of propylene with 1-octene catalyzed by rac-Me2Si(2,4,6-Me3-Ind)2ZrCl2/methyl aluminoxane. *J. Polym. Sci. Part A* **2000**, *38*, 4299. [CrossRef]
56. Brull, R.; Pasch, H.; Rauberheimer, H.G.; Sanderson, R.D.; Wahner, U.M. Polymerization of higher linear α-olefins with (CH$_3$)$_2$Si(2-methylbenz[e]indenyl)$_2$ZrCl$_2$. *J. Polym. Sci. Part A Polym. Chem.* **2000**, *38*, 2333–2339. [CrossRef]

57. Van Reenen, A.J.; Brull, R.; Wahner, U.M.; Raubenheimer, H.G.; Sanderson, R.D.; Pasch, H. The copolymerization of propylene with higher, linear α-olefins. *J. Polym. Sci. Part A Polym. Chem.* **2000**, *38*, 4110–4118. [CrossRef]
58. Lovisi, H.; Tavares, M.I.B.; da Silva, N.M.; de Menezes, S.M.C.; de Santa Maria, L.C.; Coutinho, F.M.B. Influence of comonomer content and short branch length on the physical properties of metallocene propylene copolymers. *Polymer* **2001**, *42*, 9791–9799. [CrossRef]
59. Shin, Y.-W.; Uozumi, T.; Terano, M.; Nitta, K.-H. Synthesis and characterization of ethylene–propylene random copolymers with isotactic propylene sequence. *Polymer* **2001**, *42*, 9611–9615. [CrossRef]
60. Shin, Y.-W.; Hashiguchi, H.; Terano, M.; Nitta, K. Synthesis and characterization of propylene-α-olefin random copolymers with isotactic propylene sequence. II. Propylene– hexene-1 random copolymers. *J. Appl. Polym. Sci.* **2004**, *92*, 2949. [CrossRef]
61. Hosoda, S.; Hori, H.; Yada, K.; Tsuji, M.; Nakahara, S. Degree of comonomer inclusion into lamella crystal for propylene/olefin copolymers. *Polymer* **2002**, *43*, 7451–7460. [CrossRef]
62. Fujiyama, M.; Inata, H. Crystallization and melting characteristics of metallocene isotactic polypropylenes. *J. Appl. Polym. Sci.* **2002**, *85*, 1851–1857. [CrossRef]
63. Xu, J.-T.; Xue, L.; Fan, Z.-Q. Nonisothermal crystallization of metallocene propylene–decene- 1 copolymers. *J. Appl. Polym. Sci.* **2004**, *93*, 1724–1730. [CrossRef]
64. Poon, B.; Rogunova, M.; Chum, S.P.; Hiltner, A.; Baer, E. Classification of Homogeneous Copolymers of Propylene and 1-Octene Based on Comonomer Content. *J. Polym. Sci. Polym. Phys.* **2004**, *42*, 4357–4370. [CrossRef]
65. Poon, B.; Rogunova, M.; Hiltner, A.; Baer, E.; Chum, S.P.; Galeski, A.; Piorkowska, E. Structure and Properties of Homogeneous Copolymers of Propylene and 1-Hexene. *Macromolecules* **2005**, *38*, 1232–1243. [CrossRef]
66. Palza, H.; López-Majada, J.M.; Quijada, R.; Benavente, R.; Pérez, E.; Cerrada, M.L. Metallocenic Copolymers of Isotactic Propylene and 1-Octadecene: Crystalline Structure and Mechanical Behavior. *Macromol. Chem. Phys.* **2005**, *206*, 1221–1230. [CrossRef]
67. López-Majada, J.M.; Palza, H.; Guevara, J.L.; Quijada, R.; Martinez, M.C.; Benavente, R.; Pereña, J.M.; Pérez, E.; Cerrada, M.L. Metallocene Copolymers of Propene and 1-Hexene: The Influence of the Comonomer Content and Thermal History on the Structure and Mechanical Properties. *J. Polym. Sci. Polym. Phys. Ed.* **2006**, *44*, 1253–1267. [CrossRef]
68. De Rosa, C.; Auriemma, F.; Corradini, P.; Tarallo, O.; Dello Iacono, S.; Ciaccia, E.; Resconi, L. Crystal Structure of the Trigonal Form of Isotactic Polypropylene as an Example of Density-Driven Polymer Structure. *J. Am. Chem. Soc.* **2006**, *128*, 80–81. [CrossRef]
69. De Rosa, C.; Dello Iacono, S.; Auriemma, F.; Ciaccia, E.; Resconi, L. Crystal Structure of Isotactic Propylene-Hexene Copolymers: The Trigonal Form of Isotactic Polypropylene. *Macromolecules* **2006**, *39*, 6098–6109. [CrossRef]
70. Lotz, B.; Ruan, J.; Thierry, A.; Alfonso, G.C.; Hiltner, A.; Baer, E.; Piorkowska, E.; Galeski, A. A Structure of Copolymers of Propene and Hexene Isomorphous to Isotactic Poly(1-butene) Form I. *Macromolecules* **2006**, *39*, 5777–5781. [CrossRef]
71. Toki, S.; Sics, I.; Burger, C.; Fang, D.; Liu, L.; Hsiao, B.S.; Datta, S.; Tsou, A.H. Structure Evolution during Cyclic Deformation of an Elastic Propylene-Based Ethylene-Propylene Copolymer. *Macromolecules* **2006**, *39*, 3588–3597. [CrossRef]
72. De Rosa, C.; Auriemma, F.; Ruiz de Ballesteros, O.; Resconi, L.; Camurati, I. Crystallization Behavior of Isotactic Propylene-Ethylene and Propylene-Butene Copolymers: Effect of Comonomers versus Stereodefects on Crystallization Properties of Isotactic Polypropylene. *Macromolecules* **2007**, *40*, 6600–6616. [CrossRef]
73. De Rosa, C.; Auriemma, F.; Talarico, G.; Ruiz de Ballesteros, O. Structure of Isotactic Propylene-Pentene Copolymers. *Macromolecules* **2007**, *40*, 8531–8532. [CrossRef]
74. De Rosa, C.; Auriemma, F.; Ruiz de Ballesteros, O.; De Luca, D.; Resconi, L. The double role of comonomers on the crystallization behavior of isotactic polypropylene: Propylene-hexene Copolymers. *Macromolecules* **2008**, *41*, 2172–2177. [CrossRef]
75. De Rosa, C.; Auriemma, F.; Ruiz de Ballesteros, O.; Dello Iacono, S.; De Luca, D.; Resconi, L. Stress-induced polymorphic transformations and mechanical properties of isotactic propylene-hexene copolymers. *Cryst. Grow Des.* **2009**, *9*, 165–176. [CrossRef]
76. Palza, H.; López-Majada, J.M.; Quijada, R.; Pereña, J.M.; Benavente, R.; Pérez, E.; Cerrada, M.L. Comonomer Length Influence on the Structure and Mechanical Response of Metallocenic Polypropylenic Materials. *Macromol. Chem. Phys.* **2008**, *209*, 2259–2267. [CrossRef]
77. Cerrada, M.L.; Polo-Corpa, M.J.; Benavente, R.; Pérez, E.; Velilla, T.; Quijada, R. Formation of the new trigonal polymorph in iPP-1-hexene copolymers. Competition with the mesomorphic phase. *Macromolecules* **2009**, *42*, 702–708. [CrossRef]
78. Polo-Corpa, M.J.; Benavente, R.; Velilla, T.; Quijada, R.; Pérez, E.; Cerrada, M.L. Development of the mesomorphic phase in isotactic propene/higher α-olefin copolymers at intermediate comonomer content and its effect on properties. *Eur. Polym. J.* **2010**, *46*, 1345–1354. [CrossRef]
79. De Rosa, C.; Auriemma, F.; Di Girolamo, R.; Romano, L.; De Luca, R.M. A New Mesophase of Isotactic Polypropylene in Copolymers of Propylene with Long Branched Comonomers. *Macromolecules* **2010**, *43*, 8559–8569. [CrossRef]
80. Pérez, E.; Cerrada, M.L.; Benavente, R.; Gómez-Elvira, J.M. Enhancing the formation of the new trigonal polymorph in isotactic propene-1-pentene copolymers: Determination of the X-ray crystallinity. *Macromol. Res.* **2011**, *19*, 1179–1185. [CrossRef]
81. De Rosa, C.; Auriemma, F.; Vollaro, P.; Resconi, L.; Guidotti, S.; Camurati, I. Crystallization Behavior of Propylene-Butene Copolymers: The Trigonal Form of Isotactic Polypropylene and Form I of Isotactic Poly(1-butene). *Macromolecules* **2011**, *44*, 540–549. [CrossRef]

82. De Rosa, C.; Ruiz de Ballesteros, O.; Auriemma, F.; Di Caprio, M.R. Crystal Structure of the Trigonal Form of Isotactic Propylene–Pentene Copolymers: An Example of the Principle of Entropy–Density Driven Phase Formation in Polymers. *Macromolecules* **2012**, *45*, 2749–2763. [CrossRef]
83. Pérez, E.; Gómez-Elvira, J.M.; Benavente, R.; Cerrada, M.L. Tailoring the formation rate of the mesophase in random propylene-co-1-pentene copolymers. *Macromolecules* **2012**, *45*, 6481–6490. [CrossRef]
84. Boragno, L.; Stagnaro, P.; Forlini, F.; Azzurri, F.; Alfonso, G.C. The trigonal form of i-PP in random C3/C5/C6 terpolymers. *Polymer* **2013**, *54*, 1656–1662. [CrossRef]
85. García-Peñas, A.; Gómez-Elvira, J.M.; Pérez, E.; Cerrada, M.L. Isotactic poly(propylene-co-1-pentene-co-1-hexene) terpolymers: Synthesis, molecular characterization, and evidence of the trigonal polymorph. *J. Polym. Sci. A Polym. Chem.* **2013**, *51*, 3251–3259. [CrossRef]
86. García-Peñas, A.; Gómez-Elvira, J.M.; Barranco-García, R.; Pérez, E.; Cerrada, M.L. Trigonal δ form as a tool for tuning mechanical behavior in poly(propylene-co-1-pentene-co-1-heptene) terpolymers. *Polymer* **2016**, *99*, 112–121. [CrossRef]
87. García-Peñas, A.; Gómez-Elvira, J.M.; Lorenzo, V.; Pérez, E.; Cerrada, M.L. Unprecedented dependence of stiffness parameters and crystallinity on comonomer content in rapidly cooled propylene-co-1-pentene copolymers. *Polymer* **2017**, *130*, 17–25. [CrossRef]
88. Auriemma, F.; De Rosa, C.; Di Girolamo, R.; Malafronte, A.; Scoti, M.; Cipullo, R. Yield behavior of random copolymers of isotactic polypropylene. *Polymer* **2017**, *129*, 235–246. [CrossRef]
89. Auriemma, F.; De Rosa, C.; Di Girolamo, R.; Malafronte, A.; Scoti, M.; Cioce, C. A molecular view of properties of random copolymers of isotactic polypropene. *Adv. Polym. Sci.* **2017**, *276*, 45–92.
90. De Rosa, C.; Scoti, M.; Auriemma, F.; Ruiz de Ballesteros, O.; Talarico, G.; Malafronte, A.; Di Girolamo, R. Mechanical Properties and Morphology of Propene–Pentene Isotactic Copolymers. *Macromolecules* **2018**, *51*, 3030–3040. [CrossRef]
91. De Rosa, C.; Scoti, M.; Auriemma, F.; Ruiz de Ballesteros, O.; Talarico, G.; Di Girolamo, R.; Cipullo, R. Relationships among lamellar morphology parameters, structure and thermal behavior of isotactic propene-pentene copolymers: The role of incorporation of comonomeric units in the crystals. *Eur. Polym. J.* **2018**, *103*, 251–259. [CrossRef]
92. De Rosa, C.; Scoti, M.; Ruiz de Ballesteros, O.; Di Girolamo, R.; Auriemma, F.; Malafronte, A. Propylene-Butene Copolymers: Tailoring Mechanical Properties from Isotactic Polypropylene to Polybutene. *Macromolecules* **2020**, *53*, 4407–4421. [CrossRef]
93. Scoti, M.; De Stefano, F.; Di Girolamo, R.; Talarico, G.; Malafronte, A.; De Rosa, C. Crystallization of Propene-Pentene Isotactic Copolymers as an Indicator of the General View of the Crystallization Behavior of Isotactic Polypropylene. *Macromolecules* **2022**, *55*, 241–251. [CrossRef]
94. Karger-Kocsis, J.; Bárány, T. *Polypropylene Handbook, Morphology, Blends and Composites*; Springer International Publishing: Cham, Switzerland, 2019.
95. Lotz, B.; Graff, S.; Wittmann, J.C. Crystal morphology of the γ (triclinic) phase of isotactic polypropylene and its relation to the α phase. *J. Polym. Sci. Polym. Phys. Ed.* **1986**, *24*, 2017. [CrossRef]
96. Kardos, J.L.; Christiansen, A.W.; Baer, E. Structure of pressure crystallized polypropylene. *J. Polym. Sci. Part B Polym. Phys.* **1966**, *4*, 777–788. [CrossRef]
97. Pal, K.D.; Morrow, D.R.; Sauer, J.A. Interior morphology of bulk polypropylene. *Nature* **1966**, *211*, 514–515.
98. Mezghani, K.; Phillips, P.J. The γ-phase of high molecular weight isotactic polypropylene: III. The equilibrium melting point and the phase diagram. *Polymer* **1998**, *39*, 3735. [CrossRef]
99. Mezghani, K.; Phillips, P.J. The γ-phase of high molecular weight isotactic polypropylene. II. The morphology of the γ-form crystallized at 200 MPa. *Polymer* **1997**, *38*, 5725. [CrossRef]
100. Brückner, S.; Phillips, P.J.; Mezghani, K.; Meille, S.V. On the crystallization of γ-isotactic polypropylene. A high pressure study. *Macromol. Rapid Commun.* **1997**, *18*, 1–7. [CrossRef]
101. Ewen, J.A. Mechanisms of stereochemical control in propylene polymerizations with soluble Group 4B metallocene/methylalumoxane catalysts. *J. Am. Chem. Soc.* **1984**, *106*, 6355. [CrossRef]
102. Kaminsky, W.; Kulper, K.; Brintzinger, H.H.; Wild, F.R.W. Polymerization of propene and butene with a chiral zirconocene and methylaluminoxane as cocatalyst. *Angew. Chem.* **1985**, *97*, 507–508. [CrossRef]
103. Brintzinger, H.H.; Fischer, D.; Mulhaupt, R.; Rieger, B.; Waymouth, R.M. Stereospecific Olefin Polymerization with Chiral Metallocene Catalysts. *Angew. Chem. Int. Ed. Engl.* **1995**, *34*, 1143–1170. [CrossRef]
104. Fischer, D.; Mülhaupt, R. The influence of regio-and stereoirregularities on the crystallization behavior of isotactic poly(propylene)s prepared with homogeneous group IVa metallocene/methylaluminoxane Ziegler-Natta catalysts. *Macromol. Chem. Phys.* **1994**, *195*, 1433. [CrossRef]
105. Thomann, R.; Wang, C.; Kressler, J.; Mülhaupt, R. On the γ-Phase of Isotactic Poypropylene. *Macromolecules* **1996**, *29*, 8425–8434. [CrossRef]
106. Thomann, R.; Semke, H.; Maier, R.D.; Thomann, Y.; Scherble, J.; Mülhaupt, R.; Kressler, J. Influence of stereoirregularities on the formation of the γ-phase in isotactic polypropylene. *Polymer* **2001**, *42*, 4597–4603. [CrossRef]
107. Alamo, R.G.; Kim, M.H.; Galante, M.J.; Isasi, J.R.; Mandelkern, L. Structural and Kinetic Factors Governing the Formation of the γ Polymorph of Isotactic Polypropylene. *Macromolecules* **1999**, *32*, 4050–4064. [CrossRef]
108. VanderHart, D.L.; Alamo, R.G.; Nyden, M.R.; Kim, M.H.; Mandelkern, L. Observation of Resonances Associated with Stereo and Regio Defects in the Crystalline Regions of Isotactic Polypropylene: Toward a Determination of Morphological Partioning. *Macromolecules* **2000**, *33*, 6078–6093. [CrossRef]

109. Nyden, M.R.; Vanderhart, D.L.; Alamo, R.G. The conformational structures of defect-containing chains in the crystalline regions of isotactic polypropylene. *Comput. Theor. Comput. Sci.* **2001**, *11*, 175–189. [CrossRef]
110. De Rosa, C.; Auriemma, F.; Di Capua, A.; Resconi, L.; Guidotti, S.; Camurati, I.; Nifant'ev, I.E.; Laishevtsev, I.P. Structure-property correlations in polypropylene from metallocene catalysts: Stereodefective, regioregular isotactic polypropylene. *J. Am. Chem. Soc.* **2004**, *126*, 17040–17049. [CrossRef] [PubMed]
111. De Rosa, C.; Auriemma, F.; Circelli, T.; Waymouth, R.M. Crystallization of the α and γ Forms of Isotactic Polypropylene as a Tool to Test the Degree of Segregation of Defects in the Polymer Chains. *Macromolecules* **2002**, *35*, 3622–3629. [CrossRef]
112. Auriemma, F.; De Rosa, C. Crystallization of Metallocene-Made Isotactic Polypropylene: Disordered Modifications Intermediate between the α and γ Forms. *Macromolecules* **2002**, *35*, 9057–9068. [CrossRef]
113. De Rosa, C.; Auriemma, F.; Spera, C.; Talarico, G.; Tarallo, O. Comparison between Polymorphic Behaviors of Ziegler-Natta and Metallocene-Made Isotactic Polypropylene: The Role of the Distribution of Defects in the Polymer Chains. *Macromolecules* **2004**, *37*, 1441–1454. [CrossRef]
114. De Rosa, C.; Auriemma, F.; Spera, C.; Talarico, G.; Gahleitner, M. Crystallization Properties of Elastomeric Polypropylene from Alumina-Supported Tetraalkyl Zirconium Catalysts. *Polymer* **2004**, *45*, 5875–5888. [CrossRef]
115. De Rosa, C.; Auriemma, F.; Paolillo, M.; Resconi, L.; Camurati, I. Crystallization Behavior and Mechanical Properties of Regiodefective, Highly Stereoregular Isotactic Polypropylene: Effect of Regiodefects versus Stereodefects and Influence of the Molecular Mass. *Macromolecules* **2005**, *38*, 9143–9144. [CrossRef]
116. Gahleitner, M.; Jääskeläinen, P.; Ratajski, E.; Paulik, C.; Reussner, J.; Wolfschwenger, J.; Neißl, W. Propylene-ethylene random copolymers: Comonomer effects on crystallinity and application properties. *J. Appl. Polym. Sci.* **2005**, *95*, 1073–1081. [CrossRef]
117. Caveda, S.; Pérez, E.; Blázquez-Blázquez, E.; Peña, B.; van Grieken, R.; Suárez, I.; Benavente, R. Influence of structure on the properties of polypropylene copolymers and terpolymers. *Polym.Test.* **2017**, *62*, 23–32. [CrossRef]
118. Auriemma, F.; De Rosa, C.; Di Girolamo, R.; Malafronte, A.; Scoti, M.; Mitchell, G.R.; Esposito, S. Relationship between molecular configuration and stress induced phase transitions. In *Controlling the Morphology of Polymers-Multiple Scales of Structure and Processing*; Mitchell, G.R., Tojeira, A., Eds.; Springer International Publishing: Cham, Switzerland, 2016; p. 287.
119. Auriemma, F.; De Rosa, C.; Di Girolamo, R.; Malafronte, A.; Scoti, M.; Mitchell, G.R.; Esposito, S. Deformation of Stereoirregular Isotactic Polypropylene across Length Scales. Influence of Temperature. *Macromolecules* **2017**, *50*, 2856–2870. [CrossRef]
120. Auriemma, F.; De Rosa, C.; Di Girolamo, R.; Malafronte, A.; Scoti, M.; Mitchell, G.R.; Esposito, S. Time-resolving study of stress-induced transformations of isotactic polypropylene through wide angle X-ray scattering measurements. *Polymers* **2018**, *10*, 162. [CrossRef]
121. Alamo, R.G.; Blanco, J.A.; Agarwal, P.K.; Randall, J.C. Crystallization Rates of Matched Fractions of MgCl2-Supported Ziegler Natta and Metallocene Isotactic Poly(Propylene)s. 1. The Role of Chain Microstructure. *Macromolecules* **2003**, *36*, 1559–1571. [CrossRef]
122. Randall, J.C.; Alamo, R.G.; Agarwal, P.K.; Ruff, C.J. Crystallization Rates of Matched Fractions of MgCl2-Supported Ziegler-Natta and Metallocene Isotactic Poly(propylene)s. 2. Chain Microstructures from a Supercritical Fluid Fractionation of a MgCl2-Supported Ziegler-Natta Isotactic Poly(propylene). *Macromolecules* **2003**, *36*, 1572–1584. [CrossRef]
123. De Rosa, C.; Ruiz de Ballesteros, O.; Auriemma, F.; Talarico, G.; Scoti, M.; Di Girolamo, R.; Malafronte, A.; Piemontesi, F.; Liguori, D.; Camurati, I.; et al. Crystallization Behavior of Copolymers of Isotactic Poly(1-butene) with Ethylene from Ziegler-Natta Catalyst: Evidence of the Blocky Molecular Structure. *Macromolecules* **2019**, *52*, 9114–9127. [CrossRef]
124. De Rosa, C.; Ruiz de Ballesteros, O.; Di Girolamo, R.; Malafronte, A.; Auriemma, F.; Talarico, G.; Scoti, M. The blocky structure of Ziegler-Natta "random" copolymers: Myths and experimental evidence. *Polym. Chem.* **2020**, *11*, 34–38. [CrossRef]
125. Di Girolamo, R.; Santillo, C.; Malafronte, A.; Scoti, M.; De Stefano, F.; Talarico, G.; Coates, G.W.; De Rosa, C. Structure and morphology of isotactic polypropylene-polyethylene block copolymers prepared with living and stereoselective catalyst. *Polym. Chem.* **2022**, *13*, 2950–2963. [CrossRef]
126. De Rosa, C.; Di Girolamo, R.; Auriemma, F.; Talarico, G.; Scarica, C.; Malafronte, A.; Scoti, M. Controlling Size and Orientation of Lamellar Microdomains in Crystalline Block Copolymers. *ACS Appl. Mater. Interfaces* **2017**, *9*, 31252–31259. [CrossRef] [PubMed]
127. De Rosa, C.; Di Girolamo, R.; Malafronte, A.; Scoti, M.; Talarico, G.; Auriemma, F.; Ruiz de Ballesteros, O. Polyolefins based crystalline block copolymers: Ordered nanostructures from control of crystallization. *Polymer* **2020**, *196*, 122423. [CrossRef]
128. De Rosa, C.; Malafronte, A.; Di Girolamo, R.; Auriemma, F.; Scoti, M.; Ruiz de Ballesteros, O.; Coates, G.W. Morphology of Isotactic Polypropylene-Polyethylene Block Copolymers Driven by Controlled Crystallization. *Macromolecules* **2020**, *53*, 10234–10244. [CrossRef]
129. De Rosa, C.; Di Girolamo, R.; Cicolella, A.; Talarico, G.; Scoti, M. Double Crystallization and Phase Separation in Polyethylene-Syndiotactic Polypropylene Di-Block Copolymers. *Polymers* **2021**, *13*, 2589. [CrossRef]
130. Di Girolamo, R.; Cicolella, A.; Talarico, G.; Scoti, M.; De Stefano, F.; Giordano, A.; Malafronte, A.; De Rosa, C. Structure and Morphology of Crystalline Syndiotactic Polypropylene-Polyethylene Block Copolymers. *Polymers* **2022**, *14*, 1534. [CrossRef]
131. Auriemma, F.; De Rosa, C.; Scoti, M.; Di Girolamo, R.; Malafronte, A.; Talarico, G.; Carnahan, E. Unveiling the Molecular Structure of Ethylene/1-Octene Multi-block Copolymers from Chain Shuttling Technology. *Polymer* **2018**, *154*, 298–304. [CrossRef]
132. Auriemma, F.; De Rosa, C.; Scoti, M.; Di Girolamo, R.; Malafronte, A.; Galotto Galotto, N. Structural Investigation at Nanometric Length Scale of Ethylene/1-Octene Multi-block Copolymers from Chain Shuttling Technology. *Macromolecules* **2018**, *51*, 9613. [CrossRef]

133. Auriemma, F.; De Rosa, C.; Scoti, M.; Di Girolamo, R.; Malafronte, A.; D'Alterio, M.C.; Boggioni, L.; Losio, S.; Boccia, A.C.; Tritto, I. Structure and Mechanical Properties of Ethylene/1-Octene Multi-block Copolymers from Chain Shuttling Technology. *Macromolecules* **2019**, *52*, 2669. [CrossRef]
134. Spaleck, W.; Kuber, F.; Winter, A.; Rohrmann, J.; Bachmann, B.; Antberg, M.; Dolle, V.; Paulus, E. The influence of aromatic substituents on the polymerization behavior of bridged zirconocene catalysts. *Organometallics* **1994**, *13*, 954–963. [CrossRef]
135. Kissin, Y.V.; Brandolini, A.J. 13C NMR spectra of propylene/1-hexene copolymers. *Macromolecules* **1991**, *24*, 2632–2633. [CrossRef]
136. Carman, C.J.; Harrington, R.A.; Wilkes, C.E. Monomer Sequence Distribution in Ethylene-Propylene Rubber Measured by 13C NMR. 3. Use of Reaction Probability Model. *Macromolecules* **1977**, *10*, 536–544. [CrossRef]

Article

Melt-Crystallizations of α and γ Forms of Isotactic Polypropylene in Propene-Butene Copolymers

Miriam Scoti *, Fabio De Stefano, Filomena Piscitelli †, Giovanni Talarico, Angelo Giordano and Claudio De Rosa *

Dipartimento di Scienze Chimiche, Università di Napoli Federico II, Complesso Monte S. Angelo, Via Cintia, 80126 Napoli, Italy
* Correspondence: miriam.scoti@unina.it (M.S.); claudio.derosa@unina.it (C.D.R.)
† Current address: Department of Materials and Structures, CIRA—Italian Aerospace Research Centre, Via Maiorise, 81043 Capua, Italy.

Abstract: Random isotactic propene-butene copolymers (iPPC4) of different stereoregularity have been synthesized with three different homogeneous single center metallocene catalysts having different stereoselectivity. All samples crystallize from the polymerization solution in mixtures of α and γ forms, and the relative amount of γ form increases with increasing concentrations of butene and of rr stereodefects. All samples crystallize from the melt in mixtures of α and γ forms and the fraction of γ form increases with decreasing cooling rate. At high cooling rates, the crystallization of the α form is always favored, even for samples that contain high total concentration of defects that should crystallize in the γ form. The results demonstrate that in iPPs containing significant concentrations of defects, such as stereodefects and comonomeric units, the γ form is the thermodynamically stable form of iPP and crystallizes in selective conditions of very slow crystallization, whereas the α form is the kinetically favored form and crystallizes in conditions of fast crystallization.

Keywords: propylene-butene copolymers; α and γ forms; metallocene catalysts; melt-crystallization

Citation: Scoti, M.; De Stefano, F.; Piscitelli, F.; Talarico, G.; Giordano, A.; De Rosa, C. Melt-Crystallizations of α and γ Forms of Isotactic Polypropylene in Propene-Butene Copolymers. *Polymers* **2022**, *14*, 3873. https://doi.org/10.3390/polym14183873

Academic Editors: Nikolaos Politakos and Apostolos Avgeropoulos

Received: 27 August 2022
Accepted: 13 September 2022
Published: 16 September 2022

Publisher's Note: MDPI stays neutral with regard to jurisdictional claims in published maps and institutional affiliations.

Copyright: © 2022 by the authors. Licensee MDPI, Basel, Switzerland. This article is an open access article distributed under the terms and conditions of the Creative Commons Attribution (CC BY) license (https://creativecommons.org/licenses/by/4.0/).

1. Introduction

The crystallization of α and γ forms of isotactic polypropylene (iPP) depends on the conditions of crystallization and on the molecular structure of polypropylene; the latter depends on the polymerization conditions and catalysis [1]. Different conditions of polymerization and used catalysts produce iPP macromolecules characterized by different molecular structures, because different catalysts may introduce different types and amounts of microstructural defects, such as defects of stereoregularity and regioregularity, constitutional defects and different distribution of defects along the chains [1–5].

The α form is considered the most stable form of iPP and crystallizes usually in the iPP homopoymer prepared with heterogeneous Ziegler-Natta catalysts in common crystallization conditions from the melt or from solution, and in stretched fibers [6–8]. In the same commercial iPP samples the γ form crystallizes only in special conditions, as in samples of low molecular mass [9–13] and by crystallization at high pressures [14–18]. A low amount of γ form has also been obtained in some copolymers of propylene with different comonomers synthesized with the same heterogeneous catalysts [19–47]. The γ form crystallizes, instead, easily in iPP samples [48–54] and its copolymers [55–92] synthesized with homogeneous single site metallocene catalysts [2–5], which introduce different kinds of defects depending on the catalyst structure. These defects, as stereo-defects, regio-defects and also comonomers, indeed, favors crystallization of the γ form [48–92].

The crystallization of the γ form in chains containing defects is due to the fact that the γ form crystallizes when iPP chains are characterized by short regular propene sequences, therefore, it occurs when iPP chains contain any type of defect that interrupts and shortens the regular propene sequences [51,54,62,79]. In particular, iPPs and copolymers produced with homogeneous single site metallocene catalysts are characterized by a perfectly random distribution of defects along the macromolecules, which, therefore, shortens the length of the regular propene sequences even for low concentrations of defects [51,54,62,79].

The crystallization of α form is, instead, favored when the regular propene sequences are very long, which are generated when the concentration of defects is low or when defects are segregated in blocks of the macromolecules [51,54,62,79], as, for instance, in samples synthesized with Ziegler-Natta catalysts, where the non-random distribution of defects and their segregation in blocks make the regular propene sequences always long even in the case of a high concentration of defects and in copolymers with relatively high comonomer concentration [93–96].

Understanding the conditions of crystallization of α and γ forms of iPP is of particular relevance since the two different polymorphs exhibit significant differences in mechanical behavior [7,46,47,54,76,77,79,87–91,97–101]. In general, the molecular architecture and topology of copolymers, from standard random to block or multiblock copolymers, greatly affects the crystallization behavior and properties of polymers because of the different length of crystallizable sequences [93–96,102,103]. While the effect of the molecular structure and architecture on the crystallization behavior of iPP has been extensively investigated and the effect of different kinds of defects on the crystallization of α and γ forms has been clarified, the effect of the conditions of crystallization is still unclear.

In this paper we report a study of the crystallization of α and γ forms of iPP in propene-butene copolymers in different crystallization conditions to analyze the thermodynamic and kinetics effects on the crystallization of the two polymorphic forms. Propene-butene copolymers of different stereoregularity synthesized with different metallocene catalysts have been chosen for this study because they crystallize easily in the γ form, thanks to the incorporation in the iPP chains of butene comonomeric units that shorten the length of the regular propene sequences [79]. Hence, in this system, the γ form is the thermodynamically stable form of iPP and, therefore, these copolymers represent the ideal system to study the crystallization conditions that may favor crystallization of the α form.

2. Materials and Methods

Propylene-butene isotactic copolymers (iPPC4) were synthesized with the metallocene catalysts of Scheme 1 having different stereoselectivity activated with methylalumoxane (MAO) (from Lanxess, Cologne, Germany) [79]. All operations were performed under nitrogen by using conventional Schlenk-line techniques. Toluene solvent was purified by degassing with N_2 and passing over activated Al_2O_3 (8 h, N_2 purge, 300 °C), and stored under nitrogen. The MAO cocatalyst was used as received (10 wt.%/vol. toluene solution, 1.7 M in Al). The catalyst mixture was prepared by dissolving the desired amount of the metallocene with the proper amount of the MAO solution, obtaining a solution which was stirred for 10 min at 25 °C before being injected into the reactor. All copolymerizations were run at 25 °C in a 250 mL Pyrex reactor, agitated with magnetic stirrer, containing toluene (100 mL) and MAO (2.0 mL). Gas mixtures of propene and 1-butene at the appropriate composition, prepared with vacuum line techniques in a gas cylinder at pressure of 4–5 bar and standardized by gas chromatography, were bubbled through the liquid phase at atmospheric pressure and a flow rate of 0.3 L/min. The polymerizations started by syringing in the toluene solution of the catalyst (2–3 mg) and proceeded under a constant flow of the gas mixture. Under such conditions, total monomer conversions were lower than 15%, this ensuring a nearly constant feeding ratio. The copolymers were coagulated with excess methanol acidified with enough HCl (aqueous, concentrated) to prevent the precipitation of alumina from MAO hydrolysis, filtered, washed with further methanol, and vacuum-dried. Typical yields were 2–5 g with a 120 min reaction time.

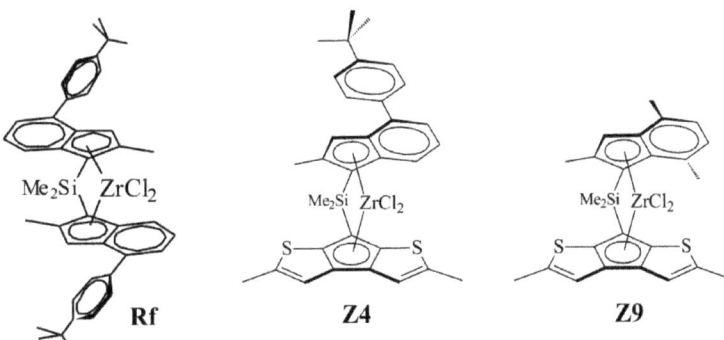

Scheme 1. Structures of the isoselective C2-symmetric (Rf) and less isoselective C1-symmetric (Z4 and Z9) metallocene catalysts used for the synthesis of iPPC4 copolymers.

The C_2-symmetric metallocene Rf is highly isospecific [104,105] and produces highly isotactic iPPC4 copolymers (samples iPPC4Rf-x, where x is the butene concentration) containing negligible amounts of stereodefects (lower than 0.1 mol% of *rr* triads) and small amount of regiodefects around 0.1–0.5 mol%, represented by secondary 2,1-erythro units (2,1e). The two C_1-symmetric metallocenes Z4 and Z9 are instead fully regioselective but introduce significant amounts of *rr* stereodefects [106–109]. Samples of iPPC4 copolymers synthesized with the catalyst Z4 (sample iPPC4Z4-x) are fully regioregular and contain 2.0–2.4 mol% of *rr* stereodefects, whereas samples synthesized with the catalyst Z9 (samples iPPC4Z9-x) contain 2.5–3.4 mol% of *rr* defects [107,108]. Consequently, the less isotactic samples synthesized with catalysts Z4 and Z9 show melting temperatures lower than those of the samples synthesized with the catalyst Rf [79]. The composition, the melting temperatures and the molecular mass of all samples are shown in Table 1.

Table 1. Catalyst, composition (mol% butene), melting temperature of as-prepared samples (T_m), molecular mass (M_w) and dispersity (M_w/M_n) of the iPPC4 copolymers [79].

Sample	Catalyst	mol% Butene	T_m (°C) [a]	M_w [b]	M_w/M_n [b]
iPPC4Rf-1	Rf	1.9	143	316,500	2.2
iPPC4Rf-2	Rf	4.3	137	228,700	2.1
iPPC4Rf-3	Rf	4.5	137	207,000	2.0
iPPC4Rf-4	Rf	8.0	125	178,500	2.0
iPPC4Rf-5	Rf	9.0	120	200,000	2.1
iPPC4Z4-1	Z4	1.3	135	172,900	2.1
iPPC4Z4-2	Z4	4.6	123	175,700	2.0
iPPC4Z4-3	Z4	8.2	112	176,700	2.0
iPPC4Z9-1	Z9	1.4	126	214,000	2.1
iPPC4Z9-2	Z9	2.2	124	214,500	2.0
iPPC4Z9-3	Z9	6.4	113	214,400	2.0

[a] Determined from DSC heating curves recorded at 10 °C/min. [b] Determined by GPC.

The composition and comonomer distribution were determined by ^{13}C NMR analysis (Table 1). All spectra were obtained using a Bruker DPX-400 spectrometer operating in the Fourier transform mode at 120 °C at 100.61 MHz (Bruker Company, Billerica, MA, USA). The samples were dissolved with 8% wt/v concentration in 1,1,2,2-tetrachloroethane-d$_2$ at 120 °C. The carbon spectra were acquired with a 90° pulse and 15 s of delay between pulses and CPD (WALTZ 16) to remove ^{1}H-^{13}C coupling. About 1500–3000 transients were stored in 32K data points using a spectral window of 6000 Hz. For all copolymer samples, the peak of the propylene methine carbon atoms was used as internal reference at 28.83 ppm. The ^{13}C NMR spectra of two samples of iPPC4 copolymers are reported in Figure S1 of the Supplementary Material. The resonances were assigned according to ref. [110] and the butene concentrations in the copolymers were evaluated from the concentrations of the constitutional diads PP, PB, BB (P = propene, B = butene), using the Equations S1–S5 of the Supplementary Material. The NMR analysis showed that all the copolymers present a random distribution of comonomers and homogeneous intermolecular composition with $r_P \times r_B \approx 1$, calculated using the Equation S6 of the Supplementary Material, according to ref. [111].

The molecular masses and the dispersity were determined by gel permeation chromatography (GPC), using a Polymer Laboratories GPC220 apparatus equipped with a differential refractive index (RI) detector and a Viscotek 220R viscometer (Agilent Company, Santa Clara, CA, USA), on polymer solutions in 1,2,4-trichlorobenzene at 135 °C of 2 mg/mL concentration. The injection volume was 300 µL with a flow rate of 1.0 mL/min. The GPC apparatus was calibrated with 12 standard samples of polystyrene having narrow dispersity and molecular masses in the range 580 and 13.2×10^6.

The calorimetry measurements were carried out with differential scanning calorimeter (DSC) Mettler Toledo DSC-822 (Columbus, OH, USA) performing scans in a flowing N_2 atmosphere and scanning rate of 10 °C/min.

X-ray powder diffraction profiles were recorded with Ni filtered Cu Kα radiation by using an Empyrean diffractometer (Malvern Panalytical, Worcestershire, UK).

All samples of iPPC4 copolymers were crystallized in DSC by cooling the melt at 180 °C down to 25 °C at different cooling rates from 1 to 40 °C/min. After the crystallization in DSC, the samples were analyzed by X-ray diffraction.

In samples that crystallized in mixtures of α and γ forms, the relative fraction of the γ form f_γ, with respect to the α form, was calculated from the intensities of the $(117)_\gamma$ and $(130)_\alpha$ reflections at $2\theta = 20.1°$ of the γ form and at $2\theta = 18.6°$ of the α form, respectively, as the ratio: $f_\gamma = I(117)_\gamma / [I(117)_\gamma + I(130)_\alpha]$.

3. Results and Discussion

The X-ray diffraction profiles of as-prepared (precipitated from the polymerization solution) samples of iPPC4 copolymers of Table 1 are reported in Figure 1. The diffraction profiles present the $(130)_\alpha$ and $(117)_\gamma$ reflections at $2\theta = 18.6$ and $20.1°$ of the α and γ forms, respectively, indicating that all as-prepared samples of iPPC4 copolymers are crystallized in mixtures of α and γ forms. The intensity of the $(117)_\gamma$ reflection increases with increasing butene concentration and, at the same butene content, is higher in the low stereoregular samples synthesized with the catalysts Z4 and Z9 (samples iPPC4Z4-x and iPPC4Z9-x, Figure 1B,C). The values of the relative amount of γ form (f_γ) calculated from the intensities of the $(117)_\gamma$ and $(130)_\alpha$ reflections in the diffraction profiles of Figure 1 are reported in Figure 2 as a function of butene concentration. Since the highly stereoregular iPP homopolymer generally crystallizes in the α form, it has been assumed $f_\gamma = 0$ for butene concentration equal to zero. It is apparent that the amount of γ form increases with increasing concentration of butene and of rr stereodefects [79] (Figure 2). The more isotactic samples iPPC4Rf-x crystallize almost in the pure γ form ($f_\gamma \approx 80\%$) for butene concentration of nearly 9 mol% (sample iPPC4Rf-5, profile e in Figures 1A and 2), whereas for the less isotactic samples iPPC4Z9-x, the highest amount of γ form ($f_\gamma \approx 88\%$) crystallizes at about 6 mol% of butene (sample iPPC4Z9-3, profile c in Figures 1C and 2).

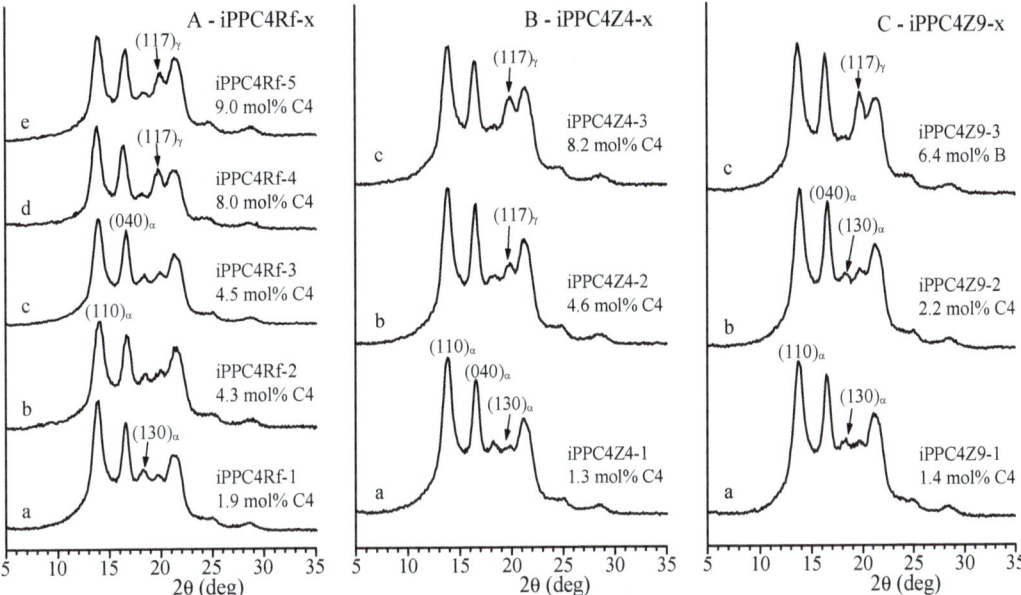

Figure 1. X-ray powder diffraction profiles of as-prepared samples of iPPC4 copolymers of the indicated butene concentration. Isotactic samples iPPC4Rf-x with [rr] < 0.1 mol% (**A**) and less isotactic samples iPPC4Z4-x with [rr] = 2.0–2.4 mol% (**B**) and iPPC4Z9-x with [rr] = 2.5–3.4 mol% (**C**).

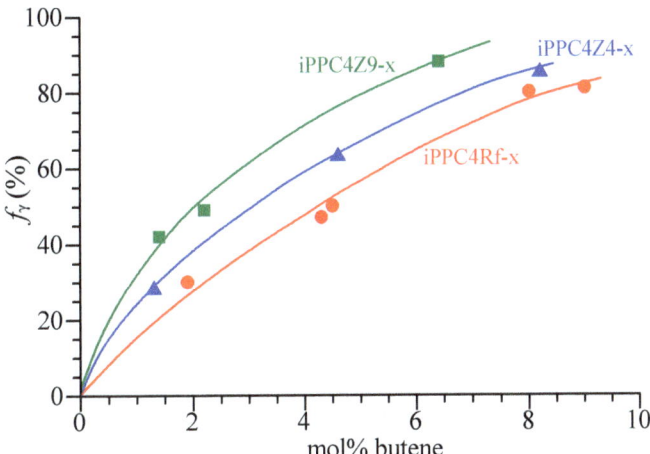

Figure 2. Values of the fraction of γ form that crystallizes in the as-prepared samples of iPPC4 copolymers as a function of butene concentration evaluated from the diffraction profiles of Figure 1. Isotactic samples iPPC4Rf-x with [rr] < 0.1 mol% (●) and less isotactic samples iPPC4Z4-x with [rr] = 2.0–2.4 mol% (▲) and iPPC4Z9-x with [rr] = 2.5–3.4 mol% (■).

It is worth reminding that in iPPC4 copolymers the further increase of butene concentration induces decrease of the amount of γ form and crystallization of the α form for concentrations higher than 15–20 mol% because butene units are included easily in the crystals of α form producing stabilization of the α form compared to the γ form [79].

The DSC heating curves of the as-polymerized samples of iPPC4 copolymers are reported in Figure 3. For the three sets of samples the melting temperature decreases with increasing butene concentration. The values of the melting temperature are reported in Table 1 and in Figure 4A as a function of butene concentration.

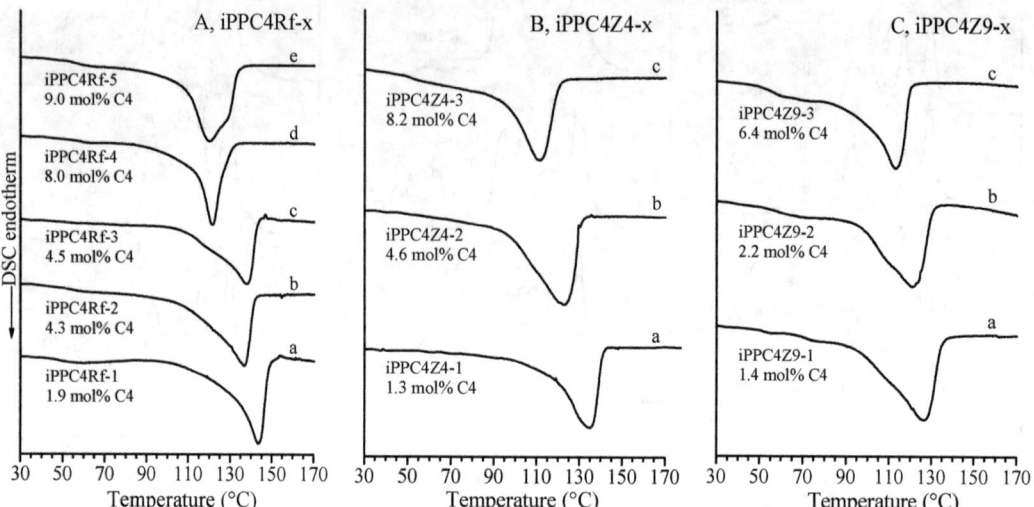

Figure 3. DSC heating curves recorded at 10 °C/min of as-prepared samples of iPPC4 copolymers of the indicated butene concentration. Isotactic samples iPPC4Rf-x with [rr] < 0.1 mol% (**A**) and less isotactic samples iPPC4Z4-x with [rr] = 2.0–2.4 mol% (**B**) and iPPC4Z9-x with [rr] = 2.5–3.4 mol% (**C**).

It is apparent from Figure 4A that the melting temperature also depends on the stereoregularity of the samples and on the concentration of *rr* defects. In fact, at the same butene concentration the more isotactic samples iPPC4Rf-x show melting temperatures higher than those of the less isotactic samples iPPC4Z4-x and iPPC4Z9-x, and the samples iPPC4Z9-x with the highest concentration of *rr* stereodefects show the lowest melting temperatures.

All samples have been crystallized in DSC from the melt by cooling at different cooling rates. The samples have been melted by heating at 10 °C/min up to 170 °C, as in Figure 3, and then cooled from 170 °C down to 25 °C at different cooling rates, from 1 °C/min to 40 °C/min. As an example, the DSC cooling curves of all samples recorded at cooling rate of 10 °C/min are reported in Figure 5. All samples crystallize during cooling and the DSC curves of Figure 5 show well-defined exothermic peaks. The values of the crystallization temperature evaluated from the DSC cooling curves of Figure 5 at cooling rate of 10 °C/min are plotted in Figure 4B as a function of butene concentration. For the three sets of samples, the crystallization temperature decreases with increasing butene concentration and, at the same butene concentration, decreasing the stereoregularity (Figure 4B). Therefore, introduction of both butene co-units and *rr* stereodefects produces a decrease of melting and crystallization temperatures (Figure 4) and an increase of the fraction of γ form (Figure 2).

After the crystallization in DSC as in Figure 5, the samples were analyzed by X-ray diffraction at 25 °C. The diffraction profiles of the samples iPPC4Rf-x, iPPC4Z4-x and iPPC4Z9-x crystallized from the melt at different cooling rates are reported in Figures 6–8, respectively. All samples crystallize into mixtures of α and γ forms and the amounts of the two forms strongly depend on the cooling rate.

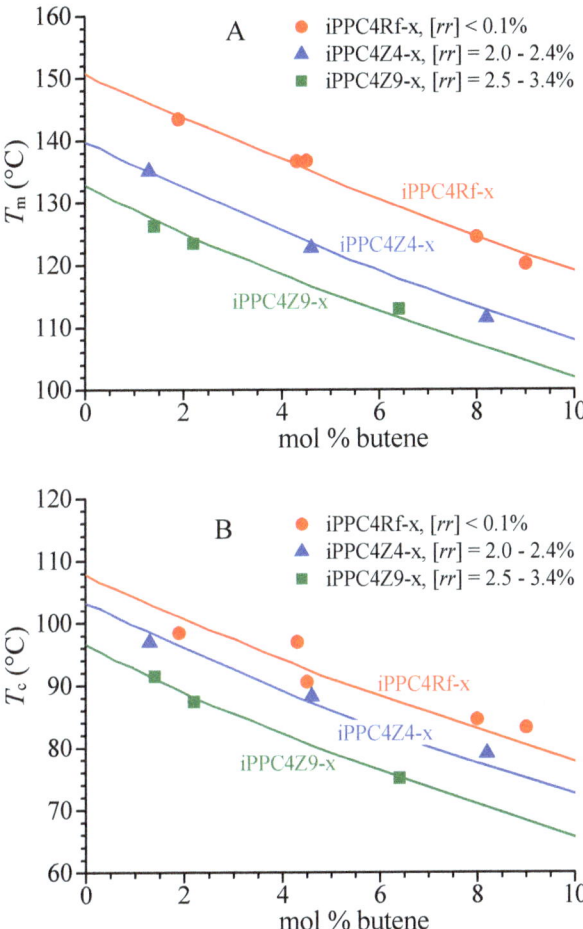

Figure 4. Melting temperature of the as-prepared samples (**A**) and crystallization temperature from the melt (**B**) of iPPC4 copolymers as a function of butene concentration evaluated from DSC thermograms recorded at scanning rates of 10 °C/min. Isotactic samples iPPC4Rf-x with [rr] < 0.1 mol% (●) and less isotactic samples iPPC4Z4-x with [rr] = 2.0–2.4 mol% (▲) and iPPC4Z9-x with [rr] = 2.5–3.4 mol% (■).

The samples displayed the same behavior regardless of the butene concentration and stereoregularity, that is, the intensity of the $(130)_\alpha$ reflection of the α form at $2\theta = 18.6°$ decreases and the intensity of the $(117)_\gamma$ reflection at $2\theta = 20.1°$ of the γ form increases with decreasing cooling rate. The amount of γ form is reported in Figure 9 as a function of the cooling rate for the three sets of samples. These data indicate that the amount of γ form increases with decreasing cooling rate. Correspondingly, the γ form almost disappears and the almost pure α form crystallizes in all samples at the highest cooling rate of 40 °C/min (profiles e of Figures 6–8). The highest concentration of γ form is always obtained at low cooling rates, whereas at high cooling rates the crystallization of the α form is always favored, even for samples that contain high total concentration of defects (high butene and high rr defects concentrations) that tend to crystallize normally in the γ form. In fact, even the less isotactic samples iPPC4Z4-3 and iPPC4Z9-3 with the highest butene concentrations of 8.2 and 6.4 mol%, respectively, that crystallize in the as-prepared samples in the almost pure γ form (f_γ = 85 and 88%, respectively) (profiles c of Figures 1B,C and 2), crystallize in the α form in the fast crystallization from the melt at high cooling rate (profiles e of Figures 7 and 8).

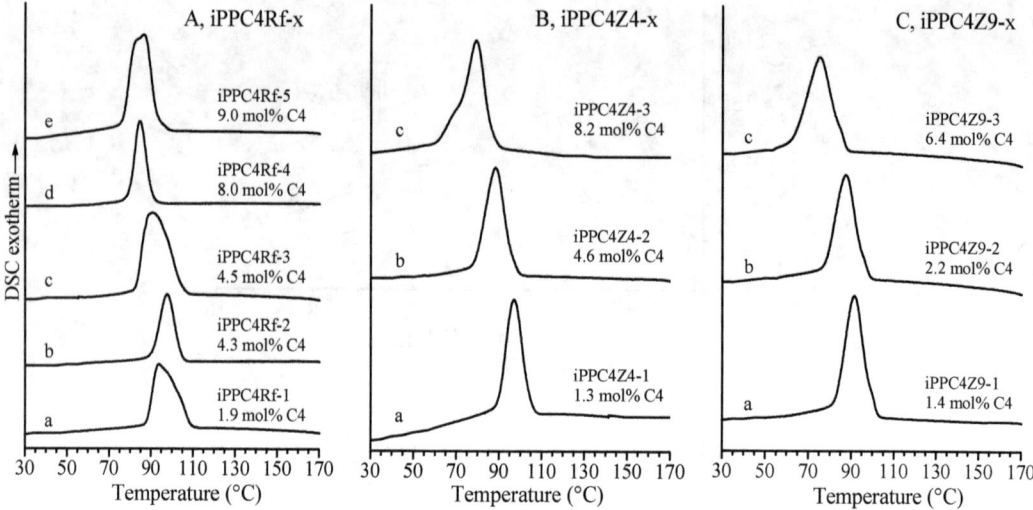

Figure 5. DSC cooling curves recorded at 10 °C/min of samples of iPPC4 copolymers of the indicated butene concentration after melting at 170 °C (Figure 3). Isotactic samples iPPC4Rf-x with [rr] < 0.1 mol% (**A**) and less isotactic samples iPPC4Z4-x with [rr] = 2.0–2.4 mol% (**B**) and iPPC4Z9-x with [rr] = 2.5–3.4 mol% (**C**).

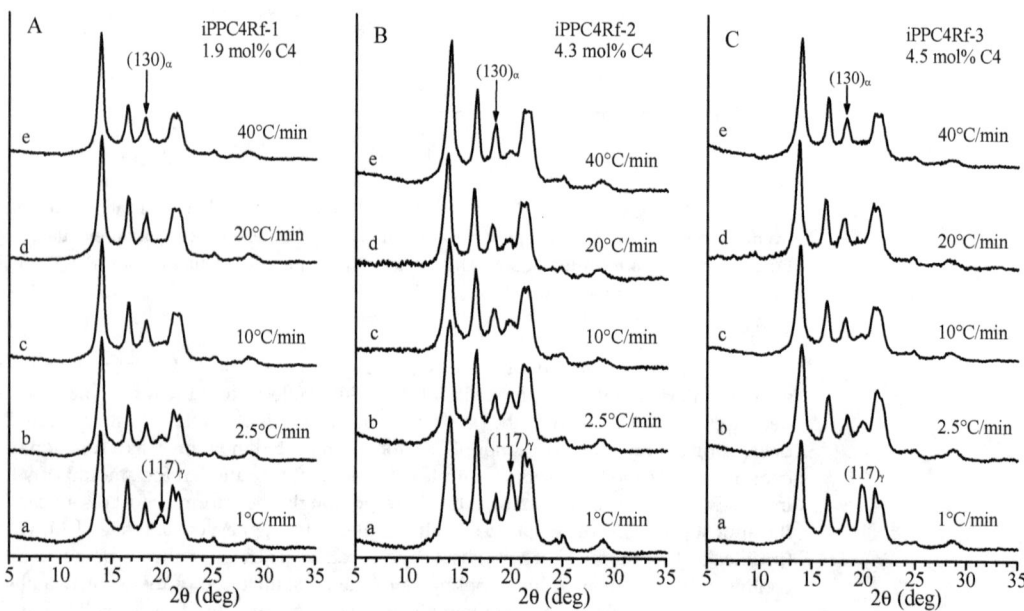

Figure 6. Cont.

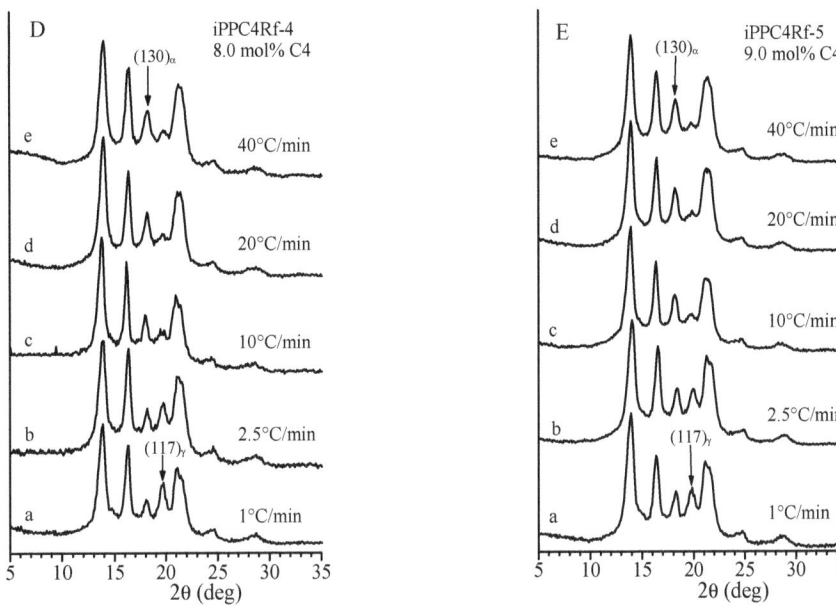

Figure 6. X-ray diffraction profiles of samples of copolymers iPPC4Rf-x crystallized from the melt by cooling the melt from 170 °C down to 25 °C at the indicated different cooling rates.

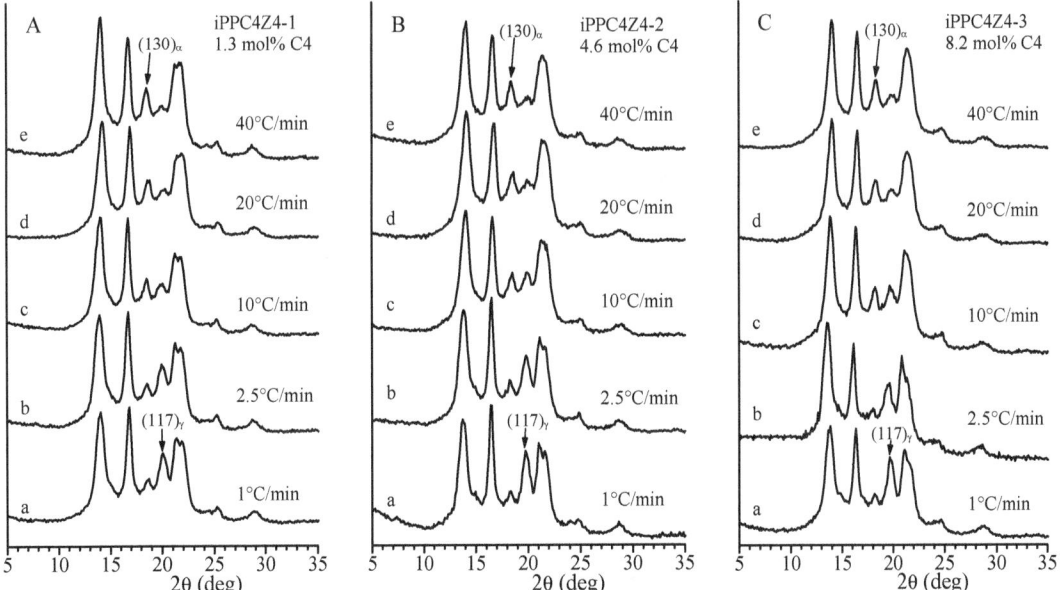

Figure 7. X-ray diffraction profiles of samples of copolymers iPPC4Z4-x crystallized from the melt by cooling the melt from 170 °C down to 25 °C at the indicated different cooling rates.

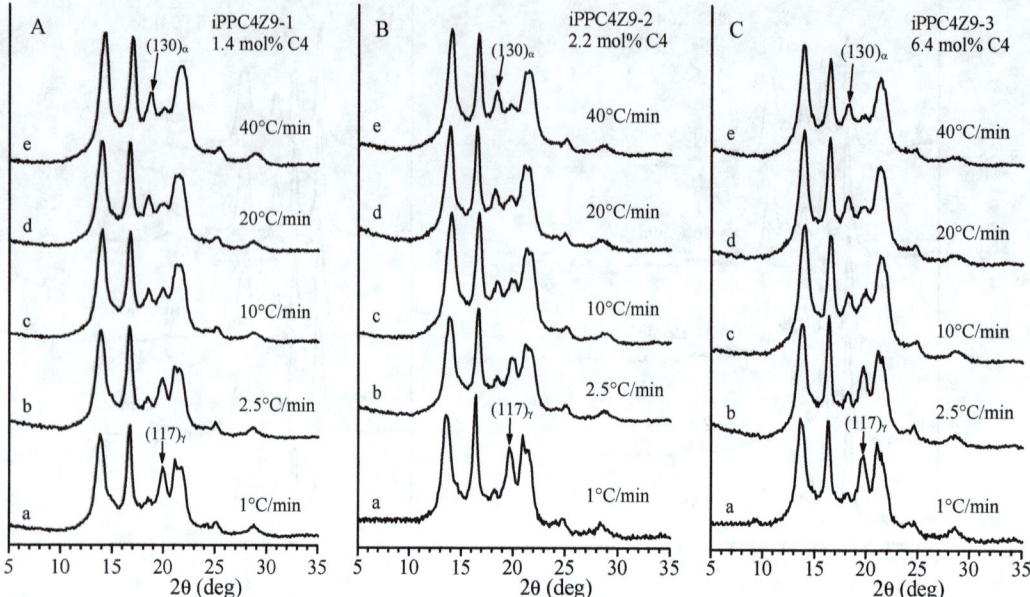

Figure 8. X-ray diffraction profiles of samples of copolymers iPPC4Z9-x crystallized from the melt by cooling the melt from 170 °C down to 25 °C at the indicated different cooling rates.

These results demonstrate that in iPP samples containing significant concentrations of defects, such as stereodefects and butene comonomeric units, the γ form is the thermodynamically stable form of iPP and crystallizes normally from polymer solution or in selective conditions of very slow crystallization, as isothermal crystallizations from the melt at high crystallization temperatures [79] or by cooling from the melt at very low cooling rates, whereas the α form is the kinetically favored form and crystallizes in conditions of fast crystallization, such as fast cooling from the melt.

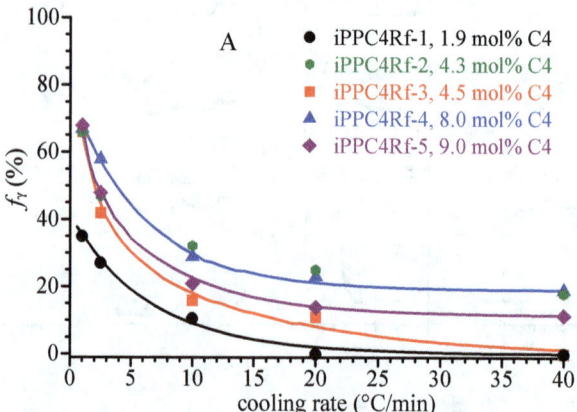

Figure 9. *Cont.*

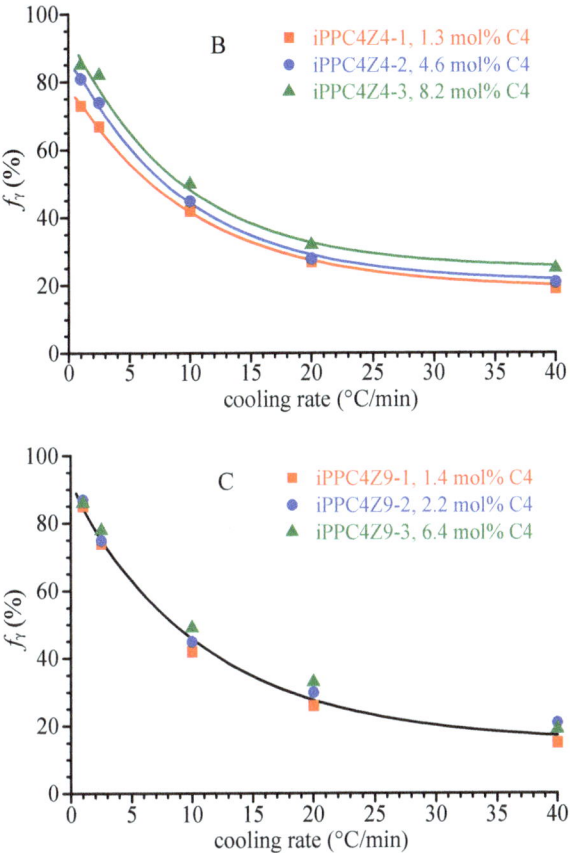

Figure 9. Values of the fraction of γ form that crystallizes in samples of iPPC4 copolymers cooled from the melt at different cooling rates as a function of the cooling rate, evaluated from the diffraction profiles of Figures 6–8. (**A**) Isotactic samples iPPC4Rf-x with [rr] < 0.1 mol%, (**B**) samples iPPC4Z4-x with [rr] = 2.0–2.4 mol%, and (**C**) samples iPPC4Z9-x with [rr] = 2.5–3.4 mol%.

4. Conclusions

Random isotactic propene-butene copolymers of different stereoregularity have been prepared with three different metallocene catalysts having different stereoselectivity. The C_2-symmetric metallocene produces highly isotactic iPPC4 copolymers containing negligible amounts of stereodefects, whereas the two C_1-symmetric metallocenes produce less isotactic copolymers containing on average 2.0–2.4 mol% and 2.5–3.4 mol% of rr stereodefects, respectively.

All as-prepared samples crystallize in mixtures of α and γ forms from the polymerization solution and the relative amount of γ form increases with increasing concentrations of butene and of rr stereodefects.

The samples have been crystallized from the melt by cooling the melt from 170 °C down to 25 °C at different cooling rates. All samples crystallize from the melt in mixtures of α and γ forms and the amount of the two forms strongly depends on the cooling rate. The amount of γ form increases with decreasing cooling rate. Correspondingly, the γ form almost disappears and the almost pure α form crystallizes in all samples at the highest cooling rate of 40 °C/min. The highest concentration of γ form is always obtained at low cooling rates, whereas at high cooling rates the crystallization of the α form is always favored, even for samples that contain high total concentration of defects that should crystallize in the γ form.

These results demonstrate that in iPP samples containing significant concentrations of defects, such as stereodefects and butene comonomeric units, the γ form is the thermodynamically stable form of iPP and crystallizes normally from polymer solution or in selective conditions of very slow crystallization, such as by cooling from the melt at very low cooling rates, whereas the α form is the kinetically favored form and crystallizes in conditions of fast crystallization, such as fast cooling from the melt.

Supplementary Materials: The following supporting information can be downloaded at: https://www.mdpi.com/article/10.3390/polym14183873/s1, Figure S1: ^{13}C NMR spectra of the iPPC4 copolymer samples iPPC4Rf-1 with 1.9 mol% of butene and iPPC4Rf-5 with 9.0 mol% of butene; Equations (S1)–(S5) used for the calculation of the concentrations of the constitutional diads PP, PP and BB (P = propene, B = butene) and the concentration of the comonomeric units from the ^{13}CNMR data; Equation S6 used for the calculation of the product of reactivity ratios $r_P \times r_B$. References [110,111] are cited in the supplementary materials.

Author Contributions: C.D.R. conceived the experiments; G.T. and F.D.S. synthesized the samples; F.P., A.G. and M.S. performed the experiments. All authors have read and agreed to the published version of the manuscript.

Funding: This research received no external funding.

Institutional Review Board Statement: Not applicable.

Informed Consent Statement: Not applicable.

Data Availability Statement: The data in this study are available on reasonable request from the corresponding author.

Acknowledgments: The task force "Polymers and biopolymers" of the University of Napoli Federico II is acknowledged.

Conflicts of Interest: The authors declare no conflict of interest.

References

1. Pasquini, N. (Ed.) *Polypropylene Handbook*; Hanser Publishers: Munich, Germany, 2005.
2. Ewen, J.A. Mechanisms of stereochemical control in propylene polymerizations with soluble Group 4B metallocene/methylalumoxane catalysts. *J. Am. Chem. Soc.* **1984**, *106*, 6355. [CrossRef]
3. Kaminsky, W.; Kulper, K.; Brintzinger, H.H.; Wild, F.R.W. Polymerization of propene and butene with a chiral zirconocene and methylaluminoxane as cocatalyst. *Angew. Chem.* **1985**, *97*, 507. [CrossRef]
4. Brintzinger, H.H.; Fischer, D.; Mulhaupt, R.; Rieger, B.; Waymouth, R.M. Stereospecific Olefin Polymerization with Chiral Metallocene Catalysts. *Angew. Chem. Int. Ed. Engl.* **1995**, *34*, 1143. [CrossRef]
5. Resconi, L.; Cavallo, L.; Fait, A.; Piemontesi, F. Selectivity in Propene Polymerization with Metallocene Catalysts. *Chem. Rev.* **2000**, *100*, 1253. [CrossRef]
6. Brückner, S.; Meille, S.V.; Petraccone, V.; Pirozzi, B. Polymorphism in Isotactic Polypropylene. *Prog. Polym. Sci.* **1991**, *16*, 361–404. [CrossRef]
7. De Rosa, C.; Scoti, M.; Di Girolamo, R.; Ruiz de Ballesteros, O.; Auriemma, F.; Malafronte, A. Polymorphism in polymers: A tool to tailor material's properties. *Polym. Cryst.* **2020**, *3*, e10101. [CrossRef]
8. Natta, G.; Corradini, P. Structure and properties of isotactic polypropylene. *Nuovo Cim. Suppl.* **1960**, *15*, 40. [CrossRef]
9. Lotz, B.; Graff, S.; Wittmann, J.C. Crystal morphology of the γ (triclinic) phase of isotactic polypropylene and its relation to the α phase. *J. Polym. Sci. Polym. Phys. Ed.* **1986**, *24*, 2017. [CrossRef]
10. Kojima, M. Solution-γ grown lamellar crystals of thermally decomposed isotactic polypropylene. *J. Polym. Sci. Part B: Polymer Letters* **1967**, *5*, 245. [CrossRef]
11. Kojima, M. Morphology of polypropylene crystals. III. Lamellar crystals of thermally decomposed polypropylene. *J. Polym. Sci. Part B Polym. Phys.* **1968**, *6*, 1255. [CrossRef]
12. Morrow, D.R.; Newman, B.A. Crystallization of low-molecular-weight polypropylene fractions. *J. Appl. Phys.* **1968**, *39*, 4944. [CrossRef]
13. Natta, G.; Mazzanti, G.; Crespi, G.; Moraglio, G. Polimeri isotattici e polimeri a stereoblocchi del propilene. *Chim. Ind. Milan* **1957**, *39*, 275.

14. Kardos, J.L.; Christiansen, A.W.; Baer, E. Structure of pressure crystallized polypropylene. *J. Polym. Sci. Part B Polym. Phys.* **1966**, *4*, 777. [CrossRef]
15. Pal, K.D.; Morrow, D.R.; Sauer, J.A. Interior morphology of bulk polypropylene. *Nature* **1966**, *211*, 514.
16. Mezghani, K.; Phillips, P.J. The γ-phase of high molecular weight isotactic polypropylene: III. The equilibrium melting point and the phase diagram. *Polymer* **1998**, *39*, 3735.
17. Mezghani, K.; Phillips, P.J. The γ-phase of high molecular weight isotactic polypropylene. II. The morphology of the γ-form crystallized at 200 MPa. *Polymer* **1997**, *38*, 5725.
18. Brückner, S.; Phillips, P.J.; Mezghani, K.; Meille, S.V. On the crystallization of γ-isotactic polypropylene. A high pressure study. *Macromol. Rapid Commun.* **1997**, *18*, 1. [CrossRef]
19. Turner-Jones, A. Development of the γ-crystal form in random copolymers of propylene and their analysis by DSC and X-ray methods. *Polymer* **1971**, *12*, 487–508. [CrossRef]
20. Cimmino, S.; Martuscelli, E.; Nicolais, L.; Silvestre, C. Thermal and mechanical properties of isotactic random propylene-butene-1 copolymers. *Polymer* **1978**, *19*, 1222. [CrossRef]
21. Crispino, L.; Martuscelli, E.; Pracella, M. Influence of composition on the melt crystallization of isotactic random propylene/1-butene copolymers. *Makromol. Chem.* **1980**, *181*, 1747. [CrossRef]
22. Cavallo, P.; Martuscelli, E.; Pracella, M. Effect of thermal treatment on solution grown crystals of isotactic propylene/butene-1 copolymers. *Polymer* **1997**, *18*, 891. [CrossRef]
23. Starkweather, H.W., Jr.; Van-Catledge, F.A.; MacDonald, R.N. Crystalline order in copolymers of ethylene and propylene. *Macromolecules* **1982**, *15*, 1600–1604. [CrossRef]
24. Guidetti, G.P.; Busi, P.; Giulianetti, I.; Zanetti, R. Structure-properties relationships in some random copolymers of propylene. *Eur. Polym. J.* **1983**, *19*, 757–759. [CrossRef]
25. Avella, M.; Martuscelli, E.; Della Volpe, G.; Segre, A.; Rossi, E.; Simonazzi, T. Composition-properties relationships in propene-ethene random copolymers obtained with high-yield Ziegler-Natta supported catalysts. *Makromol. Chem.* **1986**, *187*, 1927. [CrossRef]
26. Marigo, A.; Marega, C.; Zanetti, R.; Paganetto, G.; Canossa, E.; Coletta, F.; Gottardi, F. Crystallization of the γ-form of isotactic poly(propylene). *Makromol. Chem.* **1989**, *190*, 2805. [CrossRef]
27. Monasse, B.; Haudin, J.M. Effect of random copolymerization on growth transition and morphology change in polypropylene. *Colloid Polym. Sci.* **1988**, *266*, 679–687. [CrossRef]
28. Xu, Z.K.; Feng, L.X.; Wang, D.; Yang, S.L. Copolymerization of propene with 1-alkenes using a $MgCl_2/TiCl_4$ catalyst. *Makromol. Chem.* **1991**, *192*, 1835–1840. [CrossRef]
29. Yang, S.L.; Xu, Z.K.; Feng, L.X. Copolymerization of propene with high-1-olefin using a $MgCl_2/TiCl_4$ catalyst. *Makromol. Chem. Macromol. Symp.* **1992**, *63*, 233. [CrossRef]
30. Zimmermann, H.J. Structural analysis of random propylene-ethylene copolymers. *J. Macromol. Sci. Phys.* **1993**, *32*, 141–161. [CrossRef]
31. Sugano, T.; Gotoh, Y.; Fujita, T. Effect of catalyst isospecificity on the copolymerization of propene with 1-hexene. *Makromol. Chem.* **1992**, *193*, 43. [CrossRef]
32. Hingmann, R.; Rieger, J.; Kersting, M. Rheological Properties of a Partially Molten Polypropylene Random Copolymer during Annealing. *Macromolecules* **1995**, *28*, 3801. [CrossRef]
33. Morini, G.; Albizzati, E.; Balbontin, G.; Mingozzi, I.; Sacchi, M.C.; Forlini, F.; Tritto, I. Microstructure Distribution of Polypropylenes Obtained in the Presence of Traditional Phthalate/Silane and Novel Diether Donors: A Tool for Understanding the Role of Electron Donors in $MgCl_2$-Supported Ziegler–Natta Catalysts. *Macromolecules* **1996**, *29*, 5770. [CrossRef]
34. Pérez, E.; Benavente, R.; Bello, A.; Pereña, J.M.; Zucchi, D.; Sacchi, M.C. Crystallization behaviour of fractions of a copolymer of propene and 1-hexene. *Polymer* **1997**, *38*, 5411. [CrossRef]
35. Laihonen, S.; Gedde, U.W.; Werner, P.-E.; Martinez-Salazar, J. Crystallization kinetics and morphology of poly(propylene-*stat*-ethylene) fractions. *Polymer* **1997**, *38*, 361–369. [CrossRef]
36. Laihomen, S.; Gedde, U.W.; Werner, P.E.; Westdahl, M.; Jääskeläinen, P.; Martinez-Salazar, J. Crystal structure and morphology of melt-crystallized poly(propylene-*stat*-ethylene) fractions. *Polymer* **1997**, *38*, 371–377. [CrossRef]
37. Abiru, T.; Mizuno, A.; Weigand, F. Microstructural characterization of propylene–butene-1 copolymer using temperature rising elution fractionation. *J. Appl. Polym. Sci.* **1998**, *68*, 1493–1501. [CrossRef]
38. Feng, Y.; Jin, X.; Hay, J.N. Crystalline structure of propylene–ethylene copolymer fractions. *J. Appl. Polym. Sci.* **1998**, *68*, 381. [CrossRef]
39. Feng, Y.; Hay, J.N. The characterization of random propylene–ethylene copolymer. *Polymer* **1998**, *39*, 6589–6596. [CrossRef]
40. Xu, J.; Feng, Y. Application of temperature rising elution fractionation in polyolefins. *Eur. Polym. J.* **2000**, *36*, 867. [CrossRef]
41. Zhao, Y.; Vaughan, A.S.; Sutton, S.J.; Swingler, S.G. On nucleation and the evolution of morphology in a propylene/ethylene copolymer. *Polymer* **2001**, *42*, 6599. [CrossRef]
42. Foresta, T.; Piccarolo, S.; Goldbeck-Wood, G. Competition between α and γ phases in isotactic polypropylene: Effects of ethylene content and nucleating agents at different cooling rates. *Polymer* **2001**, *42*, 1167. [CrossRef]
43. Marega, C.; Marigo, A.; Saini, R.; Ferrari, P. The influence of thermal treatment and processing on the structure and morphology of poly(propylene-ran-1-butene) copolymers. *Polym. Int.* **2001**, *50*, 442. [CrossRef]

44. Marigo, A.; Causin, V.; Marega, C.; Ferrari, P. Crystallization of the γ form in random propylene-ethylene copolymers. *Polym. Int.* **2004**, *53*, 2001–2008. [CrossRef]
45. Dimenska, A.; Phillips, P.J. High pressure crystallization of random propylene–ethylene copolymers: α–γ Phase diagram. *Polymer* **2006**, *47*, 5445–5456. [CrossRef]
46. Gahleitner, M.; Tranninger, C.; Doshev, P. Polypropylene Copolymers. In *Polypropylene Handbook, Morphology, Blends and Composites*; Karger-Kocsis, J., Bárány, T., Eds.; Springer Nature: Cham, Switzerland, 2019; Chapter 6; p. 295.
47. Galli, P.; Haylock, J.C.; Simonazzi, T. Manufacturing and properties of polypropylene copolymers. In *Polypropylene: Structure, Blends and Composites. Vol. 2 Copolymers and Blends*; Karger-Kocsis, J., Ed.; Chapman & Hall: London, UK, 1995; p. 1.
48. Fischer, D.; Mülhaupt, R. The influence of regio- and stereoirregularities on the crystallization behaviour of isotactic poly(propylene)s prepared with homogeneous group IVa metallocene/methylaluminoxane Ziegler-Natta catalysts. *Macromol. Chem. Phys.* **1994**, *195*, 1433. [CrossRef]
49. Thomann, R.; Wang, C.; Kressler, J.; Mülhaupt, R. On the γ-Phase of Isotactic Poypropylene. *Macromolecules* **1996**, *29*, 8425–8434. [CrossRef]
50. Thomann, R.; Semke, H.; Maier, R.D.; Thomann, Y.; Scherble, J.; Mülhaupt, R.; Kressler, J. Influence of stereoirregularities on the formation of the γ-phase in isotactic polypropylene. *Polymer* **2001**, *42*, 4597–4603. [CrossRef]
51. Alamo, R.G.; Kim, M.H.; Galante, M.J.; Isasi, J.R.; Mandelkern, L. Structural and Kinetic Factors Governing the Formation of the γ Polymorph of Isotactic Polypropylene. *Macromolecules* **1999**, *32*, 4050–4064. [CrossRef]
52. VanderHart, D.L.; Alamo, R.G.; Nyden, M.R.; Kim, M.H.; Mandelkern, L. Observation of Resonances Associated with Stereo and Regio Defects in the Crystalline Regions of Isotactic Polypropylene: Toward a Determination of Morphological Partioning. *Macromolecules* **2000**, *33*, 6078–6093. [CrossRef]
53. Nyden, M.R.; Vanderhart, D.L.; Alamo, R.G. The conformational structures of defect-containing chains in the crystalline regions of isotactic polypropylene. *Comput. Theor. Comput. Sci.* **2001**, *11*, 175–189. [CrossRef]
54. De Rosa, C.; Auriemma, F.; Di Capua, A.; Resconi, L.; Guidotti, S.; Camurati, I.; Nifant'ev, I.E.; Laishevtsev, I.P. Structure-property correlations in polypropylene from metallocene catalysts: Stereodefective, regioregular isotactic polypropylene. *J. Am. Chem. Soc.* **2004**, *126*, 17040–17049. [CrossRef] [PubMed]
55. Arnold, M.; Henschke, O.; Knorr, J. Copolymerization of propene and higher α-olefins with the metallocene catalyst Et[Ind]2HfCl$_2$/methylaluminoxane. *Macromol. Chem. Phys.* **1996**, *197*, 563–573. [CrossRef]
56. Arnold, M.; Bornemann, S.; Köller, F.; Menke, T.J.; Kressler, J. Synthesis and characterization of branched polypropenes obtained by metallocene catalysis. *Macromol. Chem. Phys.* **1998**, *199*, 2647–2653. [CrossRef]
57. Busse, K.; Kressler, J.; Maier, R.D.; Scherble, J. Tailoring of the α-, β-, and γ-Modification in Isotactic Polypropene and Propene/Ethene Random Copolymers. *Macromolecules* **2000**, *33*, 8775–8780. [CrossRef]
58. Wahner, U.M.; Tincul, I.; Joubert, D.J.; Sadiku, E.R.; Forlini, F.; Losio, S.; Tritto, I.; Sacchi, M.C. 13C NMR Study of Copolymers of Propene with Higher 1-Olefins with New Microstructures by *ansa*-Zirconocene Catalysts. *Macromol. Chem. Phys.* **2003**, *204*, 1738. [CrossRef]
59. Costa, G.; Stagnaro, P.; Trefiletti, V.; Sacchi, M.C.; Forlini, F.; Alfonso, G.C.; Tincul, I.; Wahner, U.M. Thermal Behavior and Structural Features of Propene/1-Pentene Copolymers by Metallocene Catalysts. *Macromol. Chem. Phys.* **2004**, *205*, 383–389. [CrossRef]
60. Stagnaro, P.; Costa, G.; Trefiletti, V.; Canetti, M.; Forlini, F.; Alfonso, G.C. Thermal Behavior, Structure and Morphology of Propene/Higher 1-Olefin Copolymers. *Macromol. Chem. Phys.* **2006**, *207*, 2128–2141. [CrossRef]
61. Alamo, R.G.; VanderHart, D.L.; Nyden, M.R.; Mandelkern, L. Morphological Partitioning of Ethylene Defects in Random Propylene−Ethylene Copolymers. *Macromolecules* **2000**, *33*, 6094–6105. [CrossRef]
62. Hosier, I.L.; Alamo, R.G.; Esteso, P.; Isasi, G.R.; Mandelkern, L. Formation of the α and γ Polymorphs in Random Metallocene-Propylene Copolymers. Effect of Concentration and Type of Comonomer. *Macromolecules* **2003**, *36*, 5623–5636. [CrossRef]
63. Hosier, I.L.; Alamo, R.G.; Lin, J.S. Lamellar morphology of random metallocene propylene copolymers studied by atomic force microscopy. *Polymer* **2004**, *45*, 3441–3455. [CrossRef]
64. Alamo, R.G.; Ghosal, A.; Chatterjee, J.; Thompson, K.L. Linear growth rates of random propylene ethylene copolymers. The change over from γ dominated growth to mixed (α+γ) polymorphic growth. *Polymer* **2005**, *46*, 8774–8789. [CrossRef]
65. Fan, Z.; Yasin, T.; Feng, L. Copolymerization of propylene with 1-octene catalyzed by *rac*-Me$_2$Si(2,4,6-Me$_3$-Ind)$_2$ZrCl$_2$/methyl aluminoxane. *J. Polym. Sci. Part A* **2000**, *38*, 4299. [CrossRef]
66. Stagnaro, P.; Boragno, L.; Canetti, M.; Forlini, F.; Azzurri, F.; Alfonso, G.C. Crystallization and morphology of the trigonal form in random propene/1-pentene copolymers. *Polymer* **2009**, *50*, 5242. [CrossRef]
67. Brull, R.; Pasch, H.; Rauberheimer, H.G.; Sanderson, R.D.; Wahner, U.M. Polymerization of higher linear α-olefins with (CH$_3$)$_2$Si(2-methylbenz[e]indenyl)ZzrCl$_2$. *J. Polym. Sci. Part A Polym. Chem.* **2000**, *38*, 2333. [CrossRef]
68. Van Reenen, A.J.; Brull, R.; Wahner, U.M.; Raubenheimer, H.G.; Sanderson, R.D.; Pasch, H. The copolymerization of propylene with higher, linear α-olefins. *J. Polym. Sci. Part A Polym. Chem.* **2000**, *38*, 4110. [CrossRef]
69. Lovisi, H.; Tavares, M.I.B.; da Silva, N.M.; de Menezes, S.M.C.; de Santa Maria, L.C.; Coutinho, F.M.B. Influence of comonomer content and short branch length on the physical properties of metallocene propylene copolymers. *Polymer* **2001**, *42*, 9791–9799. [CrossRef]

70. Shin, Y.-W.; Uozumi, T.; Terano, M.; Nitta, K.-H. Synthesis and characterization of ethylene–propylene random copolymers with isotactic propylene sequence. *Polymer* **2001**, *42*, 9611. [CrossRef]
71. Shin, Y.-W.; Hashiguchi, H.; Terano, M.; Nitta, K. Synthesis and characterization of propylene-α-olefin random copolymers with isotactic propylene sequence. II. Propylene–hexene-1 random copolymers. *J. Appl. Polym. Sci.* **2004**, *92*, 2949.
72. Hosoda, S.; Hori, H.; Yada, K.; Tsuji, M.; Nakahara, S. Degree of comonomer inclusion into lamella crystal for propylene/olefin copolymers. *Polymer* **2002**, *43*, 7451–7460.
73. Fujiyama, M.; Inata, H. Crystallization and melting characteristics of metallocene isotactic polypropylenes. *J. Appl. Polym. Sci.* **2002**, *85*, 1851. [CrossRef]
74. Xu, J.-T.; Xue, L.; Fan, Z.-Q. Nonisothermal crystallization of metallocene propylene–decene-1 copolymers. *J. Appl. Polym. Sci.* **2004**, *93*, 1724. [CrossRef]
75. Poon, B.; Rogunova, M.; Chum, S.P.; Hiltner, A.; Baer, E. Classification of Homogeneous Copolymers of Propylene and 1-Octene Based on Comonomer Content. *J. Polym. Sci. Polym. Phys.* **2004**, *42*, 4357–4370. [CrossRef]
76. Poon, B.; Rogunova, M.; Hiltner, A.; Baer, E.; Chum, S.P.; Galeski, A.; Piorkowska, E. Structure and Properties of Homogeneous Copolymers of Propylene and 1-Hexene. *Macromolecules* **2005**, *38*, 1232–1243. [CrossRef]
77. Palza, H.; López-Majada, J.M.; Quijada, R.; Benavente, R.; Pérez, E.; Cerrada, M.L. Metallocenic Copolymers of Isotactic Propylene and 1-Octadecene: Crystalline Structure and Mechanical Behavior. *Macromol. Chem. Phys.* **2005**, *206*, 1221–1230. [CrossRef]
78. López-Majada, J.M.; Palza, H.; Guevara, J.L.; Quijada, R.; Martinez, M.C.; Benavente, R.; Pereña, J.M.; Pérez, E.; Cerrada, M.L. Metallocene Copolymers of Propene and 1-Hexene: The Influence of the Comonomer Content and Thermal History on the Structure and Mechanical Properties. *J. Polym. Sci. Polym. Phys. Ed.* **2006**, *44*, 1253–1267. [CrossRef]
79. De Rosa, C.; Auriemma, F.; Ruiz de Ballesteros, O.; Resconi, L.; Camurati, I. Crystallization Behavior of Isotactic Propylene-Ethylene and Propylene-Butene Copolymers: Effect of Comonomers versus Stereodefects on Crystallization Properties of Isotactic Polypropylene. *Macromolecules* **2007**, *40*, 6600–6616. [CrossRef]
80. Palza, H.; López-Majada, J.M.; Quijada, R.; Pereña, J.M.; Benavente, R.; Pérez, E.; Cerrada, M.L. Comonomer Length Influence on the Structure and Mechanical Response of Metallocenic Polypropylenic Materials. *Macromol. Chem. Phys.* **2008**, *209*, 2259–2267. [CrossRef]
81. Cerrada, M.L.; Polo-Corpa, M.J.; Benavente, R.; Pérez, E.; Velilla, T.; Quijada, R. Formation of the new trigonal polymorph in Ipp–1-hexene copolymers. Competition with the mesomorphic phase. *Macromolecules* **2009**, *42*, 702–708. [CrossRef]
82. Pérez, E.; Gómez-Elvira, J.M.; Benavente, R.; Cerrada, M.L. Tailoring the formation rate of the mesophase in random propylene-co-1-pentene copolymers. *Macromolecules* **2012**, *45*, 6481–6490. [CrossRef]
83. Boragno, L.; Stagnaro, P.; Forlini, F.; Azzurri, F.; Alfonso, G.C. The trigonal form of i-PP in random C3/C5/C6 terpolymers. *Polymer* **2013**, *54*, 1656–1662. [CrossRef]
84. García-Peñas, A.; Gómez-Elvira, J.M.; Pérez, E.; Cerrada, M.L. Isotactic poly(propylene-co-1-pentene-co-1-hexene) terpolymers: Synthesis, molecular characterization, and evidence of the trigonal polymorph. *J. Polym. Sci. A Polym. Chem.* **2013**, *51*, 3251–3259. [CrossRef]
85. García-Peñas, A.; Gómez-Elvira, J.M.; Barranco-García, R.; Pérez, E.; Cerrada, M.L. Trigonal δ form as a tool for tuning mechanical behavior in poly(propylene-co-1-pentene-co-1-heptene) terpolymers. *Polymer* **2016**, *99*, 112–121. [CrossRef]
86. García-Peñas, A.; Gómez-Elvira, J.M.; Lorenzo, V.; Pérez, E.; Cerrada, M.L. Unprecedented dependence of stiffness parameters and crystallinity on comonomer content in rapidly cooled propylene-co-1-pentene copolymers. *Polymer* **2017**, *130*, 17–25. [CrossRef]
87. Auriemma, F.; De Rosa, C.; Di Girolamo, R.; Malafronte, A.; Scoti, M.; Cipullo, R. Yield behavior of random copolymers of isotactic polypropylene. *Polymer* **2017**, *129*, 235–246. [CrossRef]
88. Auriemma, F.; De Rosa, C.; Di Girolamo, R.; Malafronte, A.; Scoti, M.; Cioce, C. A molecular view of properties of random copolymers of isotactic polypropene. *Adv. Polym. Sci.* **2017**, *276*, 45–92.
89. De Rosa, C.; Scoti, M.; Auriemma, F.; Ruiz de Ballesteros, O.; Talarico, G.; Malafronte, A.; Di Girolamo, R. Mechanical Properties and Morphology of Propene−Pentene Isotactic Copolymers. *Macromolecules* **2018**, *51*, 3030–3040. [CrossRef]
90. De Rosa, C.; Scoti, M.; Auriemma, F.; Ruiz de Ballesteros, O.; Talarico, G.; Di Girolamo, R.; Cipullo, R. Relationships among lamellar morphology parameters, structure and thermal behavior of isotactic propene-pentene copolymers: The role of incorporation of comonomeric units in the crystals. *Eur. Polym. J.* **2018**, *103*, 251–259. [CrossRef]
91. De Rosa, C.; Scoti, M.; Ruiz de Ballesteros, O.; Di Girolamo, R.; Auriemma, F.; Malafronte, A. Propylene-Butene Copolymers: Tailoring Mechanical Properties from Isotactic Polypropylene to Polybutene. *Macromolecules* **2020**, *53*, 4407–4421. [CrossRef]
92. Scoti, M.; De Stefano, F.; Di Girolamo, R.; Talarico, G.; Malafronte, A.; De Rosa, C. Crystallization of Propene-Pentene Isotactic Copolymers as an Indicator of the General View of the Crystallization Behavior of Isotactic Polypropylene. *Macromolecules* **2022**, *55*, 241. [CrossRef]
93. Alamo, R.G.; Blanco, J.A.; Agarwal, P.K.; Randall, J.C. Crystallization Rates of Matched Fractions of MgCl$_2$-Supported Ziegler Natta and Metallocene Isotactic Poly(Propylene)s. 1. The Role of Chain Microstructure. *Macromolecules* **2003**, *36*, 1559–1571. [CrossRef]
94. Randall, J.C.; Alamo, R.G.; Agarwal, P.K.; Ruff, C.J. Crystallization Rates of Matched Fractions of MgCl$_2$-Supported Ziegler-Natta and Metallocene Isotactic Poly(propylene)s. 2. Chain Microstructures from a Supercritical Fluid Fractionation of a MgCl$_2$-Supported Ziegler−Natta Isotactic Poly(propylene). *Macromolecules* **2003**, *36*, 1572–1584. [CrossRef]

95. De Rosa, C.; Ruiz de Ballesteros, O.; Auriemma, F.; Talarico, G.; Scoti, M.; Di Girolamo, R.; Malafronte, A.; Piemontesi, F.; Liguori, D.; Camurati, I.; et al. Crystallization Behavior of Copolymers of Isotactic Poly(1-butene) with Ethylene from Ziegler–Natta Catalyst: Evidence of the Blocky Molecular Structure. *Macromolecules* **2019**, *52*, 9114–9127. [CrossRef]
96. De Rosa, C.; Ruiz de Ballesteros, O.; Di Girolamo, R.; Malafronte, A.; Auriemma, F.; Talarico, G.; Scoti, M. The blocky structure of Ziegler–Natta "random" copolymers: Myths and experimental evidence. *Polym. Chem.* **2020**, *11*, 34–38. [CrossRef]
97. Gahleitner, M.; Jääskeläinen, P.; Ratajski, E.; Paulik, C.; Reussner, J.; Wolfschwenger, J.; Neißl, W. Propylene-ethylene random copolymers: Comonomer effects on crystallinity and application properties. *J. Appl. Polym. Sci.* **2005**, *95*, 1073–1081. [CrossRef]
98. Caveda, S.; Pérez, E.; Blázquez-Blázquez, E.; Peña, B.; van Grieken, R.; Suárez, I.; Benavente, R. Influence of structure on the properties of polypropylene copolymers and terpolymers. *Polym. Test.* **2017**, *62*, 23–32. [CrossRef]
99. Auriemma, F.; De Rosa, C.; Di Girolamo, R.; Malafronte, A.; Scoti, M.; Mitchell, G.R.; Esposito, S. Relationship between molecular configuration and stress induced phase transitions. In *Controlling the Morphology of Polymers—Multiple Scales of Structure and Processing*; Mitchell, G.R., Tojeira, A., Eds.; Springer International Publishing: Cham, Switzerland, 2016; p. 287.
100. Auriemma, F.; De Rosa, C.; Di Girolamo, R.; Malafronte, A.; Scoti, M.; Mitchell, G.R.; Esposito, S. Deformation of Stereoirregular Isotactic Polypropylene across Length Scales. Influence of Temperature. *Macromolecules* **2017**, *50*, 2856–2870. [CrossRef]
101. Auriemma, F.; De Rosa, C.; Di Girolamo, R.; Malafronte, A.; Scoti, M.; Mitchell, G.R.; Esposito, S. Time-resolving study of stress-induced transformations of isotactic polypropylene through wide angle X-ray scattering measurements. *Polymers* **2018**, *10*, 162. [CrossRef]
102. De Rosa, C.; Di Girolamo, R.; Auriemma, F.; Talarico, G.; Scarica, C.; Malafronte, A.; Scoti, M. Controlling Size and Orientation of Lamellar Microdomains in Crystalline Block Copolymers. *ACS Appl. Mater. Interfaces* **2017**, *9*, 31252. [CrossRef]
103. Auriemma, F.; De Rosa, C.; Scoti, M.; Di Girolamo, R.; Malafronte, A.; Talarico, G.; Carnahan, E. Unveiling the Molecular Structure of Ethylene/1-Octene Multi-block Copolymers from Chain Shuttling Technology. *Polymer* **2018**, *154*, 298. [CrossRef]
104. Stehling, U.; Diebold, J.; Kirsten, R.; Roell, W.; Brintzinger, H.-H.; Juengling, S.; Mülhaupt, R.; Langhauser, F. ansa-Zirconocene Polymerization Catalysts with Anelated Ring Ligands—Effects on Catalytic Activity and Polymer Chain Length. *Organometallics* **1994**, *13*, 964. [CrossRef]
105. Langhauser, F.; Kerth, J.; Kersting, M.; Koelle, P.; Lilge, D.; Mueller, P. Propylene polymerization with metallocene catalysts in industrial processes. *Angew. Makromol. Chem.* **1994**, *223*, 155. [CrossRef]
106. Resconi, L.; Guidotti, S.; Morhard, F.; Fait, A. Process for the Preparation of 1-butene/propylene Copolymers. U.S. Patent No. 7,531,609, 12 May 2009.
107. Resconi, L.; Guidotti, S.; Camurati, I.; Frabetti, R.; Focante, F.; Nifant'ev, I.E.; Laishevtsev, I.P. C1-Symmetric Heterocyclic Zirconocenes as Catalysts for Propylene Polymerization, 2. ansa-Zirconocenes with Linked Dithienocyclopentadienyl-Substituted Indenyl Ligands. *Macromol. Chem. Phys.* **2005**, *206*, 1405. [CrossRef]
108. Resconi, L.; Camurati, I.; Malizia, F. Metallocene Catalysts for 1-Butene Polymerization. *Macromol. Chem. Phys.* **2006**, *207*, 2257. [CrossRef]
109. Covezzi, M.; Fait, A. Process and Apparatus for Making Supported Catalyst Systems for Olefin Polymerization. U.S. Patent No. 7,041,750, 9 May 2006.
110. Randall, J.C. A ^{13}C NMR Determination of the Comonomer Sequence Distributions in Propylene-Butene-1 Copolymers. *Macromolecules* **1978**, *11*, 592. [CrossRef]
111. Kakugo, M.; Naito, Y.; Mizunuma, K.; Miyatake, T. Carbon-13 NMR determination of monomer sequence distribution in ethylene-propylene copolymers prepared with δ-TiCl$_3$-Al(C$_2$H$_5$)$_2$Cl. *Macromolecules* **1982**, *15*, 1150. [CrossRef]

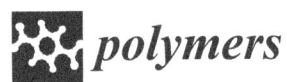

Article

The Effect of Topology on Block Copolymer Nanoparticles: Linear versus Star Block Copolymers in Toluene

Yuan Zhang *, Peng Wang, Nan Li, Chunyan Guo and Sumin Li

Research School of Polymeric Materials, School of Materials Sciences & Engineering, Jiangsu University, Zhenjiang 212013, China
* Correspondence: zhangyuan@ujs.edu.cn

Abstract: Linear and star block copolymer (BCP) nanoparticles of (polystyrene-*block*-poly(4-vinylpyridine))$_n$ (PS-*b*-P4VP)$_n$ with arm numbers of 1, 2, 3, and 4 were prepared by two methods of polymerization-induced self-assembly (PISA) and general self-assembly of block copolymers in the low-polar organic solvent, toluene. The effect of the arm number on the size and/or morphology of the (PS-*b*-P4VP)$_n$ nanoassemblies synthesized by the two methods in toluene and on the polymerization kinetics was investigated in detail. Our results show that in toluene, a low-polar solvent, the topology not only affected the morphology of the BCP nanoparticles prepared by PISA, but also influenced the BCP nanoparticles synthesized through general self-assembly.

Keywords: topology; star-block copolymer; polymerization-induced self-assembly; general self-assembly

Citation: Zhang, Y.; Wang, P.; Li, N.; Guo, C.; Li, S. The Effect of Topology on Block Copolymer Nanoparticles: Linear versus Star Block Copolymers in Toluene. *Polymers* **2022**, *14*, 3691. https://doi.org/10.3390/polym14173691

Academic Editors: Nikolaos Politakos and Apostolos Avgeropoulos

Received: 17 August 2022
Accepted: 29 August 2022
Published: 5 September 2022

Publisher's Note: MDPI stays neutral with regard to jurisdictional claims in published maps and institutional affiliations.

Copyright: © 2022 by the authors. Licensee MDPI, Basel, Switzerland. This article is an open access article distributed under the terms and conditions of the Creative Commons Attribution (CC BY) license (https://creativecommons.org/licenses/by/4.0/).

1. Introduction

Block copolymer (BCP) self-assembly has attracted much attention in the past several decades for the synthesis of polymeric nano-objects of various morphologies and their potential applications in many fields [1,2]. In general, two strategies, micellization of amphiphilic block copolymers in block-selective solvent [3–10] and polymerization-induced self-assembly (PISA) [11–19], have been adopted to synthesize/prepare BCP nano-assemblies. In the first strategy, pre-synthesized amphiphilic block copolymers are initially dissolved in a common solvent [3–5] and then block-selective solvent is added to induce the soluble-to-insoluble phase transition of the solvophobic block and therefore trigger micellization of the amphiphilic block copolymers in the solvent [6–10]. In the PISA strategy, all ingredients including a monomer, macromolecular chain transfer agent, and initiator are dissolved in solvent in the initial polymerization stage [11–13]. With the proceeding of the polymerization, the newly formed hydrophobic chain grows to a certain length, and in situ self-assembly of amphiphilic block copolymers occurs and micelles are formed [14–16]. After that, dispersion polymerization takes place dominantly in the monomer-swollen micelles, resulting in the change in size or morphology of the block copolymer nanoparticles [17–19]. Therefore, the synthesis and assembly of amphiphilic block copolymers are carried out simultaneously in one pot in the PISA method [11–19].

It is well known that the morphologies of amphiphilic BCPs are dependent on the solvent characteristics [20,21] and especially the intrinsic molecular architecture [22–27]. Star block copolymers include a single branch point from which chemically different building blocks spread out [28–31]. Owing to this exclusive structure, star block copolymers show many interesting characteristics and properties unattainable by linear polymers [22–27]. For example, Ma and coworkers successfully synthesized the novel dual-functional linear and star POSS-containing organic–inorganic hybrid block copolymer poly(glycidyl methacrylate)-*block*-poly(methacrylisobutyl polyhedral oligomeric silsesquioxane)$_{1,4,6}$ ((PGMA-*b*-PMAPOSS)$_{1,4,6}$) by the core-first ATRP method, then crosslinked the block copolymers in the presence of trimethylamine to form a 3-dimensional network. Furthermore, the self-assembly behavior of these block copolymers with similar volume fractions of insoluble block was investigated in detail in water. It was revealed that the linear block copolymers (BCPs) PGMA-*b*-PMAPOSS showed the formation of spherical micelles with a size of 140–200 nm, star four-arm BCPs (PGMA-*b*-PMAPOSS)$_4$ assembled as a multi-core core-shell morphology, and star six-arm BCPs (PGMA-*b*-PMAPOSS)$_6$ assembled as the dendritic feature [26]. Huh and his team found that self-assembling the amphiphilic star poly(ethylene glycol)-[poly(ε-caprolactone)]$_2$ and the corresponding linear counterparts with approximately the same chemical composition in selective solvent lead to cylindrical and spherical micelles, respectively [27].

The preparation of star block copolymer self-assemblies through the first strategy, that is, general self-assembly, has been widely reported [9,10]. In addition, there are also few reports on the preparation of star block copolymer assemblies via the PISA method, and most of them are carried out in a polar solvent, such as water or ethanol/water mixture solution [32–37]. As we all know, solvent properties or composition have an impact on the morphology of block copolymer nano-assemblies [20,21]. In addition, BCP nanoparticles prepared in low polar solvents could be easily used as lubrication and emulsifiers for water/oil emulsions [38,39]. However, rare reports on the effect of the arm number on block copolymer nano-assemblies, i.e., linear block copolymers versus star block copolymers in low polar organic solvent, have been reported.

The polystyrene-mediated RAFT dispersion polymerization of the 4-vinylpyridine (4VP) monomer is a typical example that uses the low-polar solvent, toluene [40,41]. In this paper, the effects of the topological structure on the morphology and evolution process of block copolymer nanoassemblies in toluene were studied using RAFT dispersion polymerization of 4VP regulated by linear or star PS macro-RAFT agents. Firstly, block copolymer nanoassemblies of (polystyrene-*block*-poly(4-vinylpyridine))$_n$ ((PS-*b*-P4VP)$_n$, n = 1, 2, 3, and 4) were formed through RAFT dispersion polymerization in toluene employing mono- and multi-functional macromolecular chain transfer agents. Then, the polymerization kinetics of these dispersion RAFT polymerizations were investigated, and the effect of the arm number or the DPs of the PS/P4VP chain on the block copolymer nanoassemblies prepared via PISA was explored by changing the length of PS or P4VP. Finally, 2-, 3-, and 4-arm star and linear block copolymers of (PS-*b*-P4VP)$_{1,2,3,4}$ with similar chemical compositions were selected to prepare the nanoassemblies in toluene by the general self-assembly method to further prove that the topology also has an important effect of on the morphology of block copolymer nanoassemblies prepared by general self-assembly.

2. Experimental Section

2.1. Materials

Styrene (St, >98%, Shanghai chemical reagent, Shanghai, China) and 4-vinylpyridine (4VP, 96%, Aladdin) were purified under reduced pressure for later use. The synthesis of the RAFT reagent of four small molecules called I, II, III, IV as shown in Scheme S1 was carried out according to the steps in our previous paper [42,43] and the detailed synthesis process and structure characterization are described in the Supporting Information (Scheme S2) and Figure S1. 2,2′-Azobis(2-methylpropionitrile) (AIBN, >99%, Tianjin Chemical Company, Tianjin, China) was recrystallized from ethanol.

2.2. Synthesis of Linear and Star macro-CTAs

Linear and star macromolecular chain transfer agents (macro-CTAs) of (PS$_m$-TTC)$_n$, in which m and n represent the arm number of the PS arms and the DP of each PS arm, with TTC representing the RAFT terminal of trithiocarbonate, were synthesized by solution RAFT polymerization of St monomers employing I-IV as chain transfer agents (CTAs). Here is a typical synthesis of 3-arm (PS$_{25}$-TTC)$_3$ under [St]:[III]:[AIBN] = 450:5:3: St (4.000 g, 0.039 mol), III (0.262 g, 0.427 mmol) and AIBN (42.05 mg, 0.2564 mmol) dissolved in 1,4-dioxane (4.000 g) were weighed into a 38 mL Schlenk flask. The mixture was degassed and ran at 70 °C for 32 h. The styrene conversion was determined by ^{1}H NMR employing 1,3,5-trioxane as an internal standard as discussed elsewhere. The synthesized (PS$_{25}$-TTC)$_3$ was precipitated into ethanol and then dried under vacuum. By changing the [St]$_0$:[CTA]$_0$:[AIBN]$_0$ molar ratio, other (PS$_m$-TTC)$_n$ macro-CTAs with different arm chain length and arm number n were also synthesized (Table 1).

Table 1. Summary of the Synthesized Macro-CTAs.

| Macro-CTA | [M]$_0$:[CTA]$_0$:[I]$_0$ | Time (h) | Conv. ^{a} (%) | M_n (kg/mol) | | | Đ ^{e} |
				$M_{n,th}$ ^{b}	$M_{n,GPC}$ ^{c}	$M_{n,NMR}$ ^{d}	
PS$_{17}$-TTC	100:5:1	32	85.0	2.02	3.1	2.1	1.11
PS$_{24}$-TTC	150:5:1	32	80.0	2.75	3.2	3.1	1.11
PS$_{61}$-TTC	500:5:1	32	61.0	6.60	6.4	6.5	1.14
(PS$_{17}$-TTC)$_2$	200:5:2	32	85.0	3.97	4.4	4.1	1.12
(PS$_{24}$-TTC)$_2$	300:5:2	32	80.4	5.43	5.5	5.38	1.13
(PS$_{57}$-TTC)$_2$	1000:5:2	32	57.2	12.29	10.6	12.4	1.18
(PS$_{17}$-TTC)$_3$	300:5:3	32	85.0	5.92	5.8	6.0	1.12
(PS$_{25}$-TTC)$_3$	450:5:3	32	83.3	8.41	7.9	8.7	1.12
(PS$_{61}$-TTC)$_3$	1500:5:3	32	59.3	19.64	21.3	19.5	1.22
(PS$_{17}$-TTC)$_4$	400:5:4	32	85.0	7.86	7.8	7.7	1.18
(PS$_{25}$-TTC)$_4$	600:5:4	32	83.3	11.19	10.9	11.0	1.20
(PS$_{60}$-TTC)$_4$	2000:5:4	32	60.0	25.75	22.6	26.1	1.26

^{a} Monomer conversion determined by ^{1}H NMR. ^{b} Theoretical molecular weight according to eq S1. ^{c} Molecular weight determined by GPC. ^{d} Molecular weight determined by ^{1}H NMR. ^{e} Đ (M_w/M_n) determined by GPC.

The styrene monomer, CTA and AIBN were dissolved in toluene to obtain a homogeneous solution with a solid content of 20 wt%, which was subjected to RAFT dispersion polymerization. Taking an example as a typical representative under [4VP]$_0$:[(PS$_{25}$-TTC)$_3$]$_0$:[AIBN]$_0$ = 2400:4:3: into a 38 mL Schlenk flask, 4VP (0.300 g, 2.86 mmol), (PS$_{25}$-TTC)$_3$ (38.571 mg, 0.0048 mmol) and AIBN (0.59 mg, 0.0036 mmol) dissolved in toluene (1.350 g) were weighed. The mixture was degassed and then polymerization was ran at 70 °C for a given time. The 4-vinylpyridine conversion was determined by ^{1}H NMR. To check the resultant block copolymer nanoassemblies, a small drop of the block copolymer dispersion was deposited onto a piece of copper grid, dried at room temperature under vacuum, and then observed by transmission electron microscope (TEM). To collect the block copolymer for GPC analysis and ^{1}H NMR analysis, the synthesized (PS-b-P4VP)$_3$ nanoassemblies were diluted with dichloromethane and precipitated into ethanol, and finally dried at 40 °C under vacuum.

2.3. Preparation of the Block Copolymer Micelles through Self-Assembly in Toluene

The above-synthesized linear or star (PS-b-P4VP)$_n$ (n = 1, 2, 3) block copolymer nanoassemblies were centrifuged and dissolved in dichloromethane (DCM), then precipitated in toluene, centrifuged and washed with ethanol/toluene (v/v = 1:8) three times, and finally dried at 40 °C under vacuum to obtain the (PS-b-P4VP)$_n$ (n = 1, 2, 3) block copolymers. The above processes were designed to eliminate the morphology of the block copolymer nanoassemblies obtained in toluene via the PISA method. Then, (PS-b-P4VP)$_n$ was dissolved in DCM at room temperature to prepare a 0.5 mg/mL solution, adding a given volume of toluene at a rate of 1 drop (1 drop was about 7 μL) every 10 s under stirring. With the addition of toluene, the solution became turbid, indicating the formation of nanoassemblies. Toluene was then added slowly until the concentration of block copolymer nanoassemblies was about 0.2 mg/mL. Finally, the DCM in the solution was removed under vacuum at 25 °C, and a dispersion solution of block copolymers with 0.2 mg/mL concentration was obtained. TEM was used to characterize the morphology of the (PS-b-P4VP)$_n$ (n = 1, 2, 3) block copolymer nanoassemblies.

2.4. Characterization

The molecular weight (M_n) and dispersity ($Ð$, $Ð = M_w/M_n$) of the polymers were obtained using a Waters 1525 μ gel permeation chromatograph. The samples were passed through three columns using DMF solution containing 0.05 M lithium bromide as the eluent at a flow rate of 1.0 mL/min at 50 °C. The narrow-polydispersity samples of polystyrene were used as a calibration standard. The ^{1}H NMR analysis was performed on a Bruker Avance II 400 MHz NMR spectrometer using CDCl$_3$ as a solvent. The TEM observation was performed using FEI Tecnai 12 transmission electron microscopy (TEM) with 120 kV or JEM-2100 TEM with 200 kV, whereby a small drop of the block copolymer dispersion was deposited onto a piece of copper grid and then dried at room temperature under vacuum.

3. Results and Discussion

3.1. Synthesis of macro-CTAs with Linear and Star Structure

Due to the similar R and Z groups in the structure of I, II, III, IV CTAs (Scheme S1), each arm of the synthesized star macro-CTAs was assumed to have a similar degree of polymerization. The linear and star macro-CTAs of (PS)$_n$-TTC (n = 1, 2, 3, 4) (Scheme 1) were synthesized by solution RAFT polymerization employing an initiator of AIBN and chain transfer agents (CTAs) of I, II, III or IV. By varying the molar ratio of St/CTAs/AIBN (Table 1), (PS)$_n$ with a suitable DP of the PS arms was prepared. The (PS-TTC)$_n$ (n = 1, 2, 3, 4) macro-RAFT agents were characterized by ^{1}H NMR analysis and GPC analysis, and the results for the typical (PS$_{24}$-TTC)$_n$ (n = 1, 2, 3, 4) with a similar DP of the PS arms, around 25, are shown in Figures 1 and 2. The molecular weight, $M_{n,NMR}$ by ^{1}H NMR analysis was calculated by comparing the integration area of the peaks at 0.88 ppm and 6.20–7.20 ppm attributed to the RAFT terminal and benzene ring. Furthermore, the $M_{n,NMR}$ was approximately equal to the theoretical molecular weight $M_{n,th}$ calculated from the monomer conversion according to eqn S1 (Table 1).

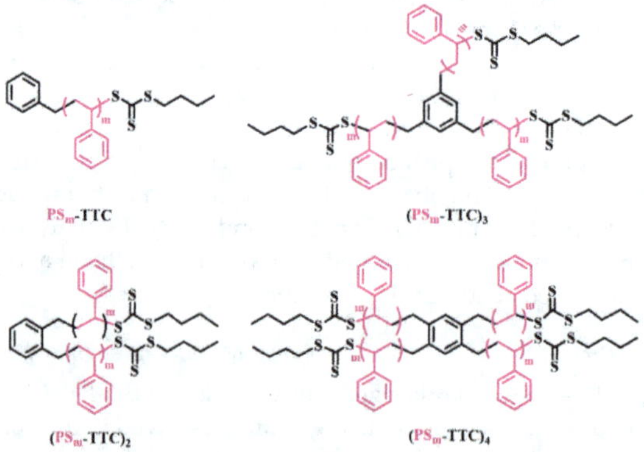

Scheme 1. (PS$_m$-TTC)$_n$ Macro-CTAs.

In Figure 2, unimodal GPC traces were observed for linear PS$_{24}$-TTC, (PS$_{24}$-TTC)$_2$, and (PS$_{25}$-TTC)$_3$, while for the 4-arm star (PS$_{25}$-TTC)$_4$, a small acromion appeared on the high-molecular-weight side. In the synthesis of star polymers, a molecule usually contains more than one propagating radical, which is more likely to lead to a bimolecular termination reaction as discussed elsewhere [44]. However, the dispersity of star (PS-TTC)$_n$ was narrow, which can be verified from Ð < 1.3 (Table 1). It should be noted that the molecular weight obtained by GPC for the star-shaped (PS-TTC)$_n$ prepared in this study was close to the amount obtained by NMR characterization, which was not quite consistent with previous results recorded in the literature, in that $M_{n,GPC}$ was less than $M_{n,NMR}$ for star-shaped polymers [45,46], which will also be discussed in the following sections. (Note: herein the DP values of PS$_{24}$-TTC, (PS$_{24}$-TTC)$_2$, (PS$_{25}$-TTC)$_3$, and (PS$_{25}$-TTC)$_4$ were calculated according to the monomer conversions. In addition, the DP values mentioned in the following sections were calculated through the monomer conversion.)

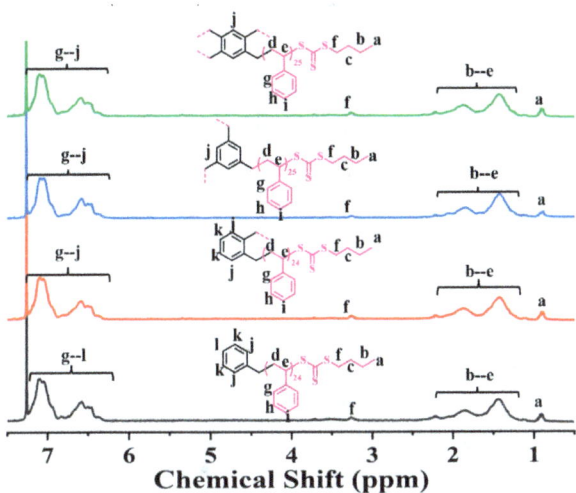

Figure 1. ^{1}H NMR spectra of (PS-TTC)n (n = 1, 2, 3, 4) with similar PS arms chain at about 25.

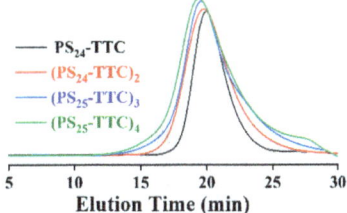

Figure 2. GPC traces of PS-TTC)n (n = 1, 2, 3, 4) with similar PS arms chain at about 25.

3.2. Effect of the Arm Number on BCP Nanoassemblies Prepared through PISA

It has been reported that the arm number of star polymers has an important effect on the morphology of block copolymer nanoassemblies in ethanol/water. However, the topology of star polymers on the morphology of nanoassemblies formed through PISA in toluene has not been studied. To achieve this, we synthesized a series of (PS-b-P4VP)$_n$ nanoparticles with different compositions by varying the length of the PS or P4VP chain segments according to Scheme S3, and characterized their morphologies by TEM.

Figure 3 lists the (PS-b-P4VP)$_n$ nanoparticles containing similar P4VP segments but different PS segments. The results indicate that all the linear and star (PS-b-P4VP)$_n$ (n = 1, 2, 3, 4) with the P4VP block chain at about 270 and the long PS arm block chain at about 60 formed discrete nanoassemblies of 32.5 ± 4.6 nm PS$_{61}$-b-P4VP$_{272}$ nanospheres, 33.2 ± 2.3 nm (PS$_{57}$-b-P4VP$_{274}$)$_2$ nanospheres, 40.1 ± 3.6 nm (PS$_{61}$-b-P4VP$_{269}$)$_3$ nanospheres and 44.3 ± 2.4 nm (PS$_{60}$-b-P4VP$_{274}$)$_4$ nanospheres, respectively. When the polymerization degree of PS arms is short, such as at about 17 or 24, linear block copolymers still form discrete nanoassemblies, while star block copolymers n = 2, 3, and 4 all formed aggregates, especially the aggregates of the 3- and 4-armed block copolymers. The self-assembly of star block copolymers or linear BAB (B is insoluble chain segment) polymers can also form large aggregates, which may be due to the formation of bridging interactions between nanoparticles [47–50].

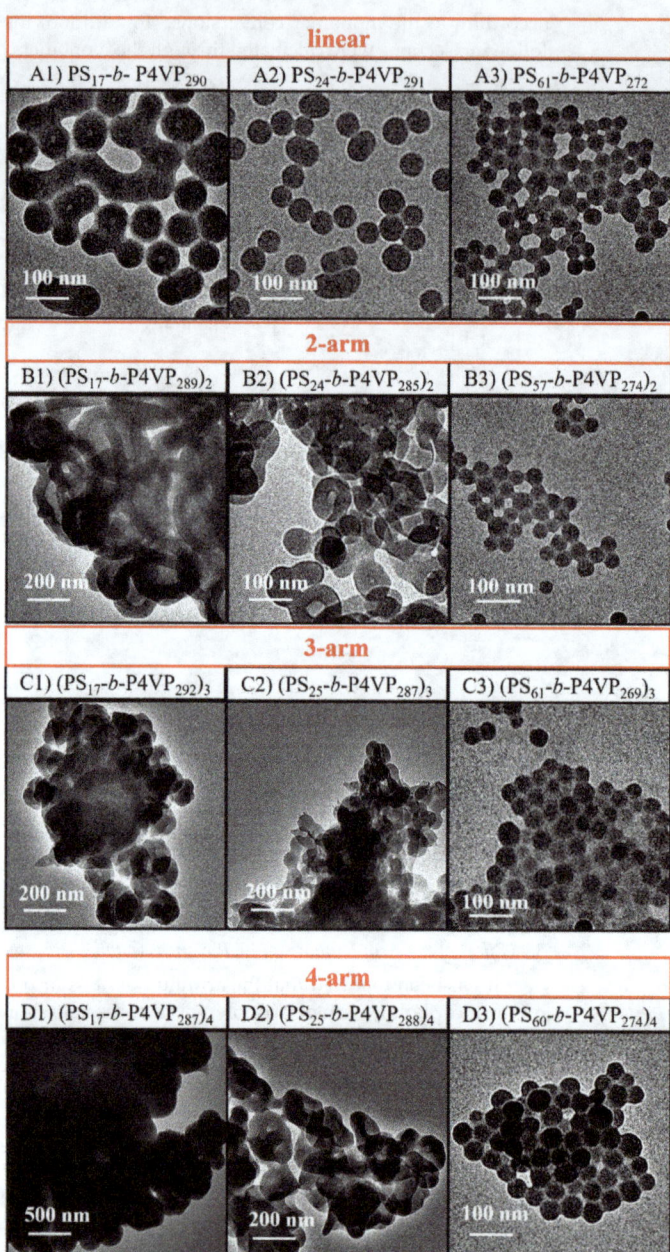

Figure 3. TEM images of (PS-*b*-P4VP)$_n$ nanoparticles containing similar P4VP segments but different PS segments.

Figure 4 lists the different case of (PS-*b*-P4VP)$_n$ nanoparticles with a similar DP of the PS segments at about 25, while increasing the DP of P4VP segments. Correspondingly, (PS-*b*-P4VP)$_n$ (n = 1−4) with a short block chain of P4VP arms (DP < 200) formed discrete nanoassemblies, and star (PS-*b*-P4VP)$_n$ (n = 2, 3, 4) with long block chain of P4VP arms at about 285 formed bridged aggregates, while linear PS-*b*-P4VP with long block chain of

P4VP still formed discrete nanoassemblies. It was found that both the linear and star (PS-*b*-P4VP)$_n$ followed similar rules, i.e., the diameter of their nanoparticles increased with the DP of P4VP block increasing. This rule is consistent with the previous morphologic change in linear BCP nanoparticles [51–54]. In addition, the morphology of star (PS-*b*-P4VP)$_{2-3}$ nanoparticles is much more complex than that of linear PS-*b*-P4VP.

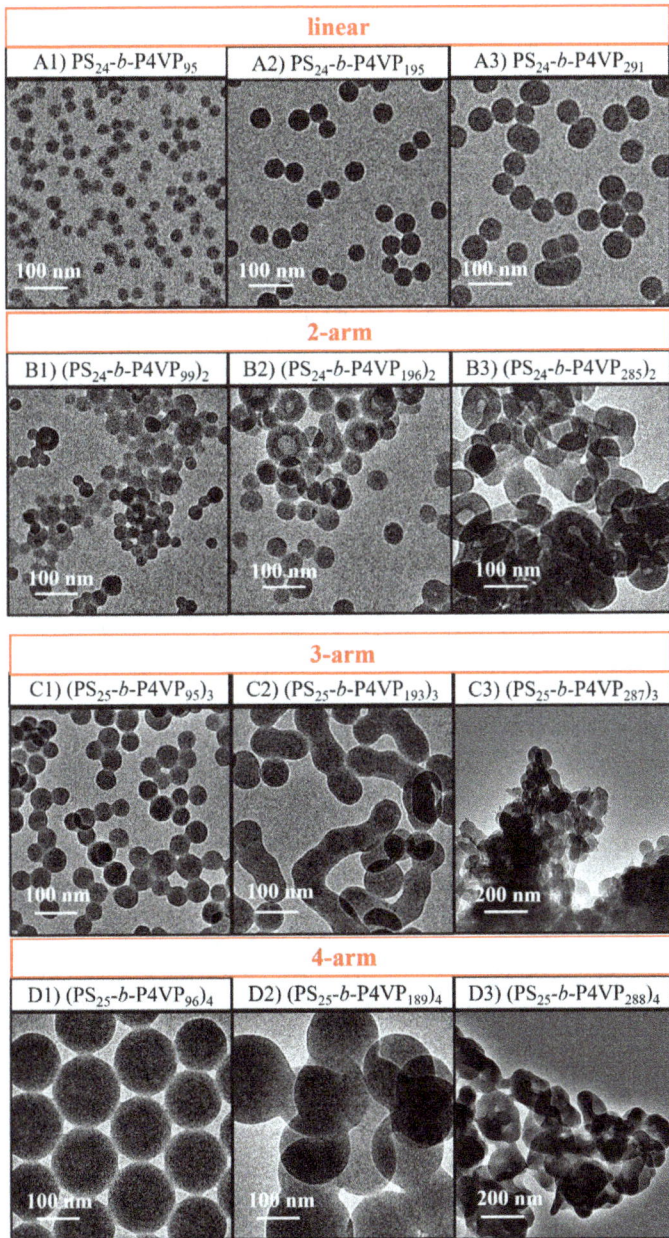

Figure 4. TEM images of (PS-*b*-P4VP)$_n$ nanoparticles containing similar PS segments but different P4VP segments.

3.3. Effect of the Arm Number on Polymerization Kinetics and the Evolution of (PS-b-P4VP)$_n$ Nanoassemblies

To further study the effect of the arm number on the polymerization kinetics and the evolution of (PS-*b*-P4VP)$_n$ nanoassemblies, the (PS-TTC)$_n$ (n = 1–3) with a similar arm length of about 25 mediated PISA of 4VP was studied in detail. In order to ensure that the polymerization conditions were similar, [4VP]$_0$:[trithiocarbonate]$_0$:[AIBN]$_0$ were designed as a constant in all PISA reactions.

As can be seen in Figure 5A, the monomer conversion-time in the three RAFT dispersion polymerization reactions was very similar, indicating that the three reactions had similar polymerization kinetics. This may be due to the fact that the (PS-TTC)$_n$ macromolecular chain transfer agents used had similar R and Z groups and close arm lengths. As shown in Figure 5B, the (PS-TTC)$_n$ (n = 1, 2, 3) mediated PISA of 4VP firstly underwent a slow polymerization process, followed by an accelerated polymerization after about 4 h reaction (Figure 5B), which had similar polymerization kinetics as commonly reported for RAFT dispersion polymerization [55–61]. Generally, the synthesized 3-armed stars (PS$_{25}$-*b*-P4VP)$_3$ were characterized by GPC (Figure 5C) and NMR to obtain their $M_{n,\,GPC}$ and $M_{n,NMR}$, respectively, and the results are listed in Figure 5D. A small acromion appeared when the monomer conversion was greater than 87% in the GPC curve of (PS$_{25}$-*b*-P4VP)$_3$, which was also reflected in the increase in the Đ value in Figure 5D. Moreover, the RAFT synthesis was controllable given that (PS$_{25}$-*b*-P4VP)$_3$ had a star structure and Đ was around 1.3. It is worth mentioning that when monomer conversion was low, the molecular weight obtained by NMR was close to the molecular weight obtained by GPC, which is in agreement with that described above. When the monomer conversion was greater than 20%, the molecular weight obtained by GPC was smaller than that obtained by NMR, which is consistent with previous results [45,46].

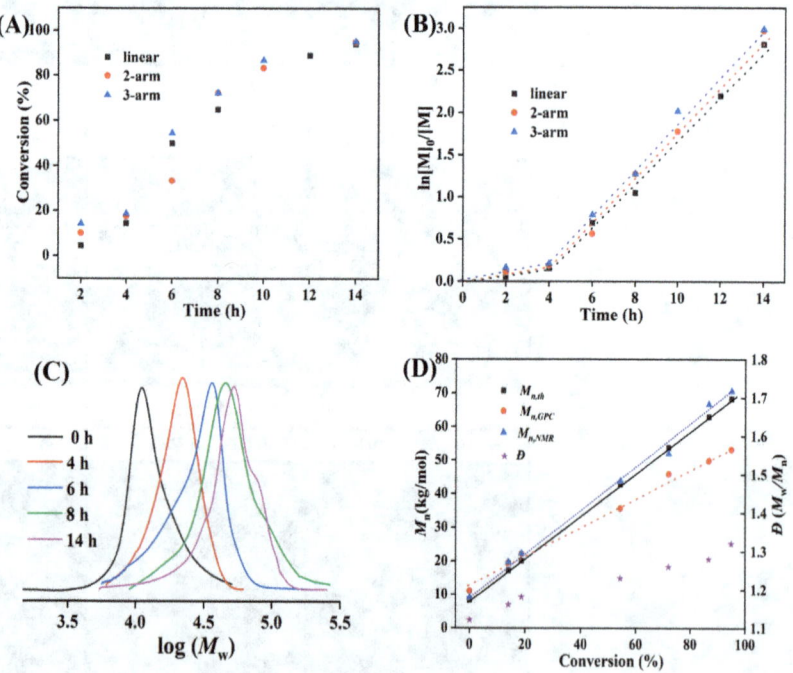

Figure 5. Polymerization kinetics (**A**) and semilogarithmic plots (**B**) of RAFT dispersion polymerization mediated by linear and 2-, and 3-armed star (PS-TTC)$_n$. GPC traces (**C**) and evolution of $M_{n,th}$, $M_{n,NMR}$, $M_{n,GPC}$, and Đ values (**D**) in (PS$_{25}$-TTC)$_3$-mediated PISA of 4VP.

In addition, the formation process of the 3-armed star-shaped $(PS_{25}\text{-}b\text{-}P4VP)_3$ nanoparticles was recorded by TEM as shown in Figure 6. With the increase in P4VP, $(PS_{25}\text{-}b\text{-}P4VP)_3$ nanoparticles changed from 16.3 ± 1.4 nm nanospheres at the beginning of 4 h to 53.8 ± 3.9 nm nanospheres at 8 h, and finally to worms of $(PS_{25}\text{-}b\text{-}P4VP_{190})_3$. However, the linear $PS_{24}\text{-}b\text{-}P4VP$ was still a solid sphere when the length of P4VP was 290 (Figure 4). Figure 6F summarizes the average diameter (D) of the $(PS_{25}\text{-}b\text{-}P4VP)_3$ and linear $PS_{24}\text{-}b\text{-}P4VP$ nanoparticles with the DP of P4VP arms increasing, showing that the average diameter of $(PS_{25}\text{-}b\text{-}P4VP)_3$ changed very little after 10 h, which may be related to the change in morphology from nanospheres to worms. Moreover, the rate of change with the DP was different for the 3-armed $(PS_{25}\text{-}b\text{-}P4VP)_3$ and linear $PS_{24}\text{-}b\text{-}P4VP$ block copolymer nanoparticles.

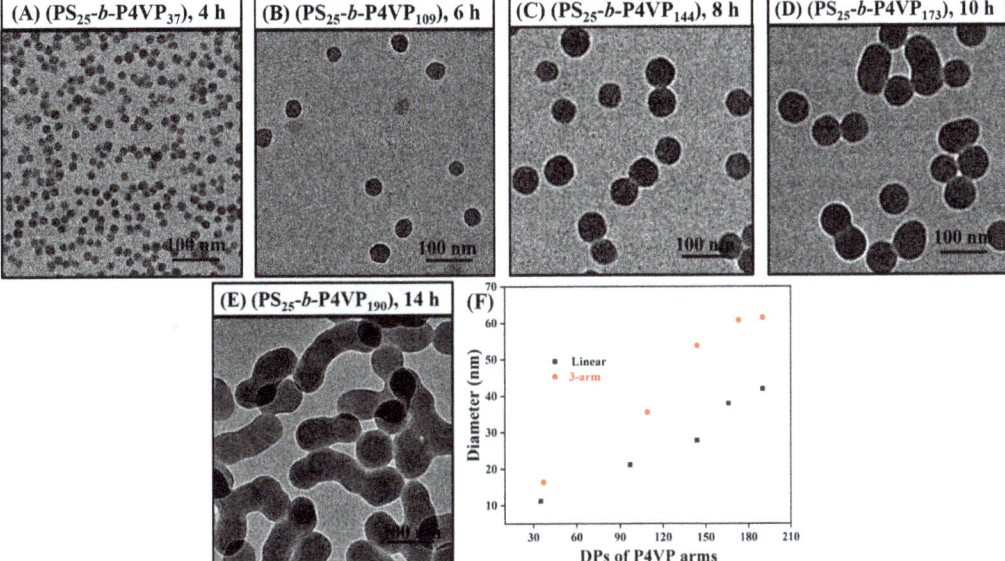

Figure 6. TEM images of 3-armed star $(PS_{25}\text{-}b\text{-}P4VP)_3$ nanoparticles formed by RAFT dispersion polymerization at 4 (**A**), 6 (**B**), 8 (**C**), 10 (**D**),14 (**E**), and average diameter (D) of $(PS_{25}\text{-}b\text{-}P4VP)_3$ and linear $PS_{24}\text{-}b\text{-}P4VP$ nanoparticles with the DP of P4VP arms increasing (**F**).

3.4. Effect of the Arm Number on BCP Nanoassemblies Prepared through General Self-Assembly in Toluene

As introduced above, the polymerization-induced self-assembly of block copolymers and the self-assembly of block copolymers in block-selective solvent represent two strategies to prepare block copolymer nano-objects. Herein, the effect of the arm number on BCP nanoassemblies prepared through general self-assembly in toluene was investigated by checking the morphology of the linear and star block copolymer nanoassemblies with similar chemical compositions. To depress the residual monomer effect, the block copolymer nanoassemblies prepared at high monomer conversion through the polymerization-induced self-assembly were chosen. The preparation of the $(PS\text{-}b\text{-}P4VP)_n$ (n = 1, 2, 3) block copolymer nanoassemblies through general self-assembly in the block-selective solvent was achieved by initially dissolving the block copolymer in DCM, and then adding toluene slowly, finally removing the excess DCM under vacuum at room temperature as discussed elsewhere [42]. Whereas, differently from the highly concentrated block copolymer in the polymerization-induced self-assembly (~20 wt %), very diluted block copolymer (0.2 mg/mL) was employed in the present self-assembly strategy, since destabilization of the (PS-*b*-P4VP)n (n = 1, 2, 3) block copolymer dispersion was found when the block copolymer concentration was above 1 wt %. Under this diluted block copolymer concentration at 0.2 mg/mL, all the block copolymer nanoassemblies were of great stability.

Figure 7 shows the TEM images of the linear or star (PS-*b*-P4VP)$_3$ block copolymer nanoassemblies with similar PS block chains at about 25 and similar P4VP block chains at about 140 (top in figure) and with similar PS block chains at about 25 and similar P4VP block chains at about 190 (bottom in figure) prepared through general self-assembly in toluene. It indicates that the PS$_{24}$-*b*-P4VP$_{139}$ formed vesicles (Figure 7A1), whereas the similar chemical composition of star block copolymers of (PS$_{24}$-*b*-P4VP$_{144}$)$_2$ self-assembled into vesicles and bicontinuous nanospheres (Figure 7B1). Similar bicontinuous nanospheres of the mixture of poly-(ethylene glycol)-*b*-polystyrene/polystyrene-*b*-poly(ethylene glycol)-*b*-polystyrene [62] and poly(ethylene oxide)-*b*-poly(octadecyl methacrylate) containing a long side octadecyl chain [63] and amphiphilic polynorbornene block copolymer [64] were also prepared, and star block copolymers of (PS$_{25}$-*b*-P4VP$_{144}$)$_3$ self-assembled into aggregates (Figure 7C1). The self-assembly of PS$_{24}$-*b*-P4VP$_{188}$ in toluene resulted in large nanospheres (Figure 7A2), and the self-assembly of star block copolymers of (PS$_{24}$-*b*-P4VP$_{190}$)$_2$ and (PS$_{25}$-*b*-P4VP$_{190}$)$_3$ with similar chemical composition lead to multilayered vesicles (Figure 7B2), and bicontinuous nanospheres (Figure 7C2). These results clearly demonstrate that the topology of block copolymers also influenced the nanoassemblies prepared through general self-assembly in toluene.

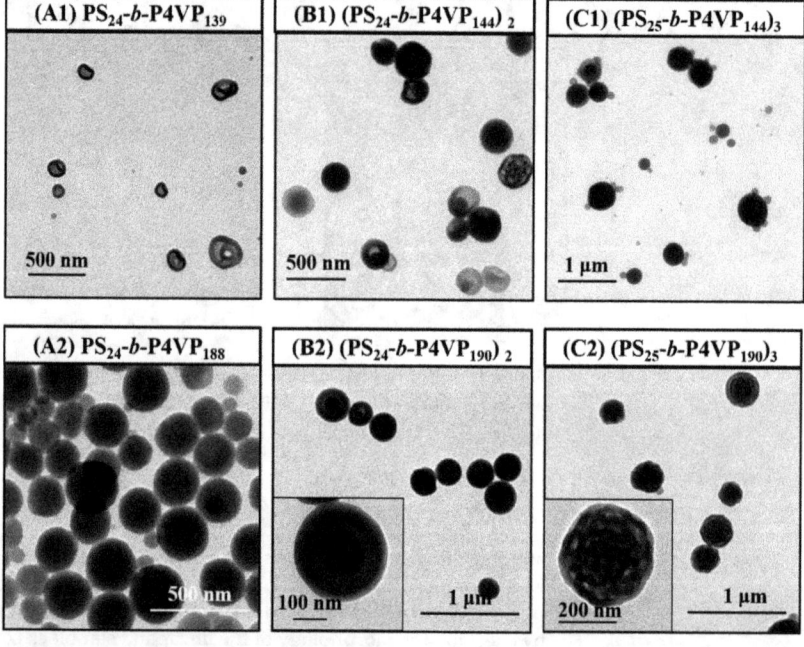

Figure 7. The TEM images of the linear or star block copolymer nanoassemblies of PS$_{24}$-*b*-P4VP$_{139}$ (**A1**) (PS$_{24}$-*b*-P4VP$_{144}$)$_2$ (**B1**) (PS$_{25}$-*b*-P4VP$_{144}$)$_3$ (**C1**) with similar PS block chains at about 25 and similar P4VP block chains at about 140 (top), and PS$_{24}$-*b*-P4VP$_{188}$ (**A2**), (PS$_{24}$-*b*-P4VP$_{190}$)$_2$ (**B2**), (PS$_{25}$-*b*-P4VP$_{190}$)$_3$ (**C2**) with similar PS block chains at about 25 and similar P4VP block chains at about 190 (bottom) prepared through general self-assembly in toluene.

4. Conclusions

In conclusion, linear and star BCP nanoparticles of (PS-*b*-P4VP)$_n$ with numbers of 1, 2, 3, and 4 were prepared by two methods of polymerization-induced self-assembly using

(PS-TTC)$_n$ (n = 1, 2, 3, 4) macro-RAFT agents and self-assembly of block copolymers in the low-polar organic solvent, toluene. Furthermore, the effect of the topology on the size and/or morphology of the (PS-b-P4VP)$_n$ nanoassemblies synthesized by the two methods in toluene was investigated in detail, and star (PS-b-P4VP)$_n$ had a more complex morphology than its linear counterpart. In addition, the PISA process of linear or star polymers with different topologies had similar polymerization kinetics. The possible reason is that all the macro-CTAs had similar R and Z groups and similar arm lengths. Our results show that in toluene, a low polar solvent, the topology not only influenced the morphology of the BCP nanoparticles synthesized by PISA, but also affected their nanoassemblies prepared through general self-assembly.

Supplementary Materials: The following supporting information can be downloaded at: https://www.mdpi.com/article/10.3390/polym14173691/s1, Scheme S1: Mono- and Multifunctional Trithiocarbonates; Scheme S2: Synthesis of linear and 2, 3, 4-arm CTAs; Figure S1: ^{1}H NMR spectra (A) and ^{13}C NMR spectra (B) of mono- and multifunctional macro-CTAs of trithiocarbonate. Note: peak e at 220 ppm is out the range of test in the 13 C NMR spectra and Scheme S3: Synthesis of the (PSm-TTC)n (n = 1, 2, 3, 4) macro-RAFT agents and the dispersion RAFT polymerization of 4-vinylpyridine in the presence of (PSm-TTC)n.

Author Contributions: Methodology, Y.Z.; validation, Y.Z., and P.W.; formal analysis, N.L. and C.G.; investigation, Y.Z., and P.W.; data curation, Y.Z., and P.W.; writing—original draft preparation, P.W.; writing—review and editing, Y.Z. and S.L.; supervision, Y.Z.; project administration, Y.Z.; funding acquisition, Li, S. All authors have read and agreed to the published version of the manuscript.

Funding: This research was funded by the National Science Foundation of China (no. 52073126).

Institutional Review Board Statement: Not applicable.

Informed Consent Statement: Not applicable.

Data Availability Statement: Not applicable.

Conflicts of Interest: The authors declare that they have no known competing financial interests or personal relationships that could have appeared to influence the work reported in this paper.

References

1. Mai, Y.; Eisenberg, A. Self-assembly of block copolymers. *Chem. Soc. Rev.* **2012**, *41*, 5969–5985. [CrossRef]
2. Rodríguez-Hernández, J.; Chécot, F.; Gnanou, Y.; Lecommandoux, S. Toward 'smart' nano-objects by self-assembly of block copolymers in solution. *Prog. Polym. Sci.* **2005**, *30*, 691–724. [CrossRef]
3. Lomas, H.; Canton, I.; MacNeil, S.; Du, J.; Armes, S.P.; Ryan, A.J.; Lewis, A.L.; Battaglia, G. Biomimetic pH Sensitive Polymersomes for Efficient DNA Encapsulation and Delivery. *Adv. Mater.* **2007**, *19*, 4238–4243. [CrossRef]
4. Bockstaller, M.R.; Lapetnikov, Y.; Margel, S.; Thomas, E.L. Size-selective organization of enthalpic compatibilized nanocrystals in ternary block copolymer/particle mixtures. *J. Am. Chem. Soc.* **2003**, *125*, 5276–5277. [CrossRef]
5. Chen, S.; Jiang, F.; Cao, Z.; Wang, G.; Dang, Z.M. Photo, pH, and thermo triple-responsive spiropyran-based copolymer nanoparticles for controlled release. *Chem. Commun.* **2015**, *51*, 12633–12636. [CrossRef]
6. Borchert, U.; Lipprandt, U.; Bilang, M.; Kimpfler, A.; Rank, A.; Peschka-Suss, R.; Schubert, R.; Lindner, P.; Forster, S. pH-induced release from P2VP-PEO block copolymer vesicles. *Langmuir* **2006**, *22*, 5843–5847. [CrossRef]
7. Khan, H.; Cao, M.; Duan, W.; Ying, T.; Zhang, W. Synthesis of diblock copolymer nano-assemblies: Comparison between PISA and micellization. *Polymer* **2018**, *150*, 204–213. [CrossRef]
8. Rodichkin, I.D.; Gumerov, R.A.; Potemkin, I.I. Self-assembly of miktoarm palm tree-like star copolymers in a selective solvent. *J. Colloid Interface Sci.* **2022**, *606*, 1966–1973. [CrossRef]
9. Xiao, J.; He, Q.; Yang, M.; Li, H.; Qiu, X.; Wang, B.; Zhang, B.; Bu, W. Hierarchical self-assembly of miktoarm star copolymers with pathway complexity. *Polym. Chem.* **2021**, *12*, 1476–1486. [CrossRef]
10. Liu, R.; Rong, Z.; Han, G.; Yang, X.; Zhang, W. Synthesis and self-assembly of star multiple block copolymer of poly(4-vinylpyridine)-block-polystyrene. *Polymer* **2021**, *215*, 123431. [CrossRef]
11. D'Agosto, F.; Rieger, J.; Lansalot, M. RAFT-Mediated Polymerization-Induced Self-Assembly. *Angew. Chem. Int. Ed. Engl.* **2020**, *59*, 8368–8392. [CrossRef]
12. Wang, X.; An, Z. New Insights into RAFT Dispersion Polymerization-Induced Self-Assembly: From Monomer Library, Morphological Control, and Stability to Driving Forces. *Macromol. Rapid Commun.* **2019**, *40*, e1800325. [CrossRef]

13. Penfold, N.J.W.; Yeow, J.; Boyer, C.; Armes, S.P. Emerging Trends in Polymerization-Induced Self-Assembly. *ACS Macro Lett* **2019**, *8*, 1029–1054. [CrossRef]
14. An, N.; Chen, X.; Yuan, J. Non-thermally initiated RAFT polymerization-induced self-assembly. *Polym. Chem.* **2021**, *12*, 3220–3232. [CrossRef]
15. Liu, D.; Cai, W.; Zhang, L.; Boyer, C.; Tan, J. Efficient Photoinitiated Polymerization-Induced Self-Assembly with Oxygen Tolerance through Dual-Wavelength Type I Photoinitiation and Photoinduced Deoxygenation. *Macromolecules* **2020**, *53*, 1212–1223. [CrossRef]
16. Chen, X.; Liu, L.; Huo, M.; Zeng, M.; Peng, L.; Feng, A.; Wang, X.; Yuan, J. Direct Synthesis of Polymer Nanotubes by Aqueous Dispersion Polymerization of a Cyclodextrin/Styrene Complex. *Angew. Chem. Int. Ed. Engl.* **2017**, *56*, 16541–16545. [CrossRef]
17. Wang, X.; Man, S.; Zheng, J.; An, Z. Alkyl alpha-Hydroxymethyl Acrylate Monomers for Aqueous Dispersion Polymerization-Induced Self-Assembly. *ACS Macro Lett.* **2018**, *7*, 1461–1467. [CrossRef]
18. Ma, Y.; Gao, P.; Ding, Y.; Huang, L.; Wang, L.; Lu, X.; Cai, Y. Visible Light Initiated Thermoresponsive Aqueous Dispersion Polymerization-Induced Self-Assembly. *Macromolecules* **2019**, *52*, 1033–1041. [CrossRef]
19. Luo, X.; Zhao, S.; Chen, Y.; Zhang, L.; Tan, J. Switching between Thermal Initiation and Photoinitiation Redirects RAFT-Mediated Polymerization-Induced Self-Assembly. *Macromolecules* **2021**, *54*, 2948–2959. [CrossRef]
20. Zhou, H.; Liu, C.; Gao, C.; Qu, Y.; Shi, K.; Zhang, W. Polymerization-induced self-assembly of block copolymer through dispersion RAFT polymerization in ionic liquid. *J. Polym. Sci. Part A Polym. Chem.* **2016**, *54*, 1517–1525. [CrossRef]
21. Liu, H.; Gao, C.; Ding, Z.; Zhang, W. Synthesis of Polystyrene-block-Poly(4-vinylpyridine) Ellipsoids through Macro-RAFT-Agent-Mediated Dispersion Polymerization: The Solvent Effect on the Morphology of the In Situ Synthesized Block Copolymer Nanoobjects. *Macromol. Chem. Phys.* **2016**, *217*, 467–476. [CrossRef]
22. Bezik, C.T.; Mysona, J.A.; Schneider, L.; Ramírez-Hernández, A.; Müller, M.; de Pablo, J.J. Is the "Bricks-and-Mortar" Mesophase Bicontinuous? Dynamic Simulations of Miktoarm Block Copolymer/Homopolymer Blends. *Macromolecules* **2022**, *55*, 745–758. [CrossRef]
23. Velychkivska, N.; Sedláček, O.; Shatan, A.B.; Spasovová, M.; Filippov, S.K.; Chahal, M.K.; Janisova, L.; Brus, J.; Hanyková, L.; Hill, J.P.; et al. Phase Separation and pH-Dependent Behavior of Four-Arm Star-Shaped Porphyrin-PNIPAM4 Conjugates. *Macromolecules* **2022**, *55*, 2109–2122. [CrossRef]
24. Kim, H.; Kang, B.-G.; Choi, J.; Sun, Z.; Yu, D.M.; Mays, J.; Russell, T.P. Morphological Behavior of A2B Block Copolymers in Thin Films. *Macromolecules* **2018**, *51*, 1181–1188. [CrossRef]
25. Sheng, Y.J.; Nung, C.H.; Tsao, H.K. Morphologies of star-block copolymers in dilute solutions. *J. Phys. Chem. B* **2006**, *110*, 21643–21650. [CrossRef] [PubMed]
26. Ma, Y.; Wu, H.; Shen, Y. Dual-functional linear and star POSS-containing organic–inorganic hybrid block copolymers: Synthesis, self-assembly, and film property. *J. Mater. Sci.* **2022**, *57*, 7791–7803. [CrossRef]
27. Yoon, K.; Kang, H.C.; Li, L.; Cho, H.; Park, M.-K.; Lee, E.; Bae, Y.H.; Huh, K.M. Amphiphilic poly(ethylene glycol)-poly(ε-caprolactone) AB$_2$ miktoarm copolymers for self-assembled nanocarrier systems: Synthesis, characterization, and effects of morphology on antitumor activity. *Polym. Chem.* **2015**, *6*, 531–542. [CrossRef]
28. Yang, J.; Dong, Q.; Liu, M.; Li, W. Universality and Specificity in the Self-Assembly of Cylinder-Forming Block Copolymers under Cylindrical Confinement. *Macromolecules* **2022**, *55*, 2171–2181. [CrossRef]
29. Deng, R.; Wang, C.; Weck, M. Supramolecular Helical Miktoarm Star Polymers. *ACS Macro. Lett.* **2022**, *11*, 336–341. [CrossRef]
30. Baulin, V.A. Topological Changes in Telechelic Micelles: Flowers versus Stars. *Macromolecules* **2022**, *55*, 517–522. [CrossRef]
31. Ren, J.M.; McKenzie, T.G.; Fu, Q.; Wong, E.H.; Xu, J.; An, Z.; Shanmugam, S.; Davis, T.P.; Boyer, C.; Qiao, G.G. Star Polymers. *Chem. Rev.* **2016**, *116*, 6743–6836. [CrossRef] [PubMed]
32. Zhang, Y.; Cao, M.; Han, G.; Guo, T.; Ying, T.; Zhang, W. Topology Affecting Block Copolymer Nanoassemblies: Linear Block Copolymers versus Star Block Copolymers under PISA Conditions. *Macromolecules* **2018**, *51*, 5440–5449. [CrossRef]
33. Qu, Y.; Chang, X.; Chen, S.; Zhang, W. In situ synthesis of thermoresponsive 4-arm star block copolymer nano-assemblies by dispersion RAFT polymerization. *Polym. Chem.* **2017**, *8*, 3485–3496. [CrossRef]
34. Cao, M.; Nie, H.; Hou, Y.; Han, G.; Zhang, W. Synthesis of star thermoresponsive amphiphilic block copolymer nano-assemblies and the effect of topology on their thermoresponse. *Polym. Chem.* **2019**, *10*, 403–411. [CrossRef]
35. Wang, X.; Figg, C.A.; Lv, X.; Yang, Y.; Sumerlin, B.S.; An, Z. Star Architecture Promoting Morphological Transitions during Polymerization-Induced Self-Assembly. *ACS Macro Lett.* **2017**, *6*, 337–342. [CrossRef]
36. Zeng, R.; Chen, Y.; Zhang, L.; Tan, J. R-RAFT or Z-RAFT? Well-Defined Star Block Copolymer Nano-Objects Prepared by RAFT-Mediated Polymerization-Induced Self-Assembly. *Macromolecules* **2020**, *53*, 1557–1566. [CrossRef]
37. Wu, J.; Zhang, L.; Chen, Y.; Tan, J. Linear and Star Block Copolymer Nanoparticles Prepared by Heterogeneous RAFT Polymerization Using an omega, omega-Heterodifunctional Macro-RAFT Agent. *ACS Macro Lett.* **2022**, *11*, 910–918. [CrossRef]
38. Thompson, K.L.; Mable, C.J.; Lane, J.A.; Derry, M.J.; Fielding, L.A.; Armes, S.P. Preparation of Pickering Double Emulsions Using Block Copolymer Worms. *Langmuir* **2015**, *31*, 4137–4144. [CrossRef]
39. Docherty, P.J.; Girou, C.; Derry, M.J.; Armes, S.P. Epoxy-functional diblock copolymer spheres, worms and vesiclesviapolymerization-induced self-assembly in mineral oil. *Polym. Chem.* **2020**, *11*, 3332–3339. [CrossRef]
40. Luo, X. A morphological transition of poly(ethylene glycol)-block-polystyrene with polymerization-induced self-assembly guided by using cosolvents. *Eur. Polym. J.* **2021**, *158*, 110639. [CrossRef]

41. Dan, M.; Huo, F.; Zhang, X.; Wang, X.; Zhang, W. Dispersion RAFT polymerization of 4-vinylpyridine in toluene mediated with the macro-RAFT agent of polystyrene dithiobenzoate: Effect of the macro-RAFT agent chain length and growth of the block copolymer nano-objects. *J. Polym. Sci. Part A Polym. Chem.* **2013**, *51*, 1573–1584. [CrossRef]
42. Zhang, Y.; Guan, T.; Han, G.; Guo, T.; Zhang, W. Star Block Copolymer Nanoassemblies: Block Sequence is All-Important. *Macromolecules* **2018**, *52*, 718–728. [CrossRef]
43. Cao, M.; Han, G.; Duan, W.; Zhang, W. Synthesis of multi-arm star thermo-responsive polymers and topology effects on phase transition. *Polym. Chem.* **2018**, *9*, 2625–2633. [CrossRef]
44. Stenzel-Rosenbaum, M.; Davis, T.P.; Chen, V.; Fane, A.G. Star-polymer synthesis via radical reversible addition–fragmentation chain-transfer polymerization. *J. Polym. Sci. Part A Polym. Chem.* **2001**, *39*, 2777–2783. [CrossRef]
45. Pang, X.; Zhao, L.; Akinc, M.; Kim, J.K.; Lin, Z. Novel Amphiphilic Multi-Arm, Star-Like Block Copolymers as Unimolecular Micelles. *Macromolecules* **2011**, *44*, 3746–3752. [CrossRef]
46. Whittaker, M.R.; Monteiro, M.J. Synthesis and aggregation behavior of four-arm star amphiphilic block copolymers in water. *Langmuir* **2006**, *22*, 9746–9752. [CrossRef]
47. Skrabania, K.; Li, W.; Laschewsky, A. Synthesis of Double-Hydrophilic BAB Triblock Copolymers via RAFT Polymerisation and their Thermoresponsive Self-Assembly in Water. *Macromol. Chem. Phys.* **2008**, *209*, 1389–1403. [CrossRef]
48. Papagiannopoulos, A.; Zhao, J.; Zhang, G.; Pispas, S.; Radulescu, A. Thermoresponsive aggregation of PS-PNIPAM-PS triblock copolymer: A combined study of light scattering and small angle neutron scattering. *Eur. Polym. J.* **2014**, *56*, 59–68. [CrossRef]
49. Kim, S.H.; Jo, W.H. A Monte Carlo Simulation for the Micellization of ABA- and BAB-Type Triblock Copolymers in a Selective Solvent. *Macromolecules* **2001**, *34*, 7210–7218. [CrossRef]
50. Lee, D.S.; Shim, M.S.; Kim, S.W.; Lee, H.; Park, I.; Chang, T. Novel Thermoreversible Gelation of Biodegradable PLGA-block-PEO-block-PLGA Triblock Copolymers in Aqueous Solution. *Macromol. Rapid Commun.* **2001**, *22*, 587–592. [CrossRef]
51. Zhang, W.-J.; Hong, C.-Y.; Pan, C.-Y. Fabrication of Spaced Concentric Vesicles and Polymerizations in RAFT Dispersion Polymerization. *Macromolecules* **2014**, *47*, 1664–1671. [CrossRef]
52. Qu, Y.; Wang, S.; Khan, H.; Gao, C.; Zhou, H.; Zhang, W. One-pot preparation of BAB triblock copolymer nano-objects through bifunctional macromolecular RAFT agent mediated dispersion polymerization. *Polym. Chem.* **2016**, *7*, 1953–1962. [CrossRef]
53. Dong, S.; Zhao, W.; Lucien, F.P.; Perrier, S.; Zetterlund, P.B. Polymerization induced self-assembly: Tuning of nano-object morphology by use of CO_2. *Polym. Chem.* **2015**, *6*, 2249–2254. [CrossRef]
54. Sahoo, S.; Gordievskaya, Y.D.; Bauri, K.; Gavrilov, A.A.; Kramarenko, E.Y.; De, P. Polymerization-Induced Self-Assembly (PISA) Generated Cholesterol-Based Block Copolymer Nano-Objects in a Nonpolar Solvent: Combined Experimental and Simulation Study. *Macromolecules* **2022**, *55*, 1139–1152. [CrossRef]
55. Gao, P.; Cao, H.; Ding, Y.; Cai, M.; Cui, Z.; Lu, X.; Cai, Y. Synthesis of Hydrogen-Bonded Pore-Switchable Cylindrical Vesicles via Visible-Light-Mediated RAFT Room-Temperature Aqueous Dispersion Polymerization. *ACS Macro Lett.* **2016**, *5*, 1327–1331. [CrossRef]
56. Ng, G.; Yeow, J.; Xu, J.; Boyer, C. Application of oxygen tolerant PET-RAFT to polymerization-induced self-assembly. *Polym. Chem.* **2017**, *8*, 2841–2851. [CrossRef]
57. Pei, Y.; Lowe, A.B. Polymerization-induced self-assembly: Ethanolic RAFT dispersion polymerization of 2-phenylethyl methacrylate. *Polym. Chem.* **2014**, *5*, 2342–2351. [CrossRef]
58. Kang, Y.; Pitto-Barry, A.; Maitland, A.; O'Reilly, R.K. RAFT dispersion polymerization: A method to tune the morphology of thymine-containing self-assemblies. *Polym. Chem.* **2015**, *6*, 4984–4992. [CrossRef]
59. Zhang, S.; Ma, X.; Zhu, Y.; Guo, R. Dispersion polymerization of styrene/acrylonitrile in polyether stabilized by macro-RAFT agents. *Colloids Surf. A* **2022**, *647*, 129166. [CrossRef]
60. Dhiraj, H.S.; Ishizuka, F.; Elshaer, A.; Zetterlund, P.B.; Aldabbagh, F. RAFT dispersion polymerization induced self-assembly (PISA) of boronic acid-substituted acrylamides. *Polym. Chem.* **2022**, *13*, 3750–3755. [CrossRef]
61. Li, J.-W.; Chen, M.; Zhou, J.-M.; Pan, C.-Y.; Zhang, W.-J.; Hong, C.-Y. RAFT dispersion copolymerization of styrene and N-methacryloxysuccinimide: Promoted morphology transition and post-polymerization cross-linking. *Polymer* **2021**, *221*, 123589. [CrossRef]
62. McKenzie, B.E.; Friedrich, H.; Wirix, M.J.M.; de Visser, J.F.; Monaghan, O.R.; Bomans, P.H.H.; Nudelman, F.; Holder, S.J.; Sommerdijk, N.A.J.M. Controlling Internal Pore Sizes in Bicontinuous Polymeric Nanospheres. *Angew. Chem. Int. Ed.* **2015**, *54*, 2457–2461. [CrossRef] [PubMed]
63. Barnhill, S.A.; Bell, N.C.; Patterson, J.P.; Olds, D.P.; Gianneschi, N.C. Phase Diagrams of Polynorbornene Amphiphilic Block Copolymers in Solution. *Macromolecules* **2015**, *48*, 1152–1161. [CrossRef]
64. Gao, C.; Wu, J.; Zhou, H.; Qu, Y.; Li, B.; Zhang, W. Self-Assembled Blends of AB/BAB Block Copolymers Prepared through Dispersion RAFT Polymerization. *Macromolecules* **2016**, *49*, 4490–4500. [CrossRef]

Article

Simultaneous Removal of Cr(VI) and Phenol from Water Using Silica-di-Block Polymer Hybrids: Adsorption Kinetics and Thermodynamics

Jia Qu *, Qiang Yang, Wei Gong, Meilan Li and Baoyue Cao

Shaanxi Key Laboratory of Comprehensive Utilization of Tailings Resources, Shaanxi Engineering Research Center for Mineral Resources Clean & Efficient Conversion and New Materials, Shangluo University, Shangluo 726000, China; yq_sust@163.com (Q.Y.); gongwei-com@escience.cn (W.G.); liecho2009@163.com (M.L.); cby0406@163.com (B.C.)
* Correspondence: 231034@slxy.edu.cn; Tel.: +86-0914-2986027

Citation: Qu, J.; Yang, Q.; Gong, W.; Li, M.; Cao, B. Simultaneous Removal of Cr(VI) and Phenol from Water Using Silica-di-Block Polymer Hybrids: Adsorption Kinetics and Thermodynamics. *Polymers* 2022, 14, 2894. https://doi.org/10.3390/polym14142894

Academic Editors: Nikolaos Politakos and Apostolos Avgeropoulos

Received: 30 June 2022
Accepted: 13 July 2022
Published: 16 July 2022

Publisher's Note: MDPI stays neutral with regard to jurisdictional claims in published maps and institutional affiliations.

Copyright: © 2022 by the authors. Licensee MDPI, Basel, Switzerland. This article is an open access article distributed under the terms and conditions of the Creative Commons Attribution (CC BY) license (https://creativecommons.org/licenses/by/4.0/).

Abstract: Heavy metal ions and organic pollutants often coexist in industrial effluents. In this work, silica-di-block polymer hybrids (SiO_2-g-PBA-b-PDMAEMA) with two ratios (SiO_2/BA/DMAEMA = 1/50/250 and 1/60/240) were designed and prepared for the simultaneous removal of Cr(VI) and phenol via a surface-initiated atom-transfer radical polymerization process using butyl methacrylate (BA) as a hydrophobic monomer and 2-(Dimethylamino)ethylmethacrylate (DMAEMA) as a hydrophilic monomer. The removal efficiency of Cr(VI) and phenol by the hybrids reached 88.25% and 88.17%, respectively. The sample with a larger proportion of hydrophilic PDMAEMA showed better adsorption of Cr(VI), and the sample with a larger proportion of hydrophobic PBA showed better adsorption of phenol. In binary systems, the presence of Cr(VI) inhibited the adsorption of phenol, yet the presence of phenol had a negligible effect on the adsorption of Cr(VI). Kinetics studies showed that the adsorption of Cr(VI) and phenol fitted the pseudo-second-order model well. Thermodynamic studies showed that the adsorption behavior of Cr(VI) and phenol were better described by the Langmuir adsorption isotherm equation, and the adsorption of Cr(VI) and phenol were all spontaneous adsorptions driven by enthalpy. The adsorbent still possessed good adsorption capacity for Cr(VI) and phenol after six adsorption–desorption cycles. These findings show that SiO_2-g-PBA-b-PDMAEMA hybrids represent a satisfying adsorption material for the simultaneous removal of heavy metal ions and organic pollutants.

Keywords: silica; amphipathic; block copolymer; adsorption; Cr(VI); phenol

1. Introduction

Water pollution is a worldwide problem that requires urgent attention and prevention. Heavy metal ions [1,2] and organic pollutants [3,4] are common pollutants in water which pose a serious threat to human health and the ecological environment. The most commonly toxic heavy metal ions in water include Cr(VI) [5–7], As(III) [8], Cd(II) [9], Hg(II) [10], Pb(II) [11], Cu(II) [12], Zn(II) [13], etc. The main organic pollutants include phenolic compounds [14,15], benzene compounds [16,17], halohydrocarbons [18,19] and so on. Among various water purification and recycling technologies, adsorption is a fast, effective, easy to operate, inexpensive, and universal method [20,21]. Current research on adsorbent development mainly focuses on the adsorption of a certain type of pollutant like heavy metal ions or organic pollutants. But in practical application, adsorption materials are required to have good adsorption performance for different types of pollutants [22,23]. Therefore, it is urgent to design a new type of adsorption material which can adsorb metal ions and organic molecules simultaneously and efficiently.

Organic/inorganic nano-hybrids have significant advantages in terms of having inorganic nanoparticles with high specific surface area, high mechanical strength, high thermal stability, and durability [24,25]. What's more, by means of molecular design, the organic polymer chains of organic/inorganic nano-hybrids can adsorb different kinds of pollutants at the same time via van der Waals forces [26], hydrogen bonding [27], charge interactions [28], complexation [29], and other driving forces [30,31]. Therefore, organic/inorganic nano-hybrids are deemed to be an ideal adsorption material [32–34].

Except for the usual advantages of inorganic nanoparticles, nano-silica is low-cost and easily functionalized [35–37]. On the other hand, block co-polymerization is a useful method for getting functional co-polymers since block co-polymer has the superiorities of a clear chemical structure and narrow molecular weight distribution [38–40]. In this work, a new kind of silica-di-block polymer hybrid adsorbent, SiO_2-g-PBA-b-PDMAEMA, was prepared by surface-initiated atom-transfer radical polymerization (SI-ATRP) [41] using butyl methacrylate (BA) as a hydrophobic monomer and 2-(Dimethylamino)ethylmethacrylate (DMAEMA) as a hydrophilic monomer. Interaction between the hydrophobic block PBA and the organic pollutants was expected, and the hydrophilic functional segment PDMAEMA was introduced into the adsorbent for its affinity with heavy metal ions. The structure of SiO_2-g-PBA-b-PDMAEMA was characterized by FTIR, GPC, and TEM. The adsorption kinetics and thermodynamics of Cr(VI) and phenol in water were studied, and the competitive-adsorption behavior of binary systems of Cr(VI) and phenol was discussed.

2. Materials and Methods

2.1. Materials

Nano-silica (SiO_2, > 99%wt, a mean particle diameter of 20 nm and a specific surface area of 120 $m^2 \cdot g^{-1}$) was purchased from Hai Tai Nano (Nanjing, China) and used as received. 2-(Dimethylamino)ethylmethacrylate (DMAEMA, 99%) was purchased from Aladdin (Shanghai, China), which was dried over calcium hydride (CaH_2, 95%, Aladdin) for 24 h and distilled under reduced pressure before use. Butyl acrylate (BA, 99%, Aladdin) was rinsed with 5 wt% NaOH (97%, Aladdin) aqueous solution and dried over CaH_2 for 24 h. Tetrahydrofuran (THF, 99%, Aladdin) and cyclohexanone (CYC, 99.5%, Aladdin) were stirred over CaH_2 for 24 h at room temperature and distilled under reduced pressure prior to use. Triethylamine (TEA, 99%, Aladdin) was purified by distillation after drying over CaH_2. Cuprous chloride (CuCl, 97%, Aladdin) was purified before use [42]. (3-aminopropyl) triethoxysilane (APTES, 99%), 2-bromoisobutyrylbromide (BiBB, 98%), N,N,N',N',N''-pentamethyldiethylenetriamine (PMDETA, 99%), copper chloride ($CuCl_2$, 98%), phenol (C_6H_6O, 99%), potassium dichromate ($K_2Cr_2O_7$, 98%), hydrofluoric acid (HF, 40%), 1,5-diphenyl carbazide (98%) and ethanol (99.5%) were supplied by Aladdin and used as-received without further purification. Nitric acid (HNO_3, 68%) was supplied by Foshan Huaxisheng Chemical Co. Ltd. (Foshan, China) and used as-received without further purification.

2.2. Preparation of Silica Initiator SiO_2-Br

The first step was to prepare the amino-modified nano-silica (SiO_2-NH_2). An amount of 2.7 g SiO_2 was dispersed in a mixture of 87 mL water and 63 mL ethanol for 30 min. Then, 10.5 mL APTES was dissolved in 24 mL ethanol and then added to the above suspension. The pH of the system was adjusted to 10 by ammonium hydroxide, and the reaction was kept for 24 h at 50 °C. After centrifugation, the lower solid layer was alternately washed and centrifuged with water and ethanol. SiO_2-NH_2 was obtained after vacuum-drying at 50 °C for 24 h with a yield of 78%. The second step was to bring the bromine atoms into SiO_2-NH_2. An amount of 0.5 g SiO_2-NH_2 and 10 mL THF was added into a dried Schlenk flask and dispersed by ultrasound for 30 min. Under the condition of an ice bath, the Schlenk flask was filled with N_2 via three vacuum/N_2 cycles and the suspension was stirred for 30 min. An amount of 1.5 mL TEA was added subsequently and then mixed with the solution of 3 mL BiBB and 40 mL THF was injected dropwise into the flask to react for 4 h in an ice bath followed by 48 h of reaction in a water bath at 35 °C. After centrifugation, the lower solid layer was alternately washed and centrifuged with water and ethanol. The silica surface-initiator (SiO$_2$-Br) was obtained after vacuum-drying at 50 °C for 36 h with a yield of 72%. The synthesis scheme of the silica initiator SiO$_2$-Br is given in Scheme 1.

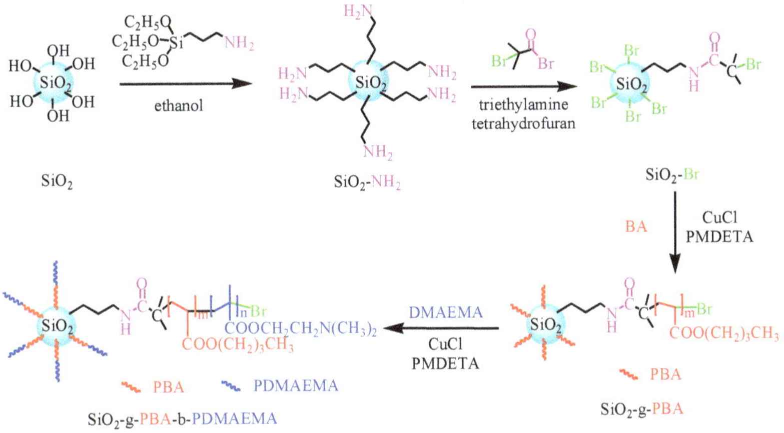

Scheme 1. Synthesis of SiO$_2$-g-PBA-b-PDMAEMA.

2.3. Preparation of SiO$_2$-g-PBA-b-PDMAEMA Hybrids by SI-ATRP

SiO$_2$-g-PBA-b-PDMAEMA hybrids were obtained via a one-step ATRP. CuCl and CuCl$_2$ were added into a dried Schlenk flask, and the flask was sealed with a rubber septum prior to three vacuum/N$_2$ cycles. BA and PMDETA were injected into the flask, followed by the suspension of SiO$_2$-Br dispersed in cyclohexanone; the reaction was performed at 90 °C for 24 h. After cooling to 70 °C, the cyclohexanone solution of DMAEMA was injected, and the reaction was performed at 70 °C for 24 h. SiO$_2$-g-PBA-b-PDMAEMA was finally obtained by centrifugation and washing with THF. The synthesis scheme of SiO$_2$-g-PBA-b-PDMAEMA is also given in Scheme 1, and the detailed recipes of polymerization are listed in Table 1.

Table 1. Detailed recipes for prepared samples.

Sample (SiO$_2$-Br:BA:DMAEMA)	SiO$_2$-Br /mmol	BA /mmol	DMAEMA /mmol	CuCl /mmol	CuCl$_2$ /mmol	PMDETA /mmol	CYC /g
S1 (1:50:250)	0.1620	8.1000	40.5000	0.1620	0.0162	0.1620	11.4773
S2 (1:60:240)	0.1620	9.7200	38.8800	0.1620	0.0162	0.1620	11.4066

Note: dosage of CYC was calculated based on a solid content of 40%.

2.4. Characterization of Initiator and Hybrids

The chemical structures for SiO$_2$-Br and SiO$_2$-g-PBA-b-PDMAEMA were characterized by Fourier transform infrared spectroscopy (FTIR, Nicolet-380, Thermo Electron Corporation, USA) in a spectral range of 4000–400 cm^{-1}. The molecular weights of PBA-b-PDMAEMA cleaved from SiO$_2$-g-PBA-b-PDMAEMA by hydrofluoric acid were measured by gel permeation chromatograph (GPC, PL-GPC50, Agilent, USA), equipped with the PL gel-MIXEDC chromatographic column and were calibrated by polystyrene standards. THF was used as the eluent, and the flow rate was 0.5 mL·min^{-1}. The particles of SiO$_2$-g-PBA-b-PDMAEMA were observed by transmission electron microscopy (TEM, Talos F200X, FEI, Czech Republic) at an acceleration voltage of 200 kV. Samples were ultrasounded in water for 30 min before measurement. The surface-grafted density of SiO$_2$-Br was obtained by thermogravimetric analyzer (TGA, STA449, NETZSCH, Germany) under N$_2$ atmosphere using a heating rate of 10 °C·min^{-1} from 25 °C to 1000 °C.

2.5. Batch Adsorption Study

The batch mode adsorption studies were carried out by adding 0.05 g SiO$_2$-g-PBA-b-PDMAEMA into 50 mL Cr(VI) aqueous solutions (prepared by dissolving K$_2$Cr$_2$O$_7$ in deionized water and mostly existing as HCrO$_4^-$ and CrO$_4^{2-}$ [43]) or 50 mL phenol aqueous solutions on a shaker at 120 rpm under controlled pH of 6. After thermostatic shaking for a certain period of time, supernatant obtained by centrifugation was analyzed to detect the concentration of Cr(VI) or phenol residues.

The concentration of residue Cr(VI) was determined using a ultraviolet-visible spectrophotometer (UV-Vis, Cary 5000, Agilent, USA) and developing a purple-violet color with 1,5-diphenyl carbazide in acidic solution as a complexing agent [44,45]. The absorbance of the purple-violet colored solution was read at 540 nm after 10 min of color development.

The concentration of phenol residue was determined using a UV-Vis spectrophotometer with maximum absorption wavelength of 270 nm [46,47].

The equilibrium adsorption amount Q_e (mg·g^{-1}) of SiO$_2$-g-PBA-b-PDMAEMA of Cr(VI) or phenol was calculated according to Equation (1) [48]. The percentage removal η(%) was calculated according to Equation (2) [4]:

$$Q = (C_0 - C_t)V/W \tag{1}$$

$$\eta = 100(C_0 - C_t)/C_0 \tag{2}$$

where C_0 (mg·L^{-1}) and C_t (mg·L^{-1}) are the initial concentrations and equilibrium concentrations of Cr(VI) or phenol, respectively, and V (L) is the volume of the solution and W (g) is the weight of the adsorbent.

2.6. Adsorption Kinetics

Adsorption kinetics were investigated to evaluate both the rate of Cr(VI) or phenol adsorption and the equilibrium time required for the adsorption isotherm. Experiments were conducted under an initial Cr(VI) or phenol concentration of 100 mg·L^{-1}, a controlled temperature of 298 K, and a controlled time t (5 min, 10 min, 20 min, 30 min, 60 min, 90 min, 120 min, 180 min, and 240 min).

Adsorption rate was analyzed using two kinetic models, i.e., the pseudo-first-order model and pseudo-second-order model. The pseudo-first-order kinetic model is expressed by Equation (3) [48]:

$$\ln(Q_e - Q_t) = -k_1 t + \ln Q_e \tag{3}$$

where Q_e (mg·g^{-1}) and Q_t (mg·g^{-1}) are the amount of Cr(VI) or phenol adsorbed at the equilibrium and time t (min), respectively, and k_1 (min^{-1}) is the pseudo-first-order rate constant for the adsorption process. Values of k_1 were calculated from the slope of the $\ln(Q_e - Q_t)$ vs. t plot. Values of Q_e were calculated from the intercept of the $\ln(Q_e - Q_t)$ vs. t plot.

The pseudo-second-order kinetic model is expressed by Equation (4) [48].

$$t/Q_t = t/Q_e + 1/(k_2 Q_e^2) \tag{4}$$

where Q_e (mg·g^{-1}) and Q_t (mg·g^{-1}) are the amount of Cr(VI) or phenol adsorbed at the equilibrium and time t (min), respectively, and k_2 (g·mg^{-1}·min^{-1}) is the pseudo-second-order rate constant for the adsorption process. Values of k_2 and Q_e were calculated from the slope and intercept of the t/Q_t vs. t plot.

2.7. Adsorption Isotherms

The adsorption isotherms were measured by controlling Cr(VI) or phenol concentration at 10 mg·L^{-1}, 20 mg·L^{-1}, 50 mg·L^{-1}, 100 mg·L^{-1}, 150 mg·L^{-1}, 200 mg·L^{-1}, and 300 mg·L^{-1}, respectively. After 120 min of equilibrium, the supernatant was analyzed for the concentration of residual Cr(VI) or phenol.

Two isotherm models were investigated, i.e., the Langmuir model and the Freundlich model. The Langmuir model is expressed by Equation (5) [48]:

$$C_e/Q_e = 1/(Q_m K_L) + C_e/Q_m \quad (5)$$

where Q_e is the equilibrium adsorption capacity (mg·g^{-1}), C_e is the equilibrium concentration in the solution (mg·L^{-1}), Q_m is the maximum adsorption capacity (mg·g^{-1}), and K_L is the Langmuir adsorption isotherm constant (L·mg^{-1}). Values of Q_m were calculated from the slope of the C_e/Q_e vs. C_e plot. Values of K_L were calculated from the intercept of the C_e/Q_e vs. C_e plot.

The Freundlich model is expressed by Equation (6) [48]:

$$\ln Q_e = \ln K_f + (1/n)\ln C_e \quad (6)$$

where Q_e is the equilibrium adsorption capacity (mg·g^{-1}) and C_e is the equilibrium concentration in the solution (mg·L^{-1}). K_f is the Freundlich adsorption isotherm constant [(mg·g^{-1})(L·mg^{-1})$^{1/n}$]. The term $1/n$ is related to the magnitude of the adsorption driving force. Values of n were calculated from the slope of the $\ln Q_e$ vs. $\ln C_e$ plot. Values of K_f were calculated from the intercept of the $\ln Q_e$ vs. $\ln C_e$ plot.

2.8. Thermodynamic Study

Three basic thermodynamic parameters were studied: the Gibbs free energy of adsorption (ΔG, J·mol^{-1}), the enthalpy change (ΔH, J·mol^{-1}), and the entropy change (ΔS, J·mol^{-1}·K^{-1}).

ΔG was calculated according to Equation (7) [48–50]:

$$\Delta G = -RT\ln K \quad (7)$$

where R is the universal gas constant (8.314 J·mol^{-1}·K^{-1}), T is the absolute temperature (K), and K is the derived Langmuir equilibrium constant.

ΔH and ΔS were calculated according to the van't Hoff equation, which is expressed by Equation (8) [48]:

$$\ln K = -\Delta H/(RT) + \Delta S/R \quad (8)$$

where K is the derived Langmuir equilibrium constant, R is the gas constant (8.314 J·mol^{-1}·K^{-1}), and T is the absolute temperature (K).

According to Equation (8), a linear relationship exists between $\ln K$ and $1/T$. ΔH and ΔS were calculated from the slope and the intercept of the $\ln K$ vs. $1/T$ plot, respectively.

2.9. Recovery Experiments

Six adsorption-desorption cycles were carried out to evaluate the recovery performance of the adsorbent. The adsorption of Cr(VI) or phenol was conducted at an initial concentration of 100 mg·L^{-1}, a controlled pH of 6, and a controlled temperature of 298 K. After adsorption equilibrium, the adsorbed Cr(VI) was removed from the adsorbent using 0.1 mol·L^{-1} nitric acid. The adsorbent was washed with deionized water and vacuum-dried before use. The adsorbed phenol was removed from the adsorbent using ethanol. The adsorbent was washed with deionized water and vacuum-dried before use.

2.10. Binary Systems Competitive Adsorption

The binary-system adsorption studies were carried out using batch mode adsorption to understand the competitive behavior between Cr(VI) and phenol onto SiO$_2$-g-PBA-b-PDMAEMA. The adsorption experiments were conducted by controlling the initial Cr(VI) concentration of 100 mg·L^{-1} and varying the initial phenol concentration stepwise or by controlling the initial phenol concentration of 100 mg·L^{-1} and varying the initial Cr(VI) concentration stepwise. The concentration of Cr(VI) residue was measured according to the diphenyl carbazide spectrophotometric method using a UV-Vis spectrophotometer at a wavelength of 540 nm. The concentration of phenol residue was determined using a UV-Vis spectrophotometer at a wavelength of 270 nm. The equilibrium adsorption amount Q_e (mg·g^{-1}) was calculated according to Equation (1).

3. Results and Discussion
3.1. Adsorbent Characterizations

The preparation of the silica initiator SiO$_2$-Br was confirmed by FTIR and TGA analysis shown in Figure 1. The new peaks in the FTIR spectrum of SiO$_2$-NH$_2$ at 2926 cm^{-1}, 2854 cm^{-1}, and 1466 cm^{-1} were attributed to the C-H vibration of APTES, which proved the grafting of APTES to SiO$_2$. Compared with the FTIR spectrum of SiO$_2$-NH$_2$, the new peak in the spectrum of SiO$_2$-Br at 1735 cm^{-1} was attributed to the C=O vibration of BiBB, which proved the grafting of BiBB onto SiO$_2$-NH$_2$. Furthermore, the TGA curves showed that the weight loss from 100 °C to 1000 °C for SiO$_2$, SiO$_2$-NH$_2$, and SiO$_2$-Br was 5.16%, 7.88%, and 21.93%, respectively; the weight loss below 100 °C was owed to the removal of the water absorbed physically. According to these results, the graft density was calculated at 0.81 mmol·g^{-1} for SiO$_2$-Br by the content of Br in SiO$_2$-Br.

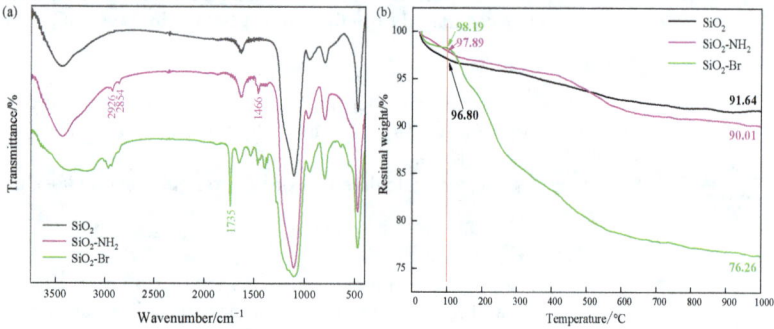

Figure 1. (a) IR spectra and (b) TGA curves of SiO$_2$, SiO$_2$-NH$_2$, and SiO$_2$-Br.

The verification of BA and DMAEMA grafting onto the silica surface was obtained by the FTIR and GPC analysis of SiO$_2$-g-PBA-b-PDMAEMA shown in Figure 2. The peaks in the FTIR spectrum at 2958 cm^{-1} and 2821 cm^{-1} were attributed to the stretching vibration of C-H in the PBA and PDMAEMA chains. The peaks at 1732 cm^{-1} were attributed to the stretching vibration of C=O. The peaks at 1105 cm^{-1} were attributed to the stretching vibration of Si-O-Si. The FTIR spectrum of SiO$_2$-g-PBA-b-PDMAEMA showed that PBA and PDMAEMA chains were successfully grafted onto the silica initiator SiO$_2$-Br. Furthermore, after etching by hydrofluoric acid, the molecular weights (M_n) of PBA-b-PDMAEMA cleaved from SiO$_2$-g-PBA-b-PDMAEMA were 32,930 g·mol^{-1} for S1 and 38,420 g·mol^{-1} for S2, which were close to the theoretical value for the single arm (45711 g·mol^{-1} for S1 and 45,421 g·mol^{-1} for S2, calculated from Table 1). The polydispersity indexes (PDIs) were 1.304 for S1 and 1.391 for S2, which revealed the narrow distribution of the molecular weights. GPC results illustrated that the polymerizations initiated by SiO$_2$-Br were all typically controllable ATRP, and the PBA-b-PDMAEMA chains were grafted as expected.

TEM images of SiO$_2$ and SiO$_2$-g-PBA-b-PDMAEMA at 100 nm are shown in Figure 3. The particle size of SiO$_2$ was about 20 nm. After grafting via PBA-b-PDMAEMA chains, the silica-di-block polymer hybrids formed spherical particles with a significantly increased diameter of 35–40 nm. Since the PBA-b-PDMAEMA chains stretched in water to some extent, the contact between the pollutants and the adsorbent in water could be improved.

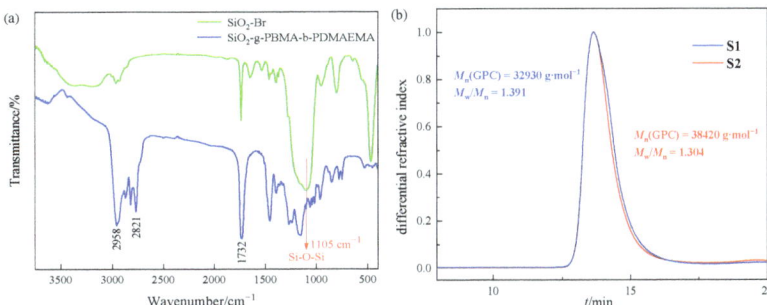

Figure 2. (**a**) IR spectra and (**b**) GPC curves of SiO$_2$-g-PBA-b-PDMAEMA.

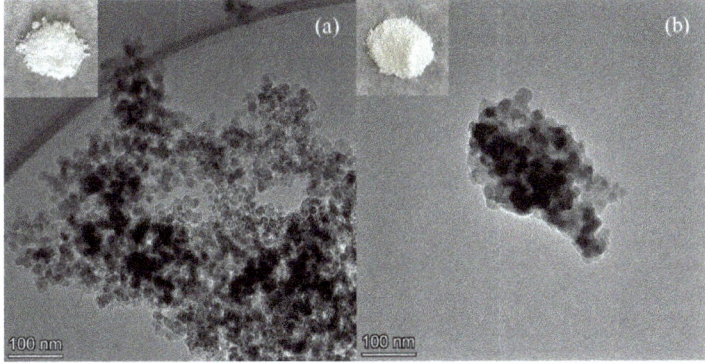

Figure 3. TEM images of (**a**) SiO$_2$ and (**b**) SiO$_2$-g-PBA-b-PDMAEMA.

3.2. Single Systems Adsorption Kinetics

Single system adsorption kinetics curves are shown in Figure 4. As a comparison, adsorption using bare nano-silica was also discussed. It could be seen from Figure 4a that Cr(VI) was adsorbed by SiO$_2$-g-PBA-b-PDMAEMA rapidly within 0–20 min, and the removal efficiency of Cr(VI) tended to be a constant after 60 min. The removal efficiency vs. t plot suggested to us that it took about 60 min for Cr(VI) to reach adsorption equilibrium on the surface of the adsorbent. The removal efficiency of Cr(VI) on S1 was 88.25%, which was more than 17-times higher than the removal efficiency using bare nano-silica (5.01%). In addition, the removal efficiency of Cr(VI) on S1 was higher than that of Cr(VI) on S2 (79.88%). This was probably because of the larger proportion of PDMAEMA in S1, which might have a stronger affinity for water-soluble ions. Figure 4b shows that phenol was adsorbed by SiO$_2$-g-PBA-b-PDMAEMA rapidly within 0–30 min, and the removal efficiency of phenol tended to be constant after 120 min. This suggested to us that the adsorption equilibrium was achieved within 120 min for phenol on the adsorbent. The removal efficiency of phenol on S2 was 88.17%, which was more than 9-times higher than the removal efficiency using bare nano-silica (9.09%) and was higher than the removal efficiency of phenol on S1 (80.44%). The adsorption capacity of SiO$_2$-g-PBA-b-PDMAEMA for phenol greatly improved after functionalization, and the larger proportion of hydrophobic PBA in S2 might be more conducive to the absorption of organic pollutants.

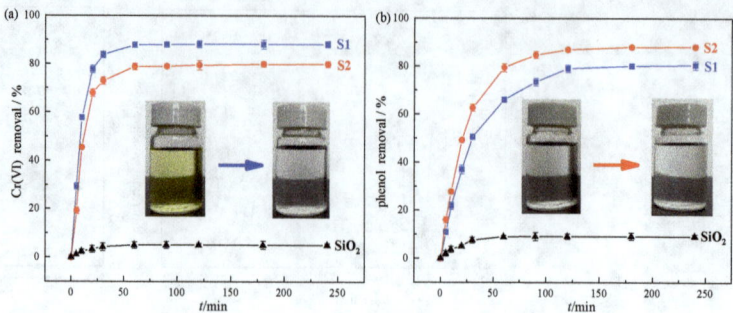

Figure 4. Adsorption kinetics of (**a**) Cr(VI) or (**b**) phenol on SiO$_2$-g-PBA-b-PDMAEMA and SiO$_2$ (C$_0$ = 100 mg·L^{-1}, pH = 6, T = 298 K, adsorbent concentration = 1 g·L^{-1}).

The adsorption kinetics data obtained experimentally were fitted to the pseudo-first-order and pseudo-second-order models. The fitting results are shown in Figure 5 and Table 2.

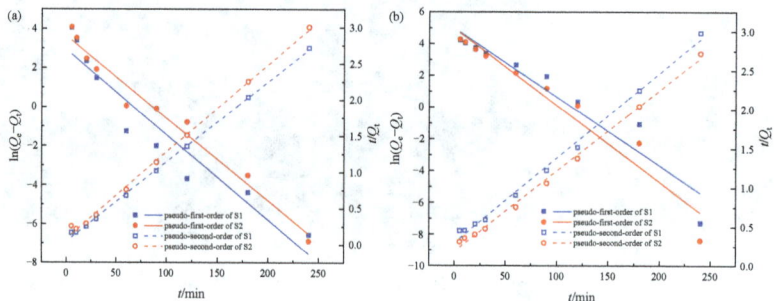

Figure 5. Fitting curves of adsorption kinetics for (**a**) Cr(VI) and (**b**) phenol on SiO$_2$-g-PBA-b-PDMAEMA.

Table 2. Fitting parameters of adsorption kinetics.

Systems	Adsorbents	$Q_{e,exp}$ /mg·g^{-1}	Pseudo-First-Order			Pseudo-Second-Order		
			$Q_{e,cal}$ /mg·g^{-1}	k_1 /min^{-1}	R^2	$Q_{e,cal}$ /mg·g^{-1}	k_2 /g·mg^{-1}·min^{-1}	R^2
Cr(VI)	S1	88.25	18.88	0.04364	0.9055	90.66	0.00242	0.9988
	S2	79.88	37.98	0.04223	0.9688	83.19	0.00167	0.9971
phenol	S1	80.44	142.62	0.04313	0.9124	92.42	0.00038	0.9944
	S2	88.17	140.87	0.04821	0.9395	97.28	0.00054	0.9957

The pseudo-second-order model showed much better fitting to Cr(VI), and phenol adsorption data with higher correlation coefficients ($R^2 > 0.99$) and a better agreement between experimental ($Q_{e,exp}$) and calculated ($Q_{e,cal}$) values is also exhibited in Table 2. This indicated that the adsorption rates of Cr(VI) and phenol onto SiO$_2$-g-PBA-b-PDMAEMA were controlled by chemical processes [9]. The rate constant k_2 of Cr(VI) on S1 (0.00242 g·mg^{-1}·min^{-1}) was relatively higher than that of Cr(VI) on S2 (0.00167 g·mg^{-1}·min^{-1}), indicating a faster uptake of Cr(VI) onto S1. Similarly, the rate constant k_2 of phenol on S2 (0.00054 g·mg^{-1}·min^{-1}) was relatively higher than that of phenol on S1 (0.00038 g·mg^{-1}·min^{-1}), indicating a faster uptake of phenol onto S2.

3.3. Single Systems Adsorption Isotherms and Thermodynamic Study

The adsorption isotherms of Cr(VI) or phenol in aqueous solution by SiO$_2$-g-PBA-b-PDMAEMA at different temperatures are shown in Figure 6. The equilibrium adsorption capacity of Cr(VI) or phenol rose with the increase of Cr(VI) or phenol equilibrium concentration and then tended to be constant, i.e., the saturated adsorption capacity. On the other hand, the equilibrium adsorption capacity of Cr(VI) or phenol decreased with the increase in adsorption temperature, indicating that the adsorption of Cr(VI) and phenol were all exothermic processes under experimental conditions. Decreasing the temperature was favorable for the adsorption of Cr(VI) or phenol on SiO$_2$-g-PBA-b-PDMAEMA.

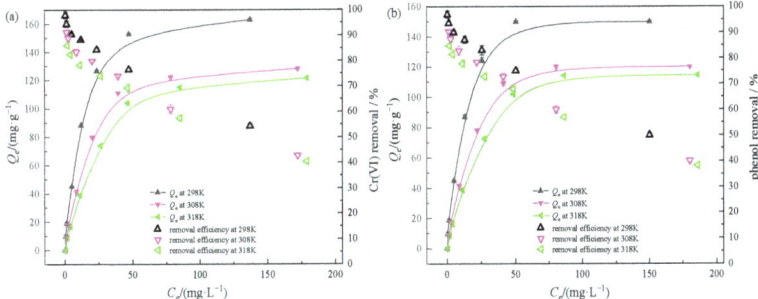

Figure 6. Adsorption isotherms of (**a**) Cr(VI) on S1 and (**b**) phenol on S2.

Fitting curves and adsorption isotherm constants for Cr(VI) adsorption on S1 are shown in Figure 7 and Table 3. According to the desired R^2 greater than 0.99, the adsorption behavior of Cr(VI) on the adsorbent surface is deemed to be better described by the Langmuir adsorption isotherm equation, indicating that Cr(VI) was mainly adsorbed by monolayer. This was because, after the monolayer adsorption of Cr(VI) on the surface of the adsorbent, the excess Cr(VI) did not easily approach the surface of the adsorbent due to the existence of electrostatic repulsion. According to the fitting parameters in Table 3, the saturated adsorption capacity of Cr(VI) at 298 K reached 174.22 mg·g^{-1}. The n values fitted by the Freundlich adsorption isotherm equation were all greater than 1, indicating that the adsorption of Cr(VI) on S1 were beneficial adsorption processes [51]. At 298 K, the n value was between 2 and 10, indicating that the adsorption of Cr(VI) on S1 more easily occurred at room temperature [52].

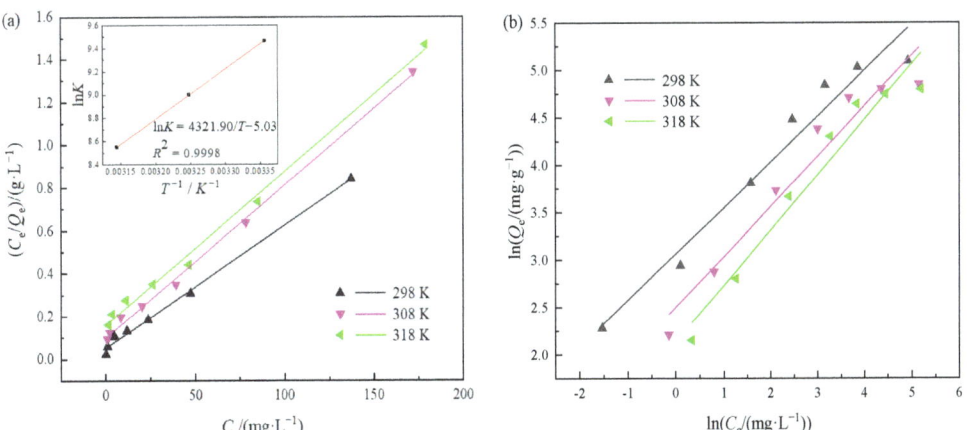

Figure 7. (**a**) Langmuir and (**b**) Freundlich adsorption isotherms for the adsorption of Cr(VI) on S1.

Table 3. Adsorption isotherm constants and thermodynamic parameters for Cr(VI) adsorption on S1 and phenol adsorption on S2.

Systems	Temperature /K	Langmuir			Freundlich			Thermodynamic Parameters		
		Q_m /mg·g^{-1}	K_L /L·mg^{-1}	R^2	n	K_f	R^2	ΔH /(kJ·mol^{-1})	ΔS /(kJ·mol^{-1}·K^{-1})	ΔG /(kJ·mol^{-1})
Cr(VI)	298	174.22	0.1114	0.9947	2.0617	21.3278	0.9541	−35.9	−0.0419	−23.4138
	308	140.06	0.0701	0.9970	1.8818	12.2125	0.9341			−22.9948
	318	139.08	0.0447	0.9940	1.7100	8.5612	0.9395			−22.5758
phenol	298	159.74	0.1191	0.9952	2.0211	19.1839	0.9385	−36.5	−0.0449	−23.1198
	308	131.58	0.0729	0.9949	1.8757	11.4701	0.9240			−22.6708
	318	130.89	0.0472	0.9906	1.7124	8.2102	0.9298			−22.2218

According to the Langmuir adsorption isotherm constant K_L, the thermodynamic equilibrium constant K was obtained by getting rid of the unit. The plot of ln K vs. $1/T$ for thermodynamic parameter calculation is shown in the inset picture of Figure 7a. The thermodynamic parameters calculated for Cr(VI) adsorption on S1 are shown in Table 3. $\Delta H < 0$ indicated that the adsorption was an exothermic process. Reducing the temperature was conducive to adsorption, and the adsorption was enthalpy-driven adsorption-type behavior. $\Delta S < 0$ indicated that the adsorption of Cr(VI) on S1 reduced the disorder degree of the system, and the adsorption was not entropy-driven adsorption-type behavior. ΔG ranged from −23.41 kJ·mol^{-1} to −22.58 kJ·mol^{-1}. The negative values of ΔG indicated that the adsorption was a spontaneous process. Moreover, it could be inferred that the adsorption driving force was more than the typical physical interactions since studies have shown that the ΔG of physical adsorption was −20–0 kJ·mol^{-1} and the ΔG of chemical adsorption was −400–−80 kJ·mol^{-1} [53]. ΔG of between −40 kJ·mol^{-1} and −20 kJ·mol^{-1} suggested to us that the main adsorption force might have been the electrostatic coulombic attraction [54]; this was consistent with our previous work on the adsorption of Cr(VI) to PDMAEMA chains [28].

Fitting curves and adsorption isotherm constants for phenol adsorption on S2 are shown in Figure 8 and Table 3. According to the desired R^2 ($R^2 > 0.99$), the adsorption behavior of phenol on the adsorbent surface could be better described by the Langmuir adsorption isotherm equation. The thermodynamic parameters calculated for phenol adsorption on S2 are shown in Table 3. $\Delta H < 0$, $\Delta S < 0$, $\Delta G < 0$, indicated that the adsorption was spontaneous adsorption, driven by enthalpy. As $\Delta H = -36.5$ kJ·mol^{-1}, which was a within the range of hydrogen bond adsorption enthalpy (2–40 kJ·mol^{-1}) [55], it could be inferred that the adsorption of S2 to phenol might be dominated by the hydrogen bond adsorption of the carbonyl group to phenol. Because of the directivity and saturation of the hydrogen bond, the adsorption of phenol on the adsorbent surface was mainly monolayered. The n values fitted by the Freundlich adsorption isotherm equation were all above 1, indicating that the adsorption of phenol on S2 was a beneficial adsorption process. At 298 K, the n value was 2.0211, indicating that the adsorption of phenol on S2 could easily occur at room temperature.

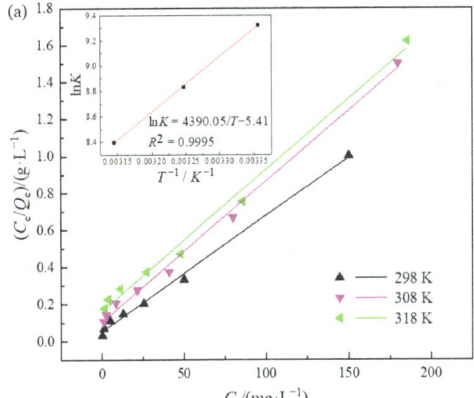 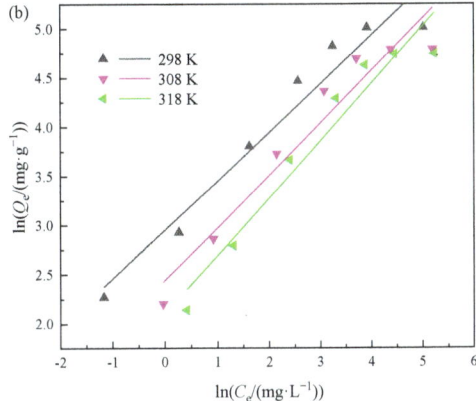

Figure 8. (a) Langmuir and (b) Freundlich adsorption isotherms for the adsorption of phenol on S2.

3.4. Recovery Performance

Six adsorption–desorption cycles were carried out to evaluate the recyclability of the adsorbent, which is an important point for practical applications. Figure 9 represents the adsorption–desorption cycle of Cr(VI) on S1 and phenol on S2. For the first three cycles, the adsorption amount of Cr(VI) was 87.92 mg·g^{-1}, 87.91 mg·g^{-1}, and 85.32 mg·g^{-1}, respectively, which showed stable adsorption performance of the hybrids to Cr(VI). After six adsorption-desorption cycles, the adsorption amount of Cr(VI) was 62.25 mg·g^{-1}, which accounted for 70.80% of the initial adsorption amount. These results indicated that the adsorbent still possessed an adequate adsorption capacity toward Cr(VI) after six cycles. For phenol adsorption, the adsorption amounts were 87.02 mg·g^{-1}, 86.98 mg·g^{-1}, and 85.43 mg·g^{-1} for the first three cycles, respectively, which also showed stable adsorption performance of the hybrids toward phenol. After six adsorption–desorption cycles, the adsorption amount for phenol was 72.79 mg·g^{-1}, which accounted for 83.65% of the initial adsorption amount. These results indicated that the adsorbent also possessed a good adsorption capacity toward phenol after six cycles.

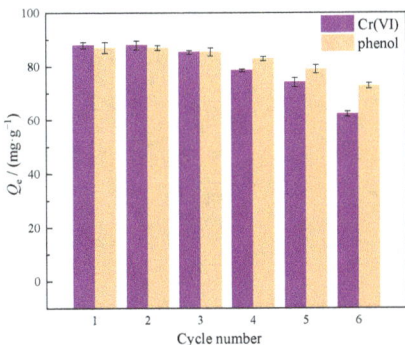

Figure 9. Adsorption–desorption cycle of Cr(VI) on S1 and phenol on S2.

3.5. Binary Systems Competitive Adsorption Behavior

The effect of the initial concentration of phenol on Cr(VI) adsorption is shown in Figure 10a. Phenol had little influence on the adsorption effect of Cr(VI). This was because when the pH < 7.8 [28,56], amino existed in the form of quaternary ammonium, and strong electrostatic attraction could have happened between the quaternary ammonium in SiO$_2$-g-PBA-b-PDMAEMA and Cr(VI) (in the forms of HCrO$_4^-$ and CrO$_4^{2-}$). The presence of phenol almost did not affect the electrostatic adsorption of Cr(VI) on the quaternary ammonium groups. The effect of the initial concentration of Cr(VI) on phenol adsorption is shown in Figure 10b. With the increase of the initial concentration of Cr(VI), the adsorption capacity of phenol decreased gradually. This might be because potassium ion in potassium dichromate can form co-ordination bonds with carbonyl groups, which compete may have competed with phenol and reduced the adsorption active site for phenol. The presumed adsorption mechanism is shown in Figure 11. Strong electrostatic attraction between the positively charged quaternary ammonium in the PDMAEMA blocks and Cr(VI) existed as oxygen anion, as the main driving force for the Cr(VI) adsorption. Hydrogen bonds between ester groups in SiO$_2$-g-PBA-b-PDMAEMA and phenol are the main driving forces for phenol adsorption. Generally, the silica-di-block polymer hybrids (SiO$_2$-g-PBA-b-PDMAEMA) showed good adsorption performance toward Cr(VI) as well as phenol in both the single and binary systems. The adsorption capacity of the hybrids compared to other materials is given in Table 4 [5–7,14,57–60].

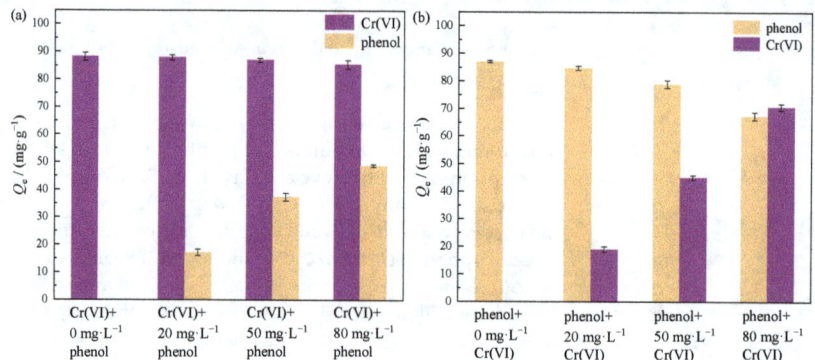

Figure 10. (a) The effect of the initial concentration of phenol on Cr(VI) adsorption and (b) the effect of the initial concentration of Cr(VI) on phenol adsorption.

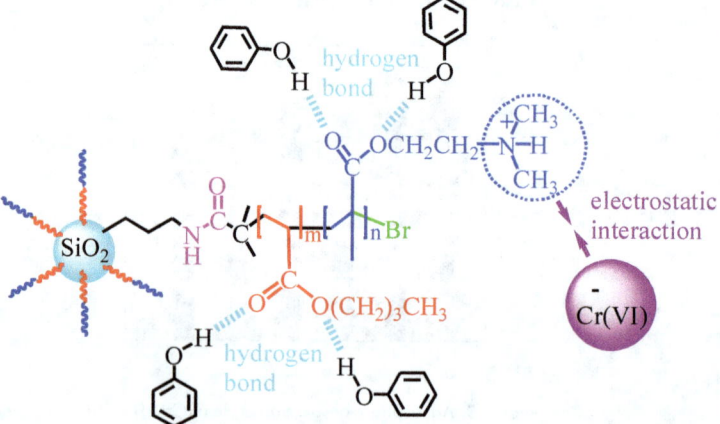

Figure 11. The proposed mechanisms for Cr(VI) and phenol adsorption.

Table 4. Comparison of different adsorbent materials.

Systems	Adsorbent Material	Q_m of Cr(VI) (mg·g^{-1})	Q_m of Phenol (mg·g^{-1})	References
Cr(VI)	divinylbenzene copolymer resin	99.91		[5]
	coffee polyphenol-formaldehyde resin	175.44		[6]
	coffee polyphenol-acetaldehyde resin	143.32		[6]
	puresorbe	76.92		[7]
	silica-di-block polymer hybrids	174.22		This study
phenol	palm-tree fruit stones		129.56	[14]
	porous acrylic ester polymer		78.70	[57]
	polymeric adsorbents IRA-96C		59.85	[58]
	silica-di-block polymer hybrids		159.74	This study
Cr(VI) and phenol	iron incorporated rice husk	36.3817	6.569	[59]
	natural red clay modified by hexadecyltrimethylammonium bromide	4.47	1.13	[60]
	silica-di-block polymer hybrids	87.02	37.36	This study

Note: Q_m refer to saturated adsorption capacity.

4. Conclusions

In this work, silica-di-block polymer hybrids, SiO$_2$-g-PBA-b-PDMAEMA, with two ratios (SiO$_2$/BA/DMAEMA = 1/50/250 and 1/60/240) were obtained via SI-ATRP methodology. All the results of the experimental studies can be summarized as follows:

(1) The results of FTIR and GPC proved that the SiO$_2$-g-PBA-b-PDMAEMA hybrids were generated as expected;

(2) The adsorbent had excellent adsorption effects for Cr(VI) as well as for phenol. Furthermore, changing the proportion of hydrophilic and hydrophobic chain segments allowed for the adjustment of the adsorption performance of the adsorbents for water-soluble ions and organic pollutants. S1 (SiO$_2$/BA/DMAEMA = 1/50/250) showed a higher removal efficiency of Cr(VI) (88.25%) than S2 (SiO$_2$/BA/DMAEMA = 1/60/240, 79.88%), and S2 showed a higher removal efficiency of phenol (88.17%) than S1 (80.44%);

(3) Kinetics studies showed that the adsorption of Cr(VI) and phenol fitted the pseudo-second-order model well;

(4) The thermodynamic studies showed that the adsorption of Cr(VI) and phenol were all exothermic processes; therefore, decreasing the temperature was favorable for the adsorption of Cr(VI) and phenol on SiO$_2$-g-PBA-b-PDMAEMA. The adsorption behavior of Cr(VI) and phenol was better described by the Langmuir adsorption isotherm equation, indicating that Cr(VI) and phenol were mainly adsorbed by monolayer. Thermodynamic parameters showed that the adsorptions were all spontaneous adsorption driven by enthalpy;

(5) The thermodynamic parameters suggested that the driving force of Cr(VI) adsorption on S1 was mainly the electrostatic attraction of anions by the quaternary ammonium of the PDMAEMA chains, while the adsorption of S2 to phenol was dominated by the hydrogen bond adsorption of carbonyl groups to phenol.

Further research will be conducted on simultaneous adsorption of various pollutants.

Author Contributions: Conceptualization, J.Q.; methodology, Q.Y. and W.G.; formal analysis, W.G. and M.L.; writing—original draft preparation, J.Q.; writing—review and editing, B.C.; supervision, Q.Y. and B.C. All authors have read and agreed to the published version of the manuscript.

Funding: This research was funded by the Special Scientific Research Plan Project of Shaanxi Provincial Department of Education, grant number 19JK0252; the Science and Technology Plan Project of Shangluo, grant number SK2019-81; and the Promotion and Application of Foamed Concrete Preparation Based on Vanadium Tailings, grant number 21HKY154.

Institutional Review Board Statement: Not applicable.

Informed Consent Statement: Not applicable.

Data Availability Statement: Data presented in this study are available on request from the first author.

Conflicts of Interest: The authors declare no conflict of interest.

References

1. Tchounwou, P.B.; Yedjou, C.G.; Patlolla, A.K.; Sutton, D.J. Heavy metal toxicity and the environment. *EXS* **2012**, *101*, 133–164. [PubMed]
2. Peralta-videa, J.R.; Lopez, M.L.; Narayan, M.; Saupe, G.; Gardea-Torresdey, J. The biochemistry of environmental heavy metal uptake by plants: Implications for the food chain. *Int. J. Biochem. Cell Biol.* **2009**, *41*, 1665–1677. [CrossRef] [PubMed]
3. Thirunavukkarasu, A.; Rajarathinam, N.; Sivashankar, R. A review on the role of nanomaterials in the removal of organic pollutants from wastewater. *Rev. Environ. Sci. Bio. Technol.* **2020**, *19*, 751–778. [CrossRef]
4. Ali, I.; Asim, M.; Khan, T.A. Low cost adsorbents for the removal of organic pollutants from wastewater. *J. Environ. Manag.* **2012**, *113*, 170–183. [CrossRef]
5. Bajpai, S.; Dey, A.; Jha, M.K.; Gupta, S.K.; Gupta, A. Removal of hazardous hexavalent chromium from aqueous solution using divinylbenzene copolymer resin. *Int. J. Environ. Sci. Technol.* **2012**, *9*, 683–690. [CrossRef]
6. Khudbudin, M.; Siona, D.; Kishor, R.; Sanjeev, T.; Nayaku, C. Adsorption of Chromium(VI) from Aqueous Solutions by Coffee Polyphenol-Formaldehyde/Acetaldehyde Resins. *J. Polym.* **2013**, 1–10.
7. Nityanandi, D.; Subbhuraam, C.V. Kinetics and thermodynamic of adsorption of chromium(VI) from aqueous solution using puresorbe. *J. Hazard. Mater.* **2009**, *170*, 876–882. [CrossRef]
8. Jagirani, M.S.; Balouch, A.; Mahesar, S.A.; Kumar, A.; Bhanger, M.L. Preparation of novel arsenic imprinted polymer for the selective extraction and enhanced adsorption of toxic As^{3+} ions from the aqueous environment. *Polym. Bull.* **2020**, *77*, 5261–5279. [CrossRef]
9. Zeng, G.M.; Liu, Y.Y.; Tang, L.; Yang, G.D.; Pang, Y.; Zhang, Y.; Zhou, Y.Y.; Li, Z.; Li, M.Y.; Lai, M.Y.; et al. Enhancement of Cd(II) adsorption by polyacrylic acid modified magnetic mesoporous carbon. *Chem. Eng.J.* **2015**, *259*, 153–160. [CrossRef]
10. Ali, S.A.; Mazumder, M.A.J. A new resin embedded with chelating motifs of biogenic methionine for the removal of Hg(II) at ppb levels. *J. Hazard. Mater.* **2018**, *350*, 169–179. [CrossRef]
11. Patil, S.A.; Suryawanshi, U.P.; Harale, N.S.; Patil, S.K.; Vadiyar, M.M.; Luwang, M.N.; Anuse, M.A.; Kim, J.H.; Kolekar, S.S. Adsorption of toxic Pb(II) on activated carbon derived from agriculture waste (Mahogany fruit shell): Isotherm, kinetic and thermodynamic study. *Int. J. Environ. Anal. Chem.* **2020**, 1–17. [CrossRef]
12. Rocha, C.G.; Zaia, D.A.M.; Alfaya, R.V.D.S.; Alfaya, A.A.D.S. Use of rice straw as biosorbent for removal of Cu(II), Zn(II), Cd(II) and Hg(II) ions in industrial effluents. *J. Hazard. Mater.* **2009**, *166*, 383–388. [CrossRef] [PubMed]
13. Dada, A.O.; Olalekan, A.P.; Olatunya, A.M.; Dada, A.O. Langmuir, Freundlich, Temkin and Dubinin–Radushkevich Isotherms Studies of Equilibrium Sorption of Zn^{2+} Unto Phosphoric Acid Modified Rice Husk. *IOSR J. Appl. Chem.* **2012**, *3*, 38–45.
14. Ahmed, M.J.; Theydan, S.K. Equilibrium isotherms, kinetics and thermodynamics studies of phenolic compounds adsorption on palm-tree fruit stones. *Ecotoxicol. Environ. Saf.* **2012**, *84*, 39–45. [CrossRef] [PubMed]
15. Soto, M.L.; Moure, A.; Dominguez, H.; Parajo, J.C. Recovery, concentration and purification of phenolic compounds by adsorption: A review. *J. Food Eng.* **2011**, *105*, 1–27. [CrossRef]
16. Karlowatz, M.; Kraft, M.; Mizaikoff, B. Simultaneous quantitative determination of benzene, toluene, and xylenes in water using mid-infrared evanescent field spectroscopy. *Anal. Chem.* **2004**, *76*, 2643–2648. [CrossRef] [PubMed]
17. Asenjo, N.G.; Alvarez, P.; Granda, M.; Blanco, C.; Santamaria, R.; Menendez, R. High performance activated carbon for benzene/toluene adsorption from industrial wastewater. *J. Hazard. Mater.* **2011**, *192*, 1525–1532. [CrossRef]
18. Adachi, A.; Ikeda, C.; Takagi, S.; Fukao, N.; Yoshie, E.; Okano, T. Efficiency of rice bran for removal of organochlorine compounds and benzene from industrial wastewater. *J. Agric. Food Chem.* **2001**, *49*, 1309–1314. [CrossRef]
19. Winid, B. Bromine and water quality-Selected aspects and future perspectives. *Appl. Geochem.* **2015**, *63*, 413–435. [CrossRef]
20. Ali, I.; Gupta, V.K. Advances in water treatment by adsorption technology. *Nat. Protoc.* **2006**, *1*, 2661–2667. [CrossRef]
21. Demirbas, A. Heavy metal adsorption onto agro-based waste materials: A review. *J. Hazard. Mater.* **2008**, *157*, 220–229. [CrossRef] [PubMed]
22. Zhou, Y.Y.; Zhang, F.F.; Tang, L.; Zhang, J.C.; Zeng, G.M.; Luo, L.; Liu, Y.Y.; Wang, P.; Peng, B.; Liu, X.C. Simultaneous removal of atrazine and copper using polyacrylic acid-functionalized magnetic ordered mesoporous carbon from water: Adsorption mechanism. *Sci. Rep.* **2017**, *7*, 43831. [CrossRef] [PubMed]
23. Jin, X.L.; Li, Y.F.; Yu, C.; Ma, Y.X.; Yang, L.Q.; Hu, H.Y. Synthesis of novel inorganic–organic hybrid materials for simultaneous adsorption of metal ions and organic molecules in aqueous solution. *J. Hazard. Mater.* **2011**, *198*, 247–256. [CrossRef] [PubMed]
24. Allahverdi, A.; Ehsani, M.; Janpour, H.; Ahmadi, S. The effect of nanosilica on mechanical, thermal and morphological properties of epoxy coating. *Prog. Org. Coat.* **2012**, *75*, 543–548. [CrossRef]
25. Jiang, J.Q.; Ashekuzzaman, S.M. Development of novel inorganic adsorbent for water treatment. *Curr. Opin. Chem. Eng.* **2012**, *1*, 191–199. [CrossRef]
26. Peng, H.; Zou, C.J.; Wang, C.J.; Tang, W.Y.; Zhou, J.X. The effective removal of phenol from aqueous solution via adsorption on CS/β-CD/CTA multicomponent adsorbent and its application for COD degradation of drilling wastewater. *Environ. Sci. Pollut. Res.* **2020**, *27*, 33668–33680. [CrossRef]

27. Thu, P.T.T.; Dieu, H.T.; Phi, H.N.; Viet, N.N.T.; Kim, S.J.; Vo, V. Synthesis, characterization and phenol adsorption of carbonyl-functionalized mesoporous silicas. *J. Porous Mater.* **2012**, *19*, 295–300. [CrossRef]
28. Huang, H.; Ren, D.Z.; Qu, J. pH and temperature-responsive POSS-based poly(2-(dimethylamino)ethyl methacrylate) for highly efficient Cr(VI) adsorption. *Colloid Polym. Sci.* **2020**, *298*, 1515–1521. [CrossRef]
29. Wang, C.; Zhao, J.L.; Wang, S.X.; Zhang, L.B.; Zhang, B. Efficient and Selective Adsorption of Gold Ions from Wastewater with Polyaniline Modified by Trimethyl Phosphate: Adsorption Mechanism and Application. *Polymers* **2019**, *11*, 652. [CrossRef]
30. Rengaraj, S.; Moon, S.H. Kinetics of adsorption of Co(II) removal from water and wastewater by ion exchange resins. *Water Res.* **2002**, *36*, 1783–1793. [CrossRef]
31. Pan, B.J.; Zhang, W.M.; Pan, B.C.; Qiu, H.; Zhang, Q.R.; Zhang, Q.X.; Zheng, S.R. Efficient removal of aromatic sulfonates from wastewater by a recyclable polymer: 2-naphthalene sulfonate as a representative pollutant. *Environ. Sci. Technol.* **2008**, *42*, 7411–7416. [CrossRef] [PubMed]
32. Moon, J.K.; Kim, K.W.; Jung, C.H.; Shul, Y.G.; Lee, E.H. Preparation of Organic-Inorganic Composite Adsorbent Beads for Removal of Radionuclides and Heavy Metal Ions. *J. Radioanal. Nucl. Chem.* **2000**, *246*, 299–307. [CrossRef]
33. Meyer, T.; Prause, S.; Spange, S.; Friedrich, M. Selective Ion Pair Adsorption of Cobalt and Copper Salts on Cationically Produced Poly(1,3-divinylimidazolid-2-one)/Silica Hybrid Particles. *J. Colloid Interface Sci.* **2001**, *236*, 335–342. [CrossRef]
34. Samiey, B.; Cheng, C.H.; Wu, J. Organic-Inorganic Hybrid Polymers as Adsorbents for Removal of Heavy Metal Ions from Solutions: A Review. *Materials* **2014**, *7*, 673–726. [CrossRef] [PubMed]
35. Huang, H.P.; Feng, Y.F.; Qu, J. Preparation and Performance of Silica-di-Block Polymer Hybrids for BSA-Resistance Coatings. *Materials* **2020**, *13*, 3478. [CrossRef]
36. Tsukagoshi, T.; Kondo, Y.; Yoshino, N. Protein adsorption on polymer-modified silica particle surface. *Colloids Surf. B Biointerfaces* **2007**, *54*, 101–107. [CrossRef]
37. Rao, A.V.; Kulkarni, M.M.; Amalnerkar, D.P.; Seth, T. Surface chemical modification of silica aerogels using various alkyl-alkoxy/chloro silanes. *Appl. Surf. Sci.* **2003**, *206*, 262–270. [CrossRef]
38. Ramakrishnan, A.; Dhamodharan, R. A novel and simple method of preparation of poly(styrene-b-2-vinylpyridine) block copolymer of narrow molecular weight distribution: Living anionic polymerization followed by mechanism transfer to controlled/"living" radical polymerization (ATRP). *J. Macromol. Sci. Part A* **2000**, *37*, 621–631. [CrossRef]
39. Leimenstoll, M.C.; Menzel, H. Behavior of ATRP-derived styrene and 4-vinylpyridine-based amphiphilic block copolymers in solution. *Colloid Polym. Sci.* **2018**, *296*, 1127–1135. [CrossRef]
40. Karkare, P.; Kumar, S.; Murthy, C.N. ARGET-ATRP using β-CD as reducing agent for the synthesis of PMMA-b-PS-b-PMMA triblock copolymers. *J. Appl. Polym. Sci.* **2019**, *136*, 47117. [CrossRef]
41. Huang, H.P.; He, L. Silica-diblock fluoropolymer hybrids synthesized by surface-initiated atom transfer radical polymerization. *RSC Adv.* **2014**, *4*, 13108–13118. [CrossRef]
42. Pan, A.Z.; He, L. Fabrication pentablock copolymer/silica hybrids as self-assembly coatings. *J. Colloid Interface Sci.* **2014**, *414*, 1–8. [CrossRef] [PubMed]
43. Wu, J.; Zhang, H.; He, P.J.; Yao, Q.; Shao, L.M. Cr(VI) removal from aqueous solution by dried activated sludge biomass. *J. Hazard. Mater.* **2010**, *176*, 697–703. [CrossRef]
44. Gan, W.E.; Li, Y.; He, Y.Z.; Zeng, R.H.; Guardia, M.D.L. Mechanism of porous core electroosmotic pump flow injection system and its application to determination of chromium(VI) in waste-water. *Talanta* **2000**, *51*, 667–675. [CrossRef]
45. Marcu, C.; Varodi, C.; Balla, A. Adsorption Kinetics of Chromium (VI) from Aqueous Solution Using an Anion Exchange Resin. *Anal. Lett.* **2020**, *54*, 140–149. [CrossRef]
46. Cherifi, H.; Hanini, S.; Bentahar, F. Adsorption of phenol from wastewater using vegetal cords as a new adsorbent. *Desalination* **2009**, *244*, 177–187. [CrossRef]
47. Ye, J.J.; Feng, W.; Tian, M.M.; Zhang, J.L.; Zhou, W.H.; Jia, Q. Spectrophotometric determination of phenol by flow injection on-line preconcentration with a micro-column containing magnetic microspheres functionalized with Cyanex272. *Anal. Methods* **2013**, *5*, 1046–1051. [CrossRef]
48. Konggidinata, M.I.; Chao, B.; Lian, Q.; Subramaniam, R.; Zappi, M.; Gang, D.D. Equilibrium, kinetic and thermodynamic studies for adsorption of BTEX onto Ordered Mesoporous Carbon (OMC). *J. Hazard. Mater.* **2017**, *336*, 249–259. [CrossRef]
49. Attia, Y.; Yu, S. Adsorption thermodynamics of a hydrophobic polymeric flocculant on hydrophobic colloidal coal particles. *Langmuir.* **1991**, *7*, 2203–2207. [CrossRef]
50. Pang, X.Y. Adsorption Kinetics and Thermodynamics Characteristics of Expanded Graphite for Polyethylene Glycol. *J. Chem.* **2010**, *7*, 1346–1358. [CrossRef]
51. Malakootian, M.; Mansoorian, H.J.; Yari, A.R. Removal of reactive dyes from aqueous solutions by a non-conventional and low cost agricultural waste: Adsorption on ash of Aloe Vera plant. *Iran. J. Health Saf. Environ.* **2014**, *1*, 117–125.
52. Yang, F.J.; Ma, C.H.; Yang, L.; Zhao, C.J.; Zhang, Y.; Zu, Y.G. Enrichment and Purification of Deoxyschizandrin and γ-Schizandrin from the Extract of Schisandra chinensis Fruit by Macroporous Resins. *Molecules* **2012**, *17*, 3510–3523. [CrossRef] [PubMed]
53. Chen, L.; Gao, X. Thermodynamic study of Th (IV) sorption on attapulgite. *Appl. Radiat. Isot.* **2009**, *67*, 1–6. [CrossRef]
54. Thanapackiam, P.; Mallaiya, K.; Rameshkumar, S.; Srikandan, S.S. Inhibition of corrosion of copper in acids by norfloxacin. *Anti-Corros. Methods Mater.* **2017**, *64*, 92–102.

55. Oepen, B.V.; Kördel, W.; Klein, W. Sorption of nonpolar and polar compounds to soils: Processes, measurements and experience with the applicability of the modified OECD-Guideline. *Chemosphere* **1991**, *22*, 285–304. [CrossRef]
56. Li, L.; Lu, B.B.; Fan, Q.K.; Wu, J.N.; Wei, L.L.; Hou, J.; Guo, X.H.; Liu, Z.Y. Synthesis and self-assembly behavior of ph-responsive starshaped POSS-(PCL-P(DMAEMA-co-PEGMA))$_{16}$ inorganic/organic hybrid block copolymer for the controlled intracellular delivery of doxorubicin. *RSC Adv.* **2016**, *6*, 61630–61640. [CrossRef]
57. Pan, B.; Pan, B.; Zhang, W.; Zhang, Q.; Zhang, Q.; Zheng, S. Adsorptive removal of phenol from aqueous phase by using a porous acrylic ester polymer. *J. Hazard. Mater.* **2008**, *157*, 293–299. [CrossRef]
58. Ming, Z.W.; Long, C.J.; Cai, P.B.; Xing, Z.Q.; Zhang, B. Synergistic adsorption of phenol from aqueous solution onto polymeric adsorbents. *J. Hazard. Mater.* **2006**, *128*, 123–129. [CrossRef]
59. Gupta, A.; Balomajumder, C. Simultaneous Adsorption of Cr(VI) and Phenol from Binary Mixture Using Iron Incorporated Rice Husk: Insight to Multicomponent Equilibrium Isotherm. *Int. J. Chem. Eng.* **2016**, 1–11. [CrossRef]
60. Gładysz-Płaska, A.; Majdan, M.; Pikus, S.; Sternik, D. Simultaneous adsorption of chromium(VI) and phenol on natural red clay modified by HDTMA. *Chem. Eng. J.* **2012**, *179*, 140–150. [CrossRef]

Article

Synthesis and Thermal Analysis of Non-Covalent PS-*b*-SC-*b*-P2VP Triblock Terpolymers via Polylactide Stereocomplexation

Ameen Arkanji, Viko Ladelta *, Konstantinos Ntetsikas * and Nikos Hadjichristidis *

Polymer Synthesis Laboratory, KAUST Catalysis Center, Physical Sciences and Engineering Division, King Abdullah University of Science and Technology (KAUST), Thuwal 23955, Saudi Arabia; ameen.arkanji@kaust.edu.sa
* Correspondence: viko.ladelta@kaust.edu.sa (V.L.); konstantinos.ntetsikas@kaust.edu.sa (K.N.); nikolaos.hadjichristidis@kaust.edu.sa (N.H.)

Citation: Arkanji, A.; Ladelta, V.; Ntetsikas, K.; Hadjichristidis, N. Synthesis and Thermal Analysis of Non-Covalent PS-*b*-SC-*b*-P2VP Triblock Terpolymers via Polylactide Stereocomplexation. *Polymers* 2022, 14, 2431. https://doi.org/10.3390/polym14122431

Academic Editors: Nikolaos Politakos and Apostolos Avgeropoulos

Received: 23 May 2022
Accepted: 13 June 2022
Published: 15 June 2022

Publisher's Note: MDPI stays neutral with regard to jurisdictional claims in published maps and institutional affiliations.

Copyright: © 2022 by the authors. Licensee MDPI, Basel, Switzerland. This article is an open access article distributed under the terms and conditions of the Creative Commons Attribution (CC BY) license (https://creativecommons.org/licenses/by/4.0/).

Abstract: Polylactides (PLAs) are thermoplastic materials known for their wide range of applications. Moreover, the equimolar mixtures of poly(L-Lactide) (PLLA) and poly(D-Lactide) (PDLA) can form stereocomplexes (SCs), which leads to the formation of new non-covalent complex macromolecular architectures. In this work, we report the synthesis and characterization of non-covalent triblock terpolymers of polystyrene-*b*-stereocomplex PLA-*b*-poly(2-vinylpyridine) (PS-*b*-SC-*b*-P2VP). Well-defined ω-hydroxy-PS and P2VP were synthesized by "living" anionic polymerization high-vacuum techniques with *sec*-BuLi as initiator, followed by termination with ethylene oxide. The resulting PS-OH and P2VP-OH were used as macroinitiators for the ring-opening polymerization (ROP) of DLA and LLA with Sn(Oct)$_2$ as a catalyst to afford PS-*b*-PDLA and P2VP-*b*-PLLA, respectively. SC formation was achieved by mixing PS-*b*-PDLA and P2VP-*b*-PLLA chloroform solutions containing equimolar PLAs segments, followed by precipitation into *n*-hexane. The molecular characteristics of the resulting block copolymers (BCPs) were determined by ^{1}H NMR, size exclusion chromatography, and Fourier-transform infrared spectroscopy. The formation of PS-*b*-SC-*b*-P2VP and the effect of molecular weight variation of PLA blocks on the resulting polymers, were investigated by differential scanning calorimetry, X-ray powder diffraction, and circular dichroism spectroscopies.

Keywords: polylactides; stereocomplexation; anionic polymerization; ring-opening polymerization; triblock terpolymers

1. Introduction

Among synthetic aliphatic polyesters, poly(lactic acids)/polylactides (PLAs) have attracted enormous interest because they are biocompatible and biodegradable [1–3]. PLAs are derived from natural renewable resources, are non-toxic to the human body, and possess good thermomechanical properties [4,5]. As a result, PLAs have been widely used in a broad range of applications, such as packaging materials [6,7] and biomedical materials (e.g., surgical sutures, implant materials, and controllable drug delivery systems) [8–10], among many other applications [11–14].

PLAs also possess several inherent defects [15,16], such as long degradation periods and slow crystallization rates, leading to inferior properties, and low heat distortion temperature, which increases the difficulty of processing [17,18]. Diverse strategies and chemical modifications have been employed to overcome such drawbacks, including the use of blends and additives [19]. Another approach is the use of PLA stereocomplex (SC)-based copolymers to enhance the properties of PLA.

Lactide (LA) exists in two optically active forms, i.e., stereoisomers, (R,R)L-lactide (LLA) and (S,S)D-lactide (DLA), and one optically inactive form, (R,S)*meso*-lactide (*m*LA) [20,21]. When LA is converted into a polymer, the stereoregularity of the chain has a strong influence on the thermomechanical properties. Consequently, PLA derived from different stereoisomers exhibits various physical and chemical properties. For example, isotactic PLLA and PDLA with high stereoregularity are semi-crystalline polymers with a melting temperature (T_m) between 170 and 190 °C [22], whereas atactic poly(*rac*-lactide) (PDLA) is amorphous due to the absence of stereoregularity. Moreover, the mixture of isotactic PLLA and PDLA can form superior material called PLA stereocomplex (SC).

The formation of SC crystallites between PLLA and PDLA was first reported by Ikada et al. in 1986 [23]. They found that an equimolar mixture of isotactic PLLA and PDLA in dichloromethane undergoes stereoselective physical association through multicenter hydrogen bonding interaction between the methyl and the carbonyl groups of the opposite configurations. This interaction was proved later by Ozaki et al. by an ultrasensitive IR spectroscopy [24,25]. These interactions result in a new arrangement of helicoidal chains (3_1) between L- and D-lactyl units interlocked side by side within the same crystal unit (i.e., racemic crystal), indicating more dense crystal packing compared to that of the homo-crystallites. Due to the strong interactions, SC exhibits exceptional physical and chemical stabilities leading to significantly enhanced properties [23]. For example, SC crystals were found to exhibit higher melting temperatures (ca. 50 °C higher than the corresponding homo-crystals), improved mechanical properties, and stronger hydrolytic resistance compared to PLA homo-crystals. Thus, stereocomplexation between PLLA and PDLA has proven to be a powerful tool for the generation of thermally and mechanically enhanced nanomaterials.

The stereocomplexation of block copolymers containing PLLA and PDLA blocks has been applied to synthesize non-covalent ABA-type triblock copolymers for various applications (A can be crystalline or amorphous block, B is PLA stereocomplex). Several examples of the A block used in such copolymers are polyethylene glycol (PEG) [26–29], poly(ε-caprolactone) (PCL) [30], polymenthide (PM) [31], poly(*N*,*N*-(dimethylamino)ethyl methacrylate) (PDMAEMA) [32], polyacrylic acid (PAA) [33], polystyrene (PS) [34], and polyisoprene (PI) [35,36]. To the best of our knowledge, P2VP-*b*-PLA systems have not been used in stereocomplex systems, even though the synthesis of such diblock copolymer has been reported via the combination of ring-opening polymerization (ROP) and reversible addition–fragmentation chain-transfer (RAFT) polymerization [37,38].

While much research has been conducted on non-covalent triblock copolymers using PLA SC, little information is known about PLA SC systems containing different block copolymers, i.e., non-covalent triblock terpolymer (ABC-type triblock terpolymer). This approach provides new insights into polymer design strategies for high-performance PLA-based materials. Therefore, a fundamental study of the synthesis and properties of such materials is required to establish the structure-property relationships.

Recently, our group reported the synthesis and characterization of non-covalent PS-*b*-SC-*b*-PI triblock terpolymers by the stereocomplex formation between well-defined PI-*b*-PLLA and PS-*b*-PDLA [39]. The synthesis of these block copolymers was accomplished by combining anionic polymerization high-vacuum techniques (HVTs) and ROP. Hydroxy-terminated PI and PS were synthesized via anionic polymerization and were used as macroinitiators for the ROP of LLA and DLA in the presence of Sn(Oct)$_2$ catalyst. The molecular characteristics, as well as the thermal properties of the precursors and the triblock terpolymers, were studied.

This study focuses on the synthesis and characterization of well-defined PS-*b*-SC-*b*-P2VP via the stereocomplex formation between P2VP-*b*-PLLA and PS-*b*-PDLA. Both BCPs were synthesized by combining anionic polymerization and ROP. First, hydroxy-functionalized P2VP and PS were synthesized by anionic polymerization HVTs, followed by termination with ethylene oxide and neutralization with methanol. The resulting polymers were then used as macroinitiators for the ROP of the corresponding LA using Sn(Oct)$_2$ as a catalyst. The synthesized BCPs were characterized by ^{1}H nuclear magnetic resonance (NMR) spectroscopy and size exclusion chromatography (SEC). Stereocomplex formation was accomplished by mixing PS-*b*-PDLA and P2VP-*b*-PLLA in chloroform, followed by precipitation in hexane. Both BCPs and their corresponding stereocomplexes were characterized by differential scanning calorimetry (DSC), Fourier-tranform infrared (FT-IR), powder X-ray diffraction (XRD), and circular dichroism (CD) spectroscopies to study the formation of PS-b-SC-b-P2VP and to evaluate the effect of the molecular weight of PLA on the resulting PS-b-SC-b-P2VP non-covalent triblock terpolymer properties.

2. Materials and Methods

2.1. For Anionic Polymerization

Benzene (VWR, Pris, France, 99%) and tetrahydrofuran (THF, VWR, Gliwice, Poland, ≥99.0%) were dried over calcium hydride (CaH_2, 95%) followed by distillation into a glass cylinder containing polystyrilithium ($PS^{(-)}Li^{(+)}$) for benzene, and sodium/potassium alloy for THF, under high vacuum. Styrene (Sigma-Aldrich, 99%) was dried over CaH_2 followed by distillation over di-n-butylmagnesium (Bu_2Mg) and stored at $-20\ °C$ in pre-calibrated ampoules. 2-Vinylpyridine (2VP,) was dried twice over CaH_2 and subsequently purified using a sodium mirror and triethylaluminum (TEA), followed by distillation into pre-calibrated ampoules. sec-Butyllithium (1.4 M in cyclohexane, Sigma-Aldrich) was diluted to the appropriate concentration in benzene for the polymerization of styrene, or in n-hexane (Sigma-Aldrich, 95%) for the polymerization of 2VP, and stored under vacuum at $-20\ °C$ within a home-made glass apparatus equipped with ampoules. Ethylene oxide (EO, Sigma-Aldrich, 99.5%) was purified by distillation over CaH_2, over n-BuLi at $0\ °C$, and stored under high vacuum in ampoules. Methanol (MeOH, Sigma-Aldrich, ≥99.9%) was purified by distillation over CaH_2 and stored in ampoules under a high vacuum.

2.2. For Ring-Opening Polymerization

Toluene, 1,4-dioxane (anhydrous, >99.9%), and benzoic acid (99.5%) were dried over CaH_2 and $PS^{(-)}Li^{(+)}$. Ethyl acetate (EtOAc) was purchased from VWR Chemicals (HiPerSolv Chromanorm) and used as received. LLA (Sigma-Aldrich, Zwijndrecht, The Netherlands, 99%) and DLA (Jinan Daigang Biomaterial Co., Ltd., Jinan, China, ≥99.5%) were recrystallized from EtOAc three times and dissolved in anhydrous 1,4-dioxane, cryo-evaporating the 1,4-dioxane, followed by drying under vacuum overnight. Stannous octoate ($Sn(Oct)_2$, Sigma-Aldrich, 95%) was distilled twice over anhydrous $MgSO_4$ and activated 4 Å molecular sieves, followed by azeotropic distillation with dry toluene. PS-OH and P2VP-OH macronitiators obtained by anionic polymerization were dried through a freeze-drying process in benzene two times. All monomers, solvents, and catalysts for polymerizations were stored under argon (Ar) in a glove box (LABmaster SP, MBraun, Stratham, NH, USA).

2.3. Instrumentation

^1H NMR measurements were performed using Bruker AVANCE III spectrometers operating at 400 or 500 MHz; chloroform-d ($CDCl_3$, 99.8% D, Sigma-Aldrich) was used as the solvent for all samples. ^1H NMR spectra were used to calculate the number-average molecular weight (M_n) of each block by using the integrals of the characteristic signals from the end-groups and repeating unit of each block. SEC measurements were performed using Agilent SEC (Agilent Technologies, Santa Clara, CA, USA) equipped with a PLgel 5 μm MIXED-C and PLgel 5 μm MIXED-D columns. THF was used as eluent at a flow rate of 1.0 mL min^{-1} at 35 °C. The instrument was calibrated with PS standards. SEC samples were prepared by dissolving 2 mg/mL solutions in THF and filtered through 0.22 μm Teflon filters before injection. DSC measurements were performed with a Mettler Toledo DSC1/TC100 under nitrogen (N_2) and calibrated with Indium (purity > 99.999%). The samples were first heated from 25 to 200 °C to erase the thermal history and then cooled to $-20\ °C$ with a heating/cooling rate of 10 °C min^{-1}. This cycle was repeated twice before the glass transition, melting, and crystallization temperatures (T_g, T_m, and T_c) were recorded. X-ray diffractograms were obtained from XRD Bruker D8 Advance using Cu Kα irradiation. The sample for XRD measurements was deposited on a glass substrate with an approximate size of 1.5 cm × 1 cm, from a chloroform solution. Circular dichroism (CD) was performed with a Jasco J-815 model, featuring a Peltier model PTC-423S/15 thermo-stabilizing system. The cell used was a 1 mm quartz suprasil cell. The solutions of the PS-*b*-PDLA, P2VP-*b*-PLLA, and their stereocomplexes were made with acetonitrile. Typical concentrations were ~0.1 mg/mL.

2.4. Synthetic Procedures

The anionic polymerization of styrene and 2VP was carried out in specific custom-made glass apparatuses, which were evacuated and washed with *n*-BuLi solution prior to polymerization. Break seals were used for the introduction of reagents. Further details regarding the polymerization techniques are provided in previous reports [40–42].

2.5. Synthesis of Hydroxy-Terminated Polystyrene (PS-OH)

Styrene (5 g) was added to the appropriate amount of solvent (benzene, 5–10% polymer concentration), followed by the addition of the initiator, *sec*-BuLi (0.833 mmol). The polymerization was left to proceed until total monomer consumption (~18 h) at room temperature. An aliquot was taken to verify the molecular characteristics (molecular weight and distribution) by heat-sealing the proper constriction. EO (~1 mL) was then added to the reaction mixture and kept for 12 h at room temperature. Finally, methanol (~0.5 mL) was added for the termination of the living polymer. The polymer solution was precipitated into a large excess of methanol. The resulting polymer was filtered and dried in a vacuum oven at 40 °C for 24 h. M_n of PS-OH was calculated to be 6300 g mol^{-1} by using ^{1}H NMR end-group analysis. SEC analysis indicated an M_n of 6200 g mol^{-1} and $Đ$ of 1.02.

2.6. Synthesis of Hydroxy-Terminated Poly(2-vinylpyridine) (P2VP-OH)

2VP (6 g) was distilled into the glass reactor containing the appropriate amount of solvent (150 mL of THF), followed by the addition of *sec*-BuLi (1.2 mmol). The polymerization was conducted at −78 °C and was left to proceed for 1 h. EO (~1 mL) was then added to the reaction mixture and kept for 12 h at room temperature. Finally, methanol (~0.5 mL) was added for the termination of the living polymer. The polymer solution was precipitated into a large excess of *n*-hexane. The resulting polymer was filtered and dried in a vacuum oven at 40 °C for 24 h. M_n of P2VP-OH was calculated to be 6000 g mol^{-1} by using ^{1}H NMR end-group analysis. SEC analysis indicated an M_n of 5500 g mol^{-1} and $Đ$ of 1.03.

2.7. Synthesis of PS-b-PDLA

In a glove box under Ar atmosphere, dry PS-OH (248 mg, 0.039 mmol), Sn(Oct)$_2$ (8.1 mg, 0.02 mmol), D-LA (288 mg, 2 mmol), and 3 mL dry toluene were added to a dry Schlenk flask equipped with a stirrer bar. The reaction mixture was stirred for 24 h at 80 °C, and the conversion was monitored by ^{1}H NMR spectroscopy. After 24 h, the reaction mixture was quenched with benzoic acid and precipitated in cold MeOH. The resulting diblock copolymer was centrifuged and dried under a vacuum for 24 h at 40 °C.

2.8. Synthesis of P2VP-b-PLLA

In a glove box under Ar atmosphere, dry P2VP-OH (240 mg, 0.04 mmol), Sn(Oct)$_2$ (8.1 mg, 0.02 mmol), L-LA (290 mg, 2 mmol), and 3 mL dry toluene were added to a dry Schlenk flask equipped with a stirrer bar. The reaction mixture was stirred for 24 h at 80 °C, and the conversion was monitored by ^{1}H NMR spectroscopy. After 24 h, the reaction mixture was quenched with benzoic acid and precipitated in cold *n*-hexane. The resulting diblock copolymer was centrifuged and dried under vacuum for 24 h at 40 °C.

2.9. Stereocomplex Formation (PS-b-SC-b-P2VP)

Equimolar solutions of PS-*b*-PDLA and P2VP-*b*-PLLA were prepared separately by dissolving each polymer (~100 mg) in chloroform (5 mL) under vigorous stirring for 15 min at 400 rpm. Subsequently, the two solutions were mixed and stirred for another 15 min at 400 rpm. Finally, the final solution was precipitated in cold n-hexane (200 mL) and stirred for 30 min at 200 rpm. The precipitate was centrifuged and dried under vacuum for 24 h at 40 °C. In the following discussion, the stereocomplex-based samples (PS-b-SC-b-P2VP) are referred to as SCPLA$_x$, where x is the calculated molecular weight of the PLA segments.

3. Results and Discussion

The anionic polymerization of 2VP and styrene was carried out in THF (at −78 °C) and benzene (at room temperature), respectively, using *sec*-BuLi as the initiator (Scheme 1). After complete consumption of monomers, the living polymers were end-capped by an excess amount of EO at room temperature. Quantitative functionalization reaction of polymeric organolithium compounds in hydrocarbon solutions with EO at room temperature proceeds in the absence of EO oligomerization [43]. The anionic polymerization of EO does not happen under these conditions, resulting in initiation without propagation. Therefore, only one monomeric unit of EO is inserted at the chain-end. This is due to the high charge density of the lithium cation resulting in the strong aggregation of terminal lithium alkoxides.

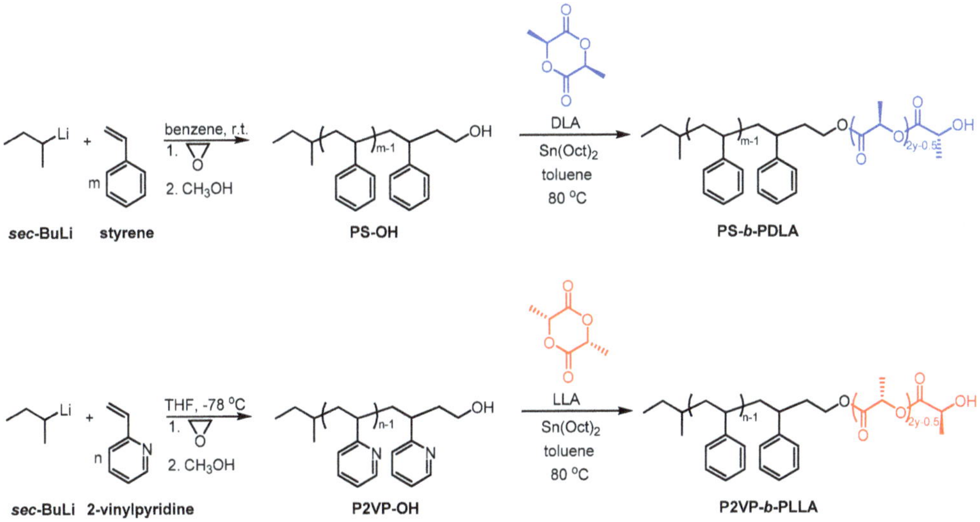

Scheme 1. General reactions for the synthesis of ω-hydroxyl functionalized PS and P2VP via anionic polymerization and the subsequent ROP of DLA/LLA.

The functionalization of both polymers (PS and P2VP) with EO was confirmed by ^{1}H NMR and by FT-IR spectrosocopies. ^{1}H NMR spectrum shows a peak around (δ = 3.2–3.7 ppm), which corresponds to the –CH$_2$ attached to the hydroxyl end-group (Figure S1). Moreover, the –OH group can be observed using FT-IR as a broad peak around 3401 cm^{-1} and 3394 cm^{-1} for PS-OH and P2VP-OH, respectively (Figure S2). Further confirmation is evident by the successful copolymerization of PLA via ROP using PS-OH and P2VP-OH as macroinitiators, as confirmed by SEC and ^{1}H NMR spectroscopy (Table 1).

PS-OH and P2VP-OH were synthesized with an M_n of 6300 and 6000 g mol^{-1}, as obtained by ^{1}H NMR end-group analysis, respectively. Their molecular characteristics are presented in Table 1. Both homopolymers have low molar-mass dispersity, as indicated by SEC, suggesting that the polymers can be considered to be well defined (Figure S3).

ROP of DLA/LLA initiated by dry PS-OH and P2VP-OH macroinitiators and catalyzed by Sn(Oct)$_2$ was performed in toluene at 80 °C to afford PS-b-PDLA and P2VP-b-PLLA diblock copolymers, respectively. Sn(Oct)$_2$, is considered one of the most effective catalysts for the ROP of lactides under a wide range of conditions [44–48]. Moreover, it is commercially available, soluble in most organic solvents, and has been approved by the United States Food and Drug Administration. The targeted molecular weights of PLAs were varied: 5000, 7000, and 10,000 g mol^{-1}. The molecular characteristics of the resulting diblock copolymers were determined by SEC and ^{1}H NMR measurements and are presented in Table 1.

Table 1. Molecular characteristics of homopolymers, diblock copolymers, and the corresponding stereocomplex.

Entry	Sample	Conv ^{c} (%)	M_n (kg mol^{-1})	Đ ^{d}
1	PS$_{6.2}$-OH ^{a}	100	6.20 ^{d}	1.02
2	P2VP$_{5.5}$-OH ^{a}	100	5.54 ^{d}	1.03
3	PS$_{6.2}$-b-PDLA$_{5.5}$ ^{b}	97	5.49 ^{e}	1.03
4	P2VP$_{5.5}$-b-PLLA$_{5.6}$ ^{b}	96	5.57 ^{e}	1.08
5	PS$_{6.2}$-b-PDLA$_{7.1}$ ^{b}	98	7.06 ^{e}	1.04
6	P2VP$_{5.5}$-b-PLLA$_7$ ^{b}	99	6.96 ^{e}	1.04
7	PS$_{6.2}$-b-PDLA$_{10.7}$ ^{b}	99	10.7 ^{e}	1.05
8	P2VP$_{5.5}$-b-PLLA$_{11}$ ^{b}	99	11.0 ^{e}	1.07

^{a} Synthesized by anionic polymerization high-vacuum techniques. ^{b} Synthesized by ROP of DLA/LLA with Sn(Oct)$_2$ as the catalyst. ^{c} Conversions of the monomers were determined by 400 MHz ^{1}H NMR spectra of crude products in CDCl$_3$ at 25 °C. ^{d} Determined by SEC in THF at 35 °C (calibrated with PS standards). ^{e} The molecular weight corresponds to the PLA block.

Figure 1 shows the SEC traces of the homopolymer precursors compared to the corresponding diblock copolymers. The SEC traces clearly show a shift towards a lower elution time, indicating an increase in molecular weight compared to the PS-OH and P2VP-OH precursors. The Đ values of all copolymers are below 1.1 (between 1.02 and 1.08), indicating that the diblock copolymers are nearly uniform (in molar mass).

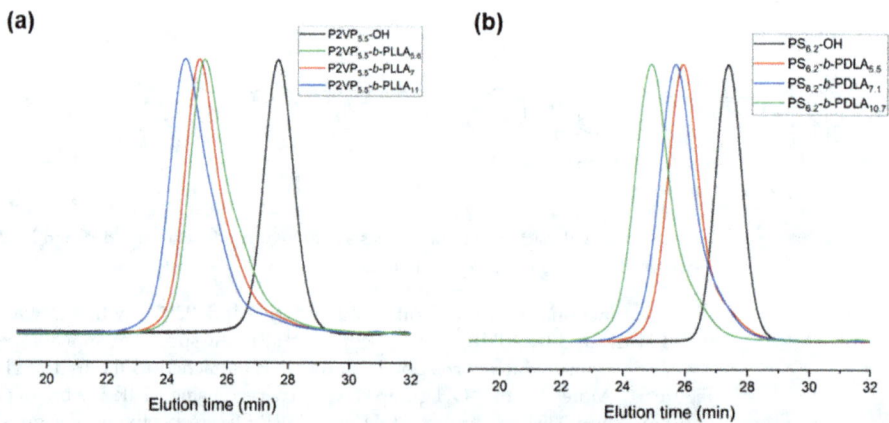

Figure 1. SEC traces of (**a**) P2VP-OH precursor and P2VP-b-PLLA diblock copolymers and (**b**) PS-OH precursor and PS-b-PDLA diblock copolymers (THF as eluent, 35 °C, PS standards).

For the following discussion on ^{1}H NMR, FT-IR, and CD results, P2VP$_{5.5}$-b-PLLA$_{5.6}$ and PS$_{6.2}$-b-PDLA$_{5.5}$ will be used as representative samples. ^{1}H NMR spectra of the diblock copolymers (Figure 2) show the characteristic peaks of methine proton from PLA main chain (c, δ = 5.2–5.3 ppm) and the terminal C–H (d, δ = 4.3–4.4 ppm). The molecular weights of the PLA blocks were determined by calculating the integral ratio of proton (c) and (d), i.e., end-group analysis.

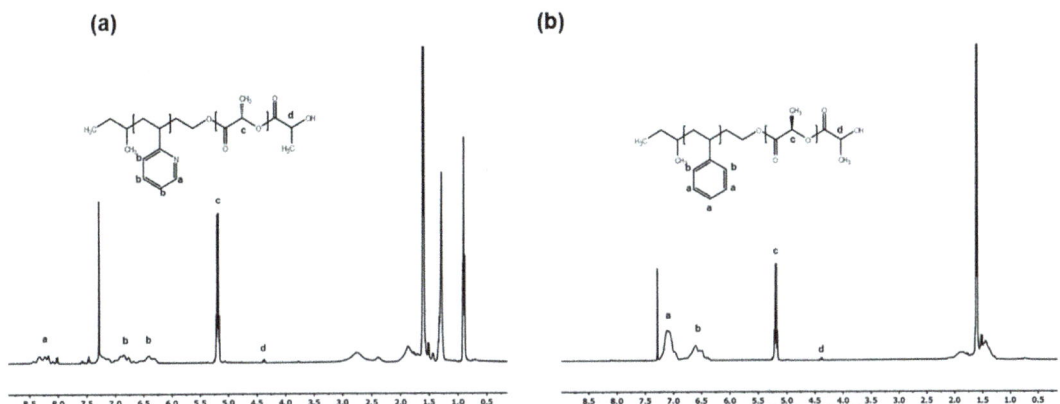

Figure 2. ^{1}H NMR spectra of (**a**) P2VP$_{5.5}$-b-PLLA$_{5.6}$, and (**b**) PS$_{6.2}$-b-PDLA$_{5.5}$ diblock copolymers (400 MHz, CDCl$_3$, 25 °C).

FT-IR spectroscopy was used to investigate the formation of the diblock copolymers and their corresponding stereocomplexes. The FT-IR spectra of the diblock copolymers reveal that a new peak is present at ~1750 cm^{-1}, which corresponds to the carbonyl (C=O) stretching, and two other peaks at ~1184 and 1088 cm^{-1} correspond to the (C–O) stretching of PLA (Figure 3).

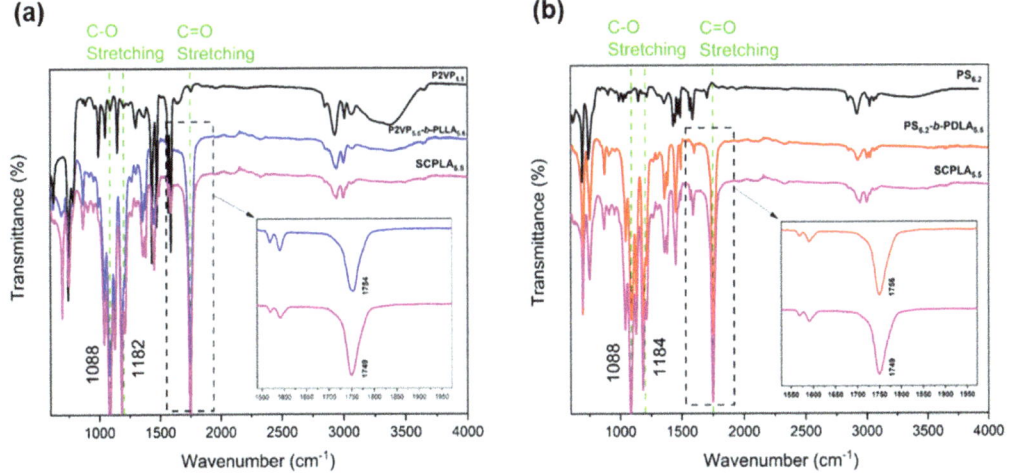

Figure 3. FT-IR spectra of (**a**) P2VP$_{5.5}$-OH, P2VP$_{5.5}$-b-PLLA$_{5.6}$, and SCPLA$_{5.5}$, and (**b**) PS$_{6.2}$-OH, PS$_{6.2}$-b-PDLA$_{5.5}$ and SCPLA$_{5.5}$.

Upon the formation of stereocomplex, the vibrational stretch of the carbonyl group of PLA, i.e., ν(C=O) band, in the SCPLA (Figure 3) shifted to a slightly lower wavenumber than that of the PS-b-PDLA (from 1756 to 1749 cm^{-1}) and P2VP-b-PLLA (from 1754 to 1749 cm^{-1}). This shift is attributed to the arrangement of the PLA chains into a more dense crystal packing due to stereocomplex formation via intermolecular H-bond interaction [49].

The specific optical rotation of PDLA/PLLA blocks in the block copolymers was evaluated by CD experiments. It is worth noting that PDLA chains take the right-handed helical conformation, whereas PLLA takes the left-handed helical conformation in acetonitrile solution. Figure S4 shows the CD spectra for both $PS_{6.2}$-b-$PDLA_{5.5}$ and $P2VP_{5.5}$-b-$PLLA_{5.6}$ in acetonitrile. The carboxylic group of PLAs with a helical conformation is accompanied by a characteristic absorption band of $n \rightarrow \pi^*$ transition. Therefore, a positive Cotton effect for $P2VP_{5.5}$-b-$PLLA_{5.6}$ and a negative Cotton effect for $PS_{6.2}$-b-$PDLA_{5.5}$ can be observed at ~233 nm. On the other hand, the solution of $SCPLA_{5.5}$ does not show such an effect, indicating that the D- and L-helical conformations complement each other due to the stereocomplex formation, resulting in zero CD response.

The influence of the molecular weight of PLA segments on the physical properties of the diblock copolymers, as well as their corresponding stereocomplexes, was investigated on the basis of DSC and XRD analyses. Figures 4 and 5 show the DSC thermograms and XRD patterns of the block copolymers and the stereocomplexes obtained by precipitation.

DSC analysis was performed in order to evaluate the thermal properties, including glass transition temperature (T_g) and melting temperature (T_m), of the homopolymers ($PS_{6.2}$-OH and $P2VP_{5.2}$-OH) and diblock copolymers ($PS_{6.2}$-b-$PDLA_x$ and $P2VP_{5.2}$-b-$PLLA_y$), as well as the corresponding SCPLAs. The DSC thermograms of the homopolymers, block copolymers, and corresponding SCPLAs are shown in Figure 4.

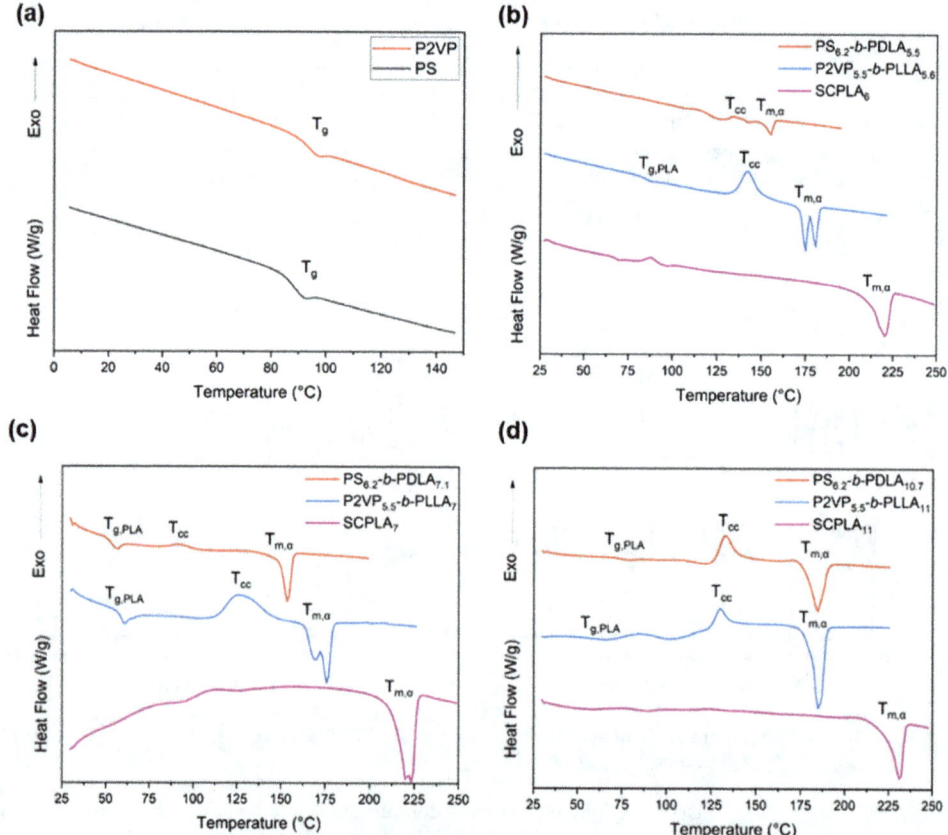

Figure 4. DSC thermograms of PS, P2VP homopolymers (**a**), PS-b-PDLAs, P2VP-b-PLLAs, and the corresponding SCPLAs with varying PLA segments with molecular weights of (**b**) 5 kg mol^{-1} (**c**) 7 kg mol^{-1}, and (**d**) 11 kg mol^{-1} (heating scan 10 °C/min, under N_2 atmosphere).

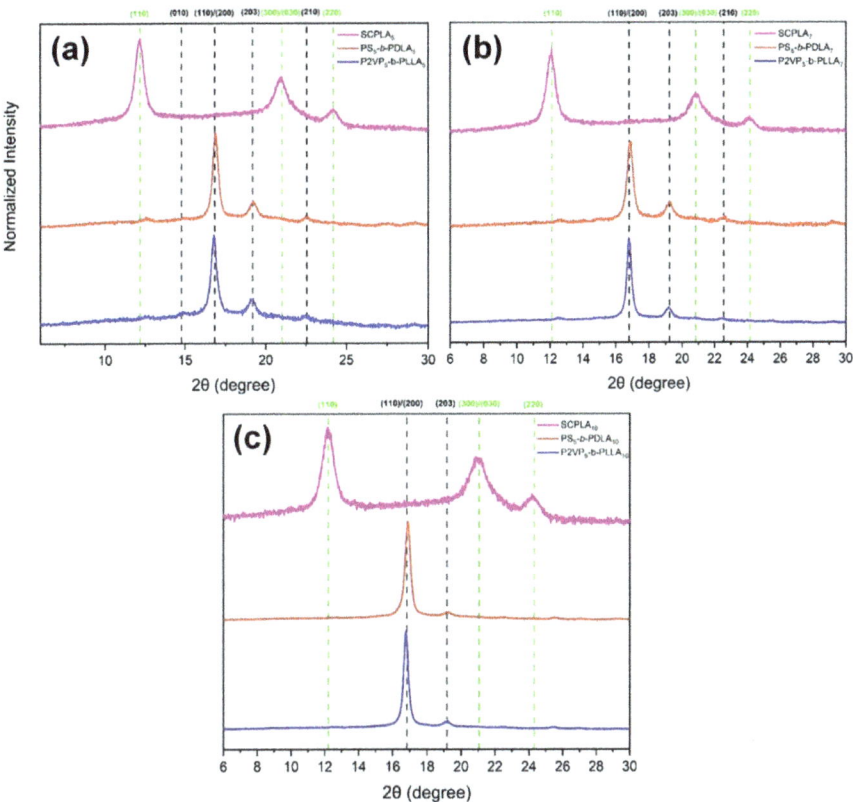

Figure 5. XRD patterns of PS-*b*-PDLAs, P2VP-*b*-PLLAs, and the corresponding SCPLAs with PLA segments with molecular weight of (**a**) 5 kg mol^{-1}, (**b**) 7 kg mol^{-1}, and (**c**) 11 kg mol^{-1} (samples were cast on top of the glass substrate).

The T_g values of the amorphous PS$_{6.2}$-OH and P2VP$_{5.2}$-OH precursors are observed to be 92.3 °C and 93.6 °C (Table 2), respectively. In the case of the diblock copolymers, the T_g values of PS and P2VP blocks cannot be observed, indicating that the PS/P2VP (amorphous blocks) and PLLA/PDLA (crystalline blocks) are miscible in the melt [50,51]. The PLLA/PDLA crystallites are well organized (as proved by the distinct ΔH_m) and limit the mobility of the amorphous blocks [50]. In addition, for PS$_{6.2}$-*b*-PDLA$_{7.1}$ (Figure 4c), the T_g of PS overlaps with the exothermic peaks from the cold-crystallization temperature (T_{cc}) of PLA. Therefore, the effect of the molecular weight of crystalline PLA blocks on the T_g of the amorphous blocks cannot be evaluated. The small endothermic humps observed between 50 and 80 °C are attributed to the T_g values of PLA.

All block copolymers exhibited a T_m of PLAs in the range of 150–180 °C, indicating the existence of crystalline PDLA/PLLA block. The PLLA block in P2VP$_{5.2}$-*b*-PLLA$_{5.6}$ and P2VP$_{5.2}$-*b*-PLLA$_7$ shows double melting (T_m) peaks. Two plausible explanations have been proposed for this observed phenomenon. The first concerns the lamellar crystal thickness [52,53], and suggests that the double endothermic behavior is the result of the existence of two kinds of crystal lamellae having different thicknesses. Consequently, the melting of the thinner lamellae would be observed at a lower temperature endotherm, whereas the thicker lamellae are related to the higher temperature endotherm. The second possible explanation is the partial melting and recrystallization process [54,55], where the lower-temperature endotherm is the result of the melting of the initial lamellae followed by recrystallization into more perfect lamellae.

Table 2. Thermal properties of homopolymers, block copolymers, and the corresponding SCPLA.

Sample	T_g (°C) [1]	T_m (°C) [1]	ΔH_m (J/g) [1]	X_c (%) [2]
PS$_{6.2}$-OH	92.3	-	-	Amorphous
P2VP$_{5.5}$-OH	93.6	-	-	Amorphous
PS$_{6.2}$-b-PDLA$_{5.5}$	-	153.0	24.4	14.5
P2VP$_{5.5}$-b-PLLA$_{5.6}$	83.0	156.0	23.5	34.0
SCPLA$_{5.5}$	-	220.3	36.1	33.0
PS$_{6.2}$-b-PDLA$_{7.1}$	54.2	154.0	25.5	37.3
P2VP$_{5.5}$-b-PLLA$_7$	58.9	152.6	27.7	39.6
SCPLA$_7$	-	223.3	41.3	38.3
PS$_{6.2}$-b-PDLA$_{10.7}$	73.9	162.6	30.0	52.4
P2VP$_{5.5}$-b-PLLA$_{11}$	59.9	162.9	43.2	51.4
SCPLA$_{11}$	-	231.1	39.9	39.3

[1] Determined by DSC (heating scan 10 °C min^{-1}, N$_2$ atmosphere). [2] Determined by XRD (samples were deposited on top of a glass substrate).

In general, the T_m values of PLLA and PDLA are affected by the increase in molecular weight. Such a trend is also observed in the PS$_{6.2}$-b-PDLA and P2VP$_{5.2}$-b-PLLA block copolymers. The higher the molecular weight of PLLA and PDLA, the higher the T_m in the diblock copolymers. When the molecular weight is increased from ~5000 g mol^{-1} to ~10,000 g mol^{-1}, the T_m values increase from 153.0 to 162.6 °C for PDLA-containing BCPs, and from 156.0 to 162.9 °C in the case of PLLA-containing BCPs. Similarly, the T_m and the melting enthalpy (ΔH_m) of SCPLA also increase with the increase in the molecular weight of PLA (from 220.3 to 230.3 °C). These results are in good agreement with our recent findings on the thermal properties of PS-b-SC-b-PI [39]. It is worth noting that the T_m of SCPLA is always ~70 °C higher than the T_m of their corresponding diblock copolymers, indicating that the effect of the amorphous PS$_{6.2}$ and P2VP$_{5.2}$ on the crystal packing of SCPLA is not significant in this case.

The investigation of crystal structure and degree of crystallinity (X_c) for the homopolymers, block copolymers, and their stereocomplexes obtained by precipitation was carried out by means of XRD analysis. The diffraction patterns are presented in Figure 5. The diblock copolymers exhibited diffraction peaks at 2θ = 16.7°, 17.6°, 19.5°, 22°, and 26°, verifying the presence of α crystals, i.e., orthorhombic unit cells. In addition, the crystal structure of SCPLAs obtained from the equimolar ratio of PLLA:PDLA was also investigated. The diffractograms (Figure 5) show diffraction peaks of triclinic crystal at 2θ = 14°, 24°, and 28°, confirming the formation of stereocomplexes. Overall, the XRD patterns of block copolymers and their stereocomplexes are in good agreement with the literature, as the α crystals show the reflection at 2θ = 16.6°, 19.1° and 17°, 19°, and the SCPLA crystals show the reflection at 2θ = 12°, 21°, 24°, and 12°, 20.9°, 24° [23].

The total degree of crystallinity (X_c) of a polymeric material can be calculated from its XRD pattern. X_c is defined as the ratio of the area of all crystalline peaks to the total area under the XRD peaks (crystalline + amorphous), as shown in Equation (1):

$$\% \, Crystallinity = \frac{I_c}{(I_c + I_a)} \times 100, \tag{1}$$

where I_a and I_c are the areas of the amorphous and crystalline domains, respectively. Based on DSC and XRD results, both PS and P2VP segments are amorphous. Therefore, the X_c of the diblock copolymers and their SCPLAs obtained from XRD can be attributed to the X_c of their PLA segments.

As can be seen in Figure 5, the X_cs of PS-b-PDLAs (14.5 < X_c < 52.4) and P2VP-b-PLLAs (34.0 < X_c < 51.4) increase with increasing molecular weight of the PLA blocks. This clearly indicates that the fraction crystalline domain increases with increasing molecular weight of PLA. A similar trend is also observed for the SCPLA (33.0 < X_c < 39.3), although the increment is insignificant.

4. Conclusions

Several well-defined diblock copolymers, $PS_{6.2}$-b-$PDLA_x$ and $P2VP_{5.5}$-b-$PLLA_y$, were successfully synthesized via the combination of anionic polymerization high-vacuum techniques and ring-opening polymerization. $PS_{6.2}$-b-$PDLA_x$ and $P2VP_{5.5}$-b-$PLLA_y$ were used as the precursors to synthesize non-covalent $PS_{6.2}$-b-SC-b-$P2VP_{5.5}$ triblock terpolymers via stereocomplexation of PDLA and PLLA blocks in chloroform. ^{1}H NMR spectroscopy and SEC confirmed the molecular characteristics of the copolymers. FT-IR, DSC, XRD, and CD spectroscopies revealed the formation of $PS_{6.2}$-b-SC-b-$P2VP_{5.5}$ as well as the effect of varying PLA molecular weights on the thermal properties of co/terpolymers. It was found that the T_m and X_c of the co/terpolymers increase with the increase of the molecular weights of PLA segments.

Comprehensive studies are necessary to further understand this system and determine a range of potential applications. Morphological and mechanical studies, including Young's modulus and tensile strength, will be conducted to fully establish the structure–properties relationship of these new non-covalent triblock terpolymers. Moreover, the presence of P2VP segments in these triblock terpolymers can be promising for biomedical applications due to their pH-sensitive nature and the ability to bind with metal cations.

Supplementary Materials: The following supporting information can be downloaded at: https://www.mdpi.com/article/10.3390/polym14122431/s1. Figure S1: ^{1}H NMR (400 MHz, CDCl$_3$) spectra of (a) PS-OH and (b) P2VP-OH; Figure S2: FT-IR spectra of (a) PS-OH and (b) P2VP-OH; Figure S3: SEC traces of (a) PS-OH and (b) P2VP-OH in THF at 35 °C; Figure S4: CD spectra of $PS_{6.2}$-b-$PDLA_{5.6}$, $P2VP_{5.5}$-b-$PLLA_{5.6}$, and $SCPLA_{5.5}$ were measured in acetonitrile with a concentration of 0.1 mg mL^{-1} at room temperature.

Author Contributions: Conceptualization was designed by N.H.; methodology and materials (homopolymers, diblock copolymers, and stereocomplexes) were synthesized by A.A., V.L. and K.N. under the supervision of N.H. writing—original draft preparation by A.A.; writing—review and editing, A.A., V.L., K.N. and N.H.; supervision by N.H; funding acquisition by KAUST. All authors have read and agreed to the published version of the manuscript.

Funding: The authors acknowledge the financial support and facilities provided by the King Abdullah University of Science and Technology (KAUST).

Institutional Review Board Statement: Not applicable.

Informed Consent Statement: Not applicable.

Data Availability Statement: Not applicable.

Conflicts of Interest: The authors declare no conflict of interest.

References

1. Ikada, Y.; Tsuji, H. Biodegradable polyesters for medical and ecological applications. *Macromol. Rapid. Comm.* **2000**, *21*, 117–132. [CrossRef]
2. Nair, L.S.; Laurencin, C.T. Biodegradable polymers as biomaterials. *Prog. Polym. Sci.* **2007**, *32*, 762–798. [CrossRef]
3. Bai, H.; Deng, S.; Bai, D.; Zhang, Q.; Fu, Q. Recent Advances in Processing of Stereocomplex-Type Polylactide. *Macromol. Rapid. Commun.* **2017**, *38*, 1700454. [CrossRef] [PubMed]
4. Nofar, M.; Sacligil, D.; Carreau, P.J.; Kamal, M.R.; Heuzey, M.-C. Poly (lactic acid) blends: Processing, properties and applications. *Int. J. Biol. Macromol.* **2018**, *125*, 307–360. [CrossRef]
5. Wu, C.-S. Renewable resource-based composites of recycled natural fibers and maleated polylactide bioplastic: Characterization and biodegradability. *Polym. Degrad. Stab.* **2009**, *94*, 1076–1084. [CrossRef]
6. Bai, H.; Huang, C.; Xiu, H.; Zhang, Q.; Deng, H.; Wang, K.; Chen, F.; Fu, Q. Significantly Improving Oxygen Barrier Properties of Polylactide via Constructing Parallel-Aligned Shish-Kebab-Like Crystals with Well-Interlocked Boundaries. *Biomacromolecules* **2014**, *15*, 1507–1514. [CrossRef]

7. Tawakkal, I.S.M.A.; Cran, M.J.; Miltz, J.; Bigger, S. A Review of Poly(Lactic Acid)-Based Materials for Antimicrobial Packaging. *J. Food Sci.* **2014**, *79*, R1477–R1490. [CrossRef]
8. Drumright, R.E.; Gruber, P.R.; Henton, D.E. Polylactic acid technology. *Adv. Mater.* **2000**, *12*, 1841–1846. [CrossRef]
9. Jacobson, G.B.; Shinde, R.; Contag, C.; Zare, R.N. Sustained Release of Drugs Dispersed in Polymer Nanoparticles. *Angew. Chem. Int. Ed.* **2008**, *47*, 7880–7882. [CrossRef]
10. Casalini, T.; Rossi, F.; Castrovinci, A.; Perale, G. A Perspective on Polylactic Acid-Based Polymers Use for Nanoparticles Synthesis and Applications. *Front. Bioeng. Biotechnol.* **2019**, *7*, 259. [CrossRef]
11. Li, Y.; Qiang, Z.; Chen, X.; Ren, J. Understanding thermal decomposition kinetics of flame-retardant thermoset polylactic acid. *RSC Adv.* **2019**, *9*, 3128–3139. [CrossRef]
12. Cai, S.; Qiang, Z.; Zeng, C.; Ren, J. Multifunctional poly(lactic acid) copolymers with room temperature self-healing and rewritable shape memory properties via Diels-Alder reaction. *Mater. Res. Express* **2019**, *6*, 045701. [CrossRef]
13. Armentano, I.; Bitinis, N.; Fortunati, E.; Mattioli, S.; Rescignano, N.; Verdejo, R.; Lopez-Manchado, M.; Kenny, J. Multifunctional nanostructured PLA materials for packaging and tissue engineering. *Prog. Polym. Sci.* **2013**, *38*, 1720–1747. [CrossRef]
14. Jing, Y.; Quan, C.; Liu, B.; Jiang, Q.; Zhang, C. A Mini Review on the Functional Biomaterials Based on Poly(lactic acid) Stereocomplex. *Polym. Rev.* **2016**, *56*, 262–286. [CrossRef]
15. Rasal, R.M.; Janorkar, A.V.; Hirt, D.E. Poly(lactic acid) modifications. *Prog. Polym. Sci.* **2010**, *35*, 338–356. [CrossRef]
16. Gao, C.; Yu, L.; Liu, H.; Chen, L. Development of self-reinforced polymer composites. *Prog. Polym. Sci.* **2012**, *37*, 767–780. [CrossRef]
17. Gupta, B.; Revagade, N.; Hilborn, J. Poly(lactic acid) fiber: An overview. *Prog. Polym. Sci.* **2007**, *32*, 455–482. [CrossRef]
18. Raquez, J.-M.; Habibi, Y.; Murariu, M.; Dubois, P. Polylactide (PLA)-based nanocomposites. *Prog. Polym. Sci.* **2013**, *38*, 1504–1542. [CrossRef]
19. Ye, S.; Lin, T.T.; Tjiu, W.W.; Wong, P.K.; He, C. Rubber toughening of poly(lactic acid): Effect of stereocomplex formation at the rubber-matrix interface. *J. Appl. Polym. Sci.* **2012**, *128*, 2541–2547. [CrossRef]
20. Garlotta, D. A Literature Review of Poly(Lactic Acid). *J. Polym. Environ.* **2001**, *9*, 63–84. [CrossRef]
21. Södergård, A.; Stolt, M. Properties of lactic acid based polymers and their correlation with composition. *Prog. Polym. Sci.* **2002**, *27*, 1123–1163. [CrossRef]
22. Jamshidi, K.; Hyon, S.-H.; Ikada, Y. Thermal characterization of polylactides. *Polymers* **1988**, *29*, 2229–2234. [CrossRef]
23. Ikada, Y.; Jamshidi, K.; Tsuji, H.; Hyon, S.H. Stereocomplex formation between enantiomeric poly(lactides). *Macromolecules* **1987**, *20*, 904–906. [CrossRef]
24. Wan, Z.-Q.; Longo, J.M.; Liang, L.-X.; Chen, H.-Y.; Hou, G.-J.; Yang, S.; Zhang, W.-P.; Coates, G.W.; Lu, X.-B. Comprehensive Understanding of Polyester Stereocomplexation. *J. Am. Chem. Soc.* **2019**, *141*, 14780–14787. [CrossRef]
25. Zhang, J.; Sato, H.; Tsuji, H.; Noda, A.I.; Ozaki, Y. Infrared spectroscopic study of CH_3 center dot center dot center dot O=C interaction during poly(L-lactide)/poly(D-lactide) stereocomplex formation. *Macromolecules* **2005**, *38*, 1822–1828. [CrossRef]
26. Fujiwara, T.; Mukose, T.; Yamaoka, T.; Yamane, H.; Sakurai, S.; Kimura, Y. Novel thermo-responsive formation of a hydrogel by stereo-complexation between PLLA-PEG-PLLA and PDLA-PEG-PDLA block copolymers. *Macromol. Biosci.* **2001**, *1*, 204–208. [CrossRef]
27. Yang, L.; El Ghzaoui, A.; Li, S. In vitro degradation behavior of poly(lactide)–poly(ethylene glycol) block copolymer micelles in aqueous solution. *Int. J. Pharm.* **2010**, *400*, 96–103. [CrossRef]
28. Chen, L.; Xie, Z.; Hu, J.; Chen, X.; Jing, X. Enantiomeric PLA–PEG block copolymers and their stereocomplex micelles used as rifampin delivery. *J. Nanopart. Res.* **2006**, *9*, 777–785. [CrossRef]
29. Song, Y.; Wang, D.; Jiang, N.; Gan, Z. Role of PEG Segment in Stereocomplex Crystallization for PLLA/PDLA-*b*-PEG-*b*-PDLA Blends. *ACS Sustain. Chem. Eng.* **2015**, *3*, 1492–1500. [CrossRef]
30. Shirahama, H.; Ichimaru, A.; Tsutsumi, C.; Nakayama, Y.; Yasuda, H. Characteristics of the biodegradability and physical properties of stereocomplexes between poly(L-lactide) and poly(D-lactide) copolymers. *J. Polym. Sci. Part A Polym. Chem.* **2004**, *43*, 438–454. [CrossRef]
31. Wanamaker, C.L.; Bluemle, M.J.; Pitet, L.; O'Leary, L.E.; Tolman, W.B.; Hillmyer, M.A. Consequences of Polylactide Stereochemistry on the Properties of Polylactide-Polymenthide-Polylactide Thermoplastic Elastomers. *Biomacromolecules* **2009**, *10*, 2904–2911. [CrossRef]
32. Li, Z.; Yuan, D.; Fan, X.; Tan, B.H.; He, C. Poly(ethylene glycol) Conjugated Poly(lactide)-Based Polyelectrolytes: Synthesis and Formation of Stable Self-Assemblies Induced by Stereocomplexation. *Langmuir* **2015**, *31*, 2321–2333. [CrossRef]
33. Fan, X.; Wang, Z.; Yuan, D.; Sun, Y.; Li, Z.; He, C. Novel linear-dendritic-like amphiphilic copolymers: Synthesis and self-assembly characteristics. *Polym. Chem.* **2014**, *5*, 4069–4075. [CrossRef]
34. Uehara, H.; Karaki, Y.; Wada, S.; Yamanobe, T. Stereo-Complex Crystallization of Poly(lactic acid)s in Block-Copolymer Phase Separation. *ACS Appl. Mater. Interfaces* **2010**, *2*, 2707–2710. [CrossRef]
35. Schmidt, S.C.; Hillmyer, M.A. Synthesis and Characterization of Model Polyisoprene−Polylactide Diblock Copolymers. *Macromolecules* **1999**, *32*, 4794–4801. [CrossRef]
36. Frick, E.M.; Hillmyer, M.A. Synthesis and characterization of polylactide-block-polyisoprene-block-polylactide triblock copolymers: New thermoplastic elastomers containing biodegradable segments. *Macromol. Rapid. Comm.* **2000**, *21*, 1317–1322. [CrossRef]

37. He, X.; He, Y.; Hsiao, M.-S.; Harniman, R.L.; Pearce, S.; Winnik, M.A.; Manners, I. Complex and Hierarchical 2D Assemblies via Crystallization-Driven Self-Assembly of Poly(l-lactide) Homopolymers with Charged Termini. *J. Am. Chem. Soc.* **2017**, *139*, 9221–9228. [CrossRef]
38. Long, J.; Azmi, A.S.; Kim, M.P.; Ali, F.B. Comparative Study of Poly(4-vinylpyridine) and Polylactic Acid-block-poly (2-vinylpyridine) Nanocomposites on Structural, Morphological and Electrochemical Properties. *Sains Malays.* **2017**, *46*, 1097–1102. [CrossRef]
39. Ladelta, V.; Ntetsikas, K.; Zapsas, G.; Hadjichristidis, N. Non-Covalent PS–SC–PI Triblock Terpolymers *via* Polylactide Stereocomplexation: Synthesis and Thermal Properties. *Macromolecules* **2022**, *55*, 2832–2843. [CrossRef]
40. Hadjichristidis, N.; Iatrou, H.; Pispas, S.; Pitsikalis, M. Anionic polymerization: High vacuum techniques. *J. Polym. Sci. Pol. Chem.* **2000**, *38*, 3211–3234. [CrossRef]
41. Uhrig, D.; Mays, J.W. Experimental techniques in high-vacuum anionic polymerization. *J. Polym. Sci. Part A Polym. Chem.* **2005**, *43*, 6179–6222. [CrossRef]
42. Bhaumik, S.; Ntetsikas, K.; Hadjichristidis, N. Noncovalent Supramolecular Diblock Copolymers: Synthesis and Microphase Separation. *Macromolecules* **2020**, *53*, 6682–6689. [CrossRef]
43. Quirk, R.P.; Ma, J.-J. Characterization of the functionalization reaction product of poly(styryl)lithium with ethylene oxide. *J. Polym. Sci. Part A Polym. Chem.* **1988**, *26*, 2031–2037. [CrossRef]
44. Kowalski, A.; Libiszowski, J.; Biela, T.; Cypryk, M.; Duda, A.A.; Penczek, S. Kinetics and Mechanism of Cyclic Esters Polymerization Initiated with Tin(II) Octoate. Polymerization of ε-Caprolactone and l,l-Lactide Co-initiated with Primary Amines. *Macromolecules* **2005**, *38*, 8170–8176. [CrossRef]
45. Grijpma, D.W.; Pennings, A.J. Polymerization Temperature Effects on the Properties of L-Lactide and Epsilon-Caprolactone Copolymers. *Polym. Bull.* **1991**, *25*, 335–341. [CrossRef]
46. Kricheldorf, H.R.; Meierhaack, J. Polylactones, 22 Aba Triblock Copolymers of L-Lactide and Poly(Ethylene Glycol). *Makromol. Chem.* **1993**, *194*, 715–725. [CrossRef]
47. Ryner, M.; Stridsberg, K.; Albertsson, A.-C.; von Schenck, H.; Svensson, M. Mechanism of Ring-Opening Polymerization of 1,5-Dioxepan-2-one and l-Lactide with Stannous 2-Ethylhexanoate. A Theoretical Study. *Macromolecules* **2001**, *34*, 3877–3881. [CrossRef]
48. Ladelta, V.; Zapsas, G.; Abou-Hamad, E.; Gnanou, Y.; Hadjichristidis, N. Tetracrystalline Tetrablock Quarterpolymers: Four Different Crystallites under the Same Roof. *Angew. Chem. Int. Ed.* **2019**, *58*, 16267–16274. [CrossRef] [PubMed]
49. Chang, Y.; Chen, Z.; Yang, Y. Benign Fabrication of Fully Stereocomplex Polylactide with High Molecular Weights via a Thermally Induced Technique. *ACS Omega* **2018**, *3*, 7979–7984. [CrossRef] [PubMed]
50. Ji, E.; Cummins, C.; Fleury, G. Precise Synthesis and Thin Film Self-Assembly of PLLA-*b*-PS Bottlebrush Block Copolymers. *Molecules* **2021**, *26*, 1412. [CrossRef]
51. Michell, R.M.; Müller, A.J.; Spasova, M.; Dubois, P.; Burattini, S.; Greenland, B.W.; Hamley, I.W.; Hermida-Merino, D.; Cheval, N.; Fahmi, A. Crystallization and Stereocomplexation Behavior of Poly(D- and L-lactide)-b-Poly(N,N-dimethylamino-2-ethyl methacrylate) Block Copolymers. *J. Polym. Sci. Part B Polym. Phys.* **2011**, *49*, 1397–1409. [CrossRef]
52. Cebe, P.; Hong, S.-D. Crystallization behaviour of poly(ether-ether-ketone). *Polymer* **1986**, *27*, 1183–1192. [CrossRef]
53. Bassett, D.; Olley, R.; Alraheil, I. On crystallization phenomena in PEEK. *Polymer* **1988**, *29*, 1745–1754. [CrossRef]
54. Lee, Y.; Porter, R.S.; Lin, J.S. On the double-melting behavior of poly(ether ether ketone). *Macromolecules* **1989**, *22*, 1756–1760. [CrossRef]
55. Jonas, A.M.; Russell, T.P.; Yoon, D.Y. Synchrotron X-ray Scattering Studies of Crystallization of Poly(ether-ether-ketone) from the Glass and Structural Changes during Subsequent Heating-Cooling Processes. *Macromolecules* **1995**, *28*, 8491–8503. [CrossRef]

Article

Block Copolymer Modified Nanonetwork Epoxy Resin for Superior Energy Dissipation

Suhail K. Siddique [1], Hassan Sadek [1], Tsung-Lun Lee [1], Cheng-Yuan Tsai [2], Shou-Yi Chang [2], Hsin-Hsien Tsai [3], Te-Shun Lin [3], Gkreti-Maria Manesi [4], Apostolos Avgeropoulos [4,5] and Rong-Ming Ho [1,*]

1 Department of Chemical Engineering, National Tsing Hua University, Hsinchu 30013, Taiwan; s105032891@m105.nthu.edu.tw (S.K.S.); hassan_sadek@azhar.edu.eg (H.S.); tsunglunlee@mx.nthu.edu.tw (T.-L.L.)
2 Department of Material Science and Engineering, National Tsing Hua University, Hsinchu 30013, Taiwan; tsai10681@gmail.com (C.-Y.T.); changsy@mx.nthu.edu.tw (S.-Y.C.)
3 Kaohsiung Factory R&D Department, Chang Chun Plastics Co., Ltd., Kaohsiung 81469, Taiwan; hsin_hsien_tsai@ccpgp.com (H.-H.T.); de_shun_lin@ccpgp.com (T.-S.L.)
4 Department of Materials Science Engineering, University Campus, University of Ioannina, 45110 Ioannina, Greece; gretimanesi@uoi.gr (G.-M.M.); aavger@uoi.gr (A.A.)
5 Faculty of Chemistry, Lomonosov Moscow State University (MSU), GSP-1, 1-3 Leninskiye Gory, 119991 Moscow, Russia
* Correspondence: rmho@mx.nthu.edu.tw; Tel.: +886-3-573-8349; Fax: +886-3-571-5408

Abstract: Herein, this work aims to fabricate well-ordered nanonetwork epoxy resin modified with poly(butyl acrylate)-b-poly(methyl methacrylate) (PBA-b-PMMA) block copolymer (BCP) for enhanced energy dissipation using a self-assembled diblock copolymer of polystyrene-b-poly(dimethylsiloxane) (PS-b-PDMS) with gyroid and diamond structures as templates. A systematic study of mechanical properties using nanoindentation of epoxy resin with gyroid- and diamond-structures after modification revealed significant enhancement in energy dissipation, with the values of 0.36 ± 0.02 nJ (gyroid) and 0.43 ± 0.03 nJ (diamond), respectively, when compared to intrinsic epoxy resin (approximately 0.02 ± 0.002 nJ) with brittle characteristics. This enhanced property is attributed to the synergic effect of the deliberate structure with well-ordered nanonetwork texture and the toughening of BCP-based modifiers at the molecular level. In addition to the deliberate structural effect from the nanonetwork texture, the BCP modifier composed of epoxy-philic hard segment and epoxy-phobic soft segment led to dispersed soft-segment domains in the nanonetwork-structured epoxy matrix with superior interfacial strength for the enhancement of applied energy dissipation.

Keywords: nanonetwork; block copolymer; modifier; templated polymerization; energy dissipation

Citation: Siddique, S.K.; Sadek, H.; Lee, T.-L.; Tsai, C.-Y.; Chang, S.-Y.; Tsai, H.-H.; Lin, T.-S.; Manesi, G.-M.; Avgeropoulos, A.; Ho, R.-M. Block Copolymer Modified Nanonetwork Epoxy Resin for Superior Energy Dissipation. *Polymers* **2022**, *14*, 1891. https://doi.org/10.3390/polym14091891

Academic Editor: Incoronata Tritto

Received: 30 March 2022
Accepted: 3 May 2022
Published: 5 May 2022

Publisher's Note: MDPI stays neutral with regard to jurisdictional claims in published maps and institutional affiliations.

Copyright: © 2022 by the authors. Licensee MDPI, Basel, Switzerland. This article is an open access article distributed under the terms and conditions of the Creative Commons Attribution (CC BY) license (https://creativecommons.org/licenses/by/4.0/).

1. Introduction

Epoxy resin is one of the most versatile thermosets due to its processability, cost-effectiveness, and superior mechanical performance, giving excellent characteristics for industrial applications. The existence of multiple oxiranes or epoxy groups in the molecular structure of epoxy resin can provide high crosslinking density for superior mechanical strength, but the excessive crosslinking of epoxy rings in the molecular structure results in the brittle or glassy characteristics, and thus limits their engineering performance [1]. As a result, the reduction in glassy characteristics of epoxy resins to enhance the toughness gained intense attention in the last century. Extensive research has been conducted to improve the energy absorption capability of epoxy resin by incorporating a variety of toughening agents including liquid rubber [2,3], thermoplastic polymers [4], core–shell particle [5,6], and ceramic fillers [7].

A well-dispersed rubber modifier without any molecular level interactions can significantly reduce the brittle characteristics of epoxy resin by lowering the crosslinking density, which enhances their toughness by enabling the internal cavitation of well-bounded rubber particles [8]. However, it might greatly affect their strength and glass transition temperature (T_g) [1]. Recent developments in the toughening of epoxy resin using amphiphilic block copolymers was reported [9,10] at which the BCPs self-assemble into spherical micelle, worm-like structures or vesicles to provide high energy dissipation with the least impact on modulus and glass transition temperature [11]. The BCPs composed of epoxy-philic hard and epoxy-phobic soft segments can self-assemble into different morphologies in epoxy resin due to their intermolecular interactions, which gives them the exceptional capability to absorb applied energy through cavitation with shear banding. The addition of a small amount of BCPs provides excellent toughening without sacrificing the strength and effective thermal properties of epoxy resin [8,12]. Most interestingly, the energy dissipation from the brittle to plastic transformation of intrinsic epoxy resin can also be achieved by well-ordered structures termed metamaterials [13–15], which offer unique emergent mechanical properties, especially for high specific energy absorption [16–19]. Mechanical metamaterials are the materials that show exceptional mechanical properties due to their deliberate structuring instead of bulk behavior [19].

Moreover, nanonetwork materials have been reported for their enhanced mechanical properties due to their network structure in the nanoscale, where the nanosize can be the secondary aspect of the mechanical property [20]. However, the fabrication of nanonetwork structures is extremely challenging due to the difficulty of controlling their structure. The block copolymers have been extensively studied recently due to their ability to self-assemble into various periodic nanostructures depending on the volume fraction of their constituent segments and molecular weight [21–24]. Due to their unique network geometry, gyroid and diamond phases are considered appealing morphologies for practical applications [25,26]. By taking advantage of the degradable segments in BCPs, polymeric templates with well-ordered periodic nanochannels can be fabricated and subsequently serve as a template for templated syntheses, giving a platform technology for the fabrication of nanonetwork functional materials [27]. The templated syntheses can be carried out by atomic layer deposition [28], electroless plating [29], sol–gel reaction [30], electrochemical deposition [31] and templated polymerization [32]. Recently, the enhanced energy dissipation from the deliberate structuring of nanonetwork textures for thermosets fabricated by templated polymerization has been demonstrated by enabling the design of mechanical metamaterials from a bottom-up approach [33].

Herein, this work aims to demonstrate the fabrication of poly(butyl acrylate)-*b*-poly(methyl methacrylate) (PBA-*b*-PMMA) modified epoxy resin with well-ordered nanonetwork structures for the enhancement of energy dissipation capability using the self-assembled BCP, followed by templated polymerization. As shown in Figure 1, a well-ordered gyroid and diamond phase with co-continuous PS and PDMS domains from self-assembly of a polystyrene-*b*-poly(dimethylsiloxane) can be fabricated by solution casting using selective solvents. Selective etching of the PDMS block from PS-*b*-PDMS can be acquired using hydrofluoric (HF) acid to give nanoporous PS with gyroid- and diamond-structured nanochannels as templates for polymerization of modified epoxy resin. The examination of the mechanical energy dissipation of these two distinct nanonetwork epoxy resins after the modification with PBA-*b*-PMMA has been carried out by using nanoindentation; the PBA-*b*-PMMA BCP can be self-assembled into spherical nanosized micelle in the epoxy matrix which acts as a toughening agent for the formation of soft domains. The synergic effect of the deliberate structuring in nanoscale and the toughening of the BCP-based modifier on a molecular level can significantly contribute to the energy dissipation capability of nanonetwork epoxy resins as compared to intrinsic brittle epoxy resin.

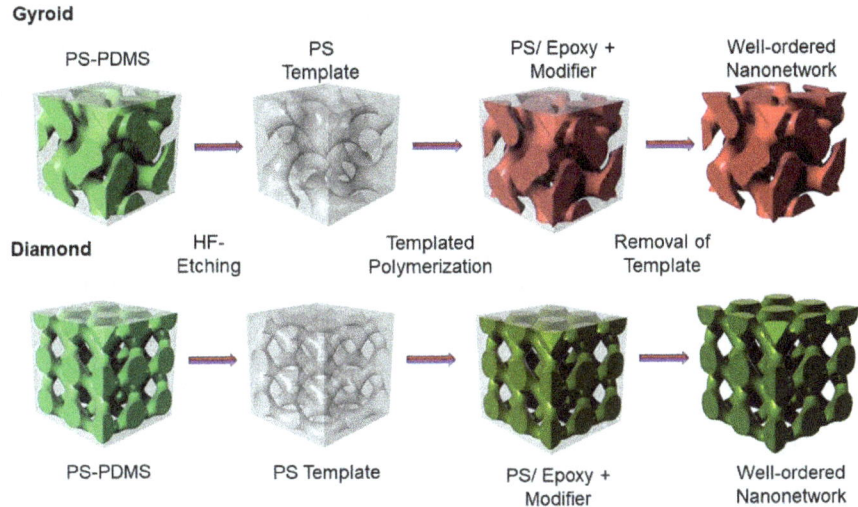

Figure 1. Schematic illustration of templated syntheses of epoxy resin modified by PBA-*b*-PMMA.

2. Materials and Methods

The detailed synthesis procedure for PS-*b*-PDMS was discussed previously [34,35]. Epoxy resin used in this study is Bisphenol-A diglycedyle ether (DGEBA) (DEH 24, Dow Chemical, Midland, MI, USA) and the hardener is Triethylenetetramine (TETA) (DEH 24, Dow Chemical). Solvents used in this study is toluene (Sigma Aldrich, St. Louis, MI, USA) and chloroform (Sigma Aldrich).

2.1. Synthesis Procedures

The total number average molecular weight of the PS-*b*-PDMS used in this study was 86,000 g/mol ($\overline{M}_n^{PS}$: 51,000 g/mol; $\overline{M}_n^{PDMS}$: 35,000 g/mol) with the volume fraction of PDMS equal to 0.42, giving lamellar morphology during self-assembly using a neutral solvent. The dispersity values of the PS precursor and the final synthesized copolymer is described in Table 1.

Table 1. Molecular characterization of the synthesized PS-*b*-PDMS sample.

Sample	$\overline{M}_n^{PS}$ (kg mol^{-1}) [a]	$\overline{M}_n^{PDMS}$ (kg mol^{-1}) [a]	$\overline{M}_n^{total}$ (kg mol^{-1}) [a]	Đ [b]	f_{PDMS}^v [c]
PS precursor	51			1.03	
PS-*b*-PDMS	51	35	86	1.05	0.42

[a] Number average molecular weight of PS and PDMS determined by membrane osmometry (MO). [b] Dispersity (Đ) measured by size exclusion chromatography (GPC). [c] Volume fraction of PDMS (f_{PDMS}^v) in the PS-*b*-PDMS as calculated from ^{1}H NMR based on ρ_{PS} = 1.04 g/cm^3, ρ_{PDMS} = 0.965 g/cm^3.

The gyroid and diamond-structures in the PS-*b*-PDMS can be fabricated by solution casting in PS selective solvents such as toluene and chloroform, respectively.

2.2. Preparation of Well-Ordered Nanoporous Template

The lamellae-forming PS-*b*-PDMS was dissolved in PS selective solvents, including toluene (Sigma Aldrich) and chloroform (Sigma Aldrich) (10 wt% concentration), in a vial with a controlled solvent evaporation rate. After drying, the samples were further dried at 60 °C in a vacuum oven. The formation of network phases from the self-assembly of lamellae-forming PS-*b*-PDMS indicates that a double gyroid structure can be formed using toluene for solution casting while a double diamond structure can be formed using chloroform as a solvent. Moreover, the affinity of PS towards the PS-selective solvents induces a reduction in the PDMS volume fraction causing the flat interfaces (lamellar morphology) to generate network or cylindrical structures. For toluene ($\delta_{toluene}$ = 8.9 cal$^{1/2}$/cm$^{3/2}$), the difference in the solubility parameters (δ_{PS} = 9.1 cal$^{1/2}$/cm$^{3/2}$, δ_{PDMS} = 7.4 cal$^{1/2}$/cm$^{3/2}$) indicates selectivity towards PS domains, giving rise to double gyroid. Furthermore, chloroform ($\delta_{chloroform}$ = 9.3 cal$^{1/2}$/cm$^{3/2}$) enables the formation of different network phases such as double diamond and double primitive (DP), as already reported in the literature [36]. Note that PDMS swollen ratio is 10% larger in chloroform than in toluene, as reported by Whitesides et al. [37]. Accordingly, higher elasticity and free stretching energy provided by the PDMS segments can contribute to the formation of kinetically trapped phases. This behavior may be attributed to PDMS blocks expansion (instead of looping) into the core of the junctions, even with the entropy loss at a higher strut number. As a result, network textures such as double diamond and double primitive can be formed due to the effect of evaporation rate control on solution casting. Subsequently, the selective etching between the PS and PDMS segments allows the formation of the nanoporous PS template from the self-assembled PS-*b*-PDMS using HF solution (HF/H$_2$O/methanol = 0.5/1/1 by volume). After the complete removal of the PDMS followed by washing with water and methanol, well-ordered nanoporous PS templates with gyroid- and diamond-structures with the corresponding nanochannels (approximately equivalent porosity) for templated polymerization can be obtained.

2.3. Templated Polymerization

For templated polymerization of epoxy resin, successful pore-filling can be achieved using methanol combined with an epoxy resin precursor for high wetting ability. The hydrophobic inner walls of the PS template can be effectively prepared for pore filling using short-chain alcohols. Note that effective pore filling must be confirmed before templated polymerization of epoxy resin; if polymerization starts before pore filling it can cause blocking of the template, which leads to incomplete networks. A mixture of dissolved epoxy resin containing Bisphenol-A type epoxy (DER 331, Dow Chemical) with 5% wt% of well soluble PBA-*b*-PMMA (see Table S1 for details) and triethylenetetramine (TETA) (DEH 24, Dow Chemical) was initially prepared. The PS templates were immersed into the precursor solution at low temperature (10 °C) for five hours to reduce the polymerization reaction and promote adequate pore filling. The epoxy resin can be pore-filled into the nano channeled templates by capillary force. The mild curing of epoxy resin inside the template gradually leads to an insufficient cross-linking reaction of the resin. A temperature increase can increase the crosslinking, but it might damage the template texture. To solve specific issues multistep curing was conducted. The temperature was gradually raised to the final setting temperature for curing through stepwise heating to provide the optimum temperature. Consequently, a higher degree of curing was acquired. Direct heating from room temperature to 150 °C led to template damage with disordered texture for templated resin. The high internal stress triggered by the fast heating causes the deformation of the resin skeleton.

2.4. Transmission Electron Microscopy (TEM)

Bright-field (BF) transmission electron microscopy (TEM) imaging was used to determine the pore-filling of epoxy resin in the PS template using JEOL JEM-2100 LaB$_6$ (Akishima, Tokyo, Japan) at an accelerating voltage of 200 kV by mass thickness contrast.

2.5. Field-Emission Scanning Electron Microscopy (FESEM)

Field emission scanning electron microscopy (FESEM) observation was performed at an accelerating voltage of 5 keV on a JEOL JSM-7401F (Akishima). The samples were collected on a silicon wafer and sputter-coated with platinum at approximately 2 to 3 nm.

2.6. Small-Angle X-ray Scattering (SAXS)

National Synchrotron Radiation Research Center (NSRRC) with synchrotron X-ray beamline X27C was used to study the SAXS where the wavelength of the beam was 0.155 nm. The two-dimensional SAXS pattern was obtained using the MAR CCD X-ray detector (Rayonix L.L.C., Evanston, IL, USA), at which the (1D) linear profile was obtained by integration of the 2D pattern. The scattering angle of the SAXS pattern was calibrated using silver behenate with the first order scattering vector $q^*(4\pi \sin\theta)/\lambda$, where 2θ is the scattering angle).

2.7. Nanoindentation Measurements

Hysitron Ti950 triboindenter (Hysitron Inc. Minneapolis, MN, USA) was used to perform the nanoindentation tests using a spherical indenter with a 2 µm diameter. The indentation measurements were conducted on a microtome film sample with 5 µm thickness in a silicon wafer as a substrate at room temperature. The load–displacement curve was recorded at the same rate of loading and unloading (60 µN/s) with a maximum load of 500 µN applied. In the nanoindentation tests, the load–displacement data were recorded continuously, while the tip was driven into the composite materials, and then smoothly removed. The load–displacement (L-D) curves were then used to calculate the mechanical energy dissipation of the fabricated materials at the same rate of loading and unloading (60 µN/s). For nanoindentation, the mechanical properties of reduced elastic modulus and hardness can be calculated from the load–displacement curve (P-h) based on the widely used Oliver-Pharr model. In the present study, the reduced elastic modulus E_r was determined from the P h curve, using the Sneddon formula for spherical indenter frictionless punch.

$$Er = \frac{\sqrt{\pi}}{2} \frac{S}{\sqrt{A_t}} \quad (1)$$

Here, E_r is the reduced elastic modulus (indentation modulus) which represents the elastic deformation that occurs in both the sample and indenter tip. S is stiffness. A_t represents the projected contact area. Note that the deformation in the diamond indenter tip is negligible. As a result, the reduced elastic modulus (indentation modulus) is a representative value for the discussion with respect to mechanical performance.

3. Results and Discussion

The study of the toughening mechanism by adding PBA-*b*-PMMA to brittle material such as epoxy resin starts from the basic characterization of the modifier (PBA-*b*-PMMA) (The detailed information about the sample is described in Table S1). As shown in Figure S1a, the glass transition temperature (T_g) of the soft PBA segment is approximately −35 °C, whereas the hard PMMA shows T_g at 87 °C (Figure S1b). Note that the T_g of the acrylic containing soft segment is significant for mechanical stability in bisphenol A diglycedyle ether (DGEBA) type of epoxy resin to achieve the desired toughness. The hard segment (PMMA) was expected to provide the molecular level association with epoxy resin in the nanoscale. Moreover, the T_g of the soft segment (PBA) was reaffirmed by DMA analysis, as shown in Figure S1c where the tan δ peak appears at approximately −37 °C. After the introduction of the PBA-*b*-PMMA modifier to the epoxy resin, the feasibility of reaction-induced phase separation was evaluated using TEM analysis of the dispersed PBA-*b*-PMMA in the epoxy matrix. As shown in Figure 2, the PBA-*b*-PMMA can be self-assembled into a spherical micelle structure dispersed in the epoxy matrix. The samples were vapor stained with 0.5 wt% OsO_4 aqueous solutions for 24 h at ambient temperature to provide adequate contrast between the two segments. In Figure 2 inset, the PBA microdomain appears as bright and PMMA as dark in a grey epoxy resin matrix, suggesting the formation of compatible PMMA in DGEBA. Note that PBA serves as the soft segment core in the dispersed micelle structure and PMMA act as the shell. Moreover, the spherical PBA-*b*-PMMA domains act as an impact modifier due to the compatibility between epoxy and PMMA, based on their interfacial strength. Namely, the incompatible

PBA block dispersed in the epoxy resin matrix can be characterized as an epoxy-phobic core and the compatible PMMA block as an epoxy-philic shell that led to the reinforcement of the interfacial strength. Additionally, as shown in Figure S1d, the glass transition temperature of the well-cured intrinsic epoxy resin was approximately 84 °C, whereas the addition of epoxy resin modified with 5% PBA-b-PMMA reduces the T_g to 83 °C; note that the reduction in T_g with respect to the addition of a modifier was negligible. After the gradual addition of PBA-b-PMMA up to 20% w/w, there is an obvious reduction in the T_g to 73 °C. Following the platform technology developed in our laboratory, the fabrication of well-ordered nanonetwork epoxy modified with PBA-b-PMMA can be successfully achieved using templated polymerization. By taking advantage of the strong segregation strength between PS and PDMS, gyroid-structured PS-b-PDMS can be fabricated from lamellae-forming BCP using toluene as a selective solvent [34]. Table 1 summarizes the molecular characterization details of the PS-b-PDMS used in this study. Figure S2a shows the TEM projection of the solution-cast PS-b-PDMS; a typical projection of a trigonal planar gyroid phase can be observed. The 1D SAXS results at the relative q values of $\sqrt{6}$, $\sqrt{8}$, $\sqrt{16}$, $\sqrt{22}$, $\sqrt{38}$, and $\sqrt{52}$ (Figure S2b) indicate the double gyroid phase (space group of $Ia\bar{3}d$). By using chloroform as a selective solvent, a typical projection of a double diamond phase with a tetrapod-like pattern can be identified by TEM analysis (Figure S2c). The corresponding 1D SAXS profile with relative q values of $\sqrt{2}$, $\sqrt{3}$, $\sqrt{6}$, $\sqrt{10}$, $\sqrt{18}$ and $\sqrt{20}$ (Figure S2d) are in good agreement with the double diamond phase in the $Pn\bar{3}m$ space group [38]. Following the experimental procedures shown in Figure 1, the pore-filling of the PBA-b-PMMA modified epoxy resin and thermal treatment at 110 °C and 150 °C for effective crosslinking of the precursor of epoxy resin can be carried out to produce PS/epoxy nanocomposites through templated polymerization. The TEM micrograph in Figure 3a shows the PS matrix as bright domain and the epoxy as dark domain due to the effective staining using OsO_4, which further confirms the formation of a gyroid-structured PS/epoxy nanocomposite. Subsequently, after removal of the PS template using organic solvents such as styrene monomer, well-ordered gyroid-structured epoxy resin can be obtained, as evidenced by FESEM (Figure 3b). Moreover, the one-dimensional small-angle X-ray scattering (1D SAXS) profile reaffirms the observed morphology.

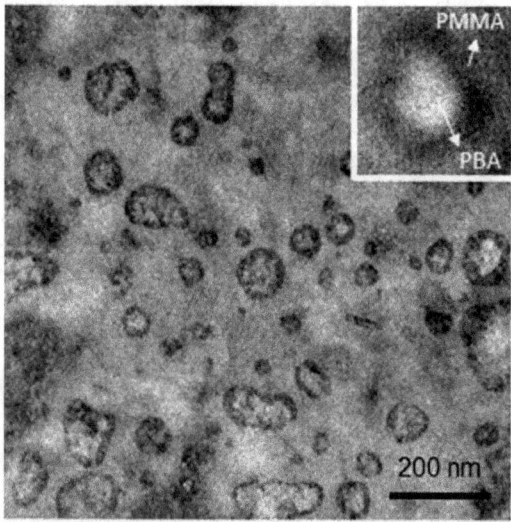

Figure 2. TEM micrograph of epoxy resin modified with 5% w/w of PBA-b-PMMA where the inset corresponds to a magnified image of the self-assembled PBA-b-PMMA and identifies the two different blocks. Vapor staining with 0.5 wt% OsO_4 aqueous solutions for 24 h at ambient temperature was performed leading to dark (PMMA) and bright (PBA) contrast.

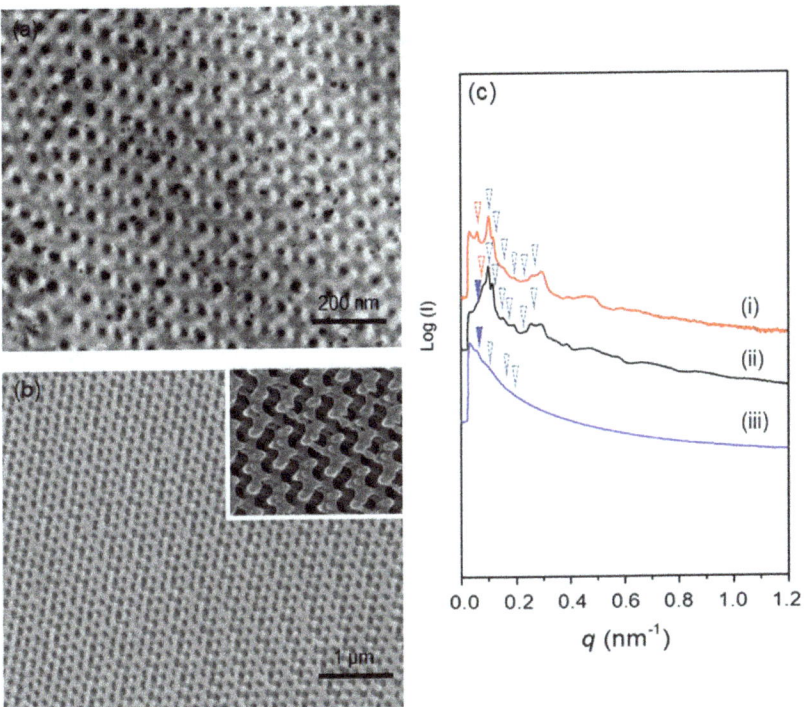

Figure 3. (a) TEM image of PS/epoxy nanocomposite in gyroid structure. (b) FESEM image of well-ordered gyroid epoxy resin, inset shows the magnified image. (c) 1D SAXS profiles of (i) PS template, (ii) PS/epoxy nanocomposites, and (iii) nanoporous epoxy resin with gyroid texture.

As shown in Figure 3c(i), the characteristic reflections corresponding to the 1D SAXS results with the relative q values of $\sqrt{6}$, $\sqrt{8}$, $\sqrt{16}$, $\sqrt{20}$, $\sqrt{30}$, and $\sqrt{38}$ and an additional reflection at $\sqrt{4}$ are recognized as the double gyroid phase (blue arrow) with an additional slight deformation peak (denoted by red arrow). The appearance of an extra peak at a relative q value of $\sqrt{2}$ can be observed for PS/epoxy nanocomposite in Figure 3c(ii), which is attributed to the (dark blue) network shifting of the gyroid nanonetworks during templated polymerization [39,40]. After the removal of the PS template, the shifting is more significant [Figure 3c(iii)]. Correspondingly, diamond-structured epoxy resin can be obtained by following the same experimental procedure. As shown in Figure 4a, the TEM projection verifies the formation of diamond network-structured PS/epoxy nanocomposites (See Figure S3 for details). Subsequently, after the removal of the PS template, nanonetwork-structured epoxy can be acquired, as evidenced by FESEM results in Figure 4b. As shown in Figure 4c(i,ii), a set of characteristic reflections for the double diamond phase at the relative q values of $\sqrt{2}$, $\sqrt{3}$, $\sqrt{4}$, $\sqrt{6}$, $\sqrt{8}$, and $\sqrt{10}$ are identified for the PS template and PS/epoxy nanocomposites. Consequently, a characteristic reflection of the nanoporous epoxy at the relative q value of $\sqrt{3}$ can be found in Figure 4c(iii) due to the network shifting after the removal of the template. The above-mentioned SAXS results (Figure 4c) confirm the successful fabrication of diamond-structured modified resin. The mechanical properties of the well-ordered nanonetwork thermosets were investigated using nanoindentation analysis. Figure 5 shows a typical load–displacement curve for intrinsic epoxy resin, as well as gyroid- and diamond-structured epoxy resins modified with PBA-*b*-PMMA. Based on the unloading curve, the reduced elastic modulus (E_r) of 4.2 GPa was calculated for intrinsic epoxy resin by the Oliver-Pharr model [41]; note that the epoxy resins are inherently brittle due to their highly cross-linked structure at which the unloading curve retracts predominantly to the initial state due to the elastic behavior. Interestingly, the modified epoxy resin with PBA-*b*-PMMA shows approximately 90% of plastic deformation behavior with minor retracting of the unloading curve (Figure S4). The reduced elastic modulus of epoxy resin (without deliberate structuring) after the addition of PBA-*b*-PMMA is approximately 3.8 GPa, indicating that there is no significant effect on the modulus after adding the modifier. In contrast to the intrinsic epoxy resin, a reduction in the modulus to 0.9 GPa for the gyroid-structured nanonetwork epoxy resin due to the introduction of porous texture. Consistently, the diamond-structured epoxy resin shows a lower modulus with a value of 0.8 GPa. It is important to note that acrylate-base block copolymer has better thermal and oxidative stability which prevents degradation during high-temperature curing (Figure S1d) due to the compatibility of PMMA towards epoxy. As a result, the dispersion of PBA-*b*-PMMA spherical domains in the matrix of epoxy resin enables better mechanical properties without compromising the thermal characteristics. For the energy dissipation measurement, the area enclosed by loading and unloading curves was calculated. For homogeneous non-structured epoxy resin, the calculated energy dissipation by area under the load–displacement curve is 0.02 ± 0.002 nJ; note that the epoxy resin with the highest crosslinking density are nearly 90% elastic at 500 μN load (Figure S4). By contrast, after being modified with 5% w/w of PBA-*b*-PMMA, a five-times higher energy dissipation can be observed which equals to 0.09 ± 0.004 nJ, which is much higher than the intrinsic brittle epoxy resin due to the plastic mode of deformation. This enhanced energy dissipation might be attributed to the stress concentration on the PBA soft segment, where the rubbery core of the self-assembled PBA-*b*-PMMA leads to a cavitate inside the epoxy matrix with shear band yielding, which accounts for the enhanced toughening. Note that the PBA in the BCP modifier incompatible to DGEBA epoxy resin that acts as a core in the self-assembled spherical micelle, where the compatibility between PMMA and epoxy allows the intermolecular interaction between the epoxy resin and the BCP-based modifier. As a result, the compatibility of PBA-*b*-PMMA controls the reduction in modulus, whereas the soft segment can enhance the energy dissipation. Most interestingly, the energy dissipation from brittle to the plastic deformation of intrinsic epoxy resin can be further enhanced by using artificially engineered structures. For gyroid-structured epoxy resin, the energy dissipation was calculated to 0.36 ± 0.02 nJ, a value quite higher than the one obtained for intrinsic epoxy resin. In contrast to gyroid-structured epoxy resin, diamond-structured epoxy resin show a large energy dissipation at a given loading with less retracting for the unloading (Figure 5); both gyroid and diamond structures give the deliberate structuring effect on energy dissipation, at which well-ordered nanonetwork structures can enhance the energy dissipation value up to six and eight times (Figure S5), compared to the intrinsic epoxy as reported in our previous publication [33]. These superior enhancements in mechanical energy dissipation is attributed to the well-ordered nanonetworks in gyroid and diamond structure with isotropic periodicity plastic deformation. Moreover, the higher strut number in the diamond network in comparison with the triagonal planar gyroid network justifies the additional energy dissipation along the struts equally and symmetrically. As a result, a recognizable increase in the energy dissipation can be found in the PBA-*b*-PMMA modified diamond epoxy resin with a value of 0.43 ± 0.03 nJ (more than twenty (20) times of the non-structured, intrinsic epoxy resin). The enhancement of the energy dissipation explicitly indicates the synergic effect of the deliberate structuring of network texture in the nanoscale and the toughening of self-assembled modifiers (BCPs) in the epoxy matrix at the molecular level.

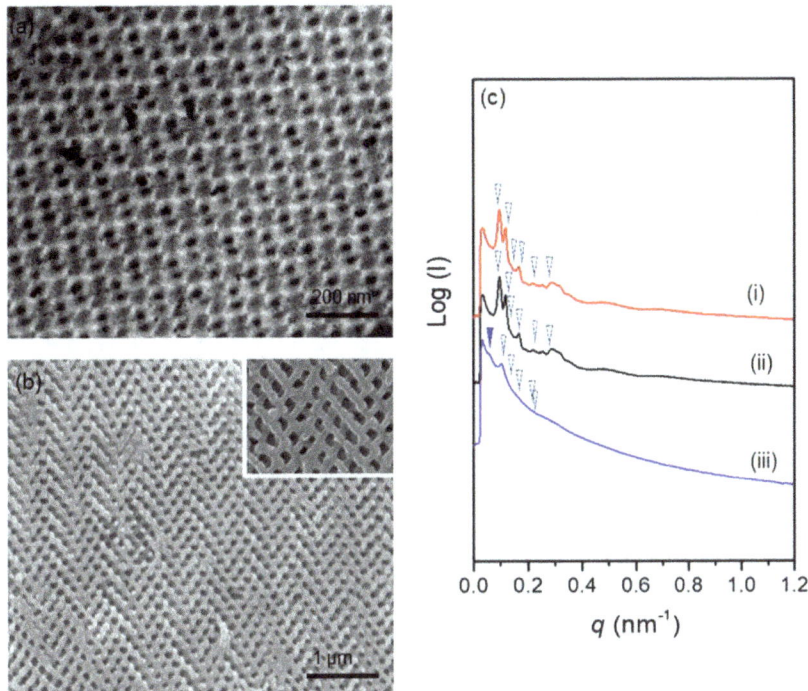

Figure 4. (**a**) TEM image of PS/epoxy nanocomposite in the diamond structure. (**b**) FESEM image of well-ordered diamond epoxy resin, inset shows the magnified image. (**c**) 1D SAXS profiles of (i) PS template, (ii) PS/epoxy nanocomposites, and (iii) nanoporous epoxy resin with diamond structure.

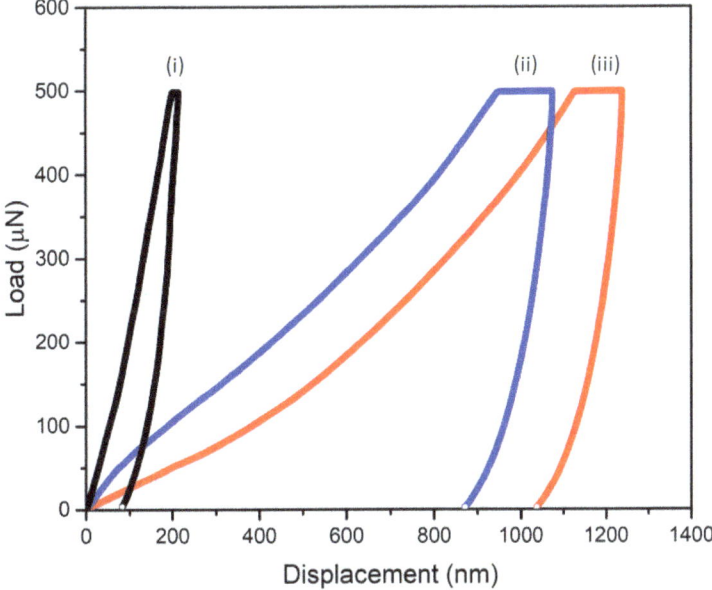

Figure 5. Load–displacement tests of (i) non-structured epoxy without a modifier, (ii) gyroid-structured epoxy, and (iii) diamond-structured epoxy resin with PBA-*b*-PMMA under 500 μN.

4. Conclusions

In conclusion, well-ordered nanonetwork-structured epoxy resin with gyroid (trigonal planar) and diamond (tetrapod) structures modified with a BCP based modifier, PBA-*b*-PMMA, can be successfully fabricated by templated polymerization, using PS templates. The periodic structured templates were acquired from PS-*b*-PDMS self-assembled samples, followed by preferential removal of PDMS through HF etching. The incorporation of PBA-*b*-PMMA in the matrix of epoxy resin imparts self-assembled spherical micelles. The PBA core serves as an energy absorbing soft domain due to the core–shell characteristics with reinforced interfacial strength from the association of PMMA in the matrix of epoxy resin. By taking advantage of the well-ordered nanonetwork structure, a further enhancement on plasticity can be achieved. The synergic effect of the deliberate structuring of nanonetwork texture and the toughening of the BCP-based modifier, thus, presented outstanding enhancement of plastic energy dissipation for gyroid and diamond-structured resin.

Supplementary Materials: The following supporting information can be downloaded at: https://www.mdpi.com/article/10.3390/polym14091891/s1, Figure S1: DSC analysis of PBA-*b*-PMMA. (a) The glass transition temperature of PBA. (b) Glass transition temperature of PMMA. (c) DMA analysis of PBA-*b*-PMMA with a glass transition temperature of PBA. (d) DSC analysis of epoxy resin with 0%, 5%,10%,15%, and 20% of PBA-*b*-PMMA.; Figure S2: (a) TEM micrograph of solution-cast PS-b-PDMS using toluene as solvent; (b) Corresponding 1D SAXS profile; (c) TEM micrograph of solution-cast PS-b-PDMS using chloroform as solvent; (d) Corresponding 1D SAXS profile; Figure S3: Three-dimensional TEM reconstruction images of (a) gyroid- and (b) diamond-structured epoxy resins in PS matrix from templated polymerization; Figure S4: Load–displacement curve of intrinsic epoxy resin (black) and intrinsic epoxy resin with 5% of PBA-*b*-PMMA (red); Figure S5: Load–displacement curve of Gyroid and diamond- structured epoxy resin without modifier; Table S1: Characterization of PBA-*b*-PMMA.

Author Contributions: Conceptualization, S.K.S.; Formal analysis, C.-Y.T.; Investigation, S.K.S.; Methodology, S.K.S. and T.-L.L.; Resources, H.-H.T., T.-S.L., G.-M.M. and A.A.; Supervision, R.-M.H.; Validation, H.S., S.-Y.C. and R.-M.H.; Visualization, H.S. and C.-Y.T.; Writing—original draft, S.K.S.; Writing—review and editing, T.-L.L., S.-Y.C., H.-H.T., T.-S.L., G.-M.M. and A.A. All authors have read and agreed to the published version of the manuscript.

Funding: This research work was partially funded by Ministry of Science and Technology of the Republic of China, Taiwan MOST with project number 107-2923-M-007-003-MY3 and Ministry of Science and Higher Education of the Russian Federation within State Contract 075-15-2019-1889. The research work was supported by the Hellenic Foundation for Research and Innovation (HFRI) under HFRI PhD Fellowship Grant 1651.

Institutional Review Board Statement: Not applicable.

Informed Consent Statement: Not applicable.

Data Availability Statement: Not applicable.

Acknowledgments: We thank the Ministry of Science and Technology of the Republic of China, Taiwan, for financially supporting this research MOST 107-2923-M-007-003-MY. We also thank The National Synchrotron Radiation Research Center (NSRRC, Taiwan) for its assistance in the Synchrotron SAXS experiments. This research was partially funded by the Ministry of Science and Higher Education of the Russian Federation within State Contract 075-15-2019-1889. The research work was supported by the Hellenic Foundation for Research and Innovation (HFRI) under HFRI PhD Fellowship Grant 1651.

Conflicts of Interest: The authors declare no conflict of interest.

References

1. Karak, N. Modification of Epoxies. In *Sustainable Epoxy Thermosets and Nanocomposites*; American Chemical Society: Washington, DC, USA, 2021; Volume 1385, pp. 37–68.
2. Shaw, S.J. Rubber modified epoxy resins. In *Rubber Toughened Engineering Plastics*; Collyer, A.A., Ed.; Springer: Dordrecht, The Netherlands, 1994; pp. 165–209.

3. Bucknall, C.B. Rubber toughening. In *The Physics of Glassy Polymers*; Haward, R.N., Young, R.J., Eds.; Springer: Dordrecht, The Netherlands, 1997; pp. 363–412.
4. Mimura, K.; Ito, H.; Fujioka, H. Toughening of epoxy resin modified with in situ polymerized thermoplastic polymers. *Polymer* **2001**, *42*, 9223–9233. [CrossRef]
5. Choi, J.; Yee, A.F.; Laine, R.M. Toughening of Cubic Silsesquioxane Epoxy Nanocomposites Using Core–Shell Rubber Particles: A Three-Component Hybrid System. *Macromolecules* **2004**, *37*, 3267–3276. [CrossRef]
6. Chen, J.; Kinloch, A.J.; Sprenger, S.; Taylor, A.C. The mechanical properties and toughening mechanisms of an epoxy polymer modified with polysiloxane-based core-shell particles. *Polymer* **2013**, *54*, 4276–4289. [CrossRef]
7. Walker, L.S.; Marotto, V.R.; Rafiee, M.A.; Koratkar, N.; Corral, E.L. Toughening in Graphene Ceramic Composites. *ACS Nano* **2011**, *5*, 3182–3190. [CrossRef] [PubMed]
8. Declet-Perez, C.; Francis, L.F.; Bates, F.S. Cavitation in Block Copolymer Modified Epoxy Revealed by In Situ Small-Angle X-Ray Scattering. *ACS Macro Lett.* **2013**, *2*, 939–943. [CrossRef]
9. Dean, J.M.; Lipic, P.M.; Grubbs, R.B.; Cook, R.F.; Bates, F.S. Micellar structure and mechanical properties of block copolymer-modified epoxies. *J. Polym. Sci. Part B Polym. Phys.* **2001**, *39*, 2996–3010. [CrossRef]
10. Dean, J.M.; Grubbs, R.B.; Saad, W.; Cook, R.F.; Bates, F.S. Mechanical properties of block copolymer vesicle and micelle modified epoxies. *J. Polym. Sci. Part B Polym. Phys.* **2003**, *41*, 2444–2456. [CrossRef]
11. Thompson, Z.J.; Hillmyer, M.A.; Liu, J.; Sue, H.-J.; Dettloff, M.; Bates, F.S. Block Copolymer Toughened Epoxy: Role of Cross-Link Density. *Macromolecules* **2009**, *42*, 2333–2335. [CrossRef]
12. Liu, J.; Sue, H.-J.; Thompson, Z.J.; Bates, F.S.; Dettloff, M.; Jacob, G.; Verghese, N.; Pham, H. Nanocavitation in Self-Assembled Amphiphilic Block Copolymer-Modified Epoxy. *Macromolecules* **2008**, *41*, 7616–7624. [CrossRef]
13. Lakes, R. Materials with structural hierarchy. *Nature* **1993**, *361*, 511–515. [CrossRef]
14. Cui, T.J.; Liu, R.; Smith, D.R. Introduction to Metamaterials. In *Metamaterials: Theory, Design, and Applications*; Cui, T.J., Smith, D., Liu, R., Eds.; Springer: Boston, MA, USA, 2010; pp. 1–19.
15. Jang, D.; Greer, J.R. Transition from a strong-yet-brittle to a stronger-and-ductile state by size reduction of metallic glasses. *Nat. Mater.* **2010**, *9*, 215–219. [CrossRef] [PubMed]
16. Gibson, L.J.; Ashby, M.F. *Cellular Solids: Structure and Properties*, 2nd ed.; Cambridge University Press: Cambridge, UK, 1997.
17. Schaedler, T.A.; Jacobsen, A.J.; Torrents, A.; Sorensen, A.E.; Lian, J.; Greer, J.R.; Valdevit, L.; Carter, W.B. Ultralight Metallic Microlattices. *Science* **2011**, *334*, 962. [CrossRef] [PubMed]
18. Lee, J.-H.; Singer, J.P.; Thomas, E.L. Micro-/Nanostructured Mechanical Metamaterials. *Adv. Mater.* **2012**, *24*, 4782–4810. [CrossRef] [PubMed]
19. Schaedler, T.A.; Ro, C.J.; Sorensen, A.E.; Eckel, Z.; Yang, S.S.; Carter, W.B.; Jacobsen, A.J. Designing Metallic Microlattices for Energy Absorber Applications. *Adv. Eng. Mater.* **2014**, *16*, 276–283. [CrossRef]
20. Biener, J.; Hodge, A.M.; Hamza, A.V.; Hsiung, L.M.; Satcher, J.H. Nanoporous Au: A high yield strength material. *J. Appl. Phys.* **2004**, *97*, 024301. [CrossRef]
21. Thomas, E.L.; Anderson, D.M.; Henkee, C.S.; Hoffman, D. Periodic area-minimizing surfaces in block copolymers. *Nature* **1988**, *334*, 598–601. [CrossRef]
22. Bates, F.S.; Fredrickson, G.H. Block Copolymers—Designer Soft Materials. *Phys. Today* **1999**, *52*, 32–38. [CrossRef]
23. Bates, F.S. Network Phases in Block Copolymer Melts. *MRS Bull.* **2005**, *30*, 525–532. [CrossRef]
24. Hajduk, D.A.; Harper, P.E.; Gruner, S.M.; Honeker, C.C.; Kim, G.; Thomas, E.L.; Fetters, L.J. The Gyroid: A New Equilibrium Morphology in Weakly Segregated Diblock Copolymers. *Macromolecules* **1994**, *27*, 4063–4075. [CrossRef]
25. Meng, L.; Watson, B.W.; Qin, Y. Hybrid conjugated polymer/magnetic nanoparticle composite nanofibers through cooperative non-covalent interactions. *Nanoscale Adv.* **2020**, *2*, 2462–2470. [CrossRef]
26. Darling, S.B. Directing the self-assembly of block copolymers. *Prog. Polym. Sci.* **2007**, *32*, 1152–1204. [CrossRef]
27. Hsueh, H.-Y.; Yao, C.-T.; Ho, R.-M. Well-ordered nanohybrids and nanoporous materials from gyroid block copolymer templates. *Chem. Soc. Rev.* **2015**, *44*, 1974–2018. [CrossRef] [PubMed]
28. Wang, Y.; Qin, Y.; Berger, A.; Yau, E.; He, C.; Zhang, L.; Gösele, U.; Knez, M.; Steinhart, M. Nanoscopic Morphologies in Block Copolymer Nanorods as Templates for Atomic-Layer Deposition of Semiconductors. *Adv. Mater.* **2009**, *21*, 2763–2766. [CrossRef]
29. Hsueh, H.-Y.; Huang, Y.-C.; Ho, R.-M.; Lai, C.-H.; Makida, T.; Hasegawa, H. Nanoporous Gyroid Nickel from Block Copolymer Templates via Electroless Plating. *Adv. Mater.* **2011**, *23*, 3041–3046. [CrossRef] [PubMed]
30. Hsueh, H.-Y.; Chen, H.-Y.; She, M.-S.; Chen, C.-K.; Ho, R.-M.; Gwo, S.; Hasegawa, H.; Thomas, E.L. Inorganic Gyroid with Exceptionally Low Refractive Index from Block Copolymer Templating. *Nano Lett.* **2010**, *10*, 4994–5000. [CrossRef]
31. Crossland, E.J.W.; Kamperman, M.; Nedelcu, M.; Ducati, C.; Wiesner, U.; Smilgies, D.M.; Toombes, G.E.S.; Hillmyer, M.A.; Ludwigs, S.; Steiner, U.; et al. A Bicontinuous Double Gyroid Hybrid Solar Cell. *Nano Lett.* **2009**, *9*, 2807–2812. [CrossRef] [PubMed]
32. Wang, X.-B.; Lin, T.-C.; Hsueh, H.-Y.; Lin, S.-C.; He, X.-D.; Ho, R.-M. Nanoporous Gyroid-Structured Epoxy from Block Copolymer Templates for High Protein Adsorbability. *Langmuir* **2016**, *32*, 6419–6428. [CrossRef]
33. Siddique, S.K.; Lin, T.-C.; Chang, C.-Y.; Chang, Y.-H.; Lee, C.-C.; Chang, S.-Y.; Tsai, P.-C.; Jeng, Y.-R.; Thomas, E.L.; Ho, R.-M. Nanonetwork Thermosets from Templated Polymerization for Enhanced Energy Dissipation. *Nano Lett.* **2021**, *21*, 3355–3363. [CrossRef]

34. Lo, T.-Y.; Chao, C.-C.; Ho, R.-M.; Georgopanos, P.; Avgeropoulos, A.; Thomas, E.L. Phase Transitions of Polystyrene-b-poly(dimethylsiloxane) in Solvents of Varying Selectivity. *Macromolecules* **2013**, *46*, 7513–7524. [CrossRef]
35. Georgopanos, P.; Lo, T.-Y.; Ho, R.-M.; Avgeropoulos, A. Synthesis, molecular characterization and self-assembly of (PS-b-PDMS)n type linear (n = 1, 2) and star (n = 3, 4) block copolymers. *Polym. Chem.* **2017**, *8*, 843–850. [CrossRef]
36. Chang, C.-Y.; Manesi, G.-M.; Yang, C.-Y.; Hung, Y.-C.; Yang, K.-C.; Chiu, P.-T.; Avgeropoulos, A.; Ho, R.-M. Mesoscale networks and corresponding transitions from self-assembly of block copolymers. *Proc. Natl. Acad. Sci. USA* **2021**, *118*, e2022275118. [CrossRef] [PubMed]
37. Whitesides, G.M.; Boncheva, M. Beyond molecules: Self-assembly of mesoscopic and macroscopic components. *Proc. Natl. Acad. Sci. USA* **2002**, *99*, 4769. [CrossRef] [PubMed]
38. Lin, C.-H.; Higuchi, T.; Chen, H.-L.; Tsai, J.-C.; Jinnai, H.; Hashimoto, T. Stabilizing the Ordered Bicontinuous Double Diamond Structure of Diblock Copolymer by Configurational Regularity. *Macromolecules* **2018**, *51*, 4049–4058. [CrossRef]
39. Hsueh, H.-Y.; Ling, Y.-C.; Wang, H.-F.; Chien, L.-Y.C.; Hung, Y.-C.; Thomas, E.L.; Ho, R.-M. Shifting Networks to Achieve Subgroup Symmetry Properties. *Adv. Mater.* **2014**, *26*, 3225–3229. [CrossRef]
40. Feng, X.; Burke, C.J.; Zhuo, M.; Guo, H.; Yang, K.; Reddy, A.; Prasad, I.; Ho, R.-M.; Avgeropoulos, A.; Grason, G.M.; et al. Seeing mesoatomic distortions in soft-matter crystals of a double-gyroid block copolymer. *Nature* **2019**, *575*, 175–179. [CrossRef]
41. Oliver, W.C.; Pharr, G.M. An improved technique for determining hardness and elastic modulus using load and displacement sensing indentation experiments. *J. Mater. Res.* **1992**, *7*, 1564–1583. [CrossRef]

Article

Poly(vinyl pyridine) and Its Quaternized Derivatives: Understanding Their Solvation and Solid State Properties

Katerina Mavronasou [1], Alexandra Zamboulis [2], Panagiotis Klonos [2,3], Apostolos Kyritsis [3], Dimitrios N. Bikiaris [2], Raffaello Papadakis [4] and Ioanna Deligkiozi [1,*]

[1] Creative Nano PC, 4 Leventi Street, Peristeri, 12132 Athens, Greece; k.mavronasou@creativenano.gr
[2] Laboratory of Polymer Chemistry and Technology, Department of Chemistry, Aristotle University of Thessaloniki, 54124 Thessaloniki, Greece; azampouli@chem.auth.gr (A.Z.); pklonos@central.ntua.gr (P.K.); dbic@chem.auth.gr (D.N.B.)
[3] Department of Physics, Zografou Campus, National Technical University of Athens, 15780 Athens, Greece; akyrits@central.ntua.gr
[4] TdB Labs AB, Ulls Väg 37, 75651 Uppsala, Sweden; rafpapadakis@gmail.com
* Correspondence: i.deligkiozi@creativenano.gr

Abstract: A series of N-methyl quaternized derivatives of poly(4-vinylpyridine) (PVP) were synthesized in high yields with different degrees of quaternization, obtained by varying the methyl iodide molar ratio and affording products with unexplored optical and solvation properties. The impact of quaternization on the physicochemical properties of the copolymers, and notably the solvation properties, was further studied. The structure of the synthesized polymers and the quaternization degrees were determined by infrared and nuclear magnetic spectroscopies, while their thermal characteristics were studied by differential scanning calorimetry and their thermal stability and degradation by thermogravimetric analysis (TG-DTA). Attention was given to their optical properties, where UV-Vis and diffuse reflectance spectroscopy (DRS) measurements were carried out. The optical band gap of the polymers was calculated and correlated with the degree of quaternization. The study was further orientated towards the solvation properties of the polymers in binary solvent mixtures that strongly depend on the degree of quaternization, enabling a better understanding of the key polymer (solute)-solvent interactions. The assessment of the underlying solvation phenomena was performed in a system of different ratios of DMSO/H_2O and the solvatochromic indicator used was Reichardt's dye. Solvent polarity parameters have a significant effect on the visible spectra of the nitrogen quaternization of PVP studied in this work and a detailed path towards this assessment is presented.

Keywords: poly(4-vinylpyridine); poly(N-methyl-4-vinylpyridinium iodide); quaternization; solvatochromism; preferential solvation; optical energy gap

Citation: Mavronasou, K.; Zamboulis, A.; Klonos, P.; Kyritsis, A.; Bikiaris, D.N.; Papadakis, R.; Deligkiozi, I. Poly(vinyl pyridine) and Its Quaternized Derivatives: Understanding Their Solvation and Solid State Properties. *Polymers* 2022, 14, 804. https://doi.org/10.3390/polym14040804

Academic Editors: Nikolaos Politakos and Apostolos Avgeropoulos

Received: 27 January 2022
Accepted: 17 February 2022
Published: 19 February 2022

Publisher's Note: MDPI stays neutral with regard to jurisdictional claims in published maps and institutional affiliations.

Copyright: © 2022 by the authors. Licensee MDPI, Basel, Switzerland. This article is an open access article distributed under the terms and conditions of the Creative Commons Attribution (CC BY) license (https://creativecommons.org/licenses/by/4.0/).

1. Introduction

Quaternized nitrogen involving polymers are a class of highly functional macromolecules that exhibit the properties of conventional polymers and an enhanced sensitivity towards the external environment due to the existing charges, e.g., solvent polarity, pH [1,2]. They can be classified in the broader class of polyelectrolytes (PEL) [3], displaying dual behavior and combining the properties of both polymers and electrolytes [4–6]. Unique properties are derived from their distinct nature, such as excellent water solubility, strong binding interactions with oppositely charged surfaces and molecules as well as a notable solvation and swelling behavior. Due to these features, they are widely employed as viscosity and surface modifiers as well as selective adsorbers. Consequently, these materials are useful in various industrial applications, materials and formulations from water treatment [7], membranes [8,9], sensors [10] and cosmetics to biomedicine with specific uses in bacterial and viral infections [11–19].

Poly(4-vinylpyridine) (PVP) is a linear polymer with pendant pyridine groups that can be used in a variety of applications, such as surface modification, by immobilizing atoms or particles, electrochemical sensors [20], fabrication of antibacterial surfaces, development of pH sensitive systems, 3D molecular level ordering systems, anti-corrosive coatings and even dye sensitized solar cells (DSSCs) and light-emitting diodes [21–29]. Positively charged polypyridines constitute a unique class of compounds in which quaternization enables the introduction of permanent positive charges into the backbone, resulting in electron delocalization phenomena that are attributed to the existence of electron-rich pyridines and electron-poor pyridiniums [30–33]. As a result, these polymers exhibit improved molecular properties, such as non-linear optical activity and increased conductivity [5,12,18,34–36]. Moreover, the counterion, as well as the length of the alkyl chain of the methylating agent, have a significant influence on solubility and conductivity [37–39]. Counterion exchange has been studied in order to obtain specific properties. It should be noted that Reillex ion exchange resins are commercial resins based on crosslinked and partially quaternized poly(4-vinylpyridine). Finally, amphiphilic copolymers containing a quaternized PVP block have also been reported [40–42].

Although it has been observed that 4-vinylpyridine polymerizes spontaneously upon protonation/quaternization, affording quaternized PVPs [43,44]; in the present study, quaternization was performed by a post-polymerization step, a strategy that yields well-defined polymers more readily [45,46]. Quaternized PVPs with a variety of alkyl halides have been reported, with various alkyl lengths and counterions. Most quaternization reactions afford quaternization degrees around 70%, while quantitative quaternizations have also been reported, usually achieved owing to a large excess of alkyl halide [14,15,17,47,48]. Although the kinetics of quaternization reactions and the retardation effect observed have been studied extensively ([49–51] and references therein), to the best of our knowledge, a systematic investigation on structure/properties relationship, especially focused on solvation and optical properties, regarding partially quaternized PVP derivatives is lacking. Indeed, most previous works focus on the use of a highly quaternized PVP derivative for specific applications, while the more fundamental study of the impact of quaternization on the molecular structure and resulting behavior of quaternized PVPs has not been reported yet.

In the present study, we report the targeted synthesis and characterization of fully (PVPQ) as well as three partially quaternized derivatives, produced by N-alkylation of the pyridine ring with methyl iodide. The aim is to cover a full range of quaternization degrees and perform a comparative study of the impact of quaternization on the physicochemical and optical properties as well as the solvation behavior of the resulting derivatives. Poly(4-vinylpyridine) (PVP) was quaternized in the presence of methyl iodide according to Scheme 1. The quaternization was quantified by ^{1}H NMR spectroscopy. The response of the polymers to temperature was assessed via TGA and DSC analysis. Emphasis was given on their behavior when it comes to light absorbance, reflectance and solvent interactions. Significant differences were observed between neutral PVP and the fully quaternized derivative (PVPQ), while the partially quaternized polymers displayed an intermediate behavior. The optical band gap difference and the solvation effect in binary solvent mixtures between those two polymers prove the significance of quaternization.

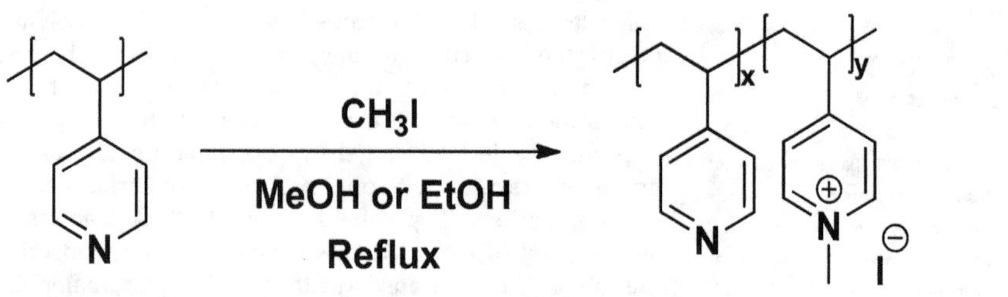

Scheme 1. Illustration of the synthetic route for the fully quaternized (x = 0) and the partially quaternized derivatives.

2. Materials and Methods

2.1. Materials

Poly(4-vinylpyridine) (Mw = 60 kDa) (PVP) and methyl iodide (CH_3I, 99.5%) were used as received from Sigma-Aldrich (St. Louis, MO, USA). Methanol (MeOH, Read, Ph.Eur) and ethanol (EtOH, 99.8%) were supplied by AppliChem GmbH (Darmstadt, Germany) and dimethyl sulfoxide (DMSO) A.G by Penta (Prague, Czech Republic). Reichardt's dye (dye content 90%) (RB) from Sigma-Aldrich (St. Louis, MO, USA) was used as the solvatochromic indicator.

2.2. Synthesis of Poly(N-methyl-4-vinyl pyridinium iodide) Derivatives

In this study, a fully quaternized (PVPQ) and three partially quaternized derivatives (PVP_Q1-3) (Table 1), were synthesized. The polyvinyl pyridine/methyl pyridinium copolymers were synthesized by dissolving 1 g (0.0167 mmol) of poly(4-vinylpyridine) (PVP) in 100 mL of ethanol and refluxing the solution in the presence of 0.35 mL (0.0055 mol), 0.47 mL (0.007 mol) and 0.7 mL (0.011 mol) of methyl iodide for 4 h so that, PVP_Q3, PVP_Q2, and PVP_Q1 were synthesized, respectively [17]. For the PVPQ derivative, 1 g of poly(4-vinylpyridine) (0.0167 mmol) was added to 50 mL of methanol and the solution was refluxed in the presence of 0.9 mL (0.014 mol) of iodomethane for 4 h. All the synthesized polymers were precipitated from the respective reaction mixtures using diethyl ether (200 mL approx.), filtered, and washed with ethanol and diethyl ether to yield a yellowish/greenish polymer product. All the resulting products were dried and fully characterized without further treatment.

Table 1. Characteristics of the polymers discussed in this study.

Sample	Calculated Degree of Quaternization (%)	Tg (°C)
PVP	0	86
PVP_Q3	35	86
PVP_Q2	46	98
PVP_Q1	85	101
PVPQ	100	114

The reaction yields for PVP_Q3, PVP_Q2, PVP_Q1 and PVPQ were 63%, 71%, 99.8% and 84%, respectively. The yields were calculated by comparing the initial PVP moles with the moles of the final product after precipitation and drying. The molecular weight of the initial PVP is 60 kDa, i.e., approximately 570 repeating units. The molecular weight of each newly synthesized polymer was calculated based on the degree of quaternization found by ^{1}H NMR (see Equation (1)), on the basis of 570 repeating units and a Mw of 247 g/mol per quaternized repeating unit.

2.3. Characterization Methods

The Fourier transform infrared (FTIR) spectra of the synthesized samples were measured on a Brucker Tension 27 FTIR spectrometer (Karlsruhe, Germany), equipped with a diamond ATR accessory at 25 °C and a spectral resolution of 4 cm^{-1} in the 4000–600 cm^{-1} range. IR spectra were analyzed with the Bruker OPUS software (version 5.2).

Nuclear magnetic resonance (NMR) spectra were recorded on an Agilent spectrometer (Agilent AM 600, Agilent Technologies, Santa Clara, CA, USA), operating at a frequency of 600 MHz for protons. Deuterated water, methanol or their mixtures were used depending on the solubility of the polymers. The spectra were internally referenced with tetramethylsilane (TMS) and calibrated using the residual solvent peak.

X-ray powder diffraction (XRD) was employed to study the structure of the synthesized polymers. The XRD spectra were recorded at room temperature through a MiniFlex II XRD system (Rigaku Co., Tokyo, Japan), with Cu Ka radiation (λ = 0.154 nm), over the 2θ range from 5° to 60° with a scanning rate of 1 °/min.

Differential scanning calorimetry (DSC) measurements were carried out in the overall temperature range from −40 to 210 °C in high purity nitrogen (99.9995%) atmosphere, utilizing a TA Q200 series DSC instrument (TA, New Castle, DE, USA), calibrated with indium for temperature and enthalpy and sapphires for heat capacity. The measurements were performed on samples of ~8–9 mg in mass closed in TA aluminum Tzero Hermetic pans. The cooling and heating rates were fixed at 10 °C/min. Overall, four (4) scans were performed in DSC, in order to follow effects on the samples as received, namely, equilibrated at environmental conditions, and upon drying. Details on the thermal profiles are given below, along with the experimental results.

The thermal stability analysis of the polymers was carried out under inert atmosphere (N_2) and air (N_2 80% O_2 20%). The measurements under air were performed using a thermogravimetric analyzer PerkinElmer Diamond Thermogravimetric/Differential Thermal Analysis (TG/DTA) (Waltham, MA, USA). The gas flow rate was 80 mL/min. The specific operation steps were as follows: the samples were heated from 25 to 700 °C at a rate of 10 °C/min and were kept at 7000 °C for 5 min. Thermogravimetric analysis under nitrogen was performed with a SETARAM SETSYS TG-DTA 16/18 instrument (Setaram instrumentation, Lyon, France). Samples were heated from ambient temperature to 600 °C in a 50 mL/min flow of N_2. A nominal heating rate of 20 °C/min was used and continuous records of sample temperature, sample weight, and heat flow were taken. All measurements were performed in triplicate.

UV-Vis spectra in solution and diffuse reflectance spectra (DRS) in the solid-state were measured between 200–800 nm using an Agilent Carry 60 spectrometer. UV-Vis spectra of PVP and its quaternized derivatives were measured in methanol and demonstrated characteristic bands at a concentration of 50 ppm. All measurements were performed at 25 ± 1 °C. The slit width and data interval were set to 1 nm, and the scan speed in all measurements was 960 nm/min. All spectra were normalized to show the same maximum intensity.

The DRS measurements were performed using a DRA Fibre Optic coupler (Harrick Agilent Barrelino). The bandgap was calculated using the Kubelka-Munk (K-M) model by plotting $[F(R) \times E]^{1/2}$ vs. E (eV), where $F(R) = (1 - R)^2/2R$ is the K-M function and R is the reflectance of the materials.

2.4. Solvatochromic Properties Study

The solvatochromic study was carried out using the following protocol: Firstly, 130 and 150 ppm stock solutions of the polymers in water (PVPQ) and ethanol (PVP) were prepared, respectively, and 0.1 mL of each solution was left to evaporate overnight in several vials. A Reichardt's dye stock solution was also prepared in acetone (150 ppm) and 1 mL of it was also left to evaporate overnight and dried under vacuum the next day. Then, to each polymer-containing vial, 1 mL of each binary solvent mixture ratio was added and after dissolution, the solvents were left to equilibrate with the polymer for 2 h. The binary mixtures of DMSO and water were prepared in multiple ratios in order to cover in detail the variation of the polymers behavior in the presence of slightly different solvent polarities. The same solvent mixture ratio was added to both polymers each time so that a comparative study would be performed. The UV spectra of each polymer-binary mixture was measured and used as a baseline for the measurement of the RB. After transferring the aforementioned quantity to the dye containing vials and dissolving the dye, RB's λmax was recorded.

3. Results

3.1. Polymer Synthesis and Structural Characterization

Poly(4-vinylpyridine) (PVP) was quaternized in the presence of methyl iodide according to Scheme 1. Progressively increasing amounts of methyl iodide afforded increasing quaternization degrees (up to 100%). Methanol was used as a solvent for the synthesis of the fully quaternized polymer (PVPQ), while ethanol was preferred for the partially quaternized derivatives (PVP_Q1-3). The success of these reactions was confirmed by FTIR and NMR spectroscopy, as discussed below. According to XRD measurements the obtained polymers were amorphous, Figure 1.

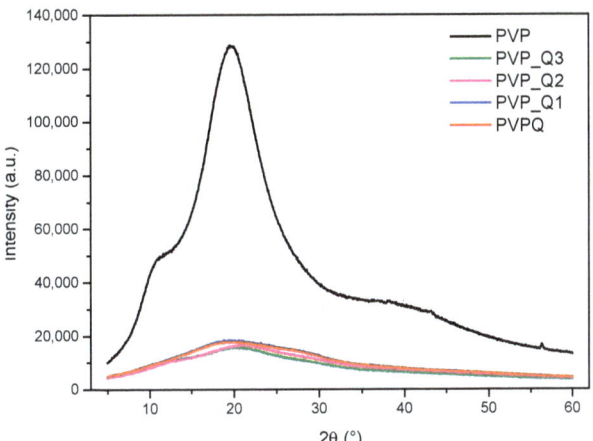

Figure 1. X-ray diffraction patterns of the synthesized polymers.

The FTIR spectrum of PVP in Figure 2a exhibits the characteristic vibrations of the pyridine ring (C=C, C–N) at 1598, 1556, 1496, 1452 and 1415 cm^{-1} [18]. The absorption bands at 1068 and 957 cm^{-1} can be assigned to the in-plane and out-of-plane C–H bending, respectively. The C-H stretching bands of the chain and the pyridine pendants are observed in the 3000 cm^{-1} region. For PVPQ, the appearance of the pyridinium bands at approximately 1640, 1570 and 1516 cm^{-1} confirm the quaternization of PVP [23,43]. It must also be noted that there is no trace of PVP, as evidenced by the absence of a band at 1598 cm^{-1}. The bands that appear at 1638, 1570, 1469 and 1300 cm^{-1} correspond to the C-C and C-N stretching vibrations of the pyridinium cation. The other peaks at 1185, 1056 and 971 cm^{-1} correspond to the symmetric bending of CH$_3$, the stretching of CH$_3$–N, and the out-of-plane stretching of H–C bond, respectively.

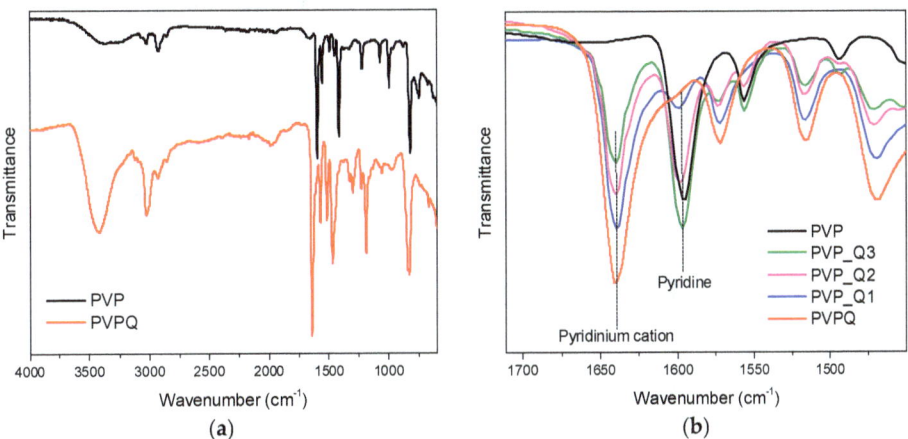

Figure 2. (**a**) Full FTIR spectra of PVP vs. the fully quaternized PVPQ derivative; (**b**) FTIR spectra of the 1400–1700 cm^{-1} region of the quaternized derivatives vs. the non-quaternized PVP.

The PVP_Q derivatives exhibit both peaks attributed to PVP and PVPQ. More specifically in Figure 2b, the peak at 1598 cm^{-1} corresponding to the non-methylated pyridine weakens when the quaternization ratio increases, while the peak attributed to the quaternized pyridine units at 1640 cm^{-1} increases accordingly. Last but not least, the peak around 3400 cm^{-1} in the full spectra is gaining intensity in the case of PVPQ when compared to PVP, due to the higher hydrophilicity and consequently, larger amount of moisture bound to the polymer [52].

The IR results are corroborated by NMR spectroscopy (Figure 3). In the ^{1}H NMR spectra, Figure 3a, the broad resonance signal around 1.5–2 ppm corresponds to the aliphatic protons of the backbone (–CH$_2$–CH–). In the quaternized polymers, the methyl group is clearly observed at ca. 4.2 ppm. The intensity of this peak increases with the quaternization degree. In the aromatic region, the protons of the pyridine and pyridinium aromatic rings are observed at 6.6 and 8.2 ppm, and 7.7 and 8.5 ppm, respectively. Similarly to the peak at 4.2 ppm, the resonance signals corresponding to the pyridinium units increase as the quaternization degree increases. Finally, in PVPQ, the peaks corresponding to PVP have completely disappeared confirming the quantitative quaternization of PVP [47,53]. Similar observations are drawn from the ^{13}C NMR spectra, Figure 3b: the resonance signals attributed to PVP (155, 150 and 124 ppm) decrease as the quaternization degree increases and are totally missing from PVPQ spectra. Correspondingly, the signals corresponding to the quaternized rings (160, 147 and 128 ppm) progressively increase.

^{1}H NMR spectra were used to calculate the quaternization degree. Due to the broadness of the peaks and some overlapping, the ratio of pyridine to pyridinium units could not be calculated by directly comparing the integrations of the corresponding peaks. It was thus calculated indirectly, according to Equation (1). On one hand, it is assumed that the integration of the peaks from 6–8 ppm corresponds to the total amount of units (quaternized and non-quaternized). On the other hand, when the peak corresponding to the methyl group at 4.2 ppm is normalized to three for all derivatives, the amount of quaternized units is equal to the amount of quaternized units in PVPQ, i.e., 4.1. The "excess" of aromatic protons is attributed to non-quaternized pyridine units. The ratio of those two integrations gives the degree of quaternization. The calculated quaternization degrees are reported in Table 1.

$$Quaternization\ degree\ (\%) = \frac{I(6-8)_{PVPQ}}{I(6-8)_{PVP\_Qx}} \times 100 \tag{1}$$

where $I(6-8)_{PVPQ}$ is the integration of the aromatic protons in for PVPQ, i.e., 4.1, and $I(6-8)_{PVP\_Qx}$ is the total integration of the aromatic protons for the partially quaternized derivatives.

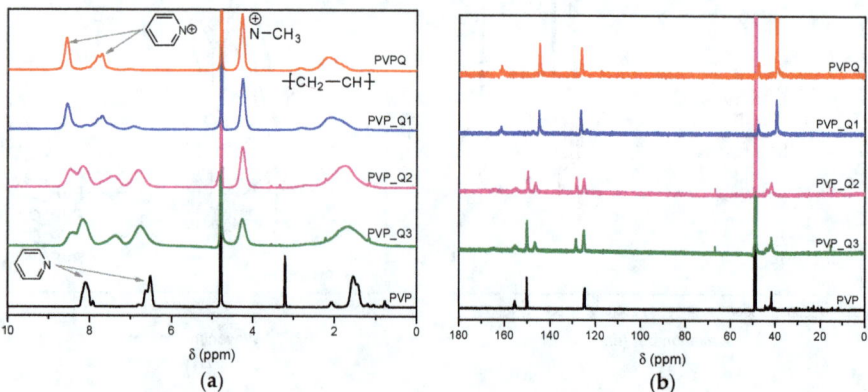

Figure 3. (a) ^{1}H NMR and (b) ^{13}C NMR spectra of poly(4-vinylpyridine) (PVP), partially (PVP_Q1–PVP_Q3) and fully (PVPQ) quaternized polymers.

3.2. Thermal Properties and Stability

The thermal behavior of PVP and all PVPQ derivatives was studied by differential scanning calorimetry (DSC). Four scans were performed by DSC. A first heating scan (scan 1), up to 150 °C, was performed to erase the thermal history in the presence of hydration water, and, subsequently (scan 2), the samples were cooled to a low temperature and heated up to 160 °C, in order to simultaneously follow the effects of structure and hydration on glass transition. Then, for scans 3 and 4, holes were made in the upper side of the hermetic pans to allow water evaporation. The samples were heated up to 160 °C (scan 3, water evaporation) and, subsequently cooled and heated up to 210 °C (scan 4). Thus, during scan 4 the direct effects of structure on glass transition could be assessed.

During scan 1 in DSC, all samples exhibit complex endothermal phenomena at temperatures above the glass transition (Figure 4a). Upon the erasing of thermal history and fixing of the polymer-pan thermal contact by the first heating, we may observe in Figure 4b that all samples demonstrate single glass transition steps. From a glance, it seems that the quaternization leads to elevation of the characteristic glass transition temperature, Tg (Table 1), and suppression of the glass transition strength (or else change in the heat capacity), Δc_p. These values were estimated and are shown in Figure 5 as a function of quaternization/modification. Beginning with PVP, which exhibits a Tg of 86 °C and Δc_p of 0.50 (±0.01) J/gK, the modification results in a sharp increase in Tg and decrease in Δc_p from the lower modification level. The changes are monotonic and suggest that the quaternization severely affects the mobility of polymer chains and particularly hinders the polymer chain diffusion (increase in Tg) and suppresses the mobile amorphous chains fraction (decrease of Δc_p) [54–57]. Both results are indicative of the transformation of the flexible polymer matrix (PVP) to a significantly more rigid one. This can also be observed from a qualitative effect in Figure 4b, namely, the change of heat flow slope (baseline) from large to gradually lower, both prior and upon glass transition [54,55].

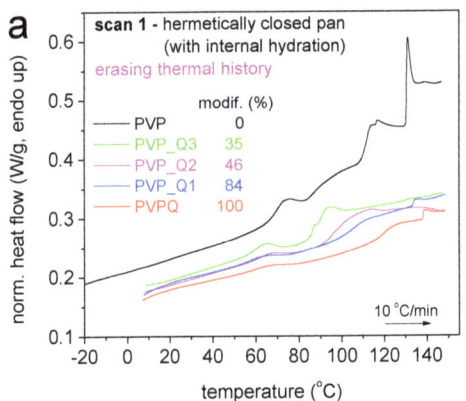 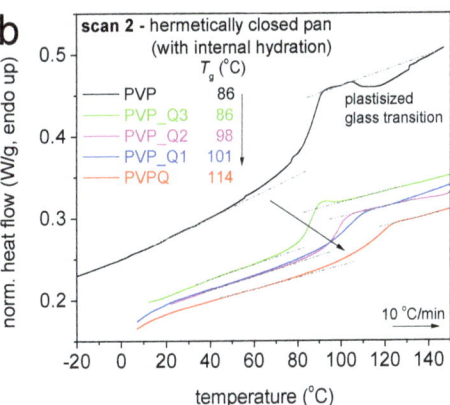

Figure 4. Heat flow curves during (**a**) scan 1 and (**b**) scan 2. The heat flow has been normalized to the sample mass. The glass transition temperatures, Tg, have been added to the plot. The added dash-dotted lines represent the baselines prior and upon glass transition.

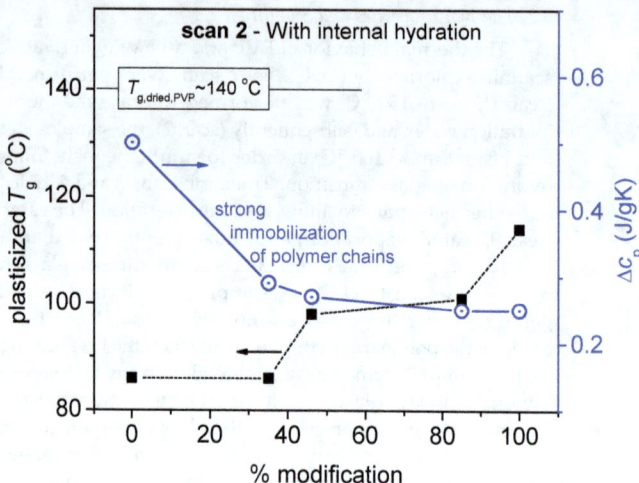

Figure 5. The modification effect on (left axis) Tg and (right axis) glass transition strength, Δc_p, in the ambient hydrated samples (as received). Included is the Tg for the initial dried PVP.

Besides these recordings, we should keep in mind that the abovementioned values refer to hydrated samples. Based on the dehydration experiments of scan 3 and 4, the Tg of the dried PVP equals ~140 °C (Figure 4). Therefore, it is essential that both the PVP and the quaternized PVPs are plasticized, i.e., exhibit lower Tg via the increased free volume involved due to the presence of water. Then, the elevated Tg in the hydrated PVPQ and PVP_Q1-3 should originate from the synergetic effect of hydration water and quaternization.

In Figure 6a, we present the DSC results for scan 3. The endothermic peak recorded between RT and 150 °C corresponds to the evaporation of 'free' and 'semi-bound' water [58]. The position of the peak (85–90 °C) barely changes between the different samples, whereas the same happens with the corresponding evaporation enthalpy. Thus, we may conclude that the polymer–water interaction degree as well as the evaporated water fraction is quite similar for the various samples. Therefore, we would expect a similar level of Tg plasticization. Despite that, the quaternization effects dominate here on the polymer mobility.

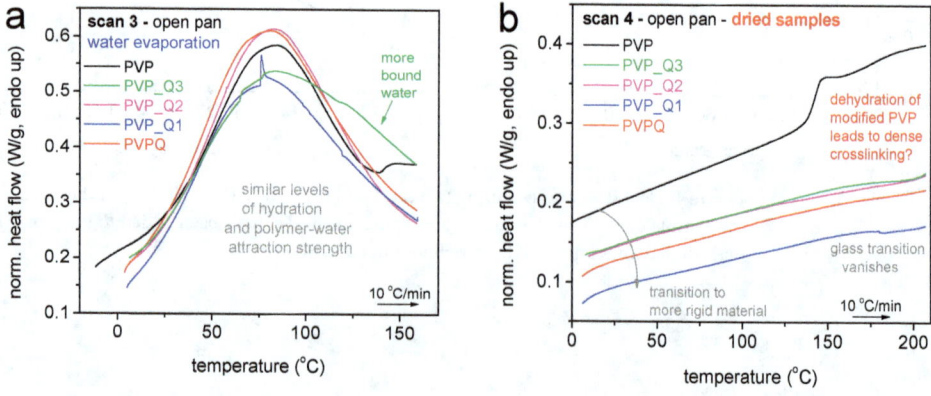

Figure 6. Comparative heat flow curves during heating for (**a**) scan 3 and (**b**) scan 4. The heat flow has been normalized to the sample mass.

A last comment on DSC refers to Figure 6b. Therein, upon dehydration, the PVP exhibits a clear glass transition step; meanwhile, on the other hand, within all quaternized samples the glass transition step has vanished. The effect suggests that the dehydration of PVPQ and PVP_Q1-3 resulted to extremely rigid structures. Upon removal of water molecules from the systems, the free volume decreases severely, whereas additionally, in the quaternized systems, dense crosslinking of the polymer chains is formed. Both parameters should be responsible for the 'elimination' of free polymer mobility. Another qualitative observation coming in to support to the latter, refers to the slopes (baselines) in Figure 6b. Comparing to PVP in the quaternized samples, the overall slope reduces. This is equivalent to the reduction of the heat capacity, c_p; the dependence from temperature. The result can be rationalized considering the reduced fraction of mobile segments that additively contribute to the transport of heat. Similar recordings have been demonstrated in polymer nanocomposites, wherein the attractive polymer–filler interactions become dominant and eliminate the free polymer mobility [55,56]. From a methodological point of view, we should report that c_p is better represented by measurements in or close to equilibrium, namely, by step-scan or temperature modulation DSC. Interestingly, these effects seem independent from the modification degree; however, they are in accordance to the similar amount of evaporated water.

The thermal stability of PVP and its quaternized derivatives was studied under both inert and oxidative conditions, Figure 7. Under inert conditions, two mass loss steps are observed. A small initial mass loss is recorded at low temperatures (up to 150 °C), which is attributed to the loss of adsorbed water. This is in agreement with the water evaporation observed by DSC (scan 3, Figure 6a). Then, the main degradation event that relates to the thermal degradation of the polymers is recorded between 300 and 400 °C for the quaternized polymers (maximum degradation rate around 365 °C) and 400–450 °C for PVP (maximum degradation rate around 420 °C). All quaternized polymers exhibit similar behavior with the two polymers containing the highest amount of quaternized units (PVPQ and PVP_Q1), demonstrating slightly lower stability compared to the two others. When it comes to air atmosphere, PVP behaves similarly by degrading at a higher temperature (between 350 and 400 °C). For the quaternized derivatives, an additional degradation step is observed between 500 and 600 °C. It can be assumed that the degradation products formed under O_2 (oxides), which differ from the ones formed under N_2, are more stable between 400 and 500 °C and result in a second mass loss between 500 and 600 °C.

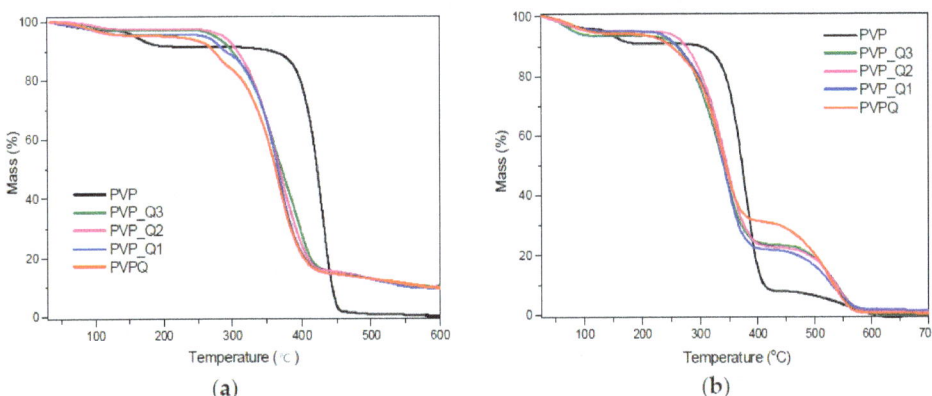

Figure 7. TGA measurements of PVP, as made, and the quaternized derivatives (**a**) under an N_2 atmosphere and (**b**) under a flow of air.

Overall, the thermal stability of PVP decreases with quaternization. It has been documented that modification by quaternization causes a decrease in the thermal stability of the polymer, as it becomes susceptible to Hofmann elimination due to the ammonium groups. Indeed, the Hofmann elimination occurs when quaternary ammonium salts are exposed to high temperatures; the reaction yields an alkene, a tertiary amine and a low molecular weight compound specific for the counterion (HI) [59]. Nevertheless, the quaternized PVP derivatives exhibit significant thermal stability.

3.3. UV-Vis Spectroscopy

The UV-Vis spectra of PVP and the partially quaternized derivatives are shown in Figure 8. Two well-defined UV bands, a strong one at 206 nm and a weaker one at 257 nm, are observed for the non-quaternized polymer. These bands are associated with n→π* transitions of N atoms with unshared electron pairs. However, in the case of the quaternized derivatives, an important shift from 206 to 217 nm was observed; this shift is associated with a bathochromic effect that occurs when the absorption wavelength shifts to longer wavelengths, indicating the formation of the pyridinium cation [23]. Moreover, it is observed that with increasing quaternization degree the intensity of the weaker band in the 260 to 280 nm region is decreased, probably due to the bonding of the nitrogen's lone pair electrons. The decreased absorbance of the PVPQ derivatives in this region is in accordance with the band gap reduction, as shown below.

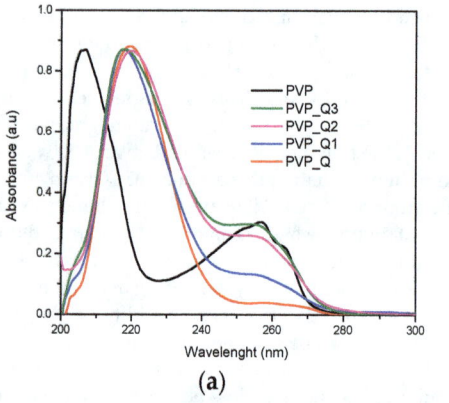 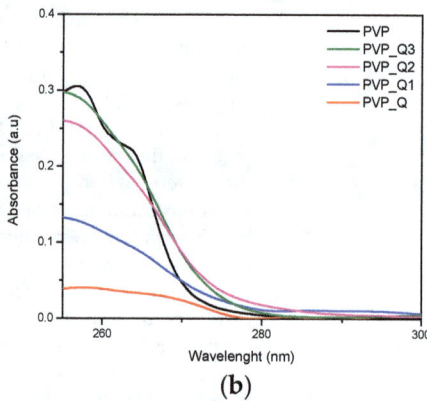

Figure 8. The UV-Vis spectra of PVP and the quaternized derivatives in methanol. (a) Full spectra. (b) Zoom in the 255–300 nm region.

Diffuse reflectance spectra (DRS) in the solid-state were also recorded between 200–800 nm. DRS calculates the absorption, which corresponds to the electron transition from the valance band to the conduction band in order to determine the bandgap of materials. The Kubelka–Munk plots of PVP and PVPQ derivatives are presented in Figure 9. The inter band electronic transitions and the absorption spectrum of all polymers represents a strong and broad absorption feature in the UV region. The bandgap values are determined from the intersection of the extrapolation of the linear part of the plots to the x-axis, indicating energy. A considerable red-shift of absorption edge in higher wavelength numbers for all quaternized polymers is observed. This shift is attributed to the reduction of the energy gap, which it is achieved for all quaternized polymers, and can be observed in Figure 9, where the absorption functions $(F \times E)^{1/2} = f(E)$ with $F = (1 - R)^2 / 2R$ corresponding to Kubelka–Munk function are presented. In Table 2, the significant difference in the energy gap between the quaternized and non-quaternized polymers is proved.

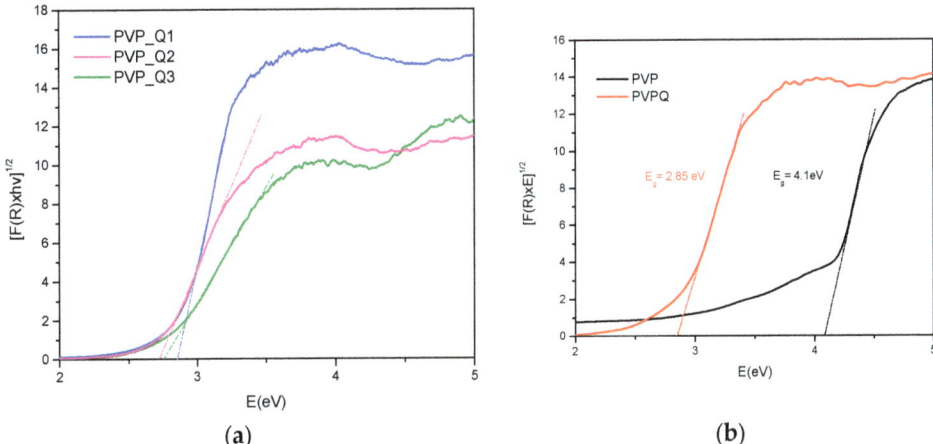

Figure 9. Kubelka–Munk plots of (**a**) PVP (Q1-Q3) and (**b**) PVP and PVPQ.

Table 2. Bandgap values (Eg) of the quaternized and non-quaternized polymers.

Sample	Eg (eV)_Indirect
PVP	4.1
PVP_Q3	2.83
PVP_Q2	2.75
PVP_Q1	2.74
PVPQ	2.85

As clearly observed, the optical bandgap is reduced from 4.1 eV for the non quaternized PVP to 2.85 eV for the fully quaternized PVPQ. The reported band gap values for PVP and PVPQ (See Table 2) correspond to the energy gaps between the higher occupied molecular orbital (HOMO) and the lower unoccupied molecular orbital (LUMO) of the corresponding polymers. More specifically, the band gap of a molecular material, i.e., the difference between its valence and conduction band, corresponds in molecular orbital theory terms to the gap between HOMO and LUMO energy levels of the molecule (the macromolecule in the case of PVP and PVPQ).

The optical bandgap (Eg) was found to decrease similarly in the partially quaternized polymers; however, it is without a well-defined relation toward the degree of quaternization in the monitored conditions. Nevertheless, a decrease can be explained for all polymers by the fact that charge transfer complexes (CTCs) in the host polymer were formed due to the quaternization of nitrogen. As a result, the lower energy transitions will be enhanced, leading to the observable optical bandgap changes [60].

Quaternization of the aromatic nitrogen by nucleophilic alkylation offers an alternative way to introduce positive charge into the backbone and provides opportunities to enhance the electron delocalization and the molecular properties of these polymers [33]. Consequently, the quaternized polymers exhibit a smaller energy gap; it should be mentioned that partial quaternization may contribute to the E_g reduction even more due to the random positioning of the positive charges and thus electron distribution.

3.4. Solvatochromic Study

The solvation effects occurring in the solutions of polymers are of high importance, as their study can reveal important information regarding key solvent–polymer interactions controlling the properties and macroscopic behavior of polymer solutions [61,62]. Studying these interactions experimentally has various limitations, many of which can be overcome using suitable probe molecules, which are sensitive to minute changes in their solvation microenvironment (cybotactic region) [63–67]. Herein, we employ the intensely sensitive solvatochromic betaine of Reichardt [68] in order to unravel the role of hydrogen bonding in solutions of PVP and PVPQ in binary solvent mixtures consisting of H_2O and DMSO. Two main criteria of choice of these two solvents were considered; firstly, the solubility of the polymers and secondly, the involvement of protic and non-protic solvents in the investigated mixtures. Water can act both as a hydrogen bond donating (HBD) as well as a hydrogen bond accepting (HBA) solvent, whereas DMSO is an HBA solvent exhibiting considerable HBA and Lewis basicity. In general terms, water molecules are expected to efficiently create H-bonds with the N atoms of PVP and PVPQ involving lone pairs of electrons, whereas DMSO is anticipated to gather around/solvate the quaternized pyridinium entities of PVPQ. In addition to these basic specific and non-specific solvent–polymer interactions, important solvent–cosolvent interactions are of high importance, as DMSO and H_2O efficiently form complexes when mixed; this happens at different extents depending on the molar ratio between the two solvents [69]. Table 3 shows the fourteen solvent ratios used as well as the recorded λmax in the presence of PVP and PVPQ. The λmax was measured three consecutive times and the average was noted.

Table 3. Solvent systems and the respective RB's absorption maxima in the visible region in the presence of PVP and PVPQ.

DMSO: H_2O Ratio	PVP (nm)	PVPQ (nm)
100:1	635	634.5
98:2	629	631.5
95:5	628	629
90:10	597	600
85:15	575	577
80:20	569	570
75:25	562	562.5
50:50	515	514
25:75	471	485.5
20:80	465	478
15:85	462	473
10:90	458	470
5:95	434.5	465
2:98	417	438

As Reichardt's betaine can act as an indicator of solvent polarity in DMSO/H_2O mixtures (the preferential solvation of Reichardt's betaine in DMSO/H_2O mixtures has been examined thoroughly in the past [68]) alterations of the solvation effect in solutions of any of the polymers PVP or PVPQ in DMSO/H_2O mixtures are expected to be detectable through the solvatochromism of Reichardt's betaine.

Even though the solvatochromic shift patterns for solutions of PVP and PVPQ in DMSO/H_2O mixtures appear to be very similar to those observed in the absence of polymer (see plots of Figures 10 and 11 and Table 4) important deviations appeared in the water-rich regions of the plots. Specifically, in the case of PVP, δE_T (PVP) differences as large as 6 kcal/mol were observed. These large differences are obviously connected to the presence of the polymer (PVP). As water is very prone to H-bond to N atoms of PVP in low DMSO molar ratios ($x_{DMSO} < 0.08$), these large deviations from linearity are attributed to synergistic solvation effects encompassing the formation of complexes of the type PVP $\cdots$ HOH $\cdots$ RB (where RB corresponds to Reichardt's betaine). The obtained microenvironment is sensed by Reichardt's betaine, as highly polar, hence the large solvatochromic shifts.

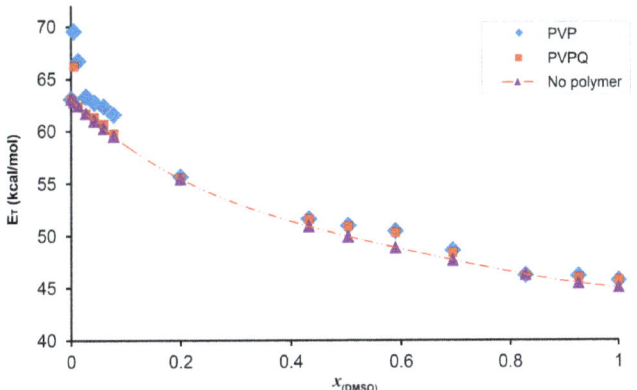

Figure 10. Plots of the measured Reichardt's betaine charge-transfer energies in DMSO/water mixtures involving PVP and PVPQ, and absence of polymer versus the molar fraction of DMSO.

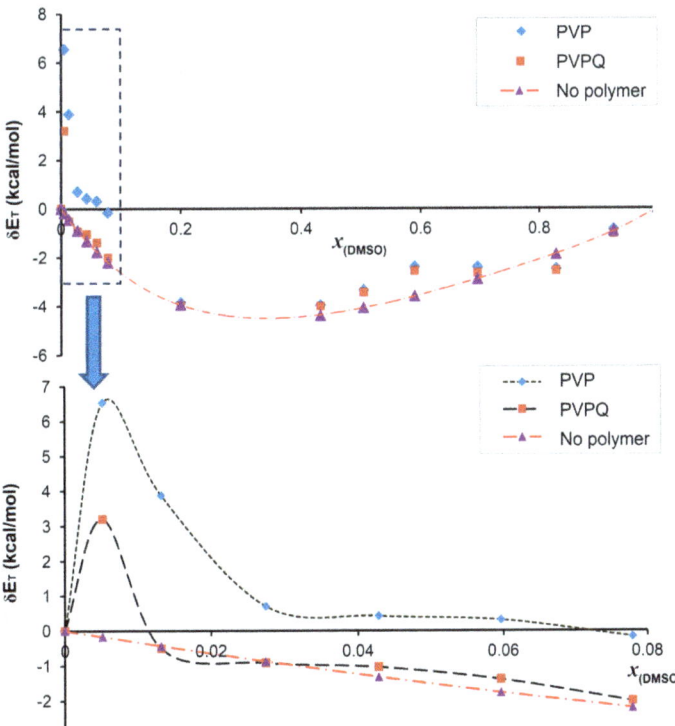

Figure 11. Plots of δE_T were observed in DMSO/H$_2$O mixtures involving PVP and PVPQ, and the absence of polymer versus the molar fraction of DMSO. Top: full DMSO fraction range; bottom magnification of the bracketed DMSO fraction range 0–0.08.

Table 4. Results of the solvatochromic study of solutions of PVP, PVPQ, and absence of polymer in binary mixtures of DMSO and H$_2$O involving Reichardt's betaine as a solvatochromic indicator.

xDMSO	E$_T$ (PVP) kcal/mol	E$_T$ (PVPQ) kcal/mol	E$_T$ (NP *) kcal/mol	δE$_T$ (PVP) ‡ kcal/mol	δE$_T$ (PVPQ) ‡ kcal/mol	δE$_T$ (NP) ‡ kcal/mol
1.000	45.7	45.7	45.0	0.0	0.0	0.0
0.926	46.1	45.9	45.4	−0.9	−1.0	−0.9
0.828	46.2	46.1	46.2	−2.5	−2.6	−1.9
0.695	48.6	48.3	47.6	−2.4	−2.6	−2.9
0.590	50.4	50.3	48.8	−2.4	−2.6	−3.6
0.504	51.0	50.9	49.9	−3.4	−3.4	−4.1
0.432	51.6	51.6	50.9	−4.0	−4.0	−4.4
0.200	55.6	55.5	55.4	−3.9	−4.0	−4.0
0.078	61.6	59.7	59.5	−0.2	−2.0	−2.2
0.060	62.4	60.7	60.2	0.3	−1.4	−1.8
0.043	62.8	61.3	61.0	0.4	−1.0	−1.3
0.027	63.3	61.7	61.7	0.7	−0.9	−0.9
0.013	66.7	62.4	62.4	3.9	−0.5	−0.4
0.005	69.5	66.2	62.8	6.5	3.2	−0.2
0.000	63.1	63.1	63.1	0.0	0.0	0.0

* NP: No polymer; ‡ δE$_T$ (PVP), δE$_T$ (PVPQ) and δE$_T$ (NP); the difference between measured charge transfer energy of Reichardt' betaine measured in PVP and PVPQ, and in absence of polymer, respectively, and the ideal (linear dependence) Reichardt' betaine charge-transfer energy are all measured in DMSO/water mixtures.

Similar effects are observed in the case of PVPQ, nevertheless, they appear to be attenuated when compared to PVP. This confirms the role of water-pyridine H-bonding in the observed solvation effect, as PVPQ contains a few non-quaternized N-atoms (i.e., N atoms with lone pairs of electrons prone to H-bonding with water molecules). Along with the remaining DMSO fraction range 1 > x > 0.08, the differences observed are less important, implying that the role of DMSO is less significantly altered in the presence of each of the polymers, PVP and PVPQ.

4. Conclusions

In this work, a series of new quaternized PVP polymers were synthesized and fully characterized. Characterization with FTIR and NMR essentially contributed to the understanding of the structure and composition (quaternization degree) of the fully (PVPQ) and partially (PVP_Q1-3) quaternized derivatives. Regarding the thermal behavior, as-made PVP as well as the quaternized PVPs exhibit single glass transition steps, with Tg systematically increasing and the heat capacity change decreasing with the modification. This suggests the transformation to a gradually more rigid matrix. In all cases the matrices are plasticized by similar amounts of hydration water (ambient), while the quaternization dominates on the hindering of the polymer chains diffusion. The thermal stability decreases with quaternization, due to the susceptibility of the quaternized polymers to Hoffman elimination reactions. Nevertheless, PVPQ and PVP_Q1-3 exhibit satisfactory thermal stability (up to 300 °C). When it comes to UV-vis absorbance, the quaternization reaction resulted in a red shift of the λmax from 206 to 217 nm for all PVPQs. Moreover, as an outcome of the quaternization, the optical energy gap of PVP was significantly reduced from 4.1 eV to 2.74–2.85 eV for PVPQ derivatives, which could be indicative of the potential of these materials in semiconducting and optoelectronic applications. Finally, the solvation behavior of the polymers was assessed in binary solvent mixtures of DMSO and water, employing a solvatochromic model dye, namely Reichardt's betaine. The impact of solvent polarity versus solvent basicity was proven to play an important role in the solvation of the studied quaternized and non PVPs. More in-depth studies on the solvatochromism of these novel polymers are in progress, broadening the understanding of the solvation behavior and chemical bonding that goes along with the optical properties related to the polymers' chemical structure.

Author Contributions: Formal analysis, K.M., A.Z., P.K. and R.P.; writing—original draft preparation, K.M., A.Z., P.K., R.P. and I.D.; writing—review and editing, K.M., A.Z., D.N.B., A.K., R.P. and I.D.; supervision, A.K., D.N.B., R.P. and I.D. All authors have read and agreed to the published version of the manuscript.

Funding: This research received no external funding.

Data Availability Statement: The data presented in this study are available on request from the corresponding author.

Acknowledgments: The authors would like to acknowledge the Center of Interdisciplinary Research and Innovation of Aristotle University of Thessaloniki (CIRI-AUTH), Greece, for access to the Large Research Infrastructure and Instrumentation of the Nuclear Magnetic Resonance Laboratory at the Center for Research of the Structure of Matter in the Chemical Engineering Department.

Conflicts of Interest: The authors declare no conflict of interest.

References

1. Malavolta, L.; Oliveira, E.; Cilli, E.M.; Nakaie, C.R. Solvation of polymers as model for solvent effect investigation: Proposition of a novel polarity scale. *Tetrahedron* **2002**, *58*, 4383–4394. [CrossRef]
2. Krokhina, L.S.; Kuleznev, V.N.; Lyusova, L.R.; Glagolev, V.A. Effect of solvent on polymer interaction in solution and properties of films obtained. *Polym. Sci. USSR* **1976**, *18*, 756–762. [CrossRef]
3. Wanasingha, N.; Dorishetty, P.; Dutta, N.K.; Choudhury, N.R. Polyelectrolyte gels: Fundamentals, fabrication and applications. *Gels* **2021**, *7*, 148. [CrossRef] [PubMed]
4. Hunkeler, D.; Wandrey, C. Polyelectrolytes: Research, Development, and Applications. *Chimia* **2001**, *55*, 223–227.
5. Vodolazkaya, N.A.; Mchedlov-Petrossyan, N.O.; Bryleva, E.Y.; Biletskaya, S.V.; Scrinner, M.; Kutuzova, L.V.; Ballauff, M. The binding ability and solvation properties of cationic spherical polyelectrolyte brushes as studied using acid-base and solvatochromic indicators. *Funct. Mater.* **2012**, *17*, 470–476.
6. Rabiee, A.; Ershad-Langroudi, A.; Zeynali, M.E. A survey on cationic polyelectrolytes and their applications: Acrylamide derivatives. *Rev. Chem. Eng.* **2015**, *31*, 239–261. [CrossRef]
7. Bucatariu, F.; Teodosiu, C.; Morosanu, I.; Fighir, D.; Ciobanu, R.; Petrila, L.M.; Mihai, M. An overview on composite sorbents based on polyelectrolytes used in advanced wastewater treatment. *Polymers* **2021**, *13*, 3963. [CrossRef]
8. Chen, N.; Lee, Y.M. Anion exchange polyelectrolytes for membranes and ionomers. *Prog. Polym. Sci.* **2021**, *113*, 101345. [CrossRef]
9. Durmaz, E.N.; Sahin, S.; Virga, E.; De Beer, S.; De Smet, L.C.P.M.; De Vos, W.M. Polyelectrolytes as Building Blocks for Next-Generation Membranes with Advanced Functionalities. *ACS Appl. Polym. Mater.* **2021**, *3*, 4347–4374. [CrossRef]
10. Ambade, A.V.; Sandanaraj, B.S.; Klaikherd, A.; Thayumanavan, S. Fluorescent polyelectrolytes as protein sensors. *Polym. Int.* **2007**, *56*, 474–481. [CrossRef]
11. Tamami, B.; Kiasat, A.R. Synthesis and application of quaternized polyvinylpyridine supported dichromate as a new polymeric oxidizing agent. *Iran. Polym. J.* **1997**, *6*, 273–279.
12. Blackmore, I.J.; Gibson, V.C.; Hitchcock, P.B.; Rees, C.W.; Williams, D.J.; White, A.J.P. Pyridine N-alkylation by lithium, magnesium, and zinc alkyl reagents: Synthetic, structural, and mechanistic studies on the bis(imino)pyridine system. *J. Am. Chem. Soc.* **2005**, *127*, 6012–6020. [CrossRef]
13. Zhang, B.Q.; Chen, G.D.; Pan, C.Y.; Luan, B.; Hong, C.Y. Preparation, characterization, and thermal properties of polystyrene-block-quaternized poly(4-vinylpyridine)/montmorillonite nanocomposites. *J. Appl. Polym. Sci.* **2006**, *102*, 1950–1958. [CrossRef]
14. Chernov'yants, M.S.; Burykin, I.V.; Pisanov, R.V.; Shalu, O.A. Synthesis and antimicrobial activity of poly(n-methyl-4-vinylpyridinium triiodide). *Pharm. Chem. J.* **2010**, *44*, 61–63. [CrossRef]
15. Singh, P.K.; Bhattacharya, B.; Nagarale, R.K.; Pandey, S.P.; Kim, K.W.; Rhee, H.W. Ionic liquid doped poly(N-methyl 4-vinylpyridine iodide) solid polymer electrolyte for dye-sensitized solar cell. *Synth. Met.* **2010**, *160*, 950–954. [CrossRef]
16. Borah, K.J.; Dutta, P.; Borah, R. Synthesis, characterization and application of poly(4-vinylpyridine)- supported Brønsted acid as reusable catalyst for acetylation reaction. *Bull. Korean Chem. Soc.* **2011**, *32*, 225–228. [CrossRef]
17. Efrati, A.; Tel-Vered, R.; Michaeli, D.; Nechushtai, R.; Willner, I. Cytochrome c-coupled photosystem i and photosystem II (PSI/PSII) photo-bioelectrochemical cells. *Energy Environ. Sci.* **2013**, *6*, 2950–2956. [CrossRef]
18. Xue, Y.; Xiao, H. Antibacterial/antiviral property and mechanism of dual-functional quaternized pyridinium-type copolymer. *Polymers* **2015**, *7*, 2290–2303. [CrossRef]
19. Manouras, T.; Platania, V.; Georgopoulou, A.; Chatzinikolaidou, M.; Vamvakaki, M. Responsive quaternized PDMAEMA copolymers with antimicrobial action. *Polymers* **2021**, *13*, 3051. [CrossRef]
20. Behzadi pour, G.; Nazarpour fard, H.; Fekri aval, L.; Esmaili, P. Polyvinylpyridine-based electrodes: Sensors and electrochemical applications. *Ionics* **2020**, *26*, 549–563. [CrossRef]
21. Raczkowska, J.; Stetsyshyn, Y.; Awsiuk, K.; Zemła, J.; Kostruba, A.; Harhay, K.; Marzec, M.; Bernasik, A.; Lishchynskyi, O.; Ohar, H.; et al. Temperature-responsive properties of poly(4-vinylpyridine) coatings: Influence of temperature on the wettability, morphology, and protein adsorption. *RSC Adv.* **2016**, *6*, 87469–87477. [CrossRef]

22. Malynych, S.; Luzinov, I.; Chumanov, G. Poly(vinyl pyridine) as a universal surface modifier for immobilization of nanoparticles. *J. Phys. Chem. B* **2002**, *106*, 1280–1285. [CrossRef]
23. Hernández-Orta, M.; Pérez, E.; Cruz-Barba, L.E.; Sánchez-Castillo, M.A. Synthesis of bactericidal polymer coatings by sequential plasma-induced polymerization of 4-vinyl pyridine and gas-phase quaternization of poly-4-vinyl pyridine. *J. Mater. Sci.* **2018**, *53*, 8766–8785. [CrossRef]
24. Shin, I.; Lee, K.; Kim, E.; Kim, T.H. Poly(ethylene glycol)-Crosslinked Poly(vinyl pyridine)-based Gel Polymer Electrolytes. *Bull. Korean Chem. Soc.* **2018**, *39*, 1058–1065. [CrossRef]
25. Urakawa, O.; Yasue, A. Glass transition behaviors of poly (vinyl pyridine)/poly (vinyl phenol) revisited. *Polymers* **2019**, *11*, 1153. [CrossRef]
26. Abed, Y.; Arrar, Z.; Hammouti, B.; Taleb, M.; Kertit, S.; Mansri, A. Poly(4-vinylpyridine) and poly(4-vinylpyridine poly-3-oxide ethylene) as corrosion inhibitors for Cu60-Zn40 in 0.5 M HNO$_3$. *Anti-Corros. Methods Mater.* **2001**, *48*, 304–308. [CrossRef]
27. Xiao, P.; Dong, T.; Xie, J.; Luo, D.; Yuan, J.; Liu, B. Emergence of white organic light-emitting diodes based on thermally activated delayed fluorescence. *Appl. Sci.* **2018**, *8*, 299. [CrossRef]
28. Panunzi, B.; Diana, R.; Caruso, U. A highly efficient white luminescent zinc (II) based metallopolymer by RGB approach. *Polymers* **2019**, *11*, 1712. [CrossRef] [PubMed]
29. Xiao, L.L.; Zhou, X.; Yue, K.; Guo, Z.H. Synthesis and self-assembly of conjugated block copolymers. *Polymers* **2021**, *13*, 110. [CrossRef] [PubMed]
30. Papadakis, R. Mono- and di-quaternized 4,4'-bipyridine derivatives as key building blocks for medium- And environment-responsive compounds and materials. *Molecules* **2020**, *25*, 1. [CrossRef] [PubMed]
31. Inuzuka, K. Electronic Properties of 4-Substituted Pyridines and Their Pyridinium Cations and Dihydropyridyl Radicals in the Ground State. *Nippon. Kagaku Kaishi* **1977**, *1977*, 355–361. [CrossRef]
32. Krygowski, T.M.; Szatyłowicz, H.; Zachara, J.E. How H-bonding modifies molecular structure and π-electron delocalization in the ring of pyridine/pyridinium derivatives involved in H-bond complexation. *J. Org. Chem.* **2005**, *70*, 8859–8865. [CrossRef] [PubMed]
33. Bunten, K.A.; Kakkar, A.K. Synthesis of Pyridine/Pyridinium-based Alkynyl Monomers, Oligomers and Polymers: Enhancing Conjugation by Pyridine N-Quaternization. *J. Mater. Chem.* **1995**, *5*, 2041–2043. [CrossRef]
34. Jonforsen, M.; Grigalevicius, S.; Andersson, M.R.; Hjertberg, T. Counter-ion induced solubility of polypyridines. *Synth. Met.* **1999**, *102*, 1200–1201. [CrossRef]
35. Aoki, A.; Rajagopalan, R.; Heller, A. Effect of quaternization on electron diffusion coefficients for redox hydrogels based on poly(4-vinylpyridine). *J. Phys. Chem.* **1995**, *99*, 5102–5110. [CrossRef]
36. Zhou, T.; He, X.; Song, F.; Xie, K. Chitosan Modified by Polymeric Reactive Dyes Containing Quanternary Ammonium Groups as a Novel Anion Exchange Membrane for Alkaline Fuel Cells. *Int. J. Electrochem. Sci.* **2016**, *11*, 590–608.
37. Zhai, L.; Li, H. Polyoxometalate-polymer hybrid materials as proton exchange membranes for fuel cell applications. *Molecules* **2019**, *24*, 3425. [CrossRef]
38. Rumyantsev, A.M.; Pan, A.; Ghosh Roy, S.; De, P.; Kramarenko, E.Y. Polyelectrolyte Gel Swelling and Conductivity vs Counterion Type, Cross-Linking Density, and Solvent Polarity. *Macromolecules* **2016**, *49*, 6630–6643. [CrossRef]
39. Bonardd, S.; Ángel, A.; Norambuena, Á.; Coll, D.; Tundidor-Camba, A.; Ortiz, P.A. Novel polyelectrolytes obtained by direct alkylation and ion replacement of a new aromatic polyamide copolymer bearing pyridinyl pendant groups. *Polymers* **2021**, *13*, 1993. [CrossRef]
40. Gokkaya, D.; Topuzogullari, M.; Arasoglu, T.; Trabzonlu, K.; Ozmen, M.M.; Abdurrahmanoğlu, S. Antibacterial properties of cationic copolymers as a function of pendant alkyl chain length and degree of quaternization. *Polym. Int.* **2021**, *70*, 829–836. [CrossRef]
41. Nagasako, T.; Ogata, T.; Kurihara, S.; Nonaka, T. Synthesis of thermosensitive copolymer beads containing pyridinium groups and their antibacterial activity. *J. Appl. Polym. Sci.* **2010**, *116*, 2580–2589. [CrossRef]
42. Luo, H.; Tang, Q.; Zhong, J.; Lei, Z.; Zhou, J.; Tong, Z. Interplay of Solvation and Size Effects Induced by the Counterions in Ionic Block Copolymers on the Basis of Hofmeister Series. *Macromol. Chem. Phys.* **2019**, *220*, 1800508. [CrossRef]
43. Mondal, P.; Saha, S.K.; Chowdhury, P. Simultaneous polymerization and quaternization of 4-vinyl pyridine. *J. Appl. Polym. Sci.* **2013**, *127*, 5045–5050. [CrossRef]
44. Chovino, C.; Gramain, P. Stereoregularity of poly(4-vinyl-n-alkyl-pyridinium) salts prepared by spontaneous polymerization. *Polymer* **1999**, *40*, 4805–4810. [CrossRef]
45. Jaeger, W.; Bohrisch, J.; Laschewsky, A. Synthetic polymers with quaternary nitrogen atoms-Synthesis and structure of the most used type of cationic polyelectrolytes. *Prog. Polym. Sci.* **2010**, *35*, 511–577. [CrossRef]
46. Laschewsky, A. Recent trends in the synthesis of polyelectrolytes. *Curr. Opin. Colloid Interface Sci.* **2012**, *17*, 56–63. [CrossRef]
47. Bicak, N.; Gazi, M. Quantitative quaternization of poly(4-vinyl pyridine). *J. Macromol. Sci.-Pure Appl. Chem.* **2003**, *40*, 585–591. [CrossRef]
48. Izumrudov, V.A.; Zhiryakova, M.V.; Melik-Nubarov, N.S. Supercharged pyridinium polycations and polyelectrolyte complexes. *Eur. Polym. J.* **2015**, *69*, 121–131. [CrossRef]
49. Boucher, E.A.; Mollett, C.C. Kinetics and Mechanism of the Quaternization of poly(4-vinyl pyridine) with Alkyl and Arylalkyl Bromides in Sulpholane. *J. Chem. Soc. Faraday Trans. 1* **1982**, *78*, 75–88. [CrossRef]

50. Chovino, C.; Gramain, P. Influence of the conformation on chemical modification of polymers: Study of the quaternization of poly(4-vinylpyridine). *Macromolecules* **1998**, *31*, 7111–7114. [CrossRef]
51. Frère, Y.; Gramain, P. Reaction Kinetics of Polymer Substituents. Macromolecular Steric Hindrance Effect in Quaternization of Poly(vinylpyridines). *Macromolecules* **1992**, *25*, 3184–3189. [CrossRef]
52. Velazquez, G.; Herrera-Gómez, A.; Martín-Polo, M.O. Identification of bound water through infrared spectroscopy in methylcellulose. *J. Food Eng.* **2003**, *59*, 79–84. [CrossRef]
53. Luo, Z.; Wang, X.; Zhang, G. Ion-specific effect on dynamics of polyelectrolyte chains. *Phys. Chem. Chem. Phys.* **2012**, *14*, 6812–6816. [CrossRef] [PubMed]
54. Sargsyan, A.; Tonoyan, A.; Davtyan, S.; Schick, C. The amount of immobilized polymer in PMMA SiO2 nanocomposites determined from calorimetric data. *Eur. Polym. J.* **2007**, *43*, 3113–3127. [CrossRef]
55. Wurm, A.; Ismail, M.; Kretzschmar, B.; Pospiech, D.; Schick, C. Retarded Crystallization in Polyamide/Layered Silicates Nanocomposites caused by an Immobilized Interphase. *Macromolecules* **2010**, *43*, 1480–1487. [CrossRef]
56. Klonos, P.; Kulyk, K.; Borysenko, M.V.; Gun'ko, V.M.; Kyritsis, A.; Pissis, P. Effects of Molecular Weight below the Entanglement Threshold on Interfacial Nanoparticles/Polymer Dynamics. *Macromolecules* **2016**, *49*, 9457–9473. [CrossRef]
57. Klonos, P.A.; Patelis, N.; Glynos, E.; Sakellariou, G.; Kyritsis, A. Molecular Dynamics in Polystyrene Single-Chain Nanoparticles. *Macromolecules* **2019**, *52*, 9334–9340. [CrossRef]
58. Pissis, P.; Kyritsis, A. Hydration studies in polymer hydrogels. *J. Polym. Sci. Part B Polym. Phys.* **2013**, *51*, 159–175. [CrossRef]
59. Szkudlarek, M.; Heine, E.; Keul, H.; Beginn, U.; Möller, M. Synthesis, characterization, and antimicrobial properties of peptides mimicking copolymers of maleic anhydride and 4-methyl-1-pentene. *Int. J. Mol. Sci.* **2018**, *19*, 2617. [CrossRef]
60. Aziz, S.B.; Rasheed, M.A.; Ahmed, H.M. Synthesis of polymer nanocomposites based on [methyl cellulose](1-x):(CuS)x (0.02M $\leq$ x $\leq$ 0.08 M) with desired optical band gaps. *Polymers* **2017**, *9*, 194. [CrossRef]
61. Nigam, S.; Rutan, S. Applications of Principles and Solvatochromism. *Appl. Spectrosc.* **2001**, *55*, 362A. [CrossRef]
62. Steven Paley, M.; Andrew Mcgill, R.; Howard, S.C.; Wallace, S.E.; Milton Harris, J. Solvatochromism. A New Method for Polymer Characterization. *Macromolecules* **1990**, *23*, 4557–4564. [CrossRef]
63. Deligkiozi, I.; Papadakis, R. Probing Solvation Effects in Binary Solvent Mixtures with the Use of Solvatochromic Dyes. In *Dyes and Pigments-Novel Applications and Waste Treatment*; Papadakis, R., Ed.; IntechOpen: London, UK, 2021.
64. Papadakis, R.; Deligkiozi, I.; Nowak, K.E. Study of the preferential solvation effects in binary solvent mixtures with the use of intensely solvatochromic azobenzene involving [2]rotaxane solutes. *J. Mol. Liq.* **2019**, *274*, 715–723. [CrossRef]
65. Papadakis, R. Solute-centric versus indicator-centric solvent polarity parameters in binary solvent mixtures. Determining the contribution of local solvent basicity to the solvatochromism of a pentacyanoferrate(II) dye. *J. Mol. Liq.* **2017**, *241*, 211–221. [CrossRef]
66. Papadakis, R. Preferential Solvation of a Highly Medium Responsive Pentacyanoferrate(II) Complex in Binary Solvent Mixtures: Understanding the Role of Dielectric Enrichment and the Specificity of Solute–Solvent Interactions. *J. Phys. Chem. B* **2016**, *120*, 9422–9433. [CrossRef] [PubMed]
67. Deligkiozi, I.; Voyiatzis, E.; Tsolomitis, A.; Papadakis, R. Synthesis and characterization of new azobenzene-containing bis pentacyanoferrate(II) stoppered push-pull [2]rotaxanes, with α- and β-cyclodextrin. Towards highly medium responsive dyes. *Dye. Pigment.* **2015**, *113*, 709–722. [CrossRef]
68. Pires, P.A.R.; El Seoud, O.A.; Machado, V.G.; De Jesus, J.C.; De Melo, C.E.A.; Buske, J.L.O.; Cardozo, A.P. Understanding Solvation: Comparison of Reichardt's Solvatochromic Probe and Related Molecular "core" Structures. *J. Chem. Eng. Data* **2019**, *64*, 2213–2220. [CrossRef]
69. Marcus, Y. *Solvent Mixtures Properties and Selective Solvation*; Marcel Dekker: New York, NY, USA, 2002.

MDPI
St. Alban-Anlage 66
4052 Basel
Switzerland
Tel. +41 61 683 77 34
Fax +41 61 302 89 18
www.mdpi.com

Polymers Editorial Office
E-mail: polymers@mdpi.com
www.mdpi.com/journal/polymers

www.ingramcontent.com/pod-product-compliance
Lightning Source LLC
LaVergne TN
LVHW082008090526
838202LV00006B/262